Environmentally Safe Approaches to CROP DISEASE CONTROL

AGRICULTURE AND ENVIRONMENT SERIES

Jack E. Rechcigl
Editor-in-Chief

Agriculture is an essential part of our economy on which we all depend for food, feed, and fiber. With the increased agricultural productivity in this country as well as abroad, the general public has taken agriculture for granted while voicing its concern and dismay over possible adverse effects of agriculture on the environment. The public debate that has ensued on the subject has been brought about, in part, by the indiscriminate use of agricultural chemicals and, in part, by misinformation, based largely on anecdotal evidence.

At the national level, recommendations have been made for increased research in this area by such bodies as the Office of Technology Assessment, the National Academy of Sciences, and the Carnegie Commission on Science, Technology, and Government. Specific issues identified for attention include: contamination of surface and groundwater by natural and chemical fertilizers, pesticides, and sediment; the continued abuse of fragile and nutrient poor soils; and suitable disposal of industrial and agricultural waste.

Although a number of publications have appeared recently on specific environmental effects of some agricultural practices, no attempt has been made to approach the subject systematically and in a comprehensive manner. The aim of this series is to fill the gap by providing the synthesis and critical analysis of the state-of-the-art in different areas of agriculture bearing on environment and vice versa. Efforts will also be made to review research in progress and comment on perspectives for the future. From time to time methodological treatises as well as compendia of important data in handbook form will also be included. The emphasis throughout the series will be on comprehensiveness, comparative aspects, alternative approaches, innovation, and worldwide orientation.

Specific topics will be selected by the Editor-in-Chief with the council of an international advisory board. Imaginative and timely suggestions for the inclusion in the series from individual scientists will be given serious consideration.

Published Titles

Environmentally Safe Approaches to Crop Disease Control
Soil Amendments and Environmental Quality
Soil Amendments: Impacts on Biotic Systems

To our parents and our family
for their love, support, and patience.

Environmentally Safe Approaches to CROP DISEASE CONTROL

Edited by
Nancy A. Rechcigl
Yoder Brothers, Inc.
Parrish, Florida
Jack E. Rechcigl
University of Florida
Soil and Water Science Department
Research and Education Center
Ona, Florida

CRC Press
Taylor & Francis Group
Boca Raton London New York

CRC Press is an imprint of the
Taylor & Francis Group, an informa business

Part I
Environmental Aspects

Table of Contents

R. S. Mehrotra
Department of Botany
Kurukshetra University
Haryana, India

Christopher C. Mundt
Department of Botany and Plant Pathology
Oregon State University
Corvallis, Oregon

John F. Murphy
Department of Plant Pathology and the Biological Control Institute
Auburn University
Auburn, Alabama

Giovanni Piccinni
Agricultural Research and Extension Center
The Texas A&M University System
Bushland, Texas

Dov Prusky
Department of Postharvest Science of Fresh Produce
The Volcani Center
Agricultural Research Organization
Bet-Dagan, Israel

Charles M. Rush
Agricultural Research and Extension Center
The Texas A&M University System
Amarillo, Texas

John L. Sherwood
Plant Pathology Department
Oklahoma State University
Stillwater, Oklahoma

Clayton A. Shillingford
Agricultural Products
Stine-Haskell Research Center
E.I. DuPont de Nemours Co., Inc.
Newark, Delaware

Constance M. Smith
Agricultural Products
Stine-Haskell Research Center
E.I. DuPont de Nemours Co., Inc.
Newark, Delaware

Franzine D. Smith
Sanford Scientific, Inc.
Waterloo, New York

Joyce M. Van Eck
Sanford Scientific, Inc.
Waterloo, New York

Mark Wilson
Department of Plant Pathology and the Biological Control Institute
Auburn University
Auburn, Alabama

Contributors

Ashok Aggarwal
Department of Botany
Kurukshetra University
Haryana, India

K. R. Aneja
Department of Botany
Kurukshetra University
Haryana, India

Paul A. Backman
Department of Plant Pathology and the Biological Control Institute
Auburn University
Auburn, Alabama

Mark A. Boudreau
Departments of Biology and Environmental Studies
Warren Wilson College
Asheville, North Carolina

John A. Bruhn
Agricultural Products
Stine-Haskell Research Center
E.I. DuPont de Nemours Co., Inc.
Newark, Delaware

Martin L. Carson
Agricultural Research Service
Department of Plant Pathology
North Carolina State University
Raleigh, North Carolina

Elizabeth K. Dann
U.S. Department of Agriculture and
Department of Botany and Plant Pathology
Michigan State University
East Lansing, Michigan

Marc Fuchs
Department of Plant Pathology
New York State Agricultural Experiment Station
Cornell University
Geneva, New York

Dennis Gonsalves
Department of Plant Pathology
New York State Agricultural Experiment Station
Cornell University
Geneva, New York

Ann B. Gould
Department of Plant Pathology
Rutgers University
New Brunswick, New Jersey

Ron A. Hamlen
Agricultural Products
Stine-Haskell Research Center
E.I. DuPont de Nemours Co., Inc.
Newark, Delaware

Ray Hammerschmidt
Department of Botany and Plant Pathology
Michigan State University
East Lansing, Michigan

Robert M. Harveson
Department of Plant Pathology
University of Florida
Gainsville, Florida

Bradley I. Hillman
Department of Plant Pathology
Rutgers University
New Brunswick, New Jersey

Mark J. Holliday
Agricultural Products
Stine-Haskell Research Center
E.I. DuPont de Nemours Co., Inc.
Newark, Delaware

Bernard Labit
Agricultural Products
Barley Mill Plaza
E.I. DuPont de Nemours Co., Inc.
Wilmington, Delaware

H. Alastair McCartney
Department of Crop and Disease Management
Institute of Arable Crops Research
Rothamsted
Harpenden, Herts, United Kingdom

Robert J. McGovern
Gulf Coast Research and Education Center
University of Florida — IFAS
Bradenton, Florida

Robert McSorley
Entomology and Nematology Department
University of Florida — IFAS
Gainesville, Florida

Preface

Plant diseases are one of the major causes of crop loss worldwide. It has been estimated that billions of dollars worth of crops are lost each year due to plant diseases. Historically, chemicals have been relied upon extensively as the primary crop management tool for preventing and controlling diseases. Recently, the use of chemicals in agriculture as well as in the home garden has come under sharp criticism by those who fear their improper use and adverse effect on the environment. This concern, coupled with the pressure from increased regulation of the pesticide industry, has prompted researchers to investigate alternative approaches for managing crop diseases. This reference book provides a balanced overview of the state-of-the-art environmentally safe approaches to crop disease control and the current research trends in this area. Each chapter was written by an expert in the respective field. This publication should be a useful resource to students and professionals in the fields of plant pathology, agronomy, horticulture, and environmental sciences, as well as those concerned with environmental issues in agriculture.

The editors wish to thank the individual contributors for the time and effort they put into the preparation of their chapters. Special thanks are also due to the CRC Press staff and editorial board for their general guidance and assistance.

Nancy A. Rechcigl
Jack E. Rechcigl

The Editors

Nancy A. Rechcigl holds the position of Technical Adviser at Yoder Bros. Inc., Parrish, Florida, specializing in plant disease and entomological problems of floricultural crops. Prior to joining Yoder Bros., Nancy worked for the University of Florida (1989–1994) as a County Horticultural Agent providing diagnostic services and information on cultural practices and pest management to the horticultural, landscape, and pest control industries. As an Extension Agent she was also responsible for supervising the County Master Gardener Program — providing instructional classes and operating a Plant Clinic which was popular with the urban community. From 1986 to 1989, she worked for Ball PanAm Inc., Parrish, Florida as a Plant Pathologist responsible for the disease certification program of ornamental plants.

Over the past 10 years, Ms. Rechcigl has given numerous lectures on the identification and control of disease and pest problems of turf and ornamentals. In addition to writing a weekly gardening column "Suncoast Gardening" for the urban community, she frequently contributes articles to local trade and professional journals.

Ms. Rechcigl received her B.S. degree (1983) in Plant Pathology from the University of Delaware, Newark. She did her graduate work at Virginia Polytechnic Institute & State University, Blacksburg, receiving her M.S. degree in 1986, specializing in Plant Virology.

Ms. Rechcigl is an active member of the American Phytopathological Society, Entomological Society of America, Florida Nurserymen and Growers Association, Czechoslovak Society of Arts and Sciences, and the Honorary Society of Phi Kappa Phi.

Jack E. Rechcigl is a Professor of Soil and Environmental Sciences at the University of Florida and is located at the Research and Education Center in Ona, Florida. He received his B.S. degree (1982) in Plant Science from the University of Delaware, Newark, DE and his M.S. (1983) and Ph.D. (1986) degrees in Soil Science from Virginia Polytechnic Institute and State University, Blacksburg. He joined the faculty of the University of Florida in 1986 as Assistant Professor. In 1991, he was promoted to Associate Professor, and in 1996 attained Full Professorship.

Dr. Rechcigl has authored over 200 publications, including contributions to books, monographs, and articles in periodicals in the fields of soil fertility, environmental quality, and water pollution. His research has been supported by research grants totaling over $2 million from both private sources and government agencies. Dr. Rechcigl has been a frequent speaker at national and international workshops and conferences and has consulted in various countries, such as Canada, Brazil, Nicaragua, Venezuela, Australia, New Zealand, Taiwan, Philippines, and the Czech Republic.

He is currently an Associate Editor for the *Journal of Environmental Quality* and is Editor-in-Chief of the Agriculture and Environment Book Series. Most recently he has edited *Soil Amendments: Impacts on Biotic Systems* (Lewis Publishers and CRC Press, 1995), *Soil Amendments and Environmental Quality* (Lewis Publishers and CRC Press, 1995), and *Agricultural Uses of By-Products and Wastes* (American Chemical Society, 1997). He is also serving as an invitational reviewer for manuscripts and grant proposals for scientific journals and granting agencies.

Dr. Rechcigl is a member of the American Society of Agronomy, American Chemical Society, Soil Science Society of America, International Soil Science Society, Czechoslovak Society of Arts and Sciences, various trade organizations, and the Honorary Societies of Sigma Xi, Gamma Sigma Delta, Phi Sigma, and Gamma Beta Phi.

CHAPTER 1

The Influence of Environment on the Development and Control of Disease

H. Alastair McCartney

I. INTRODUCTION

Crop disease epidemics are the result of the complex interaction between the pathogen (or pathogens), the host, and its environment. What is meant by "environment" in the context of disease epidemiology? Environmental factors can have either a direct effect on the pathogen, for example, many fungal spores need water to germinate, or an indirect effect through their influence on the physiological status of the crop. For soil-borne pathogens the physical and chemical nature of the soil is obviously of paramount importance, but for foliar pathogens the microclimate of the crop plays the dominant role through its effect on leaf surfaces. When considering the dispersal of pathogens we may need to include the atmosphere near the ground in our definition of the appropriate "environment." In the rest of this chapter the definition of "environment" will usually be restricted to that which has direct effects on pathogen development.

1-8493-2627-3/97/$0.00+$.50

Since the Second World War crop disease in Western agriculture has largely been controlled with fungicides, pesticides, etc. Over the last few years there has been increasing pressure to reduce the use of phytochemicals in agriculture, both from environmentalists and from governments, for example, in Scandinavia and in Holland the aim is to halve their use by the year 2000. There are several reasons for this. Fungicides are perceived to be dangerous for human health either through direct exposure (during application or by spray drift) or through residues in drinking water or food. Thus, for high-value food products there is a demand for low levels of fungicide and pesticide residues and, as detection instrumentation sensitivity increases, the acceptable levels will continue to decrease. Since 1975 the European Union has issued several directives on the reduction of pesticide levels in drinking water (Regulations 75-440/EEC, 79-869/EEC, and 80-778/EEC). Intensive use of phytochemicals is considered to be detrimental to the environment. Pimentel et al. (1991) estimated that in the U.S. the costs of indirect damage due to long-term environmental pollution from agriculture was approximately $\$10^9$ (U.S.). In Europe, where intensive farming is common, the costs are probably at least equal to if not greater than this. The environmental consequences of phytochemical use include adverse effects on nontarget organisms including beneficial fungi pathogenic to insects; direct and indirect effects on insects, fish, and mammals; subtle perturbations on natural ecosystems; and the development of resistant strains of pathogenic organisms. The pressure to reduce the use of phytochemicals, particularly in the U.S. and Europe, is likely to intensify toward the end of the millennium.

The best way to control disease in crops is to eliminate the disease by growing resistant cultivars. A significant proportion of the effort of plant breeders is devoted to developing disease resistance in crops; but even with the advent of new molecular genetic techniques, it is likely that for the foreseeable future crop protection will still rely heavily on the use of phytochemicals. Thus, crop protection strategies need to be developed which minimize the use of phytochemicals while maintaining acceptable economic returns for the farmer. This means either the development of novel control measures, for example, based on biological control, or more efficient use of control chemicals. There are several options available for developing such control strategies, each requiring knowledge of the interactions between the pathogen, host, and environment. For example, reduction in phytochemical use can be achieved by applying control measures only at critical times during the development of an epidemic. This requires an understanding of how epidemics develop, including knowledge of the relationship between microclimate and pathogen growth. Alternatively, agronomic practices may be devised based on disease avoidance, for example, by the choice of a sowing date to avoid periods of weather which favor disease development or spread. Knowledge of the interactions between the pathogen and environment may suggest methods of manipulating crop microclimate through breeding or agronomic practice, which reduces the chances of disease. The reduction in phytochemical use could also be achieved by more efficient application of sprays, for example by targeting specific parts of the crop vulnerable to attack. Environmental factors play an important role in spray application and in the retention of spray on crop surfaces.

Although there is usually little we can do to control or manipulate those environmental factors which regulate disease development in crops, except perhaps for protected crops, it is clear that an understanding of how environmental factors affect crop pathogens is essential for the development of effective control strategies. The rest of this chapter will consider the effects of the major microclimate factors on the development of plant pathogens, the crop environment, and how this information can be used to control disease. The chapter will concentrate on foliar diseases caused by fungal pathogens, but the general principles should find applications in other crop/disease systems.

II. THE PATHOGEN AND ENVIRONMENT

The physical environment affects all stages in the life cycle of plant pathogens and much of the effort in plant pathological research is devoted to the study of these changes. Temperature and moisture are the two most important microclimatic factors in the direct interaction of host and pathogen, although radiation (light) can have also have an influence, while wind and rain play dominant roles in dispersal processes (including dispersal by insect vectors). The exact relationship between the pathogen and environmental parameters differs for each pathogen/host system and a comprehensive summary would fill several volumes. This section will illustrate some of these interactions, using selected examples taken mainly from foliar fungal pathogens of arable crops.

The interaction between fungal pathogens and host broadly follows three phases: infection, development, and sporulation — each of which may be limited by environmental factors.

Infection usually includes spore germination followed by penetration and invasion of the host tissue. Some spores, referred to as resting spores, remain dormant for a period after their formation (Hawker and Madelin, 1976; Manners, 1982). In some cases, dormancy can be broken in response to environmental cues. For example, alternate wetting and drying or freezing and thawing will stimulate the germination of teliospores of *Puccinia graminis*, while spores of *Tilletia* species require exposure to temperatures below 10°C before they will germinate (Manners, 1982). Many spores do not have a dormant period as their function is to propagate the pathogen quickly. Nevertheless, their germination is often mediated by environmental factors. However, it must be remembered that spore germination can be stimulated or inhibited by other factors related to the host and the pathogen (Macko, 1981).

As with all biological functions, spore germination is temperature dependent. For all spores there is a minimum and maximum temperature range, outside of which the spores will not germinate (Table 1). The rate of germination tends to increase to a maximum at an optimum temperature, then decline as the temperature increases. For example, Harthill and Cheah (1984) found that germination of conidia of *Pyrenopeziza brassicae* on cauliflower *(Brassica oleracea)* increased almost linearly from 6°C to about 16°C before declining with higher temperatures until ceasing at 26°C; similar results were found by Figueroa et al. (1995a) on oilseed rape *(Brassica napus)*. Rust basidiospore germination usually takes place over a wide range of temperatures, but thermal death occurs in some species at temperatures greater than 30°C (Gold and Mendgen, 1991). Temperature can also affect the rate of penetration of the pathogen into the host. In epidemiology this is often quantified as the *incubation period*: the time taken between spore inoculation and appearance of visible symptoms (Manners, 1982). The temperature response of incubation periods tends to reflect that for germination, with the shortest periods found near the optimal temperatures for germination. Further development of the pathogen within the host can be quantified, in epidemiological terms, by the *latent period*: the time between inoculation and the production of spores (Manners, 1982). Both infection and latent periods are influenced by temperature (Table 1). Latent periods tend to decrease with increasing temperature, providing other factors are favorable. For example, the latent period for *Erysiphe graminis* decreases from about 14 days at 5°C to 3 days at 18 to 25°C, and that for *Puccinia hordei* decreases from 60 days at 5°C to 6 days at 25°C (Polley and Clarkson, 1978). The effects of temperature on pathogen/host interactions may be moderated by other environmental factors such as light and moisture.

Light can stimulate germination in some fungi and inhibit it others (Leach and Anderson, 1982), but the mechanisms are poorly understood. Most basidiospores are released at night or early morning; however, darkness is not necessary for germination and when other conditions are favorable light has either no effect or a marginal negative effect (Gold and Mendgen, 1991). The effects of light on germination can be complex: for example, germ-tube elongation of *Alernaria linicola* conidia on linseed under wet conditions was interrupted by exposure to light periods of up to 12 h immediately after inoculation, but resumed on return to darkness (Vloutoglou et al., 1996). However, if the light period started 6 h after inoculation germ-tube elongation was unaffected, suggesting that the germ-tubes had started penetration and were no longer sensitive to light. Similarly, exposure to light reduced the probability of infection of wheat leaves by *Mycosphaerella graminicols (Septoria tritici)* early in the infection process but stimulated it later (Shaw, 1991). Schuh (1993) found that 12 h dark/12 h light was more conducive to germination of *Cercospora kikuchii* on soybean than, in descending order, 24 h dark or 12 h light/12 h dark or 24 h light. The importance of light on the development of epidemics in the field has been little studied and is poorly understood.

Moisture is probably the most crucial environmental parameter affecting spore germination and infection of fungal pathogens. Most fungal spores require high relative humidities to germinate

Table 1a Some Recent Studies on the Effects of Environmental Factors and Pathogen Development: Infection Processes

Pathogen (Crop)	Infection/Incubation Period (ip) Temperature	Infection/Incubation Period (ip) Wetness	Latent Period Temperature	Source
Arthuriomyces peckianus (raspberry)	Germination: optimum 20°C	min 6 h @ 20°C	—	Truxall et al. (1995)
Ascochyta fabae f. sp. *lentis* (lentils)	Infection frequency > @10 and 15°C than @ 25°C	>24 h	6–7 day @ 20°C 13–14 day @ 10°C	Pedersen and Morrall (1994)
Aspergillus niger (fig)	Optimum 35°C	>97% RH	44 h @ 30°C	Subbarao and Michailides (1995)
Botrytis cinerea				
(strawberry)	—	—	5–7day @ 5°C	Sosaalvarez et al. (1995)
(black spruce)	Optimum 20°C (32 h wet) Zero @ 4 and 36°C	min 12 h	—	Zhang and Sutton (1994)
Cercospora arachidicola (groundnut)			39 day @ 12°C 13 day @ 25°C 25day @ 33°C	Wadia and Butler (1994a)
Didymella bryoniae (cucumber)	Optimum 24–25°C	—	—	Arny and Rowe (1991)
Fusarium moniliforme (fig)	Optimum 30°C	>97% RH	40 h @ 30°C (wild isolate) 60 h @ 30°C (cultivated isolate)	Subbarao and Michailides (1995)
Phaeoisariopsis personata (groundnut)	Optimum 20°C Range 8–34°C	min 20 h	38 day @ 12°C 13 day @ 25°C 26 day @ 33°C	Wadia and Butler (1994a) Butler et al. (1994)
Puccinia arachidis (groundnut)	—	—	49 day @ 12°C 12 day @ 25°C 20 day @ 33°C	Wadia and Butler (1994a)
Puccinia helianthi (sunflower)	Germination: optimum 10–24°C Range 4–30°C ip: 8–10 day @14–25°C	6–10 h	—	Shtienberg and Vintal (1995)
Pyrenopeziza brassicae (oilseed rape)	ip, opt wetness 21–11 day @ 12°C 14–18 day @ 18°C Time for 50% germination: 40h @ 5°C; 18 h @ 15–20°C	min 13 h	22day @ 5°C 14 day @ 15°C 17.5 day @ 20°C	Figueroa et al. (1995a,b)
Rhynchosporium secalis (winter barley)	—	—	(RH 100%) 24 day @ 5°C 19 day @ 10°C	Davis and Fitt (1994)

Table 1a (continued) **Some Recent Studies on the Effects of Environmental Factors and Pathogen Development: Infection Processes**

Pathogen (Crop)	Infection/Incubation Period (ip) Temperature	Infection/Incubation Period (ip) Wetness	Latent Period Temperature	Source
			16 day @ 15°C 13 day @ 20°C	
Septoria apiicola (celery)	Optimum 25°C Range at least 10–30°C	min 10 h	—	Mathiea and Kushalappa (1993)
Septoria glycines (soybean)	Optimum 25°C Range at least 15–39°C	min 6 h	—	Schuh and Adamowicz (1993)
Uromyces viciae fabae (field bean)	—	—	8–10 day @ 18°C	Sache and Zadoks (1995)

Note: See also Huber and Gillespie, 1992.

Table 1b **Some Recent Studies on the Effects of Environmental Factors and Pathogen Development: Sporulation Processes**

Pathogen (host)	Temperature	Wetness	Light	Source
Ascochyta fabae f. sp. *lentis* (lentil)	Optimum 15–25°C Maximum 30°C	—	Better in light than darkness	Kaiser (1994)
Botrytis cinerea (strawberry)	Optimum 17–18°C No sporulation @ 30°C	3–7 days	—	Sosaalvarez et al. (1995)
Cercospora carotae (carrot)	Optimum 23°C Range at least 16–32°C	>96% RH Enhanced by 12 h wetness	—	Carisse et al. (1993)
Oncobasidium theobromae (cocoa)	Maximum 26°C	Min 5 h for spore release	Alternating light and dark, release during dark	Dennis et al. (1992)
Phaeoisariopsis personata (groundnut)	Optimum 28°C Range 10–33°C	>94.5% RH Optimum 99–100% Reduced by continuous wetness, but better with intermittent wetness	Inhibited by continuous light	Butler et al. (1995)
Phoma bakerana (culture)	Range about 15–30°C No sporulation @ 10 or 35°C	—	Enhanced by near ultraviolet	Nebane and Ekpo (1992)
Puccinia helianthi (sunflower)	Optimum 20–30°C Range 4–39°C	—	—	Shtienberg and Vintal (1995)

and some need the presence of free water (Table 1). Some spores will germinate at low relative humidities, for example, *Erysiphe graminis*, but even for this pathogen optimum germination occurs when the air is close to saturation (Polley and Clarkson, 1978). The presence of free water inhibits the germination of *Erysiphe graminis*, and is generally inhibitory to basidiospores or leads to the formation of secondary basidiospores (Gold and Mendgen, 1991). The duration of wetness needed to promote germination and infection is usually dependent on temperature (Table 1) and periods as short as 0.5 h (*Phytophthora cactorum* on strawberry, Grove et al., 1985) to greater than 100 h (*Diaporthe phaseolorum* on soybean, Damicone et al., 1987) have been reported. Huber and Gillespie (1992) list the wetness requirements for a number of foliar pathogens from studies done between 1985 and 1991. In some fungal pathogens which require free water for germination and infection, discontinuous wet periods will still initiate infection (Stuckey and Zadoks, 1989; Chakrabortry et al., 1990), and in recent years there has been an increasing interest in the effects of intermittent wetness on infection and sporulation processes (e.g., Shaw, 1991; Mridha and Wheeler, 1993; Schuh, 1993; Sah, 1994; Butler et al., 1994; Wadia and Butler, 1994b; Hong and Fitt, 1995; Vallavieille-Pope et al., 1995; Vloutoglou et al., 1996). Intermittent wet periods can reduce the efficiency of infection; for example, fewer *Pyrenophora tritici-repentis* lesions developed on wheat plants exposed to different periods of dryness after an initial 6-h wetness than on plants exposed to continuous 24-h wetness (Sah, 1994). Butler et al. (1994) found that intermittent wetness enhanced infection efficiency of *Phaeoisariopsis personata* in groundnut *(Arachis hypogaea)* compared to continuous wetness. Intermittent wetness produced more germ-tubes per conidium and increased branching, which may have increased the chances of stomatal penetration (Wadia and Butler, 1994b). Interruption of wetness also enhanced infection of soybean by *Cercospora kikuchii*, but only when the relative humidity during the dry period was >95% (Schuh, 1993). There are suggestions that for some pathogens wetness periods may be partially cumulative (Stuckey and Zadoks, 1989; Wadia and Butler, 1994b). The significance of the interruption in wetness on infection processes is not known for most pathogens, and as wetness in crops is often discontinuous, especially when caused by rain, priority should be placed on studies in this area. Crop moisture status may also affect the development of disease after infection. Sah (1994) found that necrosis on wheat caused by *Pyrenophora tritici-repentis* was greater when the relative humidity during the incubation period (after a postinoculation wet period) was 95% than when it was 35 or 70%.

The first step in the spread of plant disease is the production of inoculum. For fungal pathogens inoculum consists of spores which can be dispersed in a number of ways (see below). In foliar pathogens spore production can be affected by temperature, moisture (atmospheric and surface), and light, each often interacting with the other (Rotem et al., 1978; Leach and Anderson, 1982). For soil pathogens, soil moisture and temperature are the most important environmental factors affecting sporulation (Lacey, 1986). Much work has been done on the effects of these environmental factors to sporulation *in vitro*, but there has been much less attention paid to their effects *in vivo* (Rotem et al., 1978; Huber and Gillespie, 1992) and there is concern that *in vitro* results may not be directly applicable to conditions in crops (Rotem et al., 1978; Lacey, 1986).

The effects of temperature interact with those of other environmental factors as well as those related to the host and saprophytic microflora (Rotem et al., 1978). The optimum temperature for spore production can be less than that for infection processes (Polley and Clarkson, 1978) and in some species spores may be formed at temperatures too low for lesion development (Rotem et al., 1978). Epidemiologically, the effects of temperature and moisture on sporulation are reflected by their effects on the latent period, the time between inoculation and sporulation (Table 1).

Moisture, whether humidity or free water, can be important in sporulation processes. Most leaf-infecting fungi require high relative humidities for sporulation (Lacey, 1986), although the powdery mildews, rusts, and some other pathogens sporulate over a wide range of relative humidities (Rotem et al., 1978). It is not always clear whether pathogens which sporulate under humid conditions prefer humidities close to saturation over the presence of free water (Rotem et al., 1978). Free water can inhibit sporulation in some species which require high humidities, for example, *Peronospora tabacina* and probably most powdery mildews (Lacey, 1986). Others (e.g., *Phytophthora infestans, Helminthosporium maydis*) sporulate equally well in the presence of free water or high humidity (Rotem et al., 1978). The effect of surface water varies with species and may depend in part on the type of water cover (drop size, film depth), duration of wetness, and other environmental or host factors (Lacey, 1986). For pathogens requiring surface wetness for sporulation the minimum periods can be as little as 2 to 3 h, but are generally much longer (Rotem et al., 1978) and are often longer than those needed for infection. Spore production tends to increase sigmoidally with an increasing wetness duration above the minimum needed for spore induction. For some pathogens (e.g., *Phyllosticta maydis*, *Ascochyta pisi*) the minimum wetness period may be several days, which is longer than that usually found in crops, suggesting that these fungi respond to intermittent wetness by "accumulating" wetness periods (Rotem et al., 1978). Some pathogens appear to benefit from dry interruptions: *Alternaria porri* f. sp. *solani* produced up to seven times more spores on potatoes under interrupted wetness than under continuous wetness in darkness (Bashi and Rotem, 1975); conidial production of *Phaeoisariopsis personata* on groundnut was more effective under intermittent than under continuous wetness (Butler et al., 1995). It is clear that the interactions between inoculum production and moisture are varied and complex, and that much work is still needed to quantify them for a large number of economically important pathogens.

There has been a substantial amount of work on the effect of light on inoculum production in plant pathogens (Rotem et al., 1978; Leach and Anderson, 1982), but there is little evidence that the direct effect of light on epidemic development is a substantial one once the pathogen is present in the crop (Manners, 1982). Most of the investigations of the effects of light have been made *in vitro* or in the laboratory (Leach and Anderson, 1982) and effects, particularly with shortwave (blue) or near-ultraviolet (NUV) radiation, have been demonstrated. For example, production of pycnidia in *Phoma bakeriana* in culture (Nebane and Ekpo, 1992), production of spores of *Botrytis cinerea* in culture and on leaves (Nicot et al., 1996), and the production of apothecia of *Sclerotinia sclerotiorum* (Mylchreest and Wheeler, 1987) can all be stimulated with NUV. Exposure to continuous light is inhibitory to spore production in many fungi (e.g., *Phytophthora infestans, Alternaria porri* f. sp. *solani, Peronospora destructor*), with shorter wavelengths appearing to be most effective (Rotem et al., 1978; Hildebrand and Sutton, 1984), while in others continuous light appears to have no effect (e.g., *Rhynchosporium secalis, Erysiphe cichoracearum*). *Ascochyta fabae* f. sp. *lentis* sporulated three to ten times more effectively in culture when exposed to continuous light (Kaiser, 1994) and daily sporulation rates of *Puccinia striiformis* on wheat increased with light intensity (McGregor and Manners, 1985). NUV can stimulate spore production in some species (e.g., *Leptosphaerulina trifolii, Botrytis cinerea, Alternaria solani*) (Leach and Anderson, 1982; Vakalounakis, 1991), while others (e.g., *Helminthosporium oryzae, Stemphylium botrysum*) sporulate under diurnal cycling of NUV and darkness (Leach and Anderson, 1982). Butler et al. (1995) found that *Pheaoisariopsis personata* sporulated best on groundnut under a 12-h light/dark cycle. The importance of these effects on sporulation under field conditions are not clear. A better understanding of the effects of light on pathogen inoculum production could have significant implications on disease management, especially in protected crops where control of the light climate in the crop is done routinely. For example, Vakalounakis (1991) showed that incidence of early blight of tomatoes, caused by *Alternaria solani*, could be halved when the crop was grown in greenhouses covered with a UV-absorbing vinyl film.

Spore dispersal and consequent disease spread is affected by both biological and physical processes (McCartney, 1994). Environmental factors dominate the three phases of dispersal: release, travel, and deposition. Wind obviously plays a major role in the spread of foliar diseases (Aylor, 1990), but rain, water availability, temperature, and light can also be important.

Spores and pollen can be removed from their host by simply being blown or shaken off (passive release), by an active release mechanism, or by being washed or splashed off. Many fungi have spore release mechanisms which can eject spores directly into the air (Ingold, 1971; Lacey, 1986). For example, ascospores of *Sclerotinia sclerotiorum*, can be ejected 2 to 3 cm from the apothecia (Harthill and Underhill, 1976). The mechanisms by which spores are ejected are generally not fully understood, although an adequate supply of water is needed (Gregory, 1973) and release is frequently triggered by changes in environmental factors such as humidity, temperature, or light

(Ingold, 1971; Lacey, 1986). In *Sclerotinia sclerotiorum* and *Pyrenopeziza brassicae*, ascospore release from turgid asci can occur in response to a decrease in relative humidity (McCartney and Lacey, 1990, 1992). Drying is also responsible for the release of conidia of *Peronospora tabacina*. The spores are held above the leaf surface on stalk-like structures which twist violently as they dry, throwing spores away from the leaf (Ingold, 1971).

Wind can remove spores by blowing them from surfaces or by shaking foliage or inflorescences. The aerodynamic or mechanical forces generated by wind must overcome the forces holding the particle to the surface (Aylor and Parlange, 1975). The wind speeds needed to remove spores are not known for many fungi, but, it seems likely that they can be relatively large (Grace, 1977). For example, Hammett and Manners (1974) found that conidia of *Erysiphe graminis*, which form chains above the leaf surface, are released in wind exceeding about 0.5 m s^{-1}. Likewise conidia of *Helminthosporium maydis* were only removed when the wind speed exceeded about 5 m s^{-1} (Aylor, 1975). As the mean wind speeds within plant canopies are generally smaller than those needed to release spores, it seems likely that wind gusts may play an important role in spore removal (Aylor, 1978; Aylor et al., 1981). Spores can also be dislodged by shaking (Bainbridge and Legg, 1976), thus wind gusts may indirectly remove spores by moving the crop canopy. Although the wind conditions (speed and turbulence) needed to remove spores of most pathogens are not known, the gusty nature of wind means that it is unlikely that, for many pathogens, spore removal during the day is limited by the state of the wind.

Rain or spray irrigation can remove spores and pathogenic bacteria from crop surfaces (Fitt et al., 1989) by washing them off in runoff water or in splash droplets or by the shaking caused by the impact of rain drops. The spores of many plant pathogens are held in a mucilage which prevents dispersal by wind and can only be removed by water (in runoff or in splash droplets) when the mucilage is dissolved (Gregory, 1973; Fitt et al., 1989). Splash droplets can be thrown more than a meter from the point of the splash but most only travel a few centimeters (Fitt et al., 1989). However, splash droplets, particularly those smaller than about 100 μm in diameter, have the potential to be further dispersed by wind. The effectiveness of splash in removing spores depends on the size and velocity of the drop and on the orientation and mechanical properties of the surface, but the physical mechanisms are not well understood. Drop size plays an important role in the removal of spores (Gregory,1973; Fitt and McCartney, 1986b; Fitt et al., 1989) and large drops are more effective than small one. The force of impact of the drops may determine the effectiveness of spore removal (Walklate, 1989). Thus, large slow-moving drops dripping from leaves may remove spores as effectively as small raindrops falling at their terminal velocity. Because of the importance of drop size, the potential for dispersal by splash depends partly on the size distribution of raindrops (Fitt et al., 1989; Walklate, 1989; Walklate et al., 1989), which in turn depends on the type of rainfall (Ulbricht, 1983). For example, Shaw and Royal (1993) found that *Septoria tritici* spores were only transported from the base to the top of a wheat canopy during heavy summer showers.

Wind conditions within and above crops are highly turbulent (McCartney and Fitt, 1985; Aylor, 1990). This causes individual spores to follow different paths and travel different distances, even if released from the same source under the same wind conditions. Therefore, as a spore plume is dispersed downwind of the source its concentration in the air decreases. This decrease in concentration is frequently referred to as a "concentration (or deposition) gradient" (Gregory, 1973). Deposition or concentration gradients tend to be steeper within crops than above them. Within crops the concentration is depleted not only by the expansion of the spore plume but also by deposition of spores on the crop and on the ground. Because spore-carrying water droplets are much larger than the spores they are deposited more effectively, making gradients for splash-dispersed spores steeper than for those dispersed by wind. Wind also transports spores up into the atmosphere where they have the potential to disperse over large distances. Fungal spores have been found at heights of between 500 and 1000 m in the atmosphere (Hirst et al., 1967). Dispersion gradients have been described by a number of different equations (McCartney and Fitt, 1985; Fitt et al., 1987). However, a negative exponential relationship can often be used to quantify gradients:

$$C = C_0 \exp(-\gamma \cdot x) \qquad (1)$$

where C is the concentration or deposition rate, x is the distance down wind, and C_o and γ are constants. The coefficient determines the rate of decrease in spore concentration (or deposition) with distance. Although this relationship may not be strictly valid in some situations (Aylor, 1987), γ gives a convenient method for visualizing gradients as it can be expressed as a half distance, $d_{1/2}$ (= 0.693/γ), that is, the distance in which C decreases by half. For plant pathogen spores, $d_{1/2}$ values measured within crops are usually in the order of a few meters, but when measured from the edges of the crops they tend to be larger (Fitt et al., 1987). For example, $d_{1/2}$ for 20-μm particles (similar in size to mildew spores) released within a barley crop were in the range 0.3 to 1.8 m (McCartney, and Bainbridge, 1984), whereas values of about 9 m were found for *Pyrenopeziza brassicae* ascospores (McCartney, 1990). Unfortunately, little systematic work appears to have been done in relating dispersal gradients to wind and weather conditions. Atmospheric dispersal models have also been used to assess the effects of wind on spore dispersal (McCartney and Fitt, 1985; Fitt and McCartney, 1986a; Aylor, 1990; McCartney, 1994).

Deposition of spores and pollen onto surfaces can be thought of as being a combination of two mechanisms: gravitational settling and inertial impaction (Legg and Powell, 1979). The rate at which spores settle onto surfaces, S, is proportional to the spore fall speed, V_s, and the spore concentration above the surface, C, ($S = C \cdot V_s$). V_s for most fungal spores generally lies in the range 0.1 to 3 cm s^{-1} (Gregory, 1973), but many spores and pollens have complex nonspherical shapes and usually fall more slowly than spherical ones of the same volume and density (McCartney et al., 1993). Deposition by impaction, I, is proportional to C and wind speed, u ($I = C.u.E$). The constant of proportionality, E, is known as the impaction efficiency and increases with spore size (V_s) and wind speed (u) but decreases with increasing width of the impaction surface (w) (Chamberlain, 1975; Aylor, 1982). Wind gusts can enhance the inertial impaction of spores (Aylor, 1978; Aylor et al., 1981), especially in crop canopies where mean wind speeds are small (McCartney and Bainbridge, 1987).

III. THE CROP ENVIRONMENT

Leaf wetness, temperature, wind, and to a lesser extent radiation, are the most important microclimatic variables which influence the development of epidemics of most crop pathogens. This section will briefly consider each of these variables and how they may be measured or estimated for use in disease control systems.

A. Crop Architecture

Architecture plays an important role in the microclimate of vegetative stands as it exerts a considerable influence on where the exchange processes of heat mass and momentum take place. As most exchange processes take place at leaf surfaces, the vertical variation of foliage density is the most important crop characteristic of interest. In most agricultural crops the horizontal variation in leaf density can usually be considered of secondary importance and the crop can be considered to consist of one or more uniform layers. For example, cereal crops are often modeled as a single layer while crops like maize, sunflower, or many trees may be treated as having a single canopy layer above a stem layer. Because of its influence on exchange processes crop architecture strongly influences the profiles of temperature, wind, and water vapor.

B. Temperature

Excluding soil-borne pathogens, most pathogen/crop interactions take place either within or on leaves or stems. Thus, strictly, leaf temperatures should be estimated or measured, especially when results from controlled environment studies are being used to determine pathogen behavior. Leaves have a relatively low heat storage capacity and are subject to both radiative and convective heat loss, consequently, when they are not exposed to radiative heating or cooling (i.e., in shade or under overcast conditions) their surface and internal temperatures will be close to that of the surrounding air. However, the surface of a sunlit leaf is commonly 5 to 10°C warmer than the surrounding air, while the shade side is usually 1 to 3°C cooler than the sunlit side (Oke, 1978). At night under clear skies, leaves can cool to 5 to 10°C lower than the air temperature. The effects of radiative heating and cooling could be important for pathogens which infect leaves close to the tops of crop canopies (e.g., flag leaves in cereals) as they will be subjected to much larger fluctuations in temperature than air temperature measurements would indicate. The elevation or depression of temperature of an individual leaf can be estimated by considering its energy balance. Such calculations require measurement or estimates of a number of other micrometeorological variables such as net radiation, air temperature, water vapor pressure (for transpiring leaves), and wind speed, as well as knowledge of the physiological status of the leaf (Monteith and Unsworth, 1990).

Air temperature within crop canopies is not usually uniform, but varies with height depending on the architecture of the crop, time of day, and cloud cover. Details of the variation in temperature profile depend on the nature of the crop and local climatic conditions, but a few general observations may be made. At night under clear skies, long-wave radiative cooling can give rise to a temperature minimum near the top of the canopy so that temperature increases upward into the atmosphere and downward into the crop, although the variation may only be a few degrees. Under overcast skies temperature gradients both at night and during the day tend to be relatively uniform. During the day the opposite happens, when the upper parts of the canopy may be subjected to a substantial radiation load. Then, an air temperature peak is often observed close to the top of the canopy. The difference between the maximum within-crop air temperature and that above the crop is usually about 2 to 3°C, although excess temperatures of 10°C or more are possible in cold climates when transpiration is low (Monteith and Unsworth, 1990). For many pathogens, using above-crop air temperature may be adequate for disease forecasting or modeling, especially when the temperatures are close to the optimum. When temperatures are near the upper or lower limits of the pathogen's range the use of in-crop temperatures may be more appropriate. Micrometeorological models have been developed to predict within-crop air temperatures and other microclimatic variables for a number of crops (Table 2) and these could be incorporated into disease forecasting or warning models.

Of the microclimatic factors important in disease development air temperature is probably the simplest to measure. There is a wide range of sensors (e.g., thermocouples, resistance sensors, thermistors) which are easily interfaced to automatic weather stations or data acquisition systems. Manually read maximum/minimum thermometers are readily available and could be used to provide useful information on crop temperatures. Direct measurement of leaf temperature is more difficult: the sensors need to have small thermal mass and be in good thermal contact with the leaf (fine-wire thermocouples are suitable). Because of the spatial variability of leaf temperatures a relatively large number of sensors may have to be used. Infrared thermometry allows leaf temperatures to be measured remotely and can integrate over a number of leaves. Such instruments have applications in research studies but because of their expense and complexity are currently not suitable for routine measurements.

C. Wind

Wind can influence disease epidemic development either directly through inoculum dispersal, or indirectly by affecting other environmental factors such as crop temperatures and leaf wetness duration. The measurement of mean wind speeds above crops is generally fairly straightforward and anemometers compatible with most data acquisition systems are usually available. Wind speed and direction are nearly always included in automatic weather station specifications.

Above-crop mean wind speeds increase with height, depending on the nature of the crop (height, architecture, density) and the stability of the atmosphere (temperature profile) (McCartney and Fitt, 1985). For a neutrally stratified atmosphere, when buoyancy effects can be neglected, wind speed profiles over open terrain with uniform vegetation can often be described by the "logarithmic wind profile" (Monteith, 1973; Grace, 1977; Monteith and Unsworth, 1990):

Table 2 Examples of Microclimatic Models Used to Estimated Dew or Leaf Wetness Duration in Different Crops

Source	Crop	Model	Remarks
Monteith and Butler (1979)	Cocoa pods	Heat balance including thermal lag.	Overestimated duration of wetness.
Thompson (1981)	Cereals, grass	Multilayer energy balance.	Simulation study.
Pedro and Gillespie (1982a)	Single leaf	Energy balance, micrometeorological data, exposed and shaded leaves	Dew duration within 30 min for exposed leaves and 60 min for shaded leaves.
Pedro and Gillespie (1982b)	Single leaf	As Pedro and Gillespie (1982a) but with weather station data	Dew duration within 1 h for exposed leaves and 1.5 h for shaded leaves.
Butler (1984)	Cereals	Energy balance with heat conduction, single drop micrometeorological data	Showed the effects of heat conduction from leaves.
Brain and Butler (1985)	Wheat	As Butler (1984) log-normal drop size distribution (rain)	Population of discrete drops takes longer to dry than a uniform film of the same volume.
Barthakar (1988)	Acetate film	Heat and mass transfer, single drops	Drying times <30 min of actual.
Weiss et al. (1989)	Dry bean	*CUPID* (Norman, 1982)	Good agreement with sensor measurements and observations.
Huber and Itier (1990)	Field bean	Mulilayer model with heat conduction	Good agreement with observations, especially near the base of the crop.
Jacobs et al. (1990)	Corn	Mulilayer, Bowen ratio energy balance with soil diffusivity, dew deposition	Dew deposition range +10 to −30% of observed.
Zhang and Gillespie (1990)		Energy balance, single leaf, micrometeorological or weather station data	Good agreement for shaded leaves.
Bass et al. (1991)	Corn	Zhang and Gillespie (1990)	Simulation study of geographical distribution of wetness duration.
Weihang and Goudriaan (1991)	Short grass, rice	Multilayer energy balance based on *Microweather* (Goudriaan, 1977), daily weather data.	Dew formation simulation, a simplified model works well for nighttime dew formation.
Lhomme and Jimenez (1992)	Banana, plantain	Pedro and Gillespie (1982b) applied to upper leaves, standard weather data	Predicted wetness duration to <1 h on average.
Jacobs and Nieveen (1995)	Corn, barley	Multilayer, Bowen ratio energy balance, dew deposition	Good agreement on dew deposition for corn, duration underestimated by about 1 h. Not as good for barley due to effects of runoff and dripping.
Wittich (1995)	Apple orchard	Energy balance, single leaf, different boundary conditions.	Best agreement when temperature of canopy below leaf was assumed to be the same as that of the leaf, usually <1 h difference

$$u(z) = 0.4 \cdot u_* \cdot \ln\left(\frac{z-d}{z_0}\right) \tag{2}$$

where $u(z)$ is the mean wind speed at height z above the ground. In the equation the constant u_*, the friction velocity, scales the wind speed; z_0, the roughness length, scales the height; and d is a datum level, less than the crop height, called the zero plane displacement. At height d Equation 2 predicts that $u = 0$, but the equation is not valid within the crop. For most crops z_0 is usually an order of magnitude smaller than the height of the vegetation, h, and d is between 0.6 and 0.8 h (Monteith, 1973; Thom, 1975). Values of z_0 and d for a range of surfaces have been tabulated by

McCartney and Fitt (1985). Both z_0 and d may show some dependence on wind speed if the canopy is distorted and smoothed as wind speed increases (Grace, 1977). On sunny days when there is convective activity (unstable temperature lapse rate) or in the evening when atmospheric mixing is suppressed (stable temperature lapse rate) the wind profile deviates from Equation 2 (Thom, 1975; Grace, 1977; McCartney and Fitt, 1985). The equation can be modified to account for the effects of atmospheric stability (Monteith, 1973; McCartney and Fitt, 1985), but as the modifications are not simple the reader should consult more specialized texts. The logarithmic wind profile is strictly only applicable to well-formed surface boundary layers, that is, large uniform areas. Wind profiles near obstructions such as hedges or near changes in terrain, for example woodland boundaries, may be more complex than suggested by Equation 2. However, the general form of the profile is usually similar and the equation can be used as a first approximation to estimate wind speeds at crop height from local synoptic measurements (usually made at 10 m).

The wind profiles within crops greatly depend on crop architecture, particularly the vertical distribution of foliage and the size, shape, and density of foliage elements. For crops where the foliage is relatively uniformly distributed with height, for example cereals, wind speed profiles can often be estimated using an equation suggested by Cionco et al. (1963):

$$u(z) = u(h) \cdot \exp\left[\alpha\left(\frac{z}{h} - 1\right)\right] \tag{3}$$

where h is the crop height, and the attenuation coefficient α has a value between 0.3 and 3 depending upon crop type and leaf area density. Some values of α are given by McCartney and Fitt (1985). An alternative equation, but with a similar shape, was given by Landsberg and James (1971) and Thom (1975). As a first approximation, the wind speed within the crop can be taken as a fixed fraction of the speed at the top of the crop (i.e., $u(z) = \beta.u(h)$); β usually has a value between 0.1 and 0.5 (McCartney and Fitt, 1985). In both equations $u(h)$ can be estimated from Equation 2 with z set equal to h. In crops with a "canopy and stem" structure (orchards) the wind speed profile may be "S"-shaped, with wind speeds in the "stem" layer greater than within the "canopy" layer (Grace, 1977; Oke, 1978).

D. Humidity

Relative humidity, or more strictly water vapor pressure, within the canopy can be significantly different from that of the air above the crop. The profile of vapor pressure is the consequence of the heat and water balance of the system. Understanding and modeling water vapor distribution and movement within crops has been the main concern of crop microclimatologist for many years (Monteith, 1975; Oke, 1978; Monteith and Unsworth, 1990). During the day the water vapor pressure is usually higher within the crop than above it, because the crop is transpiring, and it increases toward the soil surface. At night the differences between within- and above-crop water vapor pressures are smaller. Near the top of the canopy water vapor pressure may be less than above when dew is forming. In a transpiring crop the water vapor pressure close to the leaf surface, where the pathogen is growing, may be close to saturation.

Water vapor pressure is not normally measured directly but is calculated from relative humidity and temperature measurements. There are several techniques available to measure relative humidity including sensors compatible with most data acquisition systems (see, for example, Campbell and Madden, 1990). These include electronic (usually capacitive) sensors and wet and dry bulb psychrometers. Electronic sensors are simple to use but can suffer from calibration drift problems if they are used for long periods at high humidities. Dew-point meters give a more direct measurement of water vapor pressure, but are expensive and not usually suited for automatic use over long periods.

E. Wetness

Crop wetness duration is probably the most important environmental factor in crop disease epidemiology. Plants can become wetted directly by rain or spray irrigation or indirectly by dew formation or guttation (exudation of water from foliage). Unlike the other critical environmental parameters important in host pathogen interactions, the duration of wetness on foliar elements (including fruits, pods, flowers, etc.) is very difficult to define physically because of the spatial inhomogeneity of wet and dry areas both at leaf and crop scale (Huber and Gillespie, 1992). Thus, making meaningful measurement or estimates of wetness duration is not simple. As far as a pathogen (spore, hyphal segment) on a leaf surface is concerned, it is either covered in liquid water (wet) or not (dry). Therefore, perhaps the best way to define "wetness" in relation to foliar disease is "the area or proportion of area containing free water". Wetness duration could then be defined as the time a given proportion of the crop remained wet. Currently, none of the commonly used wetness sensors (see below) fulfil this standard and complex microclimatic models would be needed to estimate wet leaf area. Nevertheless, the measurement and estimation of wetness duration periods play essential roles in many crop disease warning and forecasting models. The ambiguousness in defining wetness duration complicates the translation of the results of laboratory or controlled environment experiments to the field. In laboratory studies leaves are usually completely dry or nearly 100% wet, therefore wetness duration is relatively easy to define. However, in the field naturally drying leaves will contain different amounts of water and some parts of the leaf will remain wet for longer periods than others, leading to a range of wetness durations for different foliar elements or areas of the crop. Huber and Itier (1990) found differences in leaf wetness duration after rain of up to 10 h in different parts of a bean canopy.

Unlike temperature and wind, it is very difficult to generalize about wetness duration in crops. Patterns of crop wetness are dependent on the complex interaction of crop and environmental factors. The method of wetting (rain, irrigation, dew, or guttation) determines both the amount of water on the crop and to some extent the pattern of leaf wetness: rainfall tends to produce a large range of water drop sizes giving a nonuniform cover on leaves (Royal and Butler, 1986), whereas dew tends to produce more uniform drop sizes and a more even cover. The nature of the foliar surface also plays an important role in determining the pattern of water on the leaf. Wettable leaves, such as tomato, will tend to have fairly uniform water films, while the water pattern on waxy leaves, oilseed brassicas for example, will consist of discrete drops. Wetness duration is a combination of the wetting period and the time taken to dry the crop. Wetting periods not only depend on the source of water but also on crop architecture as this controls the interception of rain and the pattern of dew formation on leaves. Evaporation of water from leaves depends on the radiation load, on the leaf, temperature, water vapor pressure, wind speed, leaf orientation, and water drop size and distribution on the leaf surface. Crop structure is also important in determining drying periods. Therefore, even under identical environmental conditions, wetness duration will differ for different crops or for the same crop at different growth stages.

1. Rain (Irrigation)

The length of time a crop remains wet partly depends on how much water it can hold on its foliage. Interception of rain by crops depends on rainfall duration and drop size distribution, wind speed, ground cover, crop structure, and foliage characteristics (Royal and Butler, 1986). Because of its importance in understanding the water balance of large catchments, the interception of rain by forests has been widely studied (Rutter, 1975; Huber and Gillespie, 1992). Agricultural crops, in contrast, have received relatively little attention (Clark, 1940; Appelmans et al., 1980; Butler and Huband, 1984). The type of rain determines both the rate of wetting and the efficiency with which the raindrops are captured. Small raindrops are captured efficiently but large drops will splash on leaves, redistributing the water, and may shake settled water from the plant (Fitt et al., 1989). Strong wind can also shake water from leaves, reducing the amount of water intercepted. Raindrop size distribution is generally related to rainfall intensity (Marshall and Palmer, 1948; Ulbricht, 1983; Bennett et al., 1984): light rain usually contains mostly small drops (<1 mm) but intense showers can have drops of up to about 5 mm in diameter (Willis and Tattelman, 1989). For low-intensity rainfall, crops which have complete ground cover will collect nearly all the rain which falls on them until the crop is saturated and water begins to run off the foliage (Royal and Butler, 1986). In heavy showers much of the rain falling on the crop may reach the ground as through-fall or runoff. Raindrop size distributions are also important in the dispersal of inoculum by splash (Fitt et al., 1989). Studies by Appelmans et al. (1980) suggest that the water-holding capacity of a crop and the time taken to fully "wet" it may also be dependent on rainfall intensity. There appears to be little information available on the water-holding capacity of most crops, but it depends primarily on the ground cover (leaf area index), crop architecture (leaf orientation), and the nature of the leaves (Royal and Butler, 1985). Butler and Huband's (1984) studies in wheat suggest that interception increases asymptotically with the rainfall amount until it reaches the water-holding capacity of the crop. Braden (1985, quoted in Friesland and Schrödter, 1988) has suggested the following relationship between rainfall interception I_c, rainfall amount P, leaf area index L, and ground cover b:

$$I_c = a \cdot L \cdot \left[1 - \frac{1}{1 + P \cdot b/a \cdot L} \right] \tag{4}$$

where a is the maximum water-holding capacity of the crop per unit leaf area index. Rainfall duration determines not only the amount of water held on a crop (until the maximum water-holding capacity is reached) but the distribution of water within the crop. Upper leaves will be wetted first, but after a prolonged rain the crop will become more or less uniformly wet. Mist or fog also produces uniform wetness.

Rain generally produces a large range of drop sizes on foliage, giving a relatively nonuniform cover (Brain and Butler, 1985). However, the surface characteristics of foliage also determine whether water forms films or discrete drops on leaves. Crops which hold water as discrete drops tend to have a larger water-holding capacity per unit leaf area than those which hold the water as films, although some of the leaf area may be "dry". The size and pattern of water drops on leaves affect evaporation rates and hence wetness duration.

2. Dew

At night, radiative cooling can cause the temperature of plant surfaces to fall below that of the air. Dew is formed when the surface temperature falls below that of the dewpoint of the surrounding air. Water will continue to condense as long at the surface temperature is lower than the dewpoint of the air in contact with the leaves. The effective "drying" of the air by condensation means that for dew to continue to be formed there must be a transport of water vapor to the plant surface from the atmosphere and/or from the soil or other plants. When water is transferred from the atmosphere dew formation is known as *dewfall*, otherwise it is called *distillation* (Monteith, 1957, 1963). Dew formation is therefore a fairly complex process involving not only the energy balance of individual surfaces but mechanisms of water vapor transport. For dewfall to occur wind speeds need to be within a critical range, depending on crop canopy structure and the water vapor content of the air above the crop. If wind speeds are too low the water lost by condensation cannot be replaced by turbulent transport from the more humid air above and condensation stops. When wind speeds are too high the temperature of the surface is raised above the dewpoint by turbulent heat transfer and condensation stops (Oke, 1978). For a short grass, a minimum wind speed of about 0.5 m s^{-1} at 2 m above the ground is needed (Monteith, 1957). Dewfall and distillation probably occur simultaneously in most canopies (Garrat and Segal, 1988) and the relative contributions depend on a number of factors including soil moisture availability, canopy structure, and atmospheric moisture. Under humid atmospheric conditions dewfall dominates, but distillation becomes important under semiarid conditions or at low wind speeds if there is sufficient soil moisture. Calculations suggest that under favorable conditions distillation may account for up to 80% of dew formation (Monteith, 1963; Garrat and Segal, 1988) and therefore cannot be ignored when modeling dew formation in crop canopies. Maximum rates of dewfall for saturated air have been estimated to be between 0.06 and 0.07 mm h^{-1} (Monteith, 1973) but dew formation may reach 0.09 mm h^{-1} with contributions from distillation (Garrat and Segal, 1988). Measured values of total dew formation can reach 0.6 mm per night, but values of about 0.2 mm appear to be more typical (Garrat and Segal, 1988; Huber,1992).

3. Guttation

Guttation is the exudation of water through special pores, called hydathodes, generally found on leaf margins. In grasses the hydathodes are usually at the tip of the leaf. When evaporation is low the exuded water forms drops supplied through the roots via the vascular system. Thus for guttation to occur the rate of water supply from the roots must exceed the loss by transpiration. Guttation is probably controlled by soil moisture and occurs at high moisture contents, and soil temperature may also play a regulatory role but only when soil moisture is not limiting (Hughes and Brimblecombe, 1994). In their study of short grass, Hughes and Brimblecombe (1994) found that guttation was responsible for about the same amount of water (0.1 mm per night) as dew. The average size of the guttation drops was much larger than drops formed by dew (1.49 mm compared to 0.2 mm), but they were formed mostly at the leaf tips. Guttation is probably more important in wetland crops such as a rice paddy where soil moisture is not a limitation, and when water temperature is high enough (optimum about 30°C) (Luo and Goudriann, 1997). The importance of guttation drops in disease processes is not clear, particularly since they tend to form along leaf margins.

4. Measurement of Leaf Wetness Duration

Methods to estimate or measure leaf wetness duration have been available for as long as its importance in crop disease has been recognized (Wallin, 1963, 1967) and there have been several recent reviews of methods and practices currently in use (Howard and Gillespie, 1985; Sutton et al., 1988; Campbell and Madden, 1990; Weiss, 1990; Huber and Gillespie, 1992). The ideal "wetness sensor" should measure both the duration and amount of water on the *surface of the crop* itself. This has rarely been achieved in practice, and simple-to-use relatively inexpensive sensors for general routine use which conform to this standard have still to be developed.

The amount of water deposited can be measured by weight, and this method has been used for both plant and artificial surfaces. For example, Leick (1932) used plates containing material which absorbed water and assessed dew deposition by weighing the plates before and after exposure. Mechanically recording balances have been used to automatically weigh water deposition on whole plants (Jennings and Monteith, 1954), detached plant shoots (Hirst, 1954), and on polystyrene test surfaces (Hirst, 1957). In recent years weighing methods have gone out of favor, but developments in sensor technology, especially in pressure sensors, make such methods worth reconsidering. Other indirect mechanical methods for sensing wetness have included using a water-soluble pencil to mark a rotating glass plate when it was wet (Taylor sensor) and measuring the change in length due to wetting of an absorbing material (Wallin, 1963).

Since the early 1970s, when electronic data acquisition became common, most wetness sensors have detected the presence of water by measuring the change in electrical impedance between two electrodes under wet and dry conditions. The simplest form of this type of sensor consists of an interlocking grid of electrodes on a nonconducting surface (Crossan, 1962); the surface can be painted with latex paint to enhance wettability (Davis and Hughes, 1970) and to match the color of the crop (Gillespie and Kidd, 1978). In a variation of this design McCartney and Lacey (1990) used a flat plate coated with electrically conducting paint with a "key" pattern inscribed on it; the narrow separation of the electrodes allowed the rapid detection of small droplets produced by dew. Synthetic sensors do not necessarily respond to wetting and drying in the same way as the crop, and wetness duration recorded by artificial sensors can be significantly different from that on the crop (Gillespie and Kidd, 1978; Huband and Butler, 1984; Potratz et al., 1994). To combat this, sensors have been designed to mimic the behavior of leaves; for example, "leaves" made from cotton cloth on a grid network of fine wires gave a better measure of leaf wetness on alfalfa than flat-plate painted sensors (Weiss and Lukens, 1981; Weiss and Hagen, 1983); and Gillespie and Duan (1987) used cylindrical sensors to mimic onion leaves.

The principle of measuring changes in resistance to detect surface water has also been applied directly to plant surfaces. In this type of sensor electrodes are attached directly to the leaf surface; for example, electrodes have been clipped onto leaves (Howard and Gillespie, 1985); Weiss et al.

(1989) used a grid network of independent, adjacent fine wires attached to dry bean leaves; Harold (1995) connected a circular array of fine wires to the capsules of linseed plants. McCartney and Lacey (1990) clipped oilseed rape debris between a frame consisting of two sets of electrodes in a study of *Pyrenopeziza brassicae* ascospore release and Fernandes et al. (1991) used needles inserted into wheat straw to measure water content when studying the maturation of ascospores of *Pyrenophora tritici-repentis*. Needles inserted through leaves have also been used to measure wetness in turfgrass canopies (Giesler et al., 1996). Such sensors are likely to give a more realistic measure of wetness duration than "artificial sensors": Giesler et al. (1996) found that wetness duration estimated by their sensor agreed within about 20 min of direct observation of the leaves. However, resistance sensors cannot easily be calibrated to give reliable estimates of the amount of water on the plant surface.

Indirect methods have been considered for monitoring liquid water on crop surfaces. Bunnenberg and Kühn (1977) used absorption of β-radiation to estimate dew on soil and Barthakur (1983) used this technique to measure the amount of water on individual leaves. The leaf was placed between a source (thallium-204) and a detector. The amount of radiation absorption could be directly related to the amount of water on the leaf surface. In laboratory and field tests the β-ray system gave consistently better estimates of actual wetness duration on leaves than electrical resistance sensors (Barthakur, 1985; Armstrong et al., 1993). Gillespie et al. (1990) have shown that radar backscatter measurements may have the potential to measure water deposition on a whole-crop basis. Both these approaches could have important research applications, but they are unlikely to have applications in routine disease management systems. Changes in the reflection characteristics of plant leaves or in the polarization of reflected light due to the presence of water may be potentially useful in determining crop wetness (A.F.G. Jacobs, personal communication). Microwave emissions from crops appear to be sensitive to the hydrological characteristics of the soil and crop canopy (Wigneron et al., 1996): 1.4-GHz emissions depend on soil moisture while 5-GHz emissions are sensitive to both soil moisture and crop water content. Remote sensing of microwave emissions may be potentially useful in monitoring crop wetness, at least for research purposes.

Direct monitoring of crop wetness has an important part to play in the development of disease control strategies, especially with the advent of on-farm computers and microprocessors. However, as most currently available sensors do not conform to an "ideal sensor", care must be taken with the interpretation of sensor output. This is particularly true for artificial sensors where the sensor response can be different from that of the crop. In such cases the sensor output should be "calibrated" to match the crop. Alternatively, the "disease model" should be matched to the output of specific sensors. There is a need to develop standards of wetness-sensing instruments of different types, and also to investigate and develop noncontact methods of wetness determination applicable to the whole crop or to those parts of the crop vulnerable to infection.

F. Estimating Crop Microclimate

Direct measurement of microclimatic variables is obviously the best way to obtain the information needed for disease forecasting or warning schemes or for disease control decision strategies. Automatic weather stations are available for on-farm use, but it is unlikely that for the foreseeable future direct measurement of crop environments will be a practical option for most farmers. Thus, within-crop environmental factors need to be estimated from measurements or forecasts made at sites which may be remote from the field. In most cases the only meteorological information available is from a local synoptic weather station. Measurements at such sites are usually of a restricted number of variables and are made under standard exposure conditions, usually 1 m above short grass for temperature, humidity, and radiation or 10 m for wind speed. As noted above, the presence of a crop canopy can significantly alter the local microclimate and it may not be appropriate to use synoptic measurements in disease forecasting models which have been based on controlled environment studies or direct field measurements. Mathematical modeling of crop climates offers a method to estimate critical microclimate parameters from above-crop measurements, but the development of such models is not a trivial exercise (Huber and Gillespie, 1992).

Much of the research in agricultural meteorology is aimed at understanding the interactions between plants and their soil and aerial environments, but there have been relatively few attempts to simulate the complete microclimate of a crop (e.g., *Microweather,* Goudriaan, 1977; *CUPID,* Norman, 1982; Norman and Campbell, 1983; Sauer and Norman, 1995; *AMBETI,* Braden, 1995). In these models the crop is divided into a number of layers and the microclimate (leaf temperature, air temperature, relative humidity, leaf wetness) is simulated by solving the energy balance of each layer. The forcing variables are typically air temperature, solar radiation, vapor pressure (or relative humidity), wind speed, soil temperature, and soil moisture. The models usually operate on hourly or half hourly time steps, therefore hourly values of input variables should be used, but these can be sometimes estimated from daily values (Norman, 1982). The models also require information on crop structure, usually in the form of leaf area index profiles, crop height, leaf angle distribution, and stomatal and boundary layer resistances. If this information is not available appropriate estimates can be made according to the crop (Norman, 1982). The accuracy of these models depends very much on the quality of the input information, but in the absence of direct measurements they can provide useful information.

In recent years there has been increasing interest in applying micrometeorological models to the estimation of leaf wetness duration (Huber and Gillespie, 1992). The models can use measured or estimated microclimatic inputs to simulate condensation and evaporation processes or can include these processes in complex microclimatic models (Table 2). Leaf wetness duration is calculated from evaporation rates of droplets or water films, and several models based on energy and mass balances have been devised (Butler, 1984; Brain and Butler, 1985; Barthakur, 1988; Butler, 1990; Zhang and Gillespie, 1990). The rate of evaporation depends upon the size and shape of the water drop or film, thus models must take into consideration the origin of the source of wetness. Leaf wetness duration models of various complexity have been developed for a number of crops (Table 2) and their accuracy has often been comparable with that of artificial wetness sensors (Huber and Gillespie, 1992). With the advent of personal computers the use of complex mathematical models for predicting crop microclimate is becoming a practical proposition for both regional and on-farm use.

The use of empirical models for estimating the duration of dew periods is an alternative to microclimatic modeling. Gleason et al. (1994) developed an empirical model which used relative humidity, air temperature, and wind speed to estimate occurrence and duration of dew. The model was based on data from one site and validated with hourly data from 13 weather stations in the north-central U.S. The model predicted the mean dew periods for the study to within 1 h and estimated the occurrence or absence of dew correctly 83.5% of the time. These results suggest that models of this type could provide dew period estimates over similar geographic areas where suitable weather station data are available.

IV. ENVIRONMENT AND DISEASE CONTROL

Zadoks and Schein (1979) identified six general methods for control of crop diseases: avoidance of the pathogen, exclusion of the pathogen, eradication of the pathogen, protection of the plant, development of host resistance, and therapy applied to the diseased plant. Knowledge of the interactions between host, pathogen, and environment can assist in the development of control strategies in a number of these areas (avoidance, eradication, protection, and therapy) and is one of the cornerstones of "integrated plant disease management". Disease control measures can also be grouped into "passive" measures (avoidance, exclusion, resistance) where no direct action is taken against the pathogen during the growth of the crop, and "active" measures (protection and therapy).

A. Passive Measures

Knowledge of the environmental conditions which are unsuitable for a disease to develop (e.g., lack of leaf wetness, too low or too high a temperature) can be used to devise "disease avoidance" strategies. Regional or local areas where the climate is suitable for specific diseases can be identified and the growth of susceptible crops avoided (Payen, 1983; Weltzien, 1988; Rossi et al., 1995). Such analyses nearly always have to use synoptic climatological data for the region or area. This is unlikely able to provide the microclimatic information needed to assess the impact of many crop diseases, especially those with specific leaf wetness requirements. Micrometeorological modeling can be used to provide in-crop climatic data (Huber and Gillespie, 1992). Scherm and Van Bruggen (1995) used Pedro and Gillespie's (1982b) model to study the distribution of dew periods in different climatic regions of California. Microclimatic models are also being used to develop climatic zoning for disease in Europe (Löpmeier, 1994; Orlandini et al., 1994). Additionally, knowledge of the geographic and seasonal distribution of crop microclimatic factors may allow periods when the microclimate favors disease to be avoided by adjusting the sowing date.

Passive measures also include manipulation of the microclimate to discourage disease. This is generally more difficult for field crops than for protected crops. Manipulation of the light regime in greenhouses has been used to suppress disease spread (see above, Vakalounakis, 1991). In field crops, variation of agronomic practices (e.g., row spacing, sowing density, irrigation) could be used to manipulate microclimate. For example, Giesler et al. (1996) found that in tall fescue *(Festuca arundinacea),* increasing the canopy density, by using a different seed rate or cultivars increased leaf wetness and relative humidity, except in wet years. The severity of brown patch disease *(Rhizoctonia solani)* was greater in the high-density canopies, suggesting that reduction in canopy density could reduce disease incidence. Tompkins et al. (1993) found that the microclimate in no-till winter wheat crops could be affected by changing seed rate and row spacing. High seed rates and narrow row spacing produced canopy microclimates favorable to the development of disease caused by *Septoria* spp. The duration of leaf wetness was longer and the intensity of downy mildew *(Bremia lactucae)* was greater in lettuce crops under furrow irrigation than under subsurface drip irrigation (Scherm and Van Bruggen, 1995). Detailed knowledge of crop microclimate processes, including modeling, could be effective in suggesting appropriate changes in agronomic practice. Crop microclimate can also be manipulated by altering crop architecture. This could be achieved directly by mechanical means or by plant breeding. Removal of the leaves from around berry clusters in grapevine has been shown to alter the microclimate, particularly the wind speed, and to reduce the incidence and severity of *Botrytis cinerea* infection of the fruit (Gubler et al., 1987; English et al., 1989). This practice has since been adopted by wine growers in California. Microclimatic models could be used to investigate the effects of crop architecture changes and suggest areas which could be exploited by crop breeders.

Mathematical modeling of disease epidemic development in crops has made significant advances over the last 20 years (Campbell and Madden, 1990). It is now feasible to combine disease progress, microclimate, and crop growth models to explore the potential for changing husbandry practices to minimize disease spread and development.

B. Active Measures

The majority of crop disease management measures involve taking action to either protect the crop from the pathogen or to kill or limit the growth of the pathogen on the crop. In both cases this usually involves applying the active ingredient (e.g., phytochemical, biological control agent) directly to the crop. As discussed in the Introduction to this chapter, environmental, social, and economic considerations prohibit the indiscriminate use of such measures. There is, therefore, a need to reduce the use of phytochemicals in agriculture, while maintaining an acceptable economic return for the farmer. This can be achieved in part avoiding the overuse of prophylactic measures and by taking action only when it is necessary to control the disease at an economically acceptable level. Knowledge of the relationships between pathogen, host, and environment can be used to build disease forecasting systems to aid in decisions on when or if to apply control measures. This obviously requires information on the effects of environmental factors on pathogen development.

Campbell and Madden (1990) define a *disease forecaster* as predictor of "the outbreak or increase in intensity of a disease based on information on the weather, crop or pathogen." Disease forecasters usually predict one or more critical phases in epidemic development such as initial

Table 3 Some Recent Disease Forecasting/Warning Simulation Models

Pathogen(s)	Crop	Source	Remarks
Botrytis cinerea *Peronospora destructor* *Phytophthora infestans*	Potato Onion Strawberry	Kaukoranta et al. (1993)	Device to monitor weather and calculate disease risk.
Bremoa lactucae	Lettuce	Scherm et al. (1995)	Recommends spray applications.
Colletotrichum geosporiodes	Citrus species	Timmer and Zitko (1993)	Empirical model.
Erysiphe gramminis f. sp. *tritici*	Winter wheat	Friedrich (1995)	Spore dispersal and infection model.
Leptodotium elatus *Peltaster tructicola* *Zygophiala jamaicensis*	Apple	Brown and Sutton (1995)	Empirical model to predict first symptoms based on leaf wetness duration.
Pyricularia grisea	Rice	Calvero et al. (1996)	Empirical model which uses weather factors.
Rhizoctonia solani	Perennial ryegrass	Fidauza et al. (1996)	Weather-based disease warning model.
Stagonospora nodorum	Winter wheat	Djurle et al. (1996)	Empirical weather/disease model.
Stemphyllium vesicarium	Pear	Montesinos et al. (1995)	Regression model base on wetness duration and temperature.
Uromyces appendiculatus	Phaseolus bean	Berger et al. (1995) Amorim et al. (1995)	Disease simulation model.
Venturia inaequalis	Apple	Xu et al. (1995)	Disease simulation model based on weather factors.

infection, secondary infection, or disease development. Information on environmental factors is usually essential for the disease forecaster, although some only require biological information (Campbell and Madden, 1990). The nature of disease forecasting models depends partly on the source of the environmental information needed: regional synoptic measurements, local synoptic measurements, in-crop measurements, in-crop estimates or, more rarely, weather forecasts. For example, a forecaster based on detailed studies of pathogen processes would probably require measurements or estimates of in-crop micrometeorological variables, while one based on empirical field studies would most likely use local synoptic measurements or estimates. Thus, disease forecasters vary in complexity — from disease warning schemes which identify periods when weather conditions are suitable for infection or disease spread, to systems which predict disease development and potential effects on yield (see reviews cited below). An example of a "disease warning" forecasting system is that for the apple scab *(Venturia inaequalis)* which identifies periods when the environmental conditions favor infection (leaf wetness and temperature) (Mills and LaPlante, 1951). A microprocessor-based system which monitors weather variables and calculates "infection periods" has been commercially produced and has been validated at several locations (Jones et al., 1980; Campbell and Madden, 1990). In the Netherlands a complex computer-based disease and pest management system, EPIPRE (EPIdemics PREvention) (Zadoks et al., 1984) was developed to aid pest and disease control in winter wheat. This system was field based but used a central computer to simulate disease development of a number of pathogens based on individual field observations. The system used crop and pathogen information as well as weather data.

Many different disease forecasting systems have been produced over the last 50 years and there have been several excellent reviews on their development and use (Campbell and Madden, 1990; Fry and Fohner, 1985; Jones, 1986; Pennypacker and Stevenson, 1982; Waggoner, 1984). The development of disease forecasting systems is discussed in detail by Madden and Ellis (1988). Lack of space precludes an attempt to review current and previously used disease forecasters in this chapter (some recent models are listed in Table 3). Instead, there follows an outline of three examples of disease forecasting systems which illustrate the use of environmental information in disease management.

BLIGHTCAST* (Krause et al., 1975) late blight of potato *(Phytophthora infestans) — This was a synthesis of the Hyre and the Wallin fungicide scheduling systems developed for the northeastern and midwestern U.S. *BLIGHTCAST* was based on predicting periods when infection is likely. Spray recommendations were issued in response to two different weather triggers: after 10 consecutive days when the cumulative rainfall exceeded 30 mm and the 5-day average temperature did not exceed 25.5°C; or after accumulation of 18 to 20 "severity units" (each day was scored on a 0 to 4 "severity" scale according to temperature and hours with relative humidity >90%). *BLIGHTCAST* was a computer programme which calculated the weather triggers from temperature, relative humidity, and rainfall measurements. Initially, *BLIGHTCAST* was run on a computer at Pennsylvania State University and offered as a free service. Farmers collected weather information on a recording hygrothermograph. Each week they telephoned a summary of this information to the university and received *BLIGHTCAST*'s recommendations. The free system was widely used by farmers, but a subsequent subscription service was not successful. A battery-operated microcomputer-based weather data logger and forecasting unit was developed ("Blightcaster", Campbell Scientific, Inc., Utah) for in-field use (MacKenzie, 1981) This equipment recorded the necessary weather information and displayed *BLIGHTCAST*'s fungicide recommendations. A number of growers successfully used these units, but they were not widely accepted. The experience with *BLIGHTCAST* illustrates one of the problems of using a disease forecaster. Although the program successfully predicted infection and its use generally reduced the number of fungicide applications, the direct financial saving to the farmer was relatively small compared to other production costs. Thus many farmers were not willing to risk using *BLIGHTCAST*, especially if it involved additional costs. However, as noted in the introduction, other pressures may force a reduction in the use of phytochemicals, making disease forecasters of this type more attractive to the grower.

***TOM-CAST* (Pitblado, 1992) early blight *(Alternaria solani)* on tomato** — The forecaster for tomato early blight *(FAST)* developed by Madden et al. (1978) was modified for use in Southern Ontario, Canada. Like *BLIGHTCAST*, *TOM-CAST* identifies periods when environmental conditions are favorable for disease development and recommends a fungicide application schedule. Disease severity values (0, no risk — 4, high risk) were assigned to each day according to periods of wetness and temperature during the forecast periods. The first spray was applied when at least 35 disease severity values had been accumulated. Subsequent applications were recommended after the accumulation of at leat 20 disease severity values. *TOM-CAST* required direct measurement of wetness duration in the field, which was possible using wetness sensors and data collecting systems such as those used with *BLIGHTCAST*. However, to encourage wider use of the forecaster Gillespie et al. (1993) examined methods of using synoptic weather data in place of direct wetness measurements. They concluded that with the appropriate calibrations, *TOM-CAST* could be used regionally by using locally observed daily maximum and minimum temperatures and dewpoint measurements made at two regional weather stations. This type of approach could provide a practical compromise between direct on-farm monitoring of crop microclimate (accurate but expensive) and general disease forecasts for whole regions (cheap but less accurate). Disease forecasters which use regionally estimated weather information could be operated by agricultural consultants serving a group of farmers in the same area.

***PLASMO* (Rosa et al., 1993, 1995) downy mildew *(Plasmopara viticola)* on grapevine** — This forecaster has been developed to assist growers in Northern Italy reduce the number of chemical treatments needed to control downy mildew on grapevines. *PLASMO* is a sophisticated computer programme which simulates the critical phases in the pathogen's life cycle (inoculation, incubation, and spore production) using rate factors determined by environmental parameters (temperature, relative humidity, rainfall, and leaf wetness duration). The programme has been designed to be "user friendly" and is run interactively. The programme simulates infection development and allows the grower to decide whether or not spray treatment is needed. There is also an option to input the required weather data from automatic weather stations situated within the vineyard. Conventional control of this disease is done by application of sprays at regular 7- to 10-day intervals. In field trials in Tuscany, using weather data measured in the vineyard, spraying according to *PLASMO*'s recommendations reduced the number of treatments by 25 to 35% while attaining the same degree of disease control. Forecasters of this type are at the "top end of the market" and currently are probably only commercially suited to high-value crops. However, the computer revolution of recent years has produced personal computers capable of running extremely sophisticated models. Simulation-based forecasters could easily be run on-farm or by agricultural consultant companies operating local environmental monitoring and disease prediction networks tailored to the farmers' needs.

Disease forecasting may not always improve the management of some diseases, for example, if the pathogen is always present and climatic conditions are always favorable then control measures will always be needed and forecasts may not be beneficial (Fry and Fohner, 1985). But, for diseases whose important occurrence is sporadic, disease forecasting should provide significant benefits (Fry and Fohner, 1985; Campbell and Madden, 1990). Currently, the use of disease forecasting as a tool in disease management is the exception rather than the rule. However, the need to reduce the use of phytochemicals should increase the adoption of rational decision support systems. Thus, disease forecasting incorporating environmental inputs will have an even more important role to play in crop protection programs and much effort is being devoted to developing new models for different crops (Table 3).

V. CONCLUSIONS

This chapter has highlighted the importance of environmental factors in the development of epidemics in crops and indicated how knowledge of the interactions of pathogen, crop, and environment can be of use in disease management. The are many areas which we have had to omit. For instance, the environment plays an equally important role in the epidemiology of soil-borne pathogens. Virus diseases of crops are also influenced by environmental factors, particularly though their effects on vector behavior. Knowledge of the effects of environmental factors on these pathogens is also needed to understand and control the diseases they cause. Disease can be affected by the growth of the crop and its physiological status; both are influenced by environmental factors. The effectiveness of crop spraying depends largely on the weather. Thus an understanding of the interactions between microclimate and spray droplets can help to improve the effectiveness of chemical control measures by better targeting and reduction of spray drift (see Matthews and Hislop, 1994, for a recent review of crop spraying techniques).

The environmental, social, and political pressure to reduce phytochemical use will probably continue to increase toward the millennium and beyond. It is clear that an understanding of the interactions between pathogens, crops, and environment can lead to more effective control of many crop diseases. However, for many important diseases these relationships largely remain unknown. There is, therefore, still much work to be done, and plant pathologists and environmental scientists will need to continue to work together to provide this information. Nevertheless, the potential for the development and widespread implementation of rational disease management strategies based on a sound understanding of the principles of disease processes has never been better: computing is becoming less expensive, more powerful, and more widely accepted; new methods of information handling such as decision support systems and expert systems are being developed; new methods for monitoring and estimating environmental variables are being developed. It is to be hoped that continued joint research will allow these factors to come together to produce environment- and farmer-friendly crop protection systems.

REFERENCES

Amorim, L., Berger, R. D., Bergamin, A., Hau, B., Weber, G. E., Bacchi, L. M. A., Vale, F. X. R., and Silva, M. B., A simulation model to describe epidemics of rust of Phaseolus beans. 2. Validation, *Phytopathology*, 85, 722, 1995.

Appelmans, F., Van Hove, J., and De Leenheer, L., Rain interception by wheat and sugar beet crops, in *Assessment of Soil Erosion*, de Boodt, M. and Gabriels, D., Eds., John Wiley & Sons, Chichester, 1980, 227.

Armstrong, R., Barthakur, N. N., and Norris, E., A comparative study of three leaf wetness sensors, *Int. J. Biometeorol.*, 37, 7, 1993.

Arny, C. J. and Rowe, R. C., Effects of temperature and duration of surface wetness on spore production and infection of cucumbers by *Didymella bryoniae*, *Phytopathology*, 81, 206, 1991.

Aylor, D. E., Force required to detach conidia of *Helminthosporium maydis*, *Plant Physiol.*, 55, 99, 1975.

Aylor, D. E., Dispersal in time and space: aerial pathogens, in *Plant Disease an Advanced Treatise*, Horsfall, J. G. and Cowling, E. B., Eds., Academic Press, New York, 1978, 159.

Aylor, D. E., Modelling spore dispersal in a barley crop, *Agric. Meteorol.*, 26, 215, 1982.

Aylor, D. E., Deposition gradients of *Puccinia recondita* near a source, *Phytopathology*, 77, 1442, 1987.

Aylor, D. E., The role of intermittent wind in the dispersal of fungal pathogens, *Annu. Rev. Phytopathol.*, 28, 73, 1990.

Aylor, D. E., McCartney, H. A., and Bainbridge, A., Deposition of particles liberated in gusts of wind, *J. Appl. Meteorol.*, 10, 1212, 1981.

Aylor, D. E. and Parlange, J.-Y., Ventilation required to entrain small particles from leaves, *Plant Physiol.*, 56, 97, 1975.

Bainbridge, A. and Legg, B. J., Release of barley-mildew conidia from shaken leaves, *Trans. Br. Mycolog. Soc.*, 66, 495, 1976.

Barthakur, N. N., The beta-ray gauge as a leaf surface wetness detector, *Int. J. Appl. Radiat. Isotopes*, 34, 1549, 1983.

Barthakur, N. N., A comparative study of radiometric and electronic leaf wetness sensors, *Agric. Forest Meteorol.*, 36, 83, 1985.

Barthakur, N., A non-equilibrium thermodynamic approach to surface evaporation of water drops, *Agric. Forest Meteorol.*, 42, 287, 1988.

Bashi, E. and Rotem, J., Effects of light on sporulation of *Alternaria porri* f. sp. *solani* and of *Stemphylium botryosum* f. sp. *lycopersici*, *Phytoparasitica*, 3, 63, 1975.

Bass, B., Savdie, I., and Gillespie, T. J., Simulation of leaf wetness duration of field corn, *Agric. Forest Meteorol.*, 57, 69, 1991.

Bennett, J. A., Fang, D. J., and Boston, R. C., The relationship between N_0 and Λ for Marshall-Palmer type raindrop-size distributions, *J. Climate Appl. Meteorol.*, 23, 768, 1984.

Berger, R. D., Hau, B., Weber, G. E., Bacchi, L. M. A., and Amorim, L., A simulation model to describe epidemics of rust of Phaseolus beans. 1. Development of the model and sensitivity analysis, *Phytopathology*, 85, 715, 1995.

Braden, H., Ein Energiehaushalts- und Verdunstungsmodel fur Wasser- und Stoffhaushaltsuntersuchungen landwirtschaftlich genutzer Einzugsgebiete, *Mitt. Dtsch. Bodenkundl. Ges.*, 42, 294, 1985.

Braden, H., The model AMBETI — a detailed description of a soil-plant-atmosphere model, *Berichte des Deuttscen Wetterdienstes*, 195, Deutscher Wetterdienst Offenbach., 1995.

Brain, P. and Butler, D. R., A model of drop size distribution for a system with evaporation, *Plant Cell Environ.*, 8, 247, 1985.

Bunnenberg, C. and Kühn, W., Application of the β-absorption method to measure dew on soil and plant surfaces, *Int. J. Appl. Radiat. Isotopes*, 28, 751, 1977.

Brown, E. M. and Sutton, T. B., An empirical model for predicting the first symptoms of sooty blotch and flyspeck of apples, *Plant Dis.*, 79, 1165, 1995.

Butler, D. R., The energy balance of water drops on a leaf surface, *Boundary Layer Meteorol.*, 32, 337, 1984.

Butler, D. R., Mass transfer from discrete water drops on leaves in a cereal canopy, *Boundary Layer Meteorol.*, 53, 33, 1990.

Butler, D. R. and Huband, N. D. S., Throughfall and stemflow in wheat, *Agric. Forest Meteorol.*, 35, 329, 1984.

Butler, D. R., Wadia, K. D. R., and Jadhav, D. R., Effects of leaf wetness and temperature on late leaf-spot of groundnut, *Plant Pathol.*, 43, 112, 1994.

Butler, D. R., Wadia, K. D. R., and Reddy, R. K., Effects of humidity, leaf wetness, temperature and light on conidial production by *Pheaoisariopsis personata* on groundnut, *Plant Pathol.*, 44, 662, 1995.

Calvero, S. B., Coaklet, S. M., and Teng, P. S., Development of empirical forecasting models for rice blast based on weather factors, *Plant Pathol.*, 45, 667, 1996.

Campbell, C. L. and Madden, L. V., *Introduction to Plant Disease Epidemiology*, John Wiley & Sons, New York, 1990.

Chakrabortry, S., Ratcliff, D., and McKay, F. J., Anthracnose of *Stylosanthes scabra*: effect of leaf wetness on disease severity, *Plant Dis.*, 74, 379, 1990.

Chamberlain, A. C., The movement of particles in plant communities, in *Vegetation and the Atmosphere*, Vol. 1, Monteith, J. L., Ed., Academic Press, London, 1975, 155.

Cionco, R. M., Ohmstede, W. D., and Appleby, J. F., A model for airflow in an idealized vegetative canopy, Meteorological Research Notes, Number 5, USAERDAA, Fort Huachuca, Arizona, 1963.

Clark, O. R., Interception of rainfall by Prairie grasses, weeds, and certain crop plants, *Ecol. Monogr.*, 10, 243, 1940.

Carisse, O., Kushalappa, A. C., and Cloutier, D. C., Influence of temperature, leaf wetness and high relative humidity duration on sporulation of *Cercospora carotae* on carrot leaves, *Phytopathology*, 83, 338, 1993.

Crossan, T. R., Dew Recording, Manuscript of the U.S. Weather Bureau, 1962.

Damicone, J. P., Berggren, G. T., and Snow J. P., Effects of free moisture on soybean stem canker development, *Phytopathology*, 77, 1568, 1987.

Davis, D. R. and Hughes, J. E., A new approach to recording the wetting parameter by the use of electrical resistance sensors, *Plant Dis. Rep.*, 54, 474, 1970.

Davis, H. and Fitt, B. D. L., Effects of temperature and leaf wetness on the latent period of *Rhynchosporium secalis* (leaf blotch) on leaves of winter barley, *J. Phytopathol.*, 140, 269, 1994.

Dennis, J. J. C., Holderness, M., and Keane, P. J., Weather patterns associated with sporulation of *Oncobasidium theobromae* on cocoa, *Mycol. Res.*, 96, 31, 1992.

Djurle, A., Ekbom, B., and Yuen, J. E., The relationship of leaf wetness duration and disease progress of glume blotch, caused by *Stagonospora nodorum*, in winter wheat to standard weather data, *Eur. J. Plant Pathol.*, 103, 9, 1996.

English, J. T., Thomas, C. S., Marois, J. J., and Gubler, W. D., Microclimates of grapevine canopies associated with leaf removal and control of Botrytis bunch rot, *Phytopathology*, 79, 395, 1989.

Fernandes, J. M. C., Sutton, J. C., and James, T. D. W., A sensor for monitoring moisture of wheat residues: application in ascospore maturation of *Pyrenophora tritici-repentis*, *Plant Dis.*, 75, 1011, 1991.

Fidauza, M. A., Dernoeden, P. H., and Grybauskas, A. P., Development and field validation of a brown patch warning model for perennial ryegrass turf, *Phytopathology*, 86, 385, 1996.

Figueroa, L., Fitt, B. D. L, Shaw, M. W., McCartney, H. A., and Welham, S. J., Effects of temperature on the development of light leaf spot (*Pyrenopeziza brassicae*) on oilseed rape (*Brassica napus*), *Plant Pathol.*, 44, 51, 1995a.

Figueroa, L., Fitt, B. D. L., Shaw, M. W., McCartney, H. A., and Welham, S. J., Early development of light leaf spot *Pyrenopeziza brassicae* on winter oilseed rape *Brassica napus* in relation to temperature and leaf wetness, *Plant Pathol.*, 44, 641, 1995b.

Fitt, B. D. L. and McCartney, H. A., Spore dispersal in relation to epidemic models, in *Plant Disease Epidemiology*, Vol.1: Population Dynamics and Management, Leonard, K. J. and Fry, W. E., Eds., Macmillan, New York, 1986a, 311.

Fitt, B. D. L. and McCartney, H. A., Spore dispersal in splash droplets, in *Water Fungi and Plants*, Ayres, P. G. and Boddy, L., Eds., Cambridge University Press, Cambridge, 1986b, 87.

Fitt, B. D. L., Gregory, P. H., Todd, A. D., McCartney, H. A., and Macdonald, O. C., Spore dispersal and plant disease gradients: a comparison between two empirical models, *J. Phytopathol.*, 118, 227, 1987.

Fitt, B. D. L., McCartney, H. A., and Walklate, P. J., Role of rain in the dispersal of pathogen inoculum, *Annu. Rev. Phytopathol.*, 27, 241, 1989.

Friedrich, S., Modelling infection probability of powdery mildew in winter wheat by meteorological input variables, *Z. Pflanzenkr. Pflanzenschutz*, 102, 354, 1995.

Friesland, H. and Schrödter, H., The analysis of weather factors in epidemiology, in *Experimental Techniques in Plant Disease Epidemiology*, Kranz, J. and Rotem, J., Eds., Springer-Verlag, Berlin, 1988, 116.

Fry, W. E. and Fohner, G. R., Construction of predictive models. I. Forecasting of disease development, in *Advances in Plant Pathol.*, Vol. 3: Mathematical Modelling of Crop Disease, Gilligan, C. A., Ed., Academic Press, London, 1985, 161.

Garratt, J. R. and Segal, M., On the contribution of atmospheric moisture to dew formation, *Boundary Layer Meteorol.*, 45, 209, 1988.

Gleason, M. L., Taylor, S. E., Loughin, T. M., and Koehler, K. J., Development and validation of an empirical model to estimate the duration of dew periods, *Plant Dis.*, 78, 1011, 1994.

Giesler, L. J., Horst, G. L., and Yuen, G. Y., A site specific sensor for measuring leaf wetness duration within turfgrass canopies, *Agric. Forest Meteorol.*, 81, 145, 1996.

Giesler, L. J., Yuen, G. Y., and Horst, G. L., The microclimate in tall fescue turf as affected by canopy density and its influence on brown patch disease, *Plant Dis.*, 80, 389, 1996.

Gillespie, T. R. and Kidd, G. E., Sensing duration of leaf moisture retention using electrical impedance grids, *Can. J. Plant Sci.*, 58, 179, 1978.

Gillespie T. J. and Duan, R.-X., A comparison of cylindrical and flat plate sensors for surface wetness duration, *Agric. Forest Meteorol.*, 40, 61, 1987.

Gillespie, T. J., Brisco, B., Brown, R. J., and Sofko, G. J., Radar detection of a dew event in wheat, *Remote Sensing Environ.*, 33, 151, 1990.

Gillespie, T. J., Srivastava, B., and Pitblado, R. E., Using operational weather data to schedule fungicide sprays on tomatoes in southern Ontario, Canada, *J. Appl. Meteorol.*, 32, 577, 1993.

Gold, R. E. and Mendgen, K., Rust basidiospore germlings and disease initiation, in *The Fungal Spore and Disease Initiation in Plants and Animals*, Cole, G. T. and Hoch, H. C., Eds., Plenum Press, New York, 1991, 67.

Grace, J., *Plant Response to Wind*, Academic Press, London, 1977.

Gregory, P. H., *Aerobiology of the Atmosphere*, 2nd ed., Leonard Hill, Aylesbury, 1973.

Grove, G. G., Madden, L. V., Ellis, M. A., and Schmitthener, A. F., Influence of temperature and wetness duration on the infection of immature strawberry fruit by *Phytophthora cactorum*, *Phytopathology*, 75, 165, 1985.

Goudriaan, J., *Crop Micrometeorology: A Simulation Study*, Centre for Agricultural Publishing and Documentation, Wageningen, 1977.

Gubler, W. D., Marois, J. J., Bledsoe, A. M., and Bettiga, L. J., Control of Botrytis bunch rot of grape with canopy management, *Plant Dis.*, 71, 599, 1987.

Hammett, K. R. W. and Manners, J. G., Conidium liberation in *Erysiphe graminis*. III. Wind tunnel studies, *Trans. Br. Mycol. Soc.*, 62, 267, 1974.

Harold, J. F. K., Epidemiology of Grey Mould, *Botrytis cinerea*, and Effects of Disease on Linseed, Ph.D. Thesis, University of Nottingham, U.K., 1995.

Harthill, W. F. T. and Cheah, L.-H., Some effects of climate and plant growth on the development of light leaf spot in cauliflowers, *N.Z. J. of Agric. Res.*, 27, 441, 1984.

Harthill, W. F. T. and Underhill, A. P., "Puffing" in *Sclerotinia sclerotiorum* and *S. minor*, *N.Z. J. Botany*, 14, 355, 1976.

Hawker L. E. and Madelin, M. F., The dormant spore, in *The Fungal Spore: Form and Function*, John Wiley & Sons, New York, 1976, 1.

Hildebrand, P. D. and Sutton, J. C., Interactive effects of the dark period, humid period, temperature, and light on sporulation of *Peronospora destructor*, *Phytopathology*, 74, 1444, 1984.

Hirst, J. M., A method for recording the formation and persistence of water deposits on plant shoots, *Q. J. R. Meteorol. Soc.*, 80, 227, 1954.

Hirst, J. M., A simplified surface-wetness recorder, *Plant Pathol.*, 6, 57, 1957.

Hirst, J. M., Stedman, O. J., and Hurst, G. W., Long distance spore transport: vertical sections of spore clouds over the sea, *J. Gen. Microbiol.*, 48, 357, 1967.

Hong, C. X. and Fitt, B. D. L., Effects of inoculum concentration, leaf age and wetness period on the development of dark leaf and pod spot *Alternaria brassicae* on oilseed rape *Brassica napus*, *Plant Pathol.*, 127, 283, 1995.

Howard, A. M. and Gillespie, T. J., A field guide to surface wetness duration measurement, Technical Memo 85-1, Department of Land Resources Science, University of Guelph, Ontario, Canada, 1985.

Huband, N. D. S. and Butler, D. R., A comparison of wetness sensors for use with computer or microprocessor systems designed for disease forecasting, in *1984 British Crop Protection Conference: Pests and Diseases, Vol. 2*, British Crop Protection Council, 1984, 633.

Huber, L., Déterminisme et mesure de la druée d'humectation en vue de la protection des plantes, *Agronomie*, 12, 281, 1992.

Huber, L. and Gillespie, T. J., Modelling leaf wetness in relation to plant disease epidemiology, *Annu. Rev. Phytopathol.*, 30, 553, 1992.

Huber, L. and Itier, B., Leaf wetness in a field bean canopy, *Agric. Forest Meteorol.*, 51, 281, 1990.

Hughes, R. N. and Brimblecombe, P., Dew and guttation: formation and environmental significance, *Agric. Forest Meteorol.*, 67, 173, 1994.

Ingold, C. T., *Fungal Spores Their Liberation and Dispersal*, Clarendon Press, Oxford, 1971.

Jacobs, A. F. G. and Nieveen, J. P., Formation of dew and the drying process within crop canopies, *Meteorol. Appl.*, 2, 249, 1995.

Jacobs Van Pul, W. A. J. and Van Dijken, A., Similarity of moisture dew profiles within a corn canopy, *J. Appl. Meteorol.*, 29, 1300, 1990.

Jennings, E. G. and Monteith, J. L., A sensitive recording dew balance, *Q. J. R. Meteorol. Soc.*, 80, 222, 1954.

Jones, A. L., Role of wet periods in predicting foliar disease, in *Plant Disease Epidemiology*, Vol. 1: Population Dynamics and Management, Leonard, K. J. and Fry, W. E., Eds., Macmillan, New York, 1986, 87.

Jones, A. L., Lillevik, S. L., Fisher, P. D., and Stibbens, T. C., A microcomputer-based instrument to predict primary apple scab infection periods, *Plant Dis.*, 64, 69, 1980.

Kaiser, W. J., Factors affecting growth and sporulation of *Ascochyta fabae* f.sp. *lentis*, *Plant Dis.*, 78, 374, 1994.

Kaukoranta, T., Hannukkala, A. O., and Tahvonen, R., Weather-based disease warning device, Special Report, Danish Institute of Plant and Soil Science, 7, 147, 1993.

Krause, R. A., Massie, L.B., and Hyre, R. A., BLIGHTCAST: a computerized forecast of potato late blight, *Plant Dis. Rep.*, 59, 95, 1975.

Lacey, J., Water availability and fungal reproduction: patterns of spore production, liberation and dispersal, in *Water Fungi and Plants*, Ayres, P. G. and Boddy, L., Eds., Cambridge University Press, Cambridge, 1986, 87.

Landsberg, J. J. and James, G. B., Wind profiles in plant canopies: studies on an analytical model, *J. Appl. Ecol.*, 8, 729, 1971.

Leach, C. M. and Anderson, A. J., Radiation quality and plant disease, in *Biometeorology in Integrated Pest Management*, Hatfield, J. L. and Thomason, I. J., Eds., Academic Press, New York, 1982, 267.

Legg, B. J. and Powell, F. A., Spore dispersal in a barley crop: a mathematical model, *Agric. Meteorol.*, 20, 47, 1979.

Leick, E., Zur Methodik der relativen Taumessung, *Beitr. Bot. Zentralb.*, 49, 160, 1932.

Löpmeier, F.-J., Übertragung phytopathologischer Warndienste in das orographisch gegliederte Gelände Norddeutschland, in *Computergestützte Prognose- und Entscheidungsmodelle im Pflanzenschutz, Internationales Symposium*, Bundesministerium für Ernährung, Landwirtschaft und Forsten, Bonn, 1994, 134.

Lhomme, J.-P. and Jimenez, F., Estimating dew duration on banana and plantain leaves from standard meteorological observations, *Agric. Forest Meteorol.*, 62, 263, 1992.

Luo, W. and Goudriaan, J., Effects of altering water temperature on leaf wetness in paddy rice crops, *Agric. Forest Meteorol.*, in press.

McCartney, H. A., The dispersal of plant pathogen spores and pollen from oilseed rape crops, *Aerobiologia*, 6, 147, 1990.

McCartney, H. A., Spore dispersal: environmental and biological aspects, in *Ecology of Plant Pathogens*, Blakeman, J. P. and Williamson, B., Eds., CAB International, Wallingford, 1994, 171.

McCartney, H. A. and Bainbridge, A., Deposition gradients close to a point source, *Phytopathol. Z.*, 109, 219, 1984.

McCartney, H. A. and Bainbridge, A., Deposition of *Erysiphe graminis* conidia on a barley crop. I. Sedimentation and impaction, *J. Phytopathol.*, 118, 243, 1987.

McCartney, H. A. and Fitt, B. D. L., Construction of dispersal models, in *Advances in Plant Pathol.*, Vol.3: Mathematical Modelling of Crop Disease, Gilligan, C. A., Ed., Academic Press, London, 1985, 107.

McCartney, H. A. and Lacey, M. E., The production and release of ascospores of *Pyrenopeziza brassicae* Sutton et Rawlinson on oilseed rape, *Plant Pathol.*, 39, 17, 1990.

McCartney, H. A. and Lacey, M. E., Release and dispersal of *Sclerotinia* ascospores in relation to infection, in *Brighton Crop Protection Conference: Pests and Diseases*, British Crop Protection Council, 1992, 109.

McCartney, H. A., Schmechel, D., and Lacey, M. E., Aerodynamic diameter of conidia of *Alternaria* species, *Plant Pathol.*, 42, 280, 1993.

McGregor, A. J. and Manners, J. G., The effect of temperature and light intensity on growth and sporulation of *Puccinia striiformis* on wheat, *Plant Pathol.*, 34, 263, 1985.

MacKenzie, D. R., Scheduling fungicide applications for potato late blight with BLIGHTCAST, *Plant Dis.*, 65, 394, 1981.

Macko, V., Inhibitors and stimulants of spore germination and infection structure formations, in *The Fungal Spore: Morphogenic Controls*, Turin, G. and Hohl, H. R., Eds., Academic Press, London, 1981, 564.

Madden, L. V. and Ellis, M. A., How to develop plant disease forecasters, in *Experimental Techniques in Plant Disease Epidemiology*, Kranz, J. and Rotem, J., Eds., Springer-Verlag, Heidelberg, 1988, 191.

Madden, L. V., Pennypacker, S. P., and MacNab, A. A., FAST, a forecast system for Alternaria on tomato, *Phytopathology*, 668, 1354, 1978.

Manners, J. G., *Principles of Plant Pathology*, Cambridge University Press, Cambridge, 1982.

Marshall, J. S. and Palmer, W. M., The distribution of raindrops with size, *J. Meteorol.*, 5, 165, 1948.

Mathiea, D. and Kushalappa, A. C., Effects of temperature and leaf wetness duration on the infection of celery by *Septoria apiicòla*, *Phytopathology*, 83, 1036, 1993.

Matthews, G. A. and Hislop, E. C., *Application Technology for Crop Protection*, CAB International, Wallingford, U.K., 1994.

Mills, W. D. and LaPlante, A. A., Control of diseases and insects in the orchard, *N.Y. Agric. Exp. Stn. Ithaca Bull.*, 711, 18, 1951.

Monteith, J. L., Dew, *Q. J. R. Meteorol. Soc.*, 83, 322, 1957.

Monteith, J. L., Dew: fact and fallacies, in *The Water Relations of Plants*, Rutter, J. and Whitehead, E. H., Eds., Blackwell, Scientific, London, 1963, 37.

Monteith, J. L., *Principles of Environmental Physics*, Edward Arnold, London, 1973.

Monteith, J. L. and Butler, D. R., Dew and thermal lag: a model for cocoa pods, *Q. J. R. Meteorol. Soc.*, 105, 207, 1979.

Monteith, J. L. and Unsworth, M. H., *Principles of Environmental Physics*, 2nd ed., Edward Arnold, London, 1990.

Montesinos, E., Moragrega, C., Llorente, I., Vilardell, P., Bonaterra, A., Ponti, I., Bugiani, R., Cavanni, P., and Brunelli, A., Development and evaluation of an infection model for *Stemphylium vesicarium* on pear based on temperature and wetness duration, *Phytopathology*, 85, 586, 1995.

Mridha, M. A. U. and Wheeler, B. E. J., *In vitro* effects of temperature and wet periods on the infection of oilseed rape by *Alternaria brassicae*, *Plant Pathol.*, 42, 671, 1993.

Mylchreest, S. J. and Wheeler, B. E. J., A method for inducing apothecia from sclerotia of *Sclerotinia sclerotiorum*, *Plant Pathol.*, 36, 16, 1987.

Nebane, C. L. N. and Ekpo, E. J. A., Effect of culture media, temperature and light on radial growth and pycnidium production of cowpea isolates of *Phoma bakeriana*, *Ann. Appl. Biol.*, 121, 537, 1992.

Nicot, P. C., Vaisière, B. E., and Lagier, J., Differential spore production by *Botrytis cinerea* on agar medium and plant tissue under near-ultraviolet light-absorbing polyethylene film, *Plant Dis.*, 80, 555, 1996.

Norman, J. M., Simulation of microclimates, in *Biometeorology in Integrated Pest Management*, Hatfield, J. L. and Thomason, I. J., Eds., Academic Press, New York, 1982, 65.

Norman, J. M. and Campbell, G. S., Application of a plant-environment model to problems in irrigation, in *Advances in Irrigation*, Vol. 2, Hillel, D., Ed., Academic Press, New York, 1983, 155.

Oke, T. R., *Boundary Layer Climates*, Methuen and Co., London, 1978, 372.

Orlandini, S., Bindi, M., Gozzini, B., and Seghi, L., Preliminary investigation on the spatial variability of air temperature in viticultural regions, in *Viticultural Ecological Sci.*, in press.

Payen, D., Modélisation de l'épidémiologie de *Sclerotinia sclerotiorum* sur tournesol, *Bull. Eur. Plant Prot. Organ.*, 13, 277, 1983.

Pedersen, E. A. and Morrall, R. A. A., Effects of cultivar, leaf wetness duration, temperature and growth stage on infection and development of Ascochyta blight of lentil, *Phytopathology*, 84, 1024, 1994.

Pedro, M. J. and Gillespie, T. J., Estimating dew duration. I. Utilizing micrometeorological data, *Agric. Meteorol.*, 25, 283, 1982a.

Pedro, M. J. and Gillespie, T. J., Estimating dew duration. II. Utilizing standard weather station data, *Agric. Meteorol.*, 25, 297, 1982b.

Pennypacker, S. P. and Stevenson, R. E., The management of plant pathogens, in *Biometeorology in Integrated Pest Management*, Hatfield, J. L. and Thomason, I. J., Eds., Academic Press, New York, 1982, 243.

Pimentel, D., McLaughlin, L., Zepp, A., Lakitan, B., Kraus, T., Kleinman, P., Vancini, F., Roach, W. J., Grapp, E., Keeton, W. S., and Selig, G., Environmental and economic impacts of reducing U.S. agricultural pesticide use, in *Handbook of Pest Management Agriculture*, Vol. 1, Pimentel, D., Ed., CRC Press, Boca Raton, FL, 1991, 679.

Pitblado, R. E., The Development and Implementation of TOM-CAST, Ontario Ministry of Agriculture and Food, Toronto, Canada, 1992.

Polley, R. W. and Clarkson, J. D. S., Forecasting cereal disease epidemics, in *Plant Disease Epidemiology*, Scott, P. R. and Bainbridge, A., Eds., Blackwell Scientific, Oxford, 1978, 141.

Potratz, K. J., Gleason, M. L., Hockmuth, M. L., Parker, S. K., and Pearson, G. A., Testing the accuracy and precision of wetness sensors in a tomato field and on turfgrass, *J. Iowa Acad. Sci.*, 101, 56, 1994.

Rosa, M., Genesio, R., Gozzini, B., Maracchi, G., and Orlandini, S., PLASMO: a computer program for grapevine downy mildew development forecasting, *Comput. Electron. Agric.*, 9, 205, 1993.

Rosa, M., Gozzini, B., Orlandini, S., and Seghi, L., A computer program to improve the control of grapevine downy mildew, *Comput. Electron. Agric.*, 12, 311, 1995.

Rossi, V., Racca, P., and Giosue, S., Geophytopathological analysis of Cercospora leaf spot on sugar beet in the Mediterranean area, *Phytopathol. Mediterr.*, 34, 69, 1995.

Rutter, A. J., The horological cycle in vegetation, in *Vegetation and the Atmosphere*, Monteith, J. L., Ed., Academic Press, London, 1975, 111.

Rotem, J., Cohen, Y., and Bashi, E., Host and environmental influences on sporulation *in vivo*, *Annu. Rev. Phytopathol.*, 16, 1978, 83.

Royal, D. J. and Butler, D. R., Epidemiological significance of liquid water in crop canopies and its role in disease forecasting, in *Water Fungi and Plants*, Ayers, P. G. and Boddy, L., Eds., Cambridge University Press, Cambridge, 1986, 139.

Sache, I. and Zadoks, J. C., Life-table analysis of faba bean rust, *Eur. J. Plant Pathol.*, 101, 431, 1995.

Sah, D. N., Influence of interrupted leaf wetness duration and relative humidity on development of tan spot in wheat, *Z. Pflanzenkr. Pflanzenschutz*, 101, 148, 1994.

Sauer, T. J. and Norman, J. M., Simulated canopy microclimate using estimated below-canopy soil surface transfer coefficient, *Agric. Forest Meteorol.*, 75, 135, 1995.

Schuh, W., Influence of interrupted dew periods, relative humidity, and light on disease severity and latent infections caused by *Cercospora kikuchii* of soybean, *Phytopathology*, 83, 109, 1993.

Schuh, W. and Adamowicz, A., Influence of assessment time and modeling approach on the relationship between temperature, leaf wetness periods and disease parameters of *Septoria glycines* on soybeans, *Phytopathology*, 83, 941, 1993.

Scherm, H. and Van Bruggen, A. H. C., Comparative study of microclimate and downy mildew development in subsurface drip-irrigated and furrow-irrigated lettuce fields in California, *Plant Dis.*, 79, 620, 1995.

Scherm, H., Koike, S. T., Laemmlen, F. F., and Van Bruggen, A. H. C., Field evaluation of spray advisories against lettuce downy mildew *(Bremia lactucae)* based on measured or forecast morning leaf wetness, *Plant Dis.*, 79, 511, 1995.

Shaw, M. W., Interacting effects of interrupted humid periods and light on infection of wheat leaves by *Mycosphaerella graminicola, Septoria tritici, Plant Pathol.*, 40, 595, 1991.

Shaw, M. W. and Royal, D. J., Factors determining the severity of epidemics of *Mycosphaerella graminicola, Septoria tritici* on winter wheat in the U.K., *Plant Pathol.*, 42, 882, 1993.

Shtienberg, D. and Vintal, H., Environmental influences on the development of *Puccinia helianthi* on sunflower, *Phytopathology*, 85, 1388, 1995.

Sosaalvarez, M., Madden, L. V., and Ellis, M. A., Effects of temperature and wetness duration on sporulation of *Botrytis cinerea* on strawberry leaf residues, *Plant Dis.*, 79, 609, 1995.

Stuckey, R. E. and Zadoks, J. C., Effect of interrupted leaf wetness periods on pustule development of *Puccinia recondita* f. sp. *tritici* on wheat, *Neth. J. Plant Pathol.*, Suppl. 95, 175, 1989.

Subbarao, K. V. and Michailides, T. J., Effects of temperature on isolates of *Fusarium moniliforme* causing fig endosepsis and *Aspergillus niger* causing smut, *Phytopathology*, 85, 662, 1995.

Sutton, J. C., Gillespie, T. J., and James, T. D. W., Electronic monitoring and use of microprocessors in the field, in *Experimental Techniques in Plant Disease Epidemiology*, Kranz, J. and Rotem, J., Eds., Springer, Berlin, 1988, 99.

Thom, A. S., Momentum, mass and heat exchange of plant communities, in *Vegetation and the Atmosphere* Vol. 1, Monteith, J. L., Ed., Academic Press, London, 1975, 155.

Thompson, N., The duration of leaf wetness, *Meteorol. Mag.*, 110, 1, 1981.

Timmer, L. W. and Zitko, S. E., Relationships of environmental factors and inoculum levels to the incidence of postbloom fruit drop in citrus, *Plant Dis.*, 77, 501, 1993.

Tompkins, D. K., Fowler, D. B., and Wright, A. T., Influence of agronomic practices on canopy microclimate and Septoria development in no-till winter wheat production in the parkland region of Saskatchewan, *Can. J. Plant Sci.*, 73, 331, 1993.

Truxall, D. L., Travis, J. W., and Hickey, K. D. Effect of temperature and wetness on infection of black raspberry by aeciospores of *Arthuriomyces peckianus*, *Plant Dis.*, 79, 805, 1995.

Ulbricht, C. W., Natural variations in the analytical form of the raindrop size distribution, *J. Climate Appl. Meteorol.*, 22, 1764, 1983.

Vakalounakis, D. J., Control of early blight of greenhouse tomato, caused by *Alternaria solani*, by inhibiting sporulation with ultraviolet-absorbing vinyl film, *Plant Dis.*, 75, 795, 1991.

de Vallavieille-Pope, C., Huber, L., Leconte, M., and Goyeau, H., Comparative effects of temperature and interrupted wet periods on germination, penetration, and infection of *Puccinia recindita* f. sp. *tritici* and *P. striiformis* on wheat seedlings, *Phytopathology*, 85, 409, 1995.

Vloutoglou, I., Fitt, B. D. L., and Lucas, J. A., Germination of *Alternaria linicola* conidia on linseed: effects of temperature, incubation time, leaf wetness and light regime, *Plant Pathol.*, 45, 529, 1996.

Wadia, K. D. R. and Butler, D. R., Relationship between temperature and latent periods of rust and leaf-spot disease of groundnut, *Plant Pathol.*, 43, 121, 1994a.

Wadia, K. D. R. and Butler, D. R., Infection efficiency of *Phaeoisariopsis personata* and the influence of different wetness patterns on germ-tube growth of the pathogen, *Plant Pathol.*, 43, 801, 1994b.

Waggoner, P. E., Models for forecasting disease control, in *Pest and Pathogen Control: Strategic, Tactical and Policy Models*, Conway, G. R., Ed., John Wiley & Sons, Chichester, 1984, 221.

Walklate, P. J., Vertical dispersal of plant pathogens by splashing. I. The theoretical relationship between rainfall and upward rain-splash, *Plant Pathol.*, 38, 56, 1989.

Walklate, P. J., McCartney, H. A., and Fitt, B. D. L., Vertical dispersal of plant pathogens by splashing. II. Experimental study of the relationship between rain drop size and maximum splash height, *Plant Pathol.*, 38, 64, 1989.

Wallin, J. R., Dew, its significance and measurement in phytopathology, *Phytopathology*, 53, 1210, 1963.

Wallin, J. R., Agrometeorological aspects of dew, *Agric. Meteorol.*, 4, 85, 1967.

Weihong, L. and Goudriaan, J., Analysis of leaf wetness in the rice crop caused by dew formation — a simulation study, *E. P. P. O. Bull.*, 21, 551, 1991.

Weiss, A., Leaf wetness: measurement and models, *Remote Sensing Rev.*, 5, 215, 1990.

Weiss, A. and Lukens, D. L., Electronic circuit for detecting leaf wetness and comparison of two sensors, *Plant Dis.*, 65, 41, 1981.

Weiss, A. and Hagen, A. F., Further experiments on the measurement of leaf wetness, *Agric. Meteorol.*, 29, 207, 1983.

Weiss, A., Lukens, D. L., Norman, J. M., and Steadman, J. R., Leaf wetness in dry beans under semi-arid conditions, *Agric. Forest Meteorol.*, 48, 149, 1989.

Weltzien, H. C., Use of geophytopathological information, in *Experimental Techniques in Plant Disease Epidemiology*, Kranz, J. and Rotem, J., Eds., Springer, Berlin, 1988, 238.

Wigneron, J.-P., Calvet, J.-C., and Kerr, Y., Monitoring water interception by crop fields from passive microwave observations, *Agric. Forest Meteorol.*, 80, 177, 1996.

Willis, P. T. and Tattelman, P., Drop size distribution associated with intense rainfall, *J. Appl. Meteorol.*, 28, 3, 1989.

Wittich, K.-P., Some remarks on dew duration on top of an orchard, *Agric. Forest Meteorol.*, 72, 167, 1995.

Xu, X.-M., Butt, D. J., and Van Stanten, G., A dynamic model simulating infection of apple leave by *Venturia inaequalis*, *Plant Pathol.*, 44, 865, 1995.

Zadoks, J. C. and Schein, R. D., *Epidemiology and Plant Disease Management*, Oxford University Press, Oxford, 1979.

Zadoks, J. C., Rijsdijk, F. H., and Rabbinge, R., EPIPRE: a systems approach to supervised control of pests and diseases of wheat in the Netherlands, in *Pest and Pathogen Control: Strategic, Tactical and Policy Models*, Conway, G. R., Ed., John Wiley & Sons, Chichester, 1984, 344.

Zhang, Y. and Gillespie, T. J., Estimating maximum droplet wetness duration on crops from nearby weather station data, *Agric. Forest Meteorol.*, 51, 145, 1990.

Zhang, Y. and Sutton, J. C., Effects of wetness duration, temperature and light on infection of black spruce seedlings by *Botrytis cinerea*, *Can. J. Forest Res.*, 24, 707, 1994.

CHAPTER 2

Ecological Approaches to Disease Control

Mark A. Boudreau and Christopher C. Mundt

I. INTRODUCTION

There is no question that tremendous advances have been made in the last 100 years in our understanding of plant diseases and the technology for their control. There is also little doubt that these impressive developments have prevented widespread crop losses in modern agricultural production systems, with consequences for the maintenance of the world food and fiber supply that would be difficult to exaggerate (Fry, 1982; Maloy, 1993). But disease has not gone away, nor are breeders, plant pathologists, and chemists resting on their laurels and seeking other careers; there is evidence that the force of all our efforts has merely kept pathogens at bay — losses due to disease were estimated at 10% in the 1940s, and 12% in the 1990s (Pimental et al., 1991). Not only is continued diligence obviously of great importance simply to avoid further losses, but in addition we are confronting new environmental, social, and economic risks associated with the established control methods.

1-8493-2627-3/97/$0.00+$.50

Part of the problem is that the disease control technologies we have come to depend on, particularly host resistance and chemicals, have manifested their limitations. Pathogens evolve, and so plant resistance breaks down (e.g., Green and Campbell, 1979), while pathogen resistance to pesticides develops (Dekker, 1993). Efforts to slow these processes, say by fungicide rotation or strategic regional deployment of resistance genes, suffer from lack of coordination or noncompliance (Wolfe, 1985; Maloy, 1993). Effective chemicals may represent a significant source of environmental pollution, threatening not only nontarget organisms such as wildlife and humans, but also nontarget microorganisms that act as native biocontrol agents (Blakeman and Fokkema, 1982; Vyas, 1988). Active biocontrol as an alternative control strategy has produced few commercial successes to date, and involves its own risks, particularly when nonnative agents are introduced (Andrews, 1990; Maloy, 1993). Sources of new resistance and biological control agents are being lost with the destruction of wild ecosystems, and the displacement of traditional varieties (Thurston, 1992). Indeed, whole systems of farming, which may include unique disease control strategies, disappear with the loss of primitive agriculture. Transferring new disease management techniques to traditional farmers, or old techniques to industrial farmers, has never been straightforward (Thurston, 1992; Trutmann et al., 1993). Meanwhile, available land declines due to such factors as desertification and development, while human population soars with undying hopes of ending malnutrition and famine.

We have reached an important juncture in disease management and agriculture in general. Solutions to agricultural problems involve many social issues, and there is a need to incorporate such issues into research and extension through farmer interaction and involvement of social scientists. At the biological level, we may need to take a new approach to supplement the largely trial-and-error strategy used to develop cultural, chemical, resistance, and biological control to date. A theory-based mechanistic approach is developing that may allow more effective design of disease control than strategies using empirical approaches alone.

Much is to be gained in this pursuit from the science of ecology. An ecological understanding of agricultural pathosystems is being used to analyze and optimize existing elements of control techniques, such as gene deployment, suppressive soils, and crop rotations. The potential for expansion of this effort is the topic of this chapter.

A. Learning from Natural Communities

Most ecological work has been performed in natural communities, which for our purposes means communities not significantly managed to supply a product. The study of plant pathology has been confined primarily to managed agricultural and silvicultural communities, and the literatures in ecology and plant pathology are for the most part mutually exclusive. Likewise, plant pathologists and ecologists historically have had little interaction or familiarity with each others' literatures. J. L. Harper was one of the first to recognize this and devoted a chapter of his acclaimed 1977 book to the role of pathogens in plant communities (Harper, 1977); 10 years later he declared that this remained one of the major areas of ignorance for plant population biology. He states: "It seems likely that whole blocks of fungal epidemiological theory might usefully be absorbed into plant population biology ... " (Harper, 1988). Work by individuals such as Janis Antonovics, Helen Alexander, Carol Augspurger, Jeremy Burdon, and Matthew Parker (all cited in this chapter) have pioneered studies of disease in natural communities.

Review literature on disease in natural communities generally suggests that disease levels are consistently low and that epidemic outbreaks are rare (Harlan, 1976; Harper, 1977; Burdon and Shattock, 1980; Dinoor and Eshed, 1984; Augspurger, 1989). This has led to speculation that many of the mechanisms thought to regulate disease in nature might be utilized in agricultural settings, particularly where sustainability is a major goal (Browning, 1974; Burdon, 1978; Altieri and Liebman, 1986). In contrast, Frank (1992) calls this view of benign pathogens in nature a "pervasive myth," Kranz (1990) has argued that disease is just as devastating in natural communities, and Murdoch (1975) has advised caution in applying inferences from wild conditions to agriculture. The controversy is further clouded by semantic inconsistencies in terms such as "epidemic" and "natural community."

There is, in fact, little documentation of disease levels, or particularly disease dynamics, outside of agriculture with which to assess these claims. Howard and Morrall (1975) determined detailed disease progress curves for leaf spot (caused by *Pyrenophora tritici-repens*) on graminoids in a Canadian prairie for over 3 years, and found low severity levels (<5%) throughout the study. Harlan (1976) observed low levels of rust on three wild grasses. However, a 4-year study partitioning mortality of the introduced weed cheatgrass *(Bromus tectorum)* indicated that >30% of deaths could be attributed to infection by *Ustilago bullata* at some locations (Mack and Pyke, 1984). Flax rust (*Melampsora lini*) on *Linum marginale* was severe (10 to 15% severity and almost 100% incidence) in one 2-year period, but incidence was less than 20% in another 2 years (Jarosz and Burdon, 1992). Kranz (1968) also observed severe disease with marked variability in a diverse community in Guinea, where significant thinning or death of the majority of hosts occurred in 22% of pathosystems during the wet season, and 42% of pathosystems in the dry season.

B. Limits to Application of Natural Models to Agriculture

Agroecosystems are typically highly disturbed, highly simplified communities of early successional annuals in which mechanisms regulating internal processes and interactions, both biological and physical, with the surrounding environment are severely disrupted. We must be cautious in our assumptions that a process thought to regulate disease in, say, a relatively closed, diverse, undisturbed community of indigenous vegetation with accompanying fauna, microbes, etc. will operate in a similar way in an agricultural system, and if it does, what trade-offs in yield, quality, harvestability, etc. might be involved?

In reality, there is a continuum of ecosystems entailing different levels of disturbance and complexity between the two extremes in the example above. This leads to an additional difficulty when workers state that disease levels are low in "natural communities," and that they should therefore be used as models for disease regulation. What is a natural community? The term is often taken to mean "ecologically balanced community" (Dinoor and Eshed, 1984; Kranz, 1990), but this is also a vague term. For example, the devastating epidemics caused by *Phytophthora cinnamomi* on natural eucalyptus stands in Australia, and *Cryphonectria parasitica (Endothia parasitica)* on American chestnut, have been blamed on a population out of balance — in these cases through the importation of exotic pathogens (Burdon, 1991). Pratt et al. (1973) claim that *P. cinnamomi* is in fact indigenous to southeast Australia, but only manifests itself as a serious pathogen when anthropogenic flooding occurs, such as during road construction or logging. This hypothesis has been disputed (Routley and Routley, 1974), but represents another interpretation of imbalance, that of irregular physical disturbance. An additional version of imbalance is related in a study of different-age communities on emerging islands in the Gulf of Bothnia (Carlsson et al., 1990), where disease for three pathosystems was highest in younger (i.e., more recently colonized) plant stands. Studies of disease in established systems of nonnative, naturalized plants (Mack and Pyke, 1984; Alexander, 1990) may or may not be considered natural systems in ecological balance. Clearly, qualifying the theory that disease remains at low levels in natural systems to include only ecologically balanced natural systems then requires a definition of ecological balance. Such a definition could take many forms, is rarely stated explicitly, and, when low disease levels are seen as an indicator of a balanced community (Dinoor and Eshed, 1984), becomes tautological. For the purposes of this chapter, we will include all unmanaged communities in the definition of "natural." This encompasses a great breadth of ecosystems, including old-field communities, some communities with frequent disturbance, and populations which may contain recent invaders (natural or anthropogenic) as well as indigenous species which have had ample time to coevolve. This is a practical way to distinguish agricultural and silvicultural communities from all others.

Overall, the limited evidence suggests that all levels of disease and much spatial and seasonal variability occur in natural systems, and that few generalizations may be made at this point. In any case, there are certainly mechanisms by which disease is limited in nature. Understanding them may provide clues to the variability observed, and tools for agricultural systems.

Some of these tools, such as host resistance and intrinsic biological control, will be addressed elsewhere in this volume. We will now consider, as examples, three aspects of natural communities studied in depth by ecologists, but which have largely been ignored by plant pathologists. These are the spatial and temporal occurrence of plants, plant diversity, and microbial ecology. Rather than attempt a complete review of the manipulation of these aspects for disease management, we will stress a consideration of mechanisms by which these ecological aspects contribute to disease alteration. Our focus on mechanisms is for two reasons. First, unlike traditional disease management practices such as major gene resistance and chemical control, ecological approaches to disease management have not yet benefitted from a wealth of fundamental research. Second, ecological approaches to disease management involve potentially complex interactions, with corresponding variability in their effects on disease. Thus, an understanding of mechanism may be critical to identifying the conditions under which a given ecological process can be manipulated successfully for disease management.

II. PLANT OCCURRENCE

The existence of an individual or population of plant hosts, at a particular place and time, is a fundamental aspect of ecological investigation which we may term "plant occurrence." The study of the distribution and abundance of plants indeed can be a definition of plant ecology (Begon et al., 1990), and as pointed out above, a host plant's existence may be both a result and a determinant of pathogen activity. It is therefore a reasonable starting point in this discussion to address plant occurrence in space and time and how these factors may be manipulated to manage disease.

A. Plant Occurrence in Space

Two aspects of spatial plant occurrence are density and pattern. Density is the number of individuals per unit area, acknowledging potential difficulties which may arise from clonal growth, tillering, polymorphism of growth habit, and so on. Pattern may be classified as uniform, clumped, or random (Ludwig and Reynolds, 1988). Both are highly dependent on scale and sampling strategy; a broadcast cereal may have a high density and random pattern when sampled within a 1-m^2 quadrat, a high density and uniform pattern when sampled at the field level, and a low density and aggregated pattern in a landscape.

Scale is important for interaction with pathogens as well. On a geographic scale, it is possible that occurrence of a particular plant species may be precluded by a pathogen (Weltzien, 1972). Dominant species may be reduced to low densities due to invading pathogens, with profound implications for competing species and community structure, as with *Cryphonectria parasitica (Endothia parasitica)* in eastern U.S. forests, or *Phytophthora cinnamomi* in Australian communities (Kennedy and Weste, 1986; Roame et al., 1986). The invasion of freshwater lakes in North America by Eurasian watermilfoil *(Myriophyllum spicatum)* (Grace and Wetzel, 1978), cereal fields in Australia by skeleton weed *(Chondrilla junceana)* (Burdon et al., 1981), and others, may be due in part to the absence of pathogens. However, Augspurger (1988) points out that the actual role of the pathogen in determining large-scale distribution patterns of plants is largely inferential and complicated by a number of variables. For instance, large-scale dieback of *Metrosideros polymorpha*-dominated rainforest in Hawaii in the 1960s and 1970s, for which an invading disease seemed a likely explanation for some time, may in fact be due to abiotic factors and synchronized stand-level senescence (Akashi and Mueller-Dumbois, 1995).

At the intermediate landscape scale, populations may be excluded or suppressed (and others thereby enhanced) by a pathogen or pathogens, but this has been difficult to demonstrate in natural communities, again because of the presence of confounding variables (Augspurger, 1988; Burdon et al., 1989). Agriculturalists confront this phenomenon routinely, for instance, when avoiding low, wet areas with high potential for damping-off, or planting nonsusceptible crops in fields infested with pathogenic nematodes.

In general, manipulation of host occurrence as a disease management tool is difficult on a larger-than-farm scale, requiring regional, national, or international cooperation and potentially high costs — assuming the science is sufficiently understood to pursue an effective course of action at all. Eradication of barberry, the alternate host of stem rust of wheat (caused by *Puccinia graminis* f.sp. *tritici*) in the U.S. is a rare example (Maloy, 1993). Manipulation of density and pattern for a single crop within a field is much more realistic, though poorly understood and little utilized. We will explore this scale of spatial plant occurrence in more detail.

1. Density

a. Density and Natural Communities

In natural plant populations, the interaction between plant density and disease dynamics has been difficult to assess. Augspurger (1988) stated that there was no unequivocal demonstration of a pathogen-mediated density-dependent mortality in systems studied up to that point, and little has changed since then. A more recent study of *Pinus sylvestris* naturally established at four sites in Sweden after clear cuts, found significant positive relationships between early stand density and later mortality due to *Phacidium infestans* in some of the comparisons made (Burdon et al., 1992). The authors suggest that density-dependent mortality due to pathogens may only occur during a particular stage of epidemic development when disease is most efficiently transmitted in areas of high host density, but that this is preceded and followed by stages in which density dependence will not be apparent. Mortality will increase in shore pine *(Pinus contorta* var. *contorta)* at very high density, but severity of disease (needle rust, needle cast, and stalactiform blister rust) decreases with increasing density (Xie et al., 1995).

It is clear that pathogens can reduce fecundity, growth, or winter survivorship of hosts in some wild systems (e.g., Alexander, 1984, 1990; Augspurger and Kelly, 1984; Burdon and Jarosz, 1992; Parker, 1986), but the importance of disease in determining host abundance in these cases is not clear. Moreover, the effect of plant density on these diseases has not been evaluated. Interactions of disease with other factors are particularly cumbersome in natural communities. For example, infection may make hosts less palatable to herbivores and actually protect plants, as shown in *Brassica rapa* with *Alternaria* leaf infections (Nakamura et al., 1995). One model of dynamics of plant mixtures in the *absence* of disease, as well as a number of other factors such as herbivory and allelopathy, resulted in monocultures or mixtures, stable or unstable, depending on settings for growth rates and other parameters (Gates and Westcott, 1988). Such a result again demonstrates the difficulty of inferring the role of disease from a particular pattern of community development.

b. Density and Managed Communities

In agricultural ecosystems, one of the essential features of management is the establishment of a desired density, generally at or shortly after planting. This of course exists within limits for a given species in a given environment, and will change over time (particularly in the case of perennials), but density management may continue through thinning, supplemental planting, and replanting. The value of this somewhat independent variable for disease management generally has not been exploited by growers, partly because of our poor understanding of density-disease relations, partly because other factors may dominate decisions about density, and partly, perhaps, because it is simply not considered.

Burdon and Chilvers (1982) reviewed studies on density and disease, and Power (1990) summarized work on some insect-vectored virus diseases and density. Some representative examples of density-disease relations in agricultural situations are given in Table 1 (see reviews for additional examples). Although studies are limited and varied, it is clear that we cannot subscribe to the common assumption that disease will increase with increasing host density. However, patterns are evident. In Burdon and Chilvers's (1982) survey of 46 different host-pathogen combinations, 62% showed a positive correlation between density and disease, and 27% indicated a negative correlation. As Table 1 indicates, increases in disease with density occur primarily among fungal pathogens, while virus levels tend to decrease with density. All of these viruses, and the spiroplasma, are vectored by leafhoppers or aphids.

Table 1 Influence of Host Density on Disease

Pathogen	Host	Correlation[a]	Parameter[b]	Ref.
Cercospora apii	Celery	+	S,r	Berger, 1975
Pythium irregulare	Cress	+	r	Burdon and Chilvers, 1975
Erisyphe graminis f.sp. *hordei*	Barley	+	r	Burdon and Chilvers, 1976b
Sclerotinia sclerotiorum	Sunflower	+	I,V	Huang and Hoes, 1980
Verticillium dahliae	Cotton	–	I	Minton, 1980, Huisman and Grimes, 1989
Phaeoisariopsis griseola	Bean	0–	AUDPC	Boudreau, 1993
Phytophthora colocasiae	Taro	0	r,S	Jackson et al., 1980
Corn stunt spiroplasma	Maize	–	I	Power, 1987
Beet yellows virus	Sugar beet	–	I	Heathcote, 1970
Peanut rosette virus	Peanut	–	I,N	A'Brook, 1964, Davies, 1976
Tomato curly-top virus	Tomato	–	NP[c]	Shapovalov, 1941

[a] Correlation between density and the measure(s) of disease used. "0–" = weakly suggested negative correlation.
[b] Parameter(s) used to measure disease: S = severity, I = incidence, r = apparent infection rate, AUDPC = area under the disease progress curve, V = velocity of spread.
[c] Not presented in publication.

Comparison and evaluation of these studies is complicated by methodology used (Burdon and Chilvers, 1982). When epidemics are quantified as a proportion of available tissue diseased (e.g., incidence or severity), as is usually the case, it may be that the absolute amount of disease (e.g., total number of diseased plants) will show a different trend. For instance, if incidence of disease is reduced by 25% when the density of host plants is doubled, the absolute amount of disease will increase by 50%. Some studies measure disease at one point in time, while others evaluate disease progress over time. Finally, other factors may change with host density, particularly the addition of intervening nonhost individuals. This occurs spontaneously in natural populations, or purposefully in the case of intercropping or variety mixtures.

c. Mechanism of Density Effects

The mechanisms by which density influences disease are poorly understood, but may shed light on the variety of responses observed in Table 1. Changes in host density could affect the dispersal phase of the pathogen's life cycle, the infective phase, or a dormant or overwintering phase.

Most speculation on density effects have focused on the dispersal phase, specifically the frequency of host-pathogen encounters. It is an intuitive assumption that increased density should result in increased disease due to an increased likelihood of encounter between host and pathogen (Burdon and Chilvers, 1982). In a simple Lotka-Volterra-type model, analogous to that developed for animal epidemiology (Anderson and May, 1979), a population of pathogens (P) and a population of hosts (N) randomly contact each other, with some fraction i of the contacts resulting in infections and reproduction of the pathogen. The number of infections and the population of the pathogen will increase at a rate proportional to the product iPN. Increasing N in a given area (that is, increasing density) will result in more contacts and more infections.

Failure to meet assumptions of the above oversimplification can lead to departures from a simple positive correlation between density and disease. Host and pathogen do not interact randomly; plants are not mobile like animals and nonrandom spatial patterns of disease quickly develop. The parameter i may change with density. Another complication results from an important aspect of plant demographics: increased density of individuals does not imply a proportional increase in infection sites. Above a certain density threshold, competition among plants results in smaller plants, and at sufficiently high densities a plant population will self-thin such that a predictable ratio of the log of mean plant weight to the log of density is maintained (Harper, 1977). This phenomenon, known as Yoda's self-thinning law (Yoda et al., 1963), has been demonstrated for a number of diverse species. The loss of lateral branches in high-density southern pine plantings, and the resulting loss of infection sites on needles, may be partly responsible for the negative correlation between tree density and fusiform rust incidence (Mann and Scarborough, 1948; Burdon and Chilvers, 1982). Competition in high-density arrowleaf clover stands resulting in smaller crowns probably prevents internal crown breakdown, though this may be an abiotic disease (Pratt and Knight, 1983). Also limiting the increase in available infection sites with increased density is the possibility of plant organs being protected by others, such as overlapping leaves decreasing deposition of airborne inoculum.

As in all epidemics, high levels of infection, whether on an individual plant or in a population of plants, occupy available sites on individuals and decrease the likelihood of a propagule initiating a new infection. This multiple infection phenomenon (Gregory, 1948) may be exacerbated at high densities in cases where reduced distances between hosts results in a low velocity of spread with high local levels of disease (Minogue, 1986).

None of these theoretical considerations negate a positive density-disease correlation, but each could act to diminish it significantly. Experiments to explicitly test these ideas in specific pathosystems are necessary.

Empirical evaluations of the impact of density on disease spread are rare. Velocity of spread of *Ceratocystis wagneri* on Ponderosa pine was positively correlated with stand density (Cobb et al., 1982). In another study, reduced rate of mycelial spread of Pythium damping-off in species mixtures could be attributed primarily to reduced stand density of the host (Burdon and Chilvers, 1976c).

Boudreau and Madden (1995) used simulated rain to evaluate the impact of strawberry density on spread of the splash-dispersed pathogen *Colletotrichum acutatum*, cause of strawberry anthracnose, in artificial canopies of greenhouse-grown strawberries. Spore deposition declined with increasing density for sampling locations near the source, or when integrated over areas of the plot, but this effect was not linear. That is, there was not a significant difference in deposition between the two highest densities or the two lowest densities, but a greater than fivefold decrease in deposition occurred between the second- and third-highest densities (3- and 4-cm spacing, respectively). Also notable was the effect of inoculum location: inoculum located within the row resulted in <1/6 the number of spores deposited in the plot, compared with inoculum located between rows and, therefore, not protected by the canopy. Deposition gradient slopes were not affected by plant density, but intercepts declined with increasing density.

The data were interpreted in terms of three suggested mechanisms by which density might reduce deposition: reduction in spore removal from the inoculum source, reduction in resplash of spore-carrying droplets, and interception of spores by the plant canopy. All three mechanisms appeared to impact dispersal in this system. The second and third mechanisms led to the interesting observation that density affected deposition even when inoculum was placed between rows, i.e., spores removed from an inoculum source exposed to direct rainfall were traveling into and within the canopy, where plant density influenced their movement.

Changes in density might also affect other agents of dispersal, but this remains to be experimentally evaluated. For example, density changes could influence airborne propagules by altering wind velocity and turbulence (English et al., 1989). More subtle influences on pathogen deposition may occur due to changes in plant height, leaf size, and growth habit under different density regimes (Burdon and Chilvers, 1982; Boudreau and Madden, 1995). For situations in which a soilborne pathogen remains relatively immobile and depends on root growth for contact and subsequent infection, an increase in plant density may not lead to an increase in root density. Therefore the number of root-pathogen encounters will not change, but proportion of plants infected will decrease, as seen for Verticillium wilt in Table 1 (Huisman and Grimes, 1989).

The negative correlation between density and virus incidence (Table 1) may be due primarily to dispersal phenomena. Power has studied movement patterns in homopteran virus vectors in different-density host stands. The corn leafhopper, *Dalbulus maidis* Delong and Wolcott, satisfied the proposed rule for leafhopper movement, which states that the insects are more likely to move the shorter the distance between adjacent plants (Power, 1992). Thus, in a comparison of corn *(Zea mays)* planted in rows 1 m apart, leafhoppers were more likely to move within the row than across rows, and along a high-density row (16.7 cm between plants) than a low-density row (33.3 cm between plants). The insects tended to remain on a plant longer (longer tenure time) in low-density rows, and overall leafhopper abundance (leafhoppers/plant) was higher at low density. Maize rayado fino virus, vectored by *D. maidis*, spread more quickly in the low-density stands, and disease incidence, though low (<4%), was higher at lower maize density. Maize rayado fino virus requires a minimum of 8 h for leafhopper transmission, and this may not be attained when insects move frequently, as in the high-density stands. If transmission time relative to tenure time were shorter, then one might expect higher disease incidence as density increases, assuming the movement behavior observed in the study.

Power (1990) suggests other dispersal-related mechanisms which may be responsible for the density-disease relations seen for vectored viruses. Homopteran vectors, mainly aphids and leafhoppers, account for 90% of insect spread by plant pathogens, and they generally exhibit lower abundance (insects/plant) in denser stands, and this may result in lower incidence of disease. However, in some cases the absolute number of insects, and number of plants infected, is higher in denser stands, which may indicate a limited pool of vectors diluted over more plants in a given area. Power also notes that high-density plantings may eliminate the contrast of green plants against a dark background of bare soil, which is a visual cue for aphids and leafhoppers searching for hosts to colonize.

Changes in disease with density may also be due to environmental alterations. Pathogens favored by high humidity or lower temperature may benefit from a higher density stand of hosts. Leaf removal is a practice used commonly in commercial grape *(Vitis vinifera)* vineyards for the control of bunch rot, caused by *Botrytis cinerea*. Such disease control appears to be due, at least in part, to reduced evaporative potential within the thinned canopies (English et al., 1993). The effect of reduced plant density may not always be straightforward, however. For example, a dense stand, which retains dew and maintains a lower temperature longer in the morning, may exhibit less radiative cooling at night. This would delay dew formation and reduce leaf temperature. Low densities of *Pinus contorta* are correlated with high levels of disease and insect damage, and this is thought to be due generally to temperature and light conditions which commonly favor pests (Xie et al., 1995). Competition at higher densities may alter the nutritional status or water potential of plant tissue, also influencing infection. These mechanisms have not been empirically evaluated.

There is evidence for host density effects on other species of plants which then influence disease. Oaks are alternate hosts for fusiform rust of southern pine, but are suppressed in high-density pine stands, thus limiting initial inoculum and disease development on pine (Mann and Scarborough, 1948; Burdon and Chilvers, 1982). This would only be an important mechanism in agricultural situations if a weed which was influenced by crop density was also an alternate host.

2. Pattern

The influence of plant spatial pattern on disease dynamics interacts with density in potentially complex ways, but has been studied very little even in isolation. Patchiness of hosts and pathogens in nature has been reviewed (Burdon et al., 1989). All pattern types may be seen at the field scale in agroecosystems; uniform plantings of equidistant trees in an orchard or forest plantation, random plantings of maize in a subsistence plot, or clumped hills of squash. The ubiquitous row cropping pattern is sometimes described as "linear" (e.g., Power, 1992). A row pattern amounts to a relatively high density in one dimension (within the row) and low density in another (across rows), and disease spread evaluated in each dimension might reflect the same density-disease patterns discussed earlier.

This is supported by Power (1992), in the *Dalbulus maidis*-maize-rayado fino virus system. Higher densities resulted in greater leafhopper movement. Accordingly, more insect movement was observed along rows than across rows. More than five times the number of *Colletotrichum acutatum* spores were deposited in strawberry rows containing inoculum than in the adjacent row in simulated rainfall experiments, and the row itself can form an effective barrier to cross-row transport (Boudreau and Madden, 1995; Yang and Madden, 1993).

Comparisons among planting patterns have been limited and inconclusive. Boudreau (1993) found no effect of row vs. random patterns of common bean *(Phaseolus vulgaris)* in the area under the disease progress curve for angular leaf spot caused by *Phaeoisariopsis griseola*. Incidence of maize rayado fino virus did not differ among linear, clumped, or uniform planting patterns, although abundance of the vector (leafhoppers/plant) and emigration rates were highest in the uniform planting (Power, 1989, 1992). Clear results were obtained in one study in which the rate of damping-off due to *Pythium* was reduced significantly as cress seedlings became progressively more clumped, maintaining the same overall density, probably due to the inability of hyphae to grow from clump to clump (Burdon and Chilvers, 1976a).

The effectiveness of cultivar mixtures for disease reduction, discussed in the next section, may be diminished when the size of blocks of each cultivar increases, i.e., when diversity is more "coarse-grained" *sensu* Pielou (1977). The more clumped or coarse-grained the distribution of the cultivars in such a mixture, the more disease results. This has been demonstrated in the field for oat crown rust and bean rust, and in computer epidemics, and is more apparent for uniform spatial distributions of initial inoculum than for focal epidemics (Mundt and Browning, 1985; Mundt and Leonard, 1985; Mundt et al., 1986). The phenomenon may be the result of the greater likelihood of inoculum deposition on a susceptible host where the susceptible hosts are not highly interspersed with resistant plants.

B. Plant Occurrence in Time

Temporal patterns in host occurrence and their influence on disease will be considered at the levels of the individual, population, and community. For individual plants, we will focus on phenotypic changes resulting from growth and development. Genetic changes will be discussed for populations or metapopulations. Finally, we will focus on changes in species composition at the community level. Although genetic changes may occur over millions of years, and phenotypic changes in an annual plant may occur in a few weeks, there is considerable overlap in the time scale of changes at each of these levels of organization. For example, one cohort of Douglas fir trees may grow slowly and persist for several centuries, and important genetic changes in a host population may occur in only one reproductive cycle. Also, a typical forest is a mosaic of patches with more or less stable species composition and gaps with distinct and rapidly changing populations.

1. Individual Level: Phenotypic Changes

Wide variations in the physiognomic character of a plant occurs over its life as it grows and differentiates, particularly in a long-lived perennial such as a redwood tree. In a natural community, even-aged stands with similar physiognomic character may occur (as in synchronous germination in response to a major disturbance or gap formation), or mixed-aged stands with dissimilar stature may occur, typically at later successional stages. How will this affect disease? In an even-aged stand, plants which show differential susceptibility with age (most notably in diseases such as anther smuts, which only infect plants with male inflorescences [Thrall et al., 1993]), will be either universally subject to infection, or universally incapable of infection. In a mixed-aged stand, only a fraction of the individuals will be susceptible. This could have the effect of preventing rapid disease increase and spread as some of the inoculum will be lost to nonreceptive plants; alternately, it may provide a continuous supply of initial inoculum as new individuals become susceptible, augmenting epidemic development. Which of these would occur depends on the mode of pathogen dispersal, latent and infectious periods of pathogen relative to periods of susceptibility of the host, spatial distribution of hosts, etc.

In the agricultural setting, planting and harvest date have been used to avoid coincidence of pathogen and susceptible host (Maloy, 1993). It also may be possible to utilize mixed-age stands to limit disease development, much as in a variety mixture. Differential susceptibility of plant organs with age has been demonstrated in a number of cases (e.g., Imhoff et al., 1981; Ercolani, 1991; Williams, 1995). Changes in whole-plant susceptibility are clear for infections of flowers or fruit, as discussed above, or for seedling damping-off (Neher et al., 1988), but changes in susceptibility of intermediate vegetative stages are rare or indirect (e.g., aphid vector activity changes with sugarbeet age [Williams, 1995]). This area warrants further investigation. Although there are potential logistical problems with mixed-aged stands for crop management and harvest, these are no more insurmountable than similar problems with intercrops or some variety mixtures. Straightforward application may be found in forage crops, forests, and other perennial crops.

2. Population Level: Genetic Changes

Host/pathogen coevolution can be studied primarily in wild populations or in agricultural situations with an ancient history, such as maize cultivation by the Maya or barley in the Middle East. Researchers are beginning to speculate upon and study long-term interactions and the evolution of resistance in wild populations (Browning, 1980; Burdon, 1987; Alexander, 1992; Thompson and Burdon, 1992; Leonard, 1994; Antonovics et al., 1996), but manipulation of host occurrence at this time scale is of little relevance for disease management. However, the result of this coevolution has had value for agriculture, in that resistance genes are often sought and found in centers of diversity for crops and their wild progenitors (Leppik, 1970; Harlan, 1976). In addition, patterns of resistance and virulence which evolve over long periods of coexistence, and which are then observed in a contemporary community in which disease apparently exists at low endemic levels, may offer clues to effective development and deployment of stable resistance (Browning, 1974).

An ambitious attempt at applying ideas of host/pathogen coevolution to breeding for resistance was made by Robinson (1980). Borrowing from animal behavior theory, he proposed the existence of a so-called evolutionarily stable strategy (ESS) for pathosystems, which "... is the 'best possible' strategy of (responses controlled by the genetic code), at a specified systems level. The ESS has been arrived at by the competitive elimination of all inferior, inherited strategies; and it has the maximum survival value, economy, resilience, flexibility, homeostasis and, above all, stability." He suggests that this strategy would, among other things, result in a dynamic but highly heterogeneous pattern of qualitative (gene-for-gene) resistance genes in the host (and corresponding virulence genes in the pathogen), but that this type of resistance would be of little use in a continuous pathosystem, that is, one which is spatially uniform with no "dead" season of host or parasite dormancy, such as a vegetatively propagated grass in the moist tropics infected by an imperfect fungus. In such a situation, quantitative/polygenic resistance would dominate. He provides a number of recommendations for future breeding programs based on these ideas. These include the deployment of cultivar mixtures to mimic the presumed natural heterogeneity and stability, and an emphasis on polygenic resistance development, particularly in continuous systems.

Many of Robinson's suggestions may be sound based on empirical evidence, but the theoretical foundation includes many oversimplifications and assumptions which even today are presumptuous. These include the relative prevalence of single-gene and polygenic resistance in various natural populations, the degree of importance of pathogens in influencing host populations, and the interaction of spatial patterning with resistance, as in pathogen and host refugia (Augspurger, 1989; Alexander, 1992). Nonetheless, it provides a fascinating example of how an understanding of the long-term interaction between host and pathogen can be used as a basis for methodically developing disease control strategies.

Genetic changes in hosts under the influence of disease have been observed over relatively short time periods in wild populations. A population of the herbaceous perennial *Linum marginale*, observed over a period of six years in New South Wales, Australia, underwent large changes in the pattern of resistance to the rust pathogen *Melampsora lini* (Burdon and Thompson, 1995). The period involved two substantial epidemics, one of which was associated with a change from a high-density population dominated by three resistance phenotypes, which were spatially aggregated, to a low-density population with a number of resistance phenotypes scattered throughout the population, with none dominating. However, this change did not reflect the expected selective pressures imposed by the *M. lini* pathotypes present. Instead, overall susceptibility of the population increased. Parker (1991) observed a decline in resistance of *Amphicarpaea bractata*, an annual legume, over two years in the presence of the pathogen *Synchytrium decipiens*. Burdon and colleagues (Burdon and Jarosz, 1992; Thompson and Burdon, 1992; Burdon and Thompson, 1995) suggest that these results reflect the overly simplistic nature of early models of host-pathogen coevolution, which do not incorporate the potentially large influence of the spatial pattern of the host and pathogen and its interaction with pathogen dispersal. They proposed that rather than focus on natural selection acting on an isolated local population, studies of host-pathogen coevolution must view both host and pathogen at a regional level as a patchwork of populations (i.e., a metapopulation) in which processes such as genetic drift and gene flow between populations play an important role. It is true that the limited evidence given above does not seem to support the early models of frequency-dependent selection, at least for these airborne foliar pathogens, but a metapopulation-based model of coevolution needs to be explicitly articulated and predictions tested experimentally. As Burdon and Thompson (1995) state, other explanations for these results include the linkage of resistance genes to traits for which there is more selection pressure.

The potential importance of metapopulation-level interactions would suggest that, in agriculture, resistance genes may need to be managed on a regional basis as well as at a single-field level. However, the agricultural situation is significantly different from the natural one, in that multiple generations of host plants accumulating genetic changes over time due to selective and other forces rarely occurs. Human intervention typically resets the genetic composition of a population with annual plantings of particular cultivars. Human selection occurs even when seed is saved and replanted the following year, as in many crops in less developed countries, or when quasinatural forests are thinned, burned, or replanted.

One result of long-term coevolution of host and parasite may manifest itself not in a qualitative resistance to a specific pathovar per se, but in a suite of traits which in some way minimize or avoid fitness reduction due to infection, that is, a life-history strategy. Characters such as time of emergence, longevity, flowering time, growth habit, etc., may all affect disease dynamics, though we are far from discerning the importance of disease as a selective agent for these traits. Disease has been implicated as a force maintaining sexual reproduction (Augspurger, 1988), determining the nature of vegetative reproduction (distance of clonal spread before detachment) (Wennstrom and Ericson, 1982), size of flower, and annual vs. perennial nature (Thrall et al., 1993). Again, in agricultural situations, the value of a particular life-history strategy as a disease management tool must be evaluated empirically on a case by case basis until a broader theoretical framework is developed.

3. Community Level: Species Composition Changes

Changes in species composition over time defines much of plant ecology. In simple terms, this temporal change can be described as a predictable pattern of succession for a given location, which may be arrested or redirected along a different pathway by disturbance (Pickett and White, 1985). Except in the most obvious cases (e.g., impact of chestnut blight on Eastern forests), the role of disease in either mediating the "normal" successional pattern, or as a disturbance deflecting this trajectory, is typically not addressed because of the lack of relevant studies (Harper, 1988). Augspurger (1988) suggests that pathogen effects may be largely local, perhaps creating small-scale extinctions but having less impact on a regional scale. Locally differential effects of damping-off of tropical tree seedlings (Augspurger and Kelly, 1984) and *Phellinus weirii* infection in Pacific northwest forests (Cook, 1982) support this contention.

One large-scale study evaluated 765 forest stands occurring in diverse environments over 3.38 million ha in coastal North Carolina (Wyant et al., 1991). Statistical analysis of forest service data revealed that disease did not alter species composition changes over a ten-year period from that found in undisturbed sites over the same period. The study involved managed forests, so anthropogenic causes (harvest and prescribed burning) predictably altered species composition, but the natural disturbances of wildfire and insect damage also had an impact. However, it is not clear what role disease had in regulating the development of the undisturbed stands. On the other hand, infection caused by the pathogen *Phellinus weirii* in Douglas-fir forests significantly impacted the cover, composition, and diversity of herbaceous species (Holah et al., 1993). A complex interaction between disease and disturbance regimes may also occur. For example, fire damage may create conditions favorable for some pathogens (e.g., root rot, oak wilt), and infection may then increase flammability. Witches brooms created by mistletoe infection can become detached fireballs which spread fire (Knight, 1987). Pathogen- and mycorrhiza-mediated successional changes are suggested by a study of hardwood seedling mortality under plantations of *Pinus sylvestris* and *P. resinosa* in New York (Tobiesson and Werner, 1980). The much lower establishment of hardwoods under *P. resinosa* appears to be related to the lack of endomycorrhizal infection of the hardwood seedlings.

How does an understanding of pathogen-mediated species change apply to agriculture? Species change at a site has been actively determined by growers, often to reduce disease, through the practice of crop rotation. This may be the oldest form of active disease control, alluded to in the Bible, and is effective particularly against soilborne pathogens with a limited host range, such as *Gaeumannomyces graminis* (cause of take-all of wheat) and *Heterodera glycines*, the soybean cyst nematode (Fry, 1982; Maloy, 1993; Curl, 1963). We will not review the large literature on this subject here, but some new ideas are notable. These include incorporation of crop mixtures for more efficacious disease control (Uilich, 1993), and interactions with biocontrol agents (Cook, 1990). A subset of the rotation practice includes the use of trap or decoy plants, which encourage hatching or germination of a pathogen but prevent penetration or reproduction. Marigold *(Tagetes)* and *Crotolaria* both have been used to reduce several nematodes, but are not used widely because of the loss of an economic crop during the trap-crop season (Edwards et al., 1994; Maloy, 1993).

In the case of long-term perennial crops, rotations are less relevant, but clues for disease management may be found in life-history strategies of wild perennials. Perennials tend to replace annuals during succession and may select for pathogens with certain life-history traits themselves. Thrall et al. (1993) found that anther smuts (caused by *Ustilago* spp.), which have very limited capabilities of transmission to new individuals, are five times more common on perennial then annual members of the family Caryophyllaceae. Neher and Campbell (1994) observed that perennial crops (alfalfa or tall fescue pasture) had a different community of plant parasitic nematodes than annual soybean fields, and nematode communities associated with perennial crops were more mature: long life cycles, low colonization ability, few offspring, and sensitivity to disturbance. Finally, the repeated exposure of multiple pathogen generations to genetically static perennial plantings might select for virulence in gene-for-gene resistance structures. The most effective and stable resistance, particularly for nondeciduous trees or nonfoliar pathogens, may therefore be quantitative in natural and managed systems (Robinson, 1980).

III. DIVERSITY

> I will now try to set out some of the evidence that the balance of relatively simple communities of plants and animals is more easily upset than that of richer ones; that is, more subject to direct oscillations in populations, especially of animals, and more vulnerable to invasions. For if this can be shown to be anywhere near the truth, it will have to be admitted that there is something very dangerous about handling cultivated land as we handle it now, and even more dangerous if we continue to go farther down the present road of 'simplification for efficiency'. (Elton, 1958).

With these words, written in 1958, Charles Elton articulated a concept which still finds voice in diverse circles, particularly among nonscientists and growers practicing alternative techniques. The mantra "complexity (or diversity) leads to ecosystem stability" has been used to justify and encourage inter- and intraspecific diversification of farms to minimize pest damage, including that due to disease, in hopes of mimicking the presumed dynamics of natural communities. In an excellent overview of the development of this concept, Begon et al. (1990) term Elton's view the "conventional wisdom." However, the six primary points which Elton uses to justify his concept are largely speculative, which he concedes, noting " ... it is by no means a far-fetched idea, and even if it seemed so we should still need to explore it by further research ... " (Elton, 1958).

By 1996, has the research and discussion which ensued supported the complexity-stability argument? In a word, regarding natural communities, no. The theory began to lose favor in the 1970s. Researchers pointed out that Elton's lines of evidence could be explained by alternative hypotheses, complexity led to local instability in mathematical models, the definition of "stability" was elusive, and empirical evidence for the model's predictions, such as stable populations in the diverse tropics, was not forthcoming. The reader is referred elsewhere for a discussion of this fascinating controversy (May, 1972; Bigger, 1976; Begon et al., 1990; see Murdoch, 1975 for a discussion of pests specifically). For our purposes, however, we must ask if the growers who rationalize management decisions based on Elton's principle are misguided, trapped in the ecological fashions of the 1960s? Such an accusation is harsh, for two reasons. First, the controversy is by no means dead (Tilman et al., 1996), and surprises may be in store as specific types of complexity and restricted forms of stability are evaluated in well-designed experiments with predictive and refutable hypotheses. Second, a body of evidence is accumulating on the response of various stability parameters (e.g., productivity, levels of pest occurrence, nutrient supply) to diversification in agricultural systems (Kass, 1978; Risch et al., 1983; Vandermeer, 1989; and see below). In some studies, but by no means all, disease levels are reduced by diversification. Although this phenomenon may occur in fundamentally "unnatural" systems, and sufficiently inconsistent at that to offer support for a general application of Elton's idea, it nonetheless has great potential value as an ecology-based disease management tool in many situations.

It is to this area of inquiry which we now turn. Specifically, we will discuss intercropping, the simultaneous cultivation of two or more crop species in intimate association, as an example of interspecific diversification, focusing on the mechanisms by which pathogen dynamics are affected. We will also briefly discuss the mechanisms at work during intraspecific diversification in agricultural situations, i.e., cultivar mixtures and multilines.

A. Intercropping

Intercropping is the rule rather than the exception in traditional farming systems throughout the world, and until recently was common in the U.S. (see Francis, 1986; Kass, 1978; Papendick et al., 1976; and Thurston, 1992 for general background). Although the limited research on these systems has focused on yield and agronomic features, several authors have urged intercropping as a means of low-cost, sustainable disease control (e.g., Altieri and Liebman, 1986; Browning, 1974; Burdon, 1978). Advocating such a practice may be based on presumed behavior of heterogeneous natural systems, theoretical outcomes of particular mechanisms expected to be at work (e.g., an added plant should "trap" spores), or the notion that the practice of intercropping may have evolved in most societies partly to control disease. None of these motivations is based on strong empirical evidence, however. (The last, in particular, implies the subtle prejudice that people must intercrop for some good reason, rather than practice monoculture, when *intercropping* would seem to flow more logically from the pre-agriculture condition.) We must turn to experimental comparison of disease in intercrops and monocrops for solid answers. Unfortunately, such studies are few, and much of the research has not circulated widely.

Although disease is often reduced due to intercropping, variability is quite evident, even within a particular intercrop-pathogen situation (Table 2). A case in point is bean-maize associations (perhaps the most common crop combination in the world [Papendick et al., 1976]). Van Rheenen et al. (1981), working over five years at six sites in Kenya, found that although eight of ten bean diseases were more likely to be more severe in a monocrop than a bean-maize intercrop (white mold and rust were the exceptions), there were always cases in which a given disease was worse in the intercrop, or not affected. For instance, anthracnose (caused by *Colletotricum lindemuthianum*) was less severe in an intercrop treatment in 228 cases, but more severe in 119 cases, and not affected in 173 cases. In several other studies, bean rust was reduced by intercropping in some

Table 2 Influence of Intercropping on Disease

Host Nonhost(s)	Pathogen	Effect[a]	Ref.
Barley Wheat	*Erysiphe graminis* f.sp. *hordei*	–	Burdon and Chilvers, 1977
Barley Oats	*Erysiphe graminis* f.sp. *hordei*	–	Vilich-Meller, 1992
Bean Maize	*Phaeoisariopsis griseola*	–,0,+	Moreno, 1977, Mora-E., 1978, Msuku and Edje, 1982, Boudreau, 1993,Lanter, 1990, Sengooba, 1990
Bean Maize	*Uromyces appendiculatus*	–,0	Soria et al., 1975, Monteiro et al., 1981, Moreno and Mora, 1984, Boudreau and Mundt, 1992
Cassava Bean	*Oidium manihotis*	+	Soria et al., 1975
Cassava Maize	African cassava mosaic virus	–,0,+	Fargette and Fauquet, 1988[b]
Chilli Maize or Brinjal	Chilli veinal mottle virus	–	Hussein and Samad, 1993[b]
Maize Cowpea or peanut or greengram	*Pratylenchus brachyurus*	–,0,+	Egunjobi, 1984
Maize Bean	Corn stunt spiroplasma	–	Power, 1987
Peanut Pigeonpea	*Cercospera arachidicola*, *Phaeoisariopsis personata*	–	Ghewande et al., 1993
Peanut Bean	Peanut rosette virus	–	Farrell, 1976b
Plantain Cassava	*Mycosphaerella fijiensis*	–	Emebiri and Obiefuna, 1992
Rice Maize and Cassava	*Cochliobolus miyabeanus*	+	Kass, 1978
Soybean Sorghum	Soybean mosaic virus	–	Bottenberg and Irwin, 1992
Oats Barley	*Drechslera avenae*	–	Vilich-Meller, 1992
Rye Wheat	*Rhyncosporium secalis*	–	Vilich-Meller, 1992
Wheat Rye	*Puccinia recondita* f.sp. *tritici*	–	Vilich-Meller, 1992

[a] Effect of intercropping on disease level as compared with monocrop: "+" = statistically significant increase, "0" = no significant change, "–" = statistically significant decrease. Measurement of disease level varies with study, and in some cases is not specified.

[b] Statistical analyses of data not presented.

cases, but unaffected in others (Boudreau and Mundt, 1992; Monteiro et al., 1981; Moreno and Mora, 1984; Soria et al., 1975). Angular leaf spot of bean (caused by *Phaeoisariopsis griseola*) is even more variable than bean rust from season to season or site to site, sometimes increased due to intercropping, sometimes decreased, and sometimes not affected (Boudreau, 1993; Lanter, 1990; Mora-E., 1978; Moreno, 1977; Msuku and Edje, 1982; Sengooba, 1990).

1. Mechanisms

It is clear that we cannot categorically recommend intercropping as a means of disease control for a grower, nor are we in a position to predict the net effect of combining a particular host crop with other crops, at a certain time and location, on the dynamics of a specific pathogen. Prediction is a worthy goal, however, and may be attained through a combination of the phenomenological approach above with studies on the mechanisms by which the introduction of a new crop or crops into a monoculture alters various aspects of the pathogen's life cycle. Several mechanisms have been proposed for intercrop-disease interactions (Burdon, 1978; Trenbath, 1977), but little research has actually addressed these. Focusing on short-term mechanisms (single epidemic, as opposed to long-term genetic change), these potential mechanisms may be conveniently divided into mechanisms which affect the dispersal aspect of the pathogen, and mechanisms which affect non-dispersal aspects, including (depending on the pathogen) deposition, germination, penetration, growth in/on the host, reproduction, and production of new propagules for dispersal. Change in crop density is a mechanism which may affect both dispersal and nondispersal phenomena, but is a special case which we will consider separately.

a. Density Effects

Density is an important complicating factor which must always be considered when intercropping studies are evaluated and compared. Whenever a controlled study is done comparing a monocrop with an intercrop including that species, density unavoidably changes at some level. In a replacement experiment, a fraction of the monocrop population is replaced by one or more other crops. Thus, the density of this species (the disease host) declines, though the overall plant density (host + nonhost(s)) remains constant. In an addition or additive experiment, the monocrop (host) density remains constant, but the overall density increases as a second crop is added to the community. There is also the third possibility in which *both* host and overall density change in the intercrop relative to the monocrop, as is often the case where comparisons of "realistic" cropping patterns are made. For instance, performing a direct one-to-one replacement of beans at a typical monocrop density with maize would require an unusually close spacing of maize. Therefore a more typical maize arrangement is made, and both host density and overall density is different from the monocrop. This dilemma is further complicated by potential self-thinning at high densities, and the much greater leaf area/plant of maize vs. beans. True replacement should perhaps be based on leaf area index rather than number of individuals in a situation such as this. Crops with similar physiognomy seem most amenable to replacement-type experiments, such as wheat-barley mixtures, and reduce to absurdity for mixtures of, say, herbaceous plants with trees, in which one-to-one replacement is impossible. In any case, density can influence disease as we have seen, and whether changing host density alone or overall density would affect disease depends on the mechanism at work in the particular system.

A carefully designed set of experiments can help separate density from other intercropping influences. Burdon and Chilvers (1976c, 1977) provide examples with damping-off and barley mildew, described in the previous section. Similarly, Power (1987, 1990) has found density to be largely responsible for disease reductions of maize stunt in bean-maize intercrops. Farrell (1976a,b) controlled for density in a bean-peanut study and found that the absolute level of peanut rosette virus declined due to intercropping, noting that aphids were trapped in the epidermal hairs of the beans. However, per-plant aphid abundance was reduced by increasing density, as would have disease had it been expressed as incidence. Despite these valuable efforts to separate density effects from other intercrop effects, it should be noted that it is impossible to independently alter host density, overall density, and the presence of a nonhost in a complete multifactorial experiment to truly isolate the effect of each variable.

b. Dispersal Effects

Mechanisms by which an intercrop may influence dispersal include direct interference with propagule movement by the nonhost plants, as when spores impact on a nonhost leaf and are "trapped." The dispersal medium may be altered by an intercrop. Wind velocity and quality (turbulence) may be changed. In one study of an additive intercrop with bean:maize ratios of 4:1 and 2:1, wind velocities above the bean canopy averaged 63 and 47% of the velocity above the monocrop, respectively (Boudreau, 1993). Raindrops may be intercepted and either prevented from impacting propagules or "converted" into larger, slower drops falling from leaves (Armstrong and Mitchell, 1988). Changes in the dispersal medium would be expected primarily in additive intercrops, or replacements with physiognomically different crops. Often ignored is the indirect effect intercropping may have on dispersal, in which competition or other interactions between the host and nonhost(s) could alter the host in such a way as to affect dispersal (Boudreau and Mundt, 1992). Insect vector behavior may be altered by a nonhost in all types of intercrops.

We conducted experiments over two seasons to determine the roles of two categories of effects on dispersal of the bean rust pathogen (*Uromyces appendiculatus* (Pers. ex Pers.) Unger) on additive bean-maize intercrops: effects due to competition of maize with beans, which might include reduced growth of the beans, and effects due to the physical presence of maize during dispersal, which might include trapping or wind alteration (Boudreau and Mundt, 1992). Of course, the physical presence of maize is responsible for competition, but we were able to isolate these two factors by assessing dispersal in bean plots from which maize had just been cut out (i.e., competition without physical presence), and assessing dispersal in bean plots with artificially supported cut maize stalks (physical presence without competition). Surprisingly, the slopes of the primary dispersal gradient from a focal inoculum source were generally not significantly affected by the physical presence of maize. Instead, by 50 days after planting (DAP), the slopes consistently became steeper solely due to competition of the beans with maize (average 13% increase), or when both competition and maize presence were combined (that is, an intercrop). We also estimated the relative total spore deposition in experimental plots due to these factors, which may change independent of dispersal gradient. This is because not all of the spores released from a source in a plot remain in the plot, but many may be carried up and away from the canopy by turbulent air currents at any point over the plot. The fraction of spores retained (deposited) vs. removed could have serious consequences for the level of disease which develops. For this parameter, we again found that competition, but not the presence of maize, altered spore retention significantly, but the direction of the effect changed in both years as the season progressed.

These results with bean rust are somewhat surprising and difficult to interpret without repetition and further work. They suggest that in this system, not only is spore trapping by maize of little importance, but that size reductions in bean plants (the most obvious effect of maize competition, documented by leaf area measurements [Boudreau and Mundt, 1992]), are primarily responsible for any disease reductions associated with dispersal in bean-maize intercrops. Suggestions that turbulence and resulting spore removal dominate the dispersal process for *U. appendiculatus* in sparse bean canopies (Aylor and Ferrandino, 1990) are important factors in explaining our results, but the apparent nonexistent role of the maize canopy in this process, and the change over time of season, indicate a complex set of interactions. This has implications for efforts to improve bean yields in bean-maize intercrops by, say, fertilizing beans sufficiently to compensate for reduced growth caused by competition with maize.

The only other comprehensive study of intercropping effects on dispersal regards behavior of virus vectors. We have seen that density increases tend to reduce disease for most insect-vectored diseases. Therefore density once again becomes a confounding variable when attempting to interpret the disease reductions seen in studies summarized in Table 2. In cases where the density and diversity variables both have been evaluated, density has been responsible for reducing abundance of aphids on peanuts (transmitting peanut rosette virus) (Farrell, 1976a,b), and leafhopper abundance on maize (transmitting the corn stunt spiroplasma) (Power, 1987) in intercropped systems. Disease was reduced in both studies. In the latter case, Power (1987) found that plant-to-plant movement of the leafhoppers was reduced in maize-bean intercrops, particularly across maize rows with intervening bean rows. Emigration also was higher in the intercrop than the monocrop. The reduction in corn stunt was due to both a dilution of the vector population over a greater overall number of plants (a density effect), and the reduction in vector movement. Note that reduced disease is correlated with less vector movement in the corn stunt system, but that reduced incidence of maize rayado fino virus was correlated with *greater* insect movement in high-density maize stands (Power, 1992). This discrepancy could be related to the required tenure for pathogen acquisition (short for corn stunt, long for rayado fino) (Power, 1992).

Reduction in insect pests in intercrops has been attributed to olfactory confusion due to a nonhost plant (Vandermeer, 1989), but Power (1990) suggests this is probably not true for homopterans (including aphids and leafhoppers, the major vectors of plant viruses), which do not respond to olfactory cues (Saxena and Saxena, 1975). One mechanism that may result in reduced disease is that an aphid, which must alight on a plant and probe to determine if the host is suitable, may lose a nonpersistent virus in the process (Power, 1990). Furthermore, even a persistent virus may be "trapped" as the vector alights on the added crop if the crop is a host for the insect but not the virus (Toba et al., 1977), or if the vector is physically prevented from leaving the plant, as in aphids caught on bean trichomes (Farrell, 1976b). It is clear that the influence of intercropping on vector-mediated dispersal is mechanistically complex, depending on pathogen persistence and acquisition time, host range of the pathogen, host range of the vector, sensory cues for host detection, and so on.

c. *Nondispersal Effects*

Effects of intercropping on nondispersal components of the pathogen life cycle have been dominated by suggestions of microclimate change, which may well be expected to alter some of these components (Burdon, 1978; Trenbath, 1977). Inferences may be made where sufficient data are available. For example, daytime increases in relative humidity and decreases in leaf or air temperature measured in bean-maize intercrops relative to bean monocrops (Boudreau, 1993; Stoetzer and Omunyin, 1984) may favorably influence foliar diseases that infect more readily in moist conditions, such as bean rust, angular leaf spot, or white mold (Abawi and Grogan, 1975; Cardona-Alvarez and Walker, 1956; Imhoff et al., 1981).

Other nondispersal effects of intercropping include host competition with other species which may affect susceptibility, and infection alterations mediated by a pathogen or other product originating with the nonhost. Species of *Puccinia* which cause rust on maize and sunflower can induce resistance to *U. appendiculatus* when beans are inoculated with spores, but pollen can both increase or decrease infection, depending on the pathogen and host (Allen, 1975; Allen and Skipp, 1982; Chu Chou and Preece, 1968; Yarwood, 1956).

In our study of bean-maize intercrops (Boudreau and Mundt, 1992), we directly evaluated nondispersal mechanisms by simultaneously inoculating bean plants in monocrops and intercrops and comparing latent periods and severity after approximately 14 days. In three inoculations in each of two years, the latent period was not affected by intercropping. In one season, severity was affected only late in the season (84 DAP), but here the effect was dramatic: intercropping reduced severity by 96% below the monocrop value. An additional treatment strongly suggests that this was a microclimatic effect of the maize, which may seem counterintuitive based on previous studies (see above). One explanation is that radiative cooling of bean leaves in the evening is delayed when a maize canopy is present, thus decreasing the time in which dew is present on bean leaves, essential for *U. appendiculatus* infection. (Imhoff et al., 1981).

We have discussed a few isolated cases of research evaluating some mechanisms of disease alteration in particular intercrops. The role of the many possible mechanisms in the wide range of crop combinations used has, for the most part, simply not been studied. It is at least clear that we cannot make generalizations as to how disease would behave in an intercrop from the little information available. Detailed studies specific to intercropping in particular pathosystems must be done if we are to make sense of the variability of intercrop-disease interactions which have been observed.

2. Computer Simulation

It is logistically impossible to assess the relative role of each intercrop-disease interaction in the field, or how they combine to produce the overall epidemic. This is nonetheless of great importance. For example, if dispersal-related effects dominate the role of an added species in modifying an epidemic, one would expect that situations in which dispersal gradients are steep and autoinfection (defined by Robinson [1976] as "infection in which the donor [infector] host individual is the same as the recipient [infected] host individual") is pronounced (e.g., splash-dispersed pathogens), intercrop efficacy may be low as opposed to situations in which nondispersal effects dominate. Also, perennial crops or those in continuous environments may experience disease reductions indefinitely if a mechanism acting on infection is at work, but if the mechanism acts on dispersal, the effect may be lost over time as the pathogen eventually disperses throughout the population.

Computer modeling has been a valuable component of the ecologist's efforts to understand complex natural communities, and may be useful here. Indeed, plant pathologists have developed a number of programs that model epidemics in particular crop-disease systems, and one (EPIMUL, Kampmeijer and Zadoks, 1977) has been explicitly developed to simulate disease epidemics in variety mixtures or multilines, in which different levels of resistance may be assigned to components of a monocrop. We have modified this simulator and used input variables derived from the field to study epidemics in intercrops (Boudreau and Mundt, 1994). Our objective in this study was to evaluate the relative and combined roles of the various mechanisms of intercrop influence on disease that we had studied in the bean-maize-bean rust system described earlier (Boudreau and Mundt, 1992). Specifically, we were able to simulate epidemics initiated in the center of 20 × 20 m "plots," adjusting input values to reflect effects of maize on dispersal due to competition of beans with maize, effects on dispersal due to direct interference of maize, and development of disease after deposition, each in isolation and in all possible combinations. The input values were adjusted to reflect changes observed in the field over time, and separate simulations were done with data from each year of the field studies. All simulations were also run over a range of basic infection rates (Vanderplank, 1963), reflecting the overall conduciveness of the environment for disease, regardless of the intercrop treatment.

These simulations generated disease progress curves which in many ways recreated the sort of variability observed in field studies generally. The combined effect of all intercropping mechanisms (that is, a simulated intercrop) resulted in reductions of the area under the disease progress curve (AUDPC) of 68% relative to the monocrop when input data from the 1989 season were used, with severity <10% by the end of the season in the intercrop, compared with 65% severity in the monocrop. This occurred at low basic infection rates, however, and the differences diminished as the rate increased. Data from 1990 produced almost identical disease progress curves for the monocrop and intercrop at all infection rates.

Partitioning disease reductions that occurred in the 1989-based simulations among component mechanisms revealed that dispersal alterations due to maize were responsible for only about 19% of the reduction in AUDPC, the remainder was due to reductions in postdeposition infection efficiency. In the field data on which these were based, the reduction in infection efficiency was probably due to a microclimatic effect and occurred late in the season; the simulation suggested that such late-season effects could still have a profound influence on disease severity. A final interesting outcome of the simulations was that any effects of maize on dispersal in simulations (either year, and all infection rates) were due almost entirely to changes in the fraction of spores retained (vs. removed) in the plots, but not dispersal gradients. In other words, the 19% reduction in AUDPC described above for 1989 occurred because more spores were removed from the intercrop plot than the bean monocrop, not because the dispersal gradient was steeper.

These simulations are only a first attempt to apply computer modeling to intercrop situations, but they are consistent with the variable results observed in the field and point to some reasons for it, such as sensitivity of intercrop effects to the rate of disease progress, importance of late-season microclimatic effects, and spore removal processes. This is only a beginning for an area of study which has great potential, although it must depend on a concomitant production of empirical data. Further development could someday result in models which not only allow us to understand how epidemics are conditioned by species diversity in natural and agricultural ecosystems, but also provide predictive power for the efficacy of intercropping a particular crop system under specific conditions.

B. Variety Mixtures and Multilines

Relative to the situation for intercropping, the use of variety mixtures and multilines for disease control has been studied more extensively and utilized by growers more widely. Although development of variety mixtures has been encouraged by speculations on intraspecific diversity acting to regulate disease in natural ecosystems (Browning, 1981; Robinson, 1980), we are only beginning to document the level and pattern of genetic diversity for resistance in wild systems (e.g., Alexander et al., 1993; Burdon and Jarosz, 1991; Parker, 1988; Segal et al., 1980), much less its role in pathogen dynamics (attempts include Alexander, 1991; Schmid, 1994). The commercial penetration of variety mixtures is due instead to the consistent disease reduction observed in most systems. Mixtures have been effective experimentally for barley mildew (Wolfe, 1991), rusts and Septoria on wheat (Brophy and Mundt, 1991; Jeger et al., 1981a,b; Mahmood et al., 1991), blast of rice (Bonman et al., 1986), anthracnose and rust on beans (Mundt and Leonard, 1986; Ntahimpera et al., 1996), barley yellow dwarf virus on oats (Power, 1991), and many others. Significant commercial success has been realized with small grain crops (Browning and Frey, 1981; Wolfe, 1992; Mundt, 1994a,b). The greater adoption by farmers of variety mixtures relative to intercrops may also be attributed to the amenability of same-species mixes to modern industrial farming techniques such as mechanical harvesting and specialized chemical use. This is particularly true for multilines *sensu strictu*, which are mixtures of near-isogenic lines differing only with respect to resistance genes.

In the rather voluminous literature on this subject (see reviews by Browning and Frey, 1969; Mundt and Browning, 1985; Wolfe, 1985), what can be gleaned regarding the mechanisms of disease reduction? We again will consider this in terms of mechanisms that influence the nondispersal and dispersal aspects of the pathogen life cycle.

Research and commercial application with mixtures are usually done on a replacement basis, maintaining the same overall density of the crop. The crops are physiognomically quite similar if not identical. Therefore many of the nondispersal factors potentially active in intercrop systems are eliminated. For example, one expects few microclimatic differences between single-variety and mixed-variety communities. Changes in susceptibility due to competition will be less of a factor than in an additive, or even a replacement multispecies agroecosystem, but could occur when varieties of differential competitive ability are mixed. If the nutritional status of a susceptible variety is reduced by mixing with an aggressively competitive resistant variety, and this in turn affects infection efficiency, the epidemic could be altered through this mechanism. Competitive interactions have been documented for mixtures of wheat, and susceptibility changes may mediate disease dynamics in this case (Finckh and Mundt, 1992a). Induced resistance may occur when inoculum produced on one host genotype is subsequently deposited on a different genotype and induces resistance against inoculum that would normally be virulent on the other host (Wolfe, 1985). Burdon (1978) discounted induced resistance as a significant factor unless infection levels are high, but very little field-scale work on this phenomenon has taken place (Chin and Wolfe, 1984). Competitive interactions among different virulent races can sometimes result in decreased infection on a susceptible genotype as compared to inoculation with a single race (Leonard, 1969). Such competitive interactions could be common in intraspecific crop mixtures, as diversity for resistance would select for corresponding diversity in the pathogen population. Finally, for grass species, differential tillering of host genotypes due to intergenotypic plant competition can impact disease dynamics through selection of genotypes with differing degrees of susceptibility to a pathogen and alter overall disease severity in the mixture (Finckh and Mundt, 1992a,b).

Dispersal effects function very differently with intraspecific mixtures as compared to intercrops. Because a replacement approach using similar genotypes is the norm for both experimental and practical use of intraspecific diversity, alterations in the dispersal medium (water or wind) are unlikely. Similarly, propagules are not physically intercepted on resistant plants any differently than is the case in a pure stand of susceptible plants. Rather, the mean distance between plants of like genotype is increased, effectively reducing the density of each genotype even though the overall density of the crop is unchanged. Thus, for a pathogen demonstrating race specificity, propagules deposited on intervening, incompatible plants do not infect the host and are effectively removed from propagating the epidemic. This density effect, in fact, seems to be the major mechanism by which variety mixtures and multilines reduce disease severity (Burdon and Chilvers, 1977; Chin and Wolfe, 1984; Wolfe, 1985). Further, the density of a host genotype within an intraspecific mixture is controlled very directly through the planted proportion of host genotypes in a mixture. For this reason, disease severity is reduced rather consistently in intraspecific mixtures, at least for polycyclic foliar pathogens that demonstrate race specificity (Browning and Frey, 1969; Mundt and Browning, 1985; Wolfe, 1985). This differs considerably from intercrops planted in an additive fashion, where the overall density of one or more crops may be reduced due to the effects of interspecific competition. This results in the potential for multiple impacts on both dispersal and nondispersal disease effects (see discussion above), and considerable variation in the impact of intercrops on disease severity (Table 2).

It is important to note that, with intraspecific mixtures, disease is often evaluated for the entire population, regardless of whether a given genotype is susceptible or resistant. Thus, disease progression should be measured for each genotype in mixture relative to its corresponding pure stand, or else for the mixture as a whole relative to the mean of its components grown separately in pure stands. Failure to do so can result in an artifact whereby disease severity seems reduced simply due to the presence of resistant plants within the mixture.

Leonard (1969) developed a model to predict the rate of increase of disease on susceptible plants in a mixture of near-isogenic susceptible and resistant hosts based on epidemiological considerations for a polycyclic foliar disease. This model predicts that the rate of disease increase on susceptible plants in mixture will be proportional to the logarithm of the proportion of susceptible plants and inversely proportional to the generation time of the pathogen — predictions that are supported by field experiments (Leonard, 1969). To more correctly account for disease dynamics in mixtures, however, infection rates must also be tied to models that account for the limiting carrying capacity of the host for disease (Mundt, 1990). Further, such models address only reduced density effects, and do not account for more complex dynamics, discussed above, that may occur when multiple pathogen genotypes are present and when host components have diverse genetic backgrounds. These latter considerations are important as, in practice, multiple pathogen races may be present, and each component of a host mixture may have varying reactions to each of these races. In addition, variety mixtures incorporating diverse genetic backgrounds have many advantages over multilines in practice (Mundt and Browning, 1985; Wolfe and Barrett, 1980; Wolfe, 1985) and are currently receiving much more attention than multilines at the applied level.

The relative importance of dispersal mechanisms for disease reduction in intraspecific mixtures mitigates against their efficacy in situations where epidemic development is determined more by infection processes than by dispersal, as in soil-borne or splash-dispersed pathogens (Wolfe, 1985), or when a high proportion of infections are autoinfections, as when host genotypes are aggregated (Brophy and Mundt, 1991). Indeed, use of intraspecific diversity for disease control has been limited primarily to airborne foliar pathogens of cereals. On the other hand, positive yield benefits of mixtures have been found against some soil-borne pathogens of wheat even when disease severity is not reduced (Mundt, unpublished), coffee *(Coffea arabica)* mixtures are being grown in Colombia to control rust (caused by *Hemileia vastatrix*) (Ruiz and Zapata, 1990) and modeling approaches suggest that aggregates of different host genotypes may have significant impact on epidemic progression when deployed at large spatial scales (Mundt and Brophy, 1988). How intraspecific diversity might affect dispersal by vectored pathogens is poorly understood, but Power

(1991) found that movement of the aphid *Sitobion avenae* was greater in two-variety mixtures of oats than in pure stands and that there was a reduction in barley yellow dwarf virus in the mixture, for which *S. avenae* is a vector.

Most studies of intraspecific mixtures have focused on pathogens with clear race specificity. Jeger et al. (1981a) modeled the effect of two-component intraspecific mixtures on epidemic progression for polycyclic foliar diseases caused by nonspecific pathogens. They concluded that mixtures may decrease, increase, or have no effect on epidemic progression depending on the relative values of infection efficiency and sporulation of the component host genotypes. Field studies of nonspecific pathogens in variety mixtures showed that significant disease reductions were possible in the apparent absence of pathogen specificity (Jeger et al., 1981b).

In the longer term, it is critical to consider the impacts of intraspecific mixtures on pathogen population dynamics as well as on epidemic progression. Deployment of resistance genes in mixtures has apparently increased the durability of resistance genes on a regional basis in some cases (Mundt, 1994b). On the other hand, there is concern that use of intraspecific mixtures may select for superraces able to overcome all of the resistance genes incorporated in a mixture. Some have argued that there is little field data to support this concern (Mundt and Browning, 1985; Mundt, 1994a). The question is difficult to address experimentally, however, and mathematical models (Leonard and Czochor, 1980; Kiyosawa, 1989; Marshall, 1989) have provided inconclusive results. These models all assume that a fitness cost to the pathogen is associated with virulence, an assumption that has little empirical support (Parlevliet, 1991). The models, however, generally have not accounted for other mechanisms that may contribute to maintenance of genetic polymorphism in pathogen populations (Mundt and Browning, 1985, Mundt, 1994b). Thus, at this point, it seems prudent to attempt to manage mixtures so as to reduce potential selection for superraces (Wolfe, 1985).

Finally, it should be noted that even with high levels of disease on one or more components of a variety mix, the resistant components may produce more in the less competitive environment, and compensate somewhat for this yield loss. This also applies to mixed-species stands. In any case, even without compensatory yield, diversity of species or varieties offers greater yield stability and a hedge against even a devastating epidemic on one unfortunate component of a mixture (Wolfe, 1985). This has long been realized by farmers in less developed countries, with very little land at their disposal and little willingness to risk it all to a single genotype of a single crop.

IV. ECOLOGICAL THEORY AND MICROBES

This chapter has thus far discussed how ecology of plants might influence disease and be utilized for disease control in agroecosystems. There is, of course, a large body of work regarding the ecology of microbes, including those in the rhizosphere and, to a growing extent, on the phylloplane (Blakeman, 1981; Wicklow and Carrol, 1981, 1992; Fokkema and van den Heuvel, 1986; Beattie and Lindow, 1995). At times the microbial ecology of plant pathogens is specifically addressed, and the burgeoning field of biocontrol has provoked analysis of competition, predation, and antibiosis among pathogens and nonpathogens. Microbial ecology of endophytes and suppressive soils are fascinating subsets of this work (Part III of this volume; Fokkema and van den Heuvel, 1986; Lumsden, 1981; Windels and Lindow, 1985; Cook, 1990; Thomashow and Weller, 1996).

Attempts have been made to apply or interpret specific macroecological theory in terms of microbial populations on plants. For example, island biogeography theory has been evaluated using apple leaves as islands for microorganisms (Andrews and Kinkel, 1986; Kinkel et al., 1987), succession of fungal communities on plant surfaces has been analyzed (Frankland, 1981; Mishra and Das, 1981), and concepts of fundamental and realized niche have been addressed using yeasts on cactus (Starmer, 1981). Life-history strategies have been discussed with respect to microbes and plant pathogens (Andrews and Rouse, 1982; Andrews and Harris, 1986). These studies may have application to plant disease control; for example, for ascertaining or creating the best situation in which to introduce a biological control agent which generally is most effective when added to simple microbial communities (Fravel and Engelkes, 1994).

Space limitations preclude a review of the microbial ecology literature, but we will describe one of the rare instances in which macroecological theory has been applied directly to plant pathogens. This involves the theory of life-history strategies known as *r*- and *K*-selection. The notion that certain sets of traits typically occur together, running on a continuum from extreme *r*-selection (short life, polymorphic, highly migratory, asexual reproduction, poor competitor) to extreme *K*-selection (long life, monomorphic, sedentary, biparental sexual reproduction, adaptations for competition) was first developed in the 1950s and 1960s. These combinations were thought to manifest selection pressures of variable, disturbed environments with lax competition in the former case, and constant environments near carrying capacity in the latter case (Pianka, 1970). Application of this theory generally has been limited to animals and vascular plants, and in that context has been somewhat controversial.

Andrews and Rouse (1982) furthered the controversy by discussing plant pathogens in light of *r*- and *K*-selection. They suggested that overall, plant pathogens seem to possess the characteristics of *r*-strategists, if one focuses on the parasitic phase of their life cycles. They note that pathogens typically occupy fluctuating short-lived environments (e.g., a leaf surface) which they must colonize and reproduce upon quickly before the substrate disappears. Indeed, the nature of parasitism itself dictates that the substrate will be destroyed, or at least compromised, by the pathogen. The authors note that exceptions to the pattern occur, such as the relatively stable environments of a tree trunk interior, or the highly competitive environment of multiple infection. Polymorphism in some pathogen species, such as facultative parasitism or alternate feeding modes of nematodes, may reflect adaptation to this heterogeneity of host environments. Thus, a pathogen may alter its strategy depending on environment or life stage, or may possess a mix of *r*- and *K*-strategies at one point. Do the exceptions and ambiguity preclude application of this theory to plant pathogens?

Andrews and Rouse (1982) proposed a model to clarify the placement of pathogens on the *r*-*K* continuum based on relative allocation of resources to maintenance, growth, and reproduction. They classify organisms that devote a large proportion of their resources to reproductive structures, particularly large numbers of small propagules, as *r*-selected, and pathogens which preferentially allocate resources to vegetative growth as *K*-selected. Using this criterion, fungal examples are cited along the continuum from *r* to *K* as *Puccinia striiformis* (large numbers of widely disseminated spores) → *Helminthosporium* spp. → *Pythium*/*Phytophthora* spp. → *Rhizoctonia solani* (no spore production). As a secondary measure, they propose consideration of the nature of the host-parasite relationship. Here the distinction is made between pathogens which destroy the host, thereby creating a disturbance (*r*-selected), and those which "merely" stress the host, say by reducing photosynthesis. They note that both these extremes can occur within the traditional nutritional associations such as obligate biotroph, facultative necrotroph, etc. *Venturia inaequalis* (cause of apple scab) may in some cases defoliate its host, while the slow-growing pathogen causing cedar-apple rust, *Gymnosporangium juniperi-virginianae*, will cause low-grade symptoms on its hosts over many years. The former is *r*-selected, the latter *K*-selected, though both are obligate biotrophs.

The ideas that Andrews and Rouse (1982) present embody a unique application of ecological theory to plant pathogens. Furthermore, the authors note that their approach represents an attempt to unify theories of macro- and microorganisms, and free-living and parasitic organisms, and that it suggests new systems in which to test these theories. Finally, they suggest that their ideas may "... provide a new conceptual perspective for the design of plant disease control tactics generally and biological control in particular." Unfortunately, this idea is not elaborated further, and other workers have not related this theoretical framework to disease control applications.

This is one of the rare instances in which an ecological theory developed in plant and animal communities is discussed with reference to plant pathogens. In fact, studies of macro-and micro-ecology generally have been largely independent, and little attempt has been made to synthesize or integrate the two, if that is indeed possible. Andrews (1991) has discussed the comparative ecology of macro- and microorganisms and the relatively recent notions of hierarchy in ecosystems

(O'Neill et al., 1986) may provide means to bridge these disparate scales. This should be of concern to plant pathologists, for it is at the infection court that these scales are intrinsically bridged.

V. CONCLUSIONS

The research community needs to explore the potential for utilizing the unique ecological interactions seen in natural communities for their promise in disease management. Lenné et al. (1994) wrote that " ... the evolution of pathogens in complex ecosystems has been neglected due to the immense difficulties of characterizing the multitude of components and interactions occurring in such systems." These difficulties are well known to ecologists, however, and a dialogue between pathologists and ecologists can only serve to benefit both camps. As an understanding of disease dynamics in unmanaged systems develops, coupled with the extant ecological theory derived from these systems, plant pathologists should creatively explore the specific ways in which these ideas can generate predictions for agroecosystems, then design experiments to test them. The definition of "agroecosystem" should also be creatively rendered and not confined to neat rows of one genetically uniform crop.

This chapter has consolidated some recent work on observations of plant communities which leads to unorthodox disease management strategies that are more ecologically based. These strategies and others like them could contribute significantly to sustainable, low-input farming systems throughout the world. The demands of the future and the limitations of past approaches make it clear that we can no longer ignore the potential for an ecological approach to managing disease.

REFERENCES

Abawi, G. S. and Grogan, R. G., Source of primary inoculum and effects of temperature and moisture on infection of beans by *Whetzelinia sclerotium, Phytopathology,* 65, 300, 1975.

A'Brook, J., The effect of planting date and spacing on the incidence of groundnut rosette disease and of the vector, *Aphis craccivora* Koch, at Mokwa, Northern Nigeria, *Ann. Appl. Biol.*, 54, 199, 1964.

Akashi, Y. and Mueller-Dumbois, D., A landscape perspective of the Hawaiian rain forest dieback, *J. Veg. Sci.*, 6, 449, 1995.

Alexander, H. M., The effect of disease induced by *Albugo candida* and *Peronospora parasitica* on the survival and reproduction of *Capsella bursa-pastoris*, *Oecologia*, 64, 314, 1984.

Alexander, H. M., Epidemiology of anther-smut infection of *Silene alba* caused by *Ustilago violacea.* Patterns of spore deposition and disease incidence, *J. Ecol.*, 78, 166, 1990.

Alexander, H. M., Plant population heterogeneity and pathogen and herbivore levels: a field experiment, *Oecologia,* 86, 125, 1991.

Alexander, H. M., Evolution of disease resistance in natural plant populations, in *Plant Resistance to Herbivores and Pathogens*, Fritz, R. S. and Simms, E. L., Eds., University of Chicago Press, Chicago, 1992, chap. 14.

Alexander, H. M., Antonovics, J., and Kelly, A. W., Genotypic variation in plant disease resistance — physiological resistance in relation to field disease transmission, *J. Ecol.*, 81, 325, 1993.

Allen, D. J., Induced resistance to bean rust and its possible epidemiological significance in mixed cropping, in *Intercropping in Semi-Arid Areas*, Monyo, J. H., Kerr, A. D. R., and Campbell, M., Eds., IDRC, Ottawa, Canada, 1975, 46.

Allen, D. J. and Skipp, R. A., Maize pollen alters the reaction of cowpea to pathogens, *Fla. Crops Res.*, 5, 265, 1982.

Altieri, M. A. and Liebman, M., Insect, weed, and plant disease management in multiple cropping systems, in *Multiple Cropping Systems*, Francis, C. A., Ed., Macmillan, New York, 1986, 183.

Anderson, R. M. and May, R. M., Population biology of infectious disease. I, *Nature*, 280, 361, 1979.

Andrews, J. H., Biological control in the phyllosphere — realistic goal or false hope?, *Can. J. Plant Pathol.*, 12, 300, 1990.

Andrews, J. H., *Comparative Ecology of Microorganisms and Macroorganisms*, Springer-Verlag, New York, 1991.

Andrews, J. H. and Harris, R. F., *r*- and *K*-selection and microbial ecology, *Adv. Microb. Ecol.*, 9, 99, 1986.

Andrews, J. H. and Rouse, D. I., Plant pathogens and the theory of *r*- and *K*-selection, *Am. Nat.*, 120, 283, 1982.

Andrews, J. H. and Kinkel, L. L., Colonization dynamics: the island theory, in *Microbiology of the Phyllosphere*, Fokkema, N. J. and van den Heuvel, J., Eds., Cambridge University Press, New York, 1986, 63.

Antonovics, J., Stratton, D., Thrall, P. H., and Jarosz, A. M., An anther-smut disease *(Ustilago violacea)* of fire-pink *(Silene virginica)*. Its biology and relationship to the anther-smut disease of white campion *(Silene alba)*, *Am. Mid. Nat.*, 135, 130, 1996.

Armstrong, C. L. and Mitchell, J. K., Plant canopy characteristics and processes which affect transformation of rainfall properties, *Trans. Am. Soc. Agric. Eng.*, 31, 1400, 1988.

Augspurger, C. K., Impact of pathogens on natural plant populations, in *Plant Population Ecology*, Davy, A. J., Hutchings, M. J., and Watkinson, A. R., Eds., Blackwell Scientific, Oxford, 1988, 413.

Augspurger, C. K. and Kelly, C. K., Pathogen mortality of tropical tree seedlings. Experimental studies of the effects of dispersal distance, seedling density, and light conditions, *Oecologia*, 61, 211, 1984.

Aylor, D. E. and Ferrandino, F. J., Initial spread of bean rust close to an inoculated bean leaf, *Phytopathology*, 80, 1469, 1990.

Beattie, G. A. and Lindow, S. E., The secret life of foliar bacterial pathogens on leaves, *Annu. Rev. Phytopathol.*, 33, 145, 1995.

Begon, M., Harper, J. L., and Townsend, C. R., *Ecology: Individuals, Populations, and Communities*, Blackwell Scientific, Boston, 1990.

Berger, R. D., Disease incidence and infection rates of *Cercospora apii* in plant spacing plots, *Phytopathology*, 65, 485, 1975.

Bigger, M., Oscillations of tropical insect populations, *Nature*, 259, 207, 1976.

Blakeman, J. P., Ed., *Microbial Ecology of the Phylloplane*, Academic Press, New York, 1981.

Blakeman, J. P. and Fokkema, N. J., Potential for biological control of plant diseases on the phylloplane, *Annu. Rev. Phytopathol.*, 20, 167, 1982.

Bonman, J. M., Estrada, B. A., and Denton, R. I., Blast management with upland rice cultivar mixtures, in *Progress in Upland Rice Research*, IRRI, Manila, 1986, 375.

Bottenberg, H. and Irwin, M. E., Using mixed cropping to limit seed mottling induced by soybean mosaic virus, *Plant Dis.*, 76, 304, 1992.

Boudreau, M. A., Effect of intercropping beans with maize on the severity of angular leaf spot of beans in Kenya, *Plant Pathol.*, 42, 16, 1993.

Boudreau, M. A. and Mundt, C. C., Mechanisms of alteration in bean rust epidemiology due to intercropping with maize, *Phytopathology*, 82, 1051, 1992.

Boudreau, M. A. and Mundt, C. C., Mechanisms of alteration in bean rust development due to intercropping, in computer-simulated epidemics, *Ecol. Appl.*, 4, 729, 1994.

Boudreau, M. A. and Madden, L. V., Effect of strawberry density on dispersal of *Colletotrichum acutatum* by simulated rain, *Phytopathology*, 85, 934, 1995.

Brophy, L. S. and Mundt, C. C., Influence of plant spatial patterns on disease dynamics, plant competition and grain yield in genetically diverse wheat populations, *Agric. Ecosyst. Environ.*, 35, 1, 1991.

Browning, J. A., Relevance of knowledge about natural ecosystems to the development of pest management programs for agroecosystems, *Proc. Am. Phytopathol. Soc.*, 1, 191, 1974.

Browning, J. A., Genetic protective mechanisms of plant-pathogen populations: their coevolution and use in breeding for resistance, in *Biology and Breeding for Resistance to Arthropods and Pathogens of Cultivated Plants*, Texas Ag. Exp. Stn. Misc. Publ. No. 1451, Harris, M. K., Ed., 1980, 52.

Browning, J. A., The agro-ecosystem — natural ecosystem dichotomy and its impact on phytopathological concepts, in *Pests, Pathogens, and Vegetation*, Thresh, J. M., Ed., Pitman Publishing, Boston, 1981, 159.

Browning, J. A. and Frey, K. J., Multiline cultivars as a means of disease control, *Annu. Rev. Phytopathol.*, 7, 355, 1969.

Browning, J. A., and Frey, K. J., The multiline concept in theory and practice, in *Strategies for the Control of Cereal Disease*, Jenkyn, J. F. and Plumb, R. T., Eds., Blackwell Scientific, Oxford, 1981, 37.

Burdon, J. J., Mechanisms of disease control in heterogeneous plant populations — an ecologist's view, in *Plant Disease Epidemiology,* Scott, P. R. and Bainbridge, A. W., Eds., Blackwell Scientific, Oxford, 1978, 193.

Burdon, J. J., *Disease and Plant Population Biology,* Cambridge Studies in Ecology, Cambridge University Press, Cambridge, 1987.

Burdon, J. J., Fungal pathogens as selective forces in plant populations and communities, *Aust. J. Ecol.,* 16, 423, 1991.

Burdon, J. J. and Chilvers, G. A., Epidemiology of damping-off disease *(Pythium irregulare)* in relation to density of *Lepidium sativum* seedlings, *Ann. Appl. Biol.,* 81, 135, 1975.

Burdon, J. J. and Chilvers, G. A., The effect of clumped planting patterns on epidemics of damping-off disease in cress seedlings, *Oecologia,* 23, 17, 1976a.

Burdon, J. J. and Chilvers, G. A., Controlled environment experiments on epidemics of barley mildew in different density host stands, *Oecologia,* 26, 61, 1976b.

Burdon, J. J. and Chilvers, G. A., Epidemiology of *Pythium*-induced damping-off in mixed species stands, *Ann. Appl. Biol.,* 82, 233, 1976c.

Burdon, J. J. and Chilvers, G. A., Controlled environment experiments on epidemic rates of barley mildew in different mixtures of barley and wheat, *Oecologia,* 28, 141, 1977.

Burdon, J. J. and Chilvers, G. A., Host density as a factor in plant disease ecology, *Annu. Rev. Phytopathol.,* 20, 143, 1982.

Burdon, J. J., Groves, R. H., and Cullen, J. M., The impact of biological control on the distribution and abundance of *Chondrilla juncea* in southeastern Australia, *J. Appl. Ecol.,* 18, 957, 1981.

Burdon, J. J. and Jarosz, A. M., The ecological genetics of plant-pathogen interactions in natural communities, *Philos. Trans. R. Soc. London Ser. B.,* 321, 349, 1988.

Burdon, J. J. and Jarosz, A. M., Pattern and patchiness in plant-pathogen interactions: causes and consequences, *Annu. Rev. Ecol. Syst.,* 20, 119, 1989.

Burdon, J. J. and Jarosz, A. M., Host-pathogen interactions in natural populations of *Linum marginale* and *Melampsora lini.* I. Patterns of resistance and racial variation in a large host population, *Evolution,* 45, 205, 1991.

Burdon J. J. and Jarosz A. M., Temporal variation in the racial structure of flax rust *(Melampsora lini)* populations growing on natural stands of wild flax *(Linum marginale):* local versus metapopulation dynamics, *Plant Pathol.,* 41, 165, 1992.

Burdon, J. J., Jarosz, A. M., and Kirby, G. C., Pattern and patchiness in plant-pathogen interactions: causes and consequences, *Annu. Rev. Ecol. Syst.,* 20, 119, 1989.

Burdon, J. J. and Shattock, R. C., Disease in plant communities. *Appl. Biol.,* 5, 145, 1980.

Burdon, J. J. and Thompson, J. N., Changed patterns of resistance in a population of *Linum marginale* attacked by the rust pathogen *Melampsora lini, J. Ecol.,* 83, 199, 1995.

Burdon, J. J., Wennstrom, A., Ericson, L., Miller, W. J., and Morton, R., Density-dependent mortality in *Pinus sylvestris* caused by the snow blight pathogen *Phacidium infestans, Oecologia,* 90, 74, 1992.

Cardona-Alvarez, C. and Walker, J. C., Angular leaf spot of bean, *Phytopathology,* 46, 610, 1956.

Carlsson, U., Elmquist, T., Wennstrom, A., and Ericson, L., Infection by pathogens and population age of host plants, *J. Ecol.,* 78, 1094, 1990.

Chin, K. M. and Wolfe, M. S., The spread of *Erysiphe graminis* f.sp. *hordei* in mixtures of barley varieties, *Plant Pathol.,* 33, 89, 1984.

Chu Chou, M. and Preece, T. F., The effect of pollen grains on infection caused by *Botrytis cinerea, Ann. Appl. Biol.,* 62, 11, 1968.

Cobb, F. W., Jr., Slaughter, G. W., Rowney, D. L., and DeMars, C. J., Rate of spread of *Ceratocystis wageneri* in ponderosa pine stands in the central Sierra Nevada, *Phytopathology,* 72, 1359, 1982.

Cook, R. J., Plant management in relation to biological control of plant pathogens, *UCLA Symp. Mol. Cell. Biol.,* 112, 441, 1990.

Cook, S. A., Stand development in the presence of a pathogen, *Phellinus weirii,* in *Forest Succession and Stand Development Research in the Northwest,* Means, J., Ed., USFS Forest Research Lab, Corvallis, OR, 1982, 159.

Curl, E. A., Control of plant diseases by crop rotation, *Bot. Rev.,* 29, 413, 1963.

Davies, J. C., The incidence of rosette disease in groundnut in relation to plant density and its effect on yield, *Ann. Appl. Biol.,* 82, 489, 1976.

Dekker, J., The fungicide resistance problem: current status and the role of systematics, in *Pesticide Interactions in Crop Production,* Altman, J., Ed., CRC Press, Boca Raton, FL, 1993, chap. 8.

Dinoor, A. and Eshed, N., The role and importance of pathogens in natural plant communities, *Annu. Rev. Phytopathol.,* 22, 443, 1984.

Edwards, L., Utkhede, R. S., and Vrain, T., Effect of antagonistic plants on apple replant disease, *Acta Hortic.,* 363, 135, 1994.

Egunjobi, O. A., Effects of intercropping maize with grain legumes and fertilizer treatment on populations of *Pratylenchus brachyurus* Godfrey (Nematoda) and on the yield of maize (*Zea mays* L.), *Prot. Ecol.,* 6, 153, 1984.

Elton, C. S., *The Ecology of Invasion by Animals and Plants,* Methuen, London, 1958.

Emebiri, L. C. and Obiefuna, J. C., Effects of leaf removal and intercropping on the incidence and severity of black Sigatoka disease at the establishment phase of plantains (*Musa* spp. AAB), *Agric. Ecosyst. Environ.,* 39, 213, 1992.

English, J. T., Kaps, M. L., Moore, J. F., Hill, J., and Nakova, M., Leaf removal for control of Botrytis bunch rot of wine grapes in the midwestern United States, *Plant Dis.,* 77, 1224, 1993.

English, J. T., Thomas, C. S., Marois, J. J., and Gubler, W. D., Microclimates of grapevine canopies associated with leaf removal and control of Botrytis bunch rot, *Phytopathology,* 79, 395, 1989.

Ercolani, G. L., Distribution of epiphytic bacteria on olive leaves and the influence of leaf age and sampling time, *Microb. Ecol.,* 21, 35, 1991.

Fargette, D. and Fauquet, C., A preliminary study on the influence of intercropping maize and cassava on the spread of African cassava mosaic virus by whiteflies, *Aspects Appl. Entomol.,* 17, 195, 1988.

Farrell, J. A. K., Effects of groundnut crop density on the population dynamics of *Aphis craccivora* Koch (Hemiptera, Aphididae) in Malawi, *Bull. Entomol. Res.,* 66, 317, 1976a.

Farrell, J. A. K., Effects of intersowing with beans on the spread of groundnut rosette virus by *Aphis craccivora* Koch (Hemiptera, Aphididae) in Malawi, *Bull. Entomol. Res.,* 66, 331, 1976b.

Finckh, M. R. and Mundt, C. C., Plant competition and disease in genetically diverse wheat populations, *Oecologia,* 91, 82, 1992a.

Finckh, M.R. and Mundt, C. C., Stripe rust, yield, and plant competition in wheat cultivar mixtures, *Phytopathology,* 82, 905, 1992b.

Fokkema, N. J. and van den Heuvel, J., *Microbiology of the Phyllosphere,* Cambridge University Press, Cambridge, 1986.

Francis, C. A., Ed., *Multiple Cropping Systems,* Macmillan, New York, 1986.

Frank, S. A., Models of plant-pathogen coevolution, *Trends Genet.,* 8, 213, 1992.

Frankland, J. C., Mechanisms in fungal succession, in *The Fungal Community,* Wicklow, D. T. and Carroll, G. C., Eds., Marcel Dekker, New York, 1981, chap. 22.

Fravel, D. R. and Engelkes, C. A., Biological management, in *Epidemiology and Management of Root Diseases,* Campbell, C. L. and Benson, D. M., Eds., Springer-Verlag, Berlin, 1994, chap. 10.

Fry, W. E., *Principles of Plant Disease Management,* Academic Press, New York, 1982.

Gates, D. J. and Westcott, M., Stability of plant mixtures in the absence of infection or predation, *J. Theor. Biol.,* 131, 15, 1988.

Ghewande, M. P., Desai, S., Narayan, P., and Ingle, A. P., Integrated management of foliar diseases of groundnut (*Arachis hypogaea* L.) in India, *Int. J. Pest Manage.,* 39, 375, 1993.

Grace, J. B. and Wetzel, R. G., The production biology of Eurasian watermilfoil (*Myriophyllum spicatum* L.): a review, *J. Aquat. Plant Manage.,* 16, 1, 1978.

Green, G. J. and Campbell, A. B., Wheat cultivars resistant to *Puccinia graminis tritici* in Western Canada: their development, performance, and economic value, *Can. J. Plant Pathol.,* 1, 13, 1979.

Gregory, P. H., The multiple-infection transformation, *Ann. Appl. Biol.,* 35, 412, 1948.

Harlan, J. R., Disease as a factor in plant evolution, *Annu. Rev. Phytopathol.,* 14, 31, 1976.

Harper, J. L., *Population Biology of Plants,* Academic Press, London, 1977.

Harper, J. L., An apophasis of plant population biology, in *Plant Population Ecology,* Davy, A. J., Hutchings, M. J., and Watkinson, A. R., Eds., Blackwell Scientific, Oxford, 1988, chap. 21.

Heathcote, G. D., Effect of plant spacing and time of sowing of sugar beet on aphid infestation and spread of virus yellows, *Plant Pathol.,* 19, 32, 1970.

Holah, J. C., Wilson, M. V., and Hansen, E. M., Effects of a native forest pathogen, *Phellinus weirii,* on Douglas-fir forest composition in western Oregon, *Can. J. For. Res.,* 23, 2473, 1993.

Howard, R. J. and Morrall, R. A. A., The epidemiology of leaf spot disease in a native prairie. I. The progression of disease with time, *Can. J. Bot.*, 53, 1040, 1975.

Huang, H. C. and Hoes, J. A., Importance of plant spacing and sclerotial position to development of Sclerotinia wilt of sunflower, *Plant Dis.*, 64, 81, 1980.

Huisman, O. C. and Grimes, D. W., Cultural practices: the effect of plant density and irrigation regimes on Verticillium wilt of cotton, *NATO ASI Ser., Ser. H, Cell Biol.*, 28, 537, 1989.

Hussein, M. Y. and Samad, N. A., Intercropping chilli with maize on brinjal to suppress populations of *Aphis gossypii* Glov., and transmission of chile viruses, *Int. J. Pest Manage.*, 39, 216, 1993.

Imhoff, M. W., Main, C. E., and Leonard, K. J., Effect of temperature, dew period, and age of leaves, spores, and source pustules on germination of bean rust urediospores, *Phytopathology*, 71, 577, 1981.

Jackson, G. V. H., Gollifer, D. E., and Newhook, F. J., Studies on the taro leaf blight fungus *Phytophthora colocasiae* in Soloman Islands: control by fungicides and spacing, *Ann. Appl. Biol.*, 96, 1, 1980.

Jarosz, A. M. and Burdon, J. J., Host-pathogen interactions in natural populations of *Linum marginale* and *Melampsora lini.* III. Influence of pathogen epidemics on host survivorship and flower production, *Oecologia*, 89, 53, 1992.

Jeger, M. J., Jones, D. G., and Griffiths, E., Disease progress of non-specialized fungal pathogens in intraspecific mixed stands of cereal cultivars. I. Models, *Ann. Appl. Biol.*, 98, 187, 1981a.

Jeger, M. J., Jones, D. G., and Griffiths, E., Disease progress of non-specialized fungal pathogens in intraspecific mixed stands of cereal cultivars. II. Field experiments, *Ann. Appl. Biol.*, 98, 199, 1981b.

Kampmeijer, P. and Zadoks, J. C., *EPIMUL, a Simulator of Foci and Epidemics in Mixtures of Susceptible and Resistant Plants, Mosaics, and Multilines,* Centre for Agricultural Publishing and Documentation, Wageningen, The Netherlands, 1977.

Kass, D. C. L., *Polyculture Cropping Systems: Review and Analysis,* Cornell Int. Agric. Bull. No. 32, Cornell University, Ithaca, NY, 1978.

Kennedy, J. and Weste, G., Vegetation changes associated with invasion by *Phytophthora cinnamomi* on monitored sites in the Grampians of western Victoria, Australia, *Aust. J. Bot.*, 34, 251, 1986.

Kinkel, L. L., Andrews, J. H., Berbee, F., and Nordheim, E. V., Leaves as islands for microbes, *Oecologia,* 71, 405, 1987.

Kiyosawa, S., Breakdown of blast in relation to general strategies of resistance gene deployment to prolong effectiveness of resistance in plants, in *Plant Disease Epidemiology,* Vol. 2, Leonard, K. J. and Fry, W. E., Eds., McGraw-Hill, New York, 1989, 251.

Knight, D. H., Parasites, lightning, and the vegetation mosaic in wilderness landscapes, in *Landscape Heterogeneity and Disturbance,* Turner, M. G., Ed., Springer-Verlag, New York, 1987, 59.

Kranz, J., Eine Analyse von annuellen Epidemien pilzlicher Parasiten. II. Quantitative und qualitative Merkmale von Befallskurven, *Phytopathol. Z.*, 61, 59, 1968.

Kranz, J., Fungal diseases in multispecies plant communities, *New Phytol.*, 116, 383, 1990.

Lanter, J. M., Epidemiology of Angular Leaf Spot of Bean in Monocultures and in Bean-Maize Intercrops, Ph.D. Dissertation, University of California-Berkeley, 1990.

Lenné, J. M., Teverson, D. M., and Jeger, M. J., Evaluation of plant pathogens in complex ecosystems, in *Ecology of Plant Pathogens,* Blakeman, J. P. and Williamson, B., Eds., CAB International, Oxon, 1994, chap. 4.

Leonard, K. J., Stability of equilibria in a gene-for-gene coevolution model of host-parasite interactions, *Phytopathology,* 84, 70, 1994.

Leonard, K. J., Factors affecting rates of stem rust increase in mixed plantings of susceptible and resistant oat varieties, *Phytopathology,* 59, 1845, 1969.

Leonard, K. J. and Czochor, R. J., Theory of genetic interactions among populations of plants and their parasites, *Annu. Rev. Phytopathol.*, 18, 237, 1980.

Leppik, E. E., Gene centers of plants as sources of disease resistance, *Annu. Rev. Phytopathol.*, 8, 323, 1970.

Ludwig, J. A. and Reynolds, J. F., *Statistical Ecology,* John Wiley & Sons, New York, 1988, chap. 2.

Lumsden, R. D., Ecology of mycoparasitism, in *The Fungal Community,* Wicklow, D. T. and Carroll, G. C., Eds., Marcel Dekker, New York, 1981, chap. 17.

Mack, R. N. and Pyke, D. A., The demography of *Bromus tectorum*: the role of microclimate, grazing, and disease, *J. Ecol.*, 72, 731, 1984.

Mahmood, T., Marshall, D., and McDaniel, M. E., Effect of winter wheat cultivar mixtures on leaf rust severity and grain yield, *Phytopathology,* 81, 470, 1991.

Maloy, O. C., *Plant Disease Control: Principles and Practice,* John Wiley & Sons, New York, 1993.

Mann, W. T., Jr. and Scarborough, N. M., Close spacing reduces fusiform rust, Forest Notes 53, U.S. Department of Agriculture, Washington, D.C., 1948.

Marshall, D. R., Modeling the effects of multiline varieties on the population genetics of plant pathogens, in *Plant Disease Epidemiology,* Vol. 2, Leonard, K. J. and Fry, W. E., Eds., McGraw-Hill, New York, 1989, 284.

May, R. M., Will a large complex system be stable?, *Nature,* 238, 413, 1972.

Minogue, K. P., Disease gradients and the spread of disease, in *Plant Disease Epidemiology,* Vol. 1, Leonard, K. J. and Fry, W. E., Eds., Macmillan, New York, 1986, chap. 12.

Minton, E. B., Effects of row spacing and cotton cultivars on seedling diseases, Verticillium wilt, and yield, *Crop Sci.,* 20, 347, 1980.

Mishra, R. R. and Das, P. K., Fungal succession on conifer needles, in *Microbial Ecology of the Phylloplane,* Blakeman, J. P., Ed., Academic Press, New York, 1981, 475.

Monteiro, A. A. T., Vieira, C., and da Silva, C. C., Comportamentode de cultivares de feijao (*Phaseolus vulgaris* L) na Zona da Mata de Minas Gerais *Rev. Ceres (Brazil),* 28, 588, 1981.

Mora-E., L. E., Effecto de labranzas de suelo en la incidencia y severidad de enfermedades foliares del maiz (*Zea mays* L.) y frijol comun (*Phaseolus vulgaris* L.) en diferentes sistemas de cultivos, Thesis, University of Costa Rica-CATIE, Turrialba, Costa Rica, 1978.

Moreno, R. A., Efecto de diferentes sistemas de cultivo sobre la severidad de la mancha angular del frijol (*Phaseolus vulgaris* L.), causada por *Isariopsis griseola* Sacc., *Agron. Costarricense,* 1, 39, 1977.

Moreno, R. A. and Mora-E., L. E., Cropping pattern and soil management influence on plant disease. II. Bean rust epidemiology, *Turrialba,* 34, 41, 1984.

Msuku, W. A. B. and Edje, O. T., Effect of mixed cropping with maize and bean on bean diseases, *Annu. Rep. Bean Improve. Coop.,* 26, 16, 1982.

Mundt, C. C., Disease dynamics in agroecosystems, in *Agroecology,* Carroll, C. R., Vandermeer, J. H., and Rossett, P. M., Eds., McGraw-Hill, New York, 1990, 263.

Mundt, C. C., Use of host genetic diversity to control cereal diseases: implications for rice blast, in *Rice Blast Disease,* Zeigler, R. S., Leong, S. A., and Teng, P. S., Eds., CAB International/International Rice Research Institute, Oxon/Manila, 1994a, 293.

Mundt, C. C., Techniques to manage pathogen coevolution with host plants to prolong resistance, in *Rice Pest Science and Management,* Teng, P. S., Heong, K. L., and Moody, K., Eds., International Rice Research Institute, Manila, 1994b, 193.

Mundt, C. C. and Brophy, L. S., Influence of number of host genotype units on the effectiveness of host mixtures for disease control: a modeling approach, *Phytopathology,* 78, 1087, 1988.

Mundt, C. C. and Browning, J. A., Genetic diversity and cereal rust management, in *The Cereal Rusts,* Vol. 2, Roelfs, A. P. and Bushnell, W. R., Eds., Academic Press, Orlando, FL, 1985, 527.

Mundt, C. C., Leonard, K. J., Thal, W. M., and Fulton, J. H., Computerized simulation of crown rust epidemics in mixtures of immune and susceptible oat plants with different genotype unit areas and spatial distributions of initial disease, *Phytopathology,* 76, 590, 1986.

Mundt, C. C. and Leonard, K. J., Effect of host genotype unit area on epidemic development of crown rust following focal and general inoculations of mixtures of immune and susceptible oat plants, *Phytopathology,* 75, 1141, 1985.

Mundt, C. C. and Leonard, K. J., Effect of host genotype unit area on development of focal epidemics of bean rust and common maize rust in mixtures of resistant and susceptible plants, *Phytopathology,* 76, 895, 1986.

Murdoch, W. W., Diversity, complexity, stability, and pest control, *J. Appl. Ecol.,* 2, 795, 1975.

Nakamura, R. R., Mitchell-Olds, T., Manasse, R. S., and Lello, D., Seed predation, pathogen infection and life-history traits in *Brassica rapa, Oecologia,* 102, 324, 1995.

Neher, D. A., Augspurger, C. K., and Wilkinson, H. T., Influence of age structure of plant population on damping-off epidemics, *Oecologia,* 74, 419, 1988.

Neher, D. A. and Campbell, C. L., Nematode communities and microbial biomass in soils with annual and perennial crops, *Appl. Soil Ecol.,* 1, 17, 1994.

Ntahimpera, N., Dillard, H. R., Cobb, A. C., and Seem, R. C., Anthracnose development in mixtures of resistant and susceptible dry bean cultivars, *Phytopathology,* 86, 668, 1996.

O'Neill, R. V., DeAngelis, D. L., Waide, J. B., and Allen, T. F. H., *A Hierarchical Concept of Ecosystems,* Princeton University Press, Princeton, NJ, 1986.
Papendick, R. I., Sanchez, P. A., and Triplett, G. B., Eds., *Multiple Cropping,* Spec. Publ. No. 27, American Society of Agronomy, Madison, WI, 1976.
Parker, M. A., Individual variation in pathogen attack and differential reproductive success in the annual legume, *Amphicarpaea bracteata, Oecologia,* 69, 253, 1986.
Parker, M. A., Polymorphism for disease resistance in the annual legume *Amphicarpaea bracteata, Heredity,* 60, 27, 1988.
Parker, M. A., Nonadaptive evolution of disease resistance in an annual legume, *Evolution,* 45, 1209, 1991.
Parlevliet, J. E., Stabilizing selection in crop pathosystems: an empty concept or a reality?, *Euphytica,* 30, 259, 1991.
Pianka, E. R., On *r*- and *K*-selection, *Am. Nat.,* 104, 592, 1970.
Pielou, E. C., *Mathematical Ecology,* John Wiley & Sons, New York, 1977, 195.
Pickett, S. T. A. and White, P. S., Eds., *The Ecology of Natural Disturbance and Patch Dynamics,* Academic Press, New York, 1985.
Pimental, D., McLaughlin, L., Zepp, A., Lakitan, B., Kraus, T., Kleinman, P., Vancini, F., Roach, W. J., Graap, E., Keeton, W. S., and Selig, G., Environmental and economic effects of reducing pesticide use, *BioScience,* 41, 402, 1991.
Power, A. G., Plant community diversity, herbivore movement, and an insect-transmitted disease of maize, *Ecology,* 68, 1658, 1987.
Power, A. G., Influence of plant spacing and nitrogen fertilization in maize on *Dalbulus maidis* (Homoptera: Cicadellidae), vector of corn stunt, *Environ. Entomol.,* 18, 494, 1989.
Power, A. G., Cropping systems, insect movement, and the spread of insect-transmitted diseases in crops, in *Agroecology: Researching the Ecological Basis for Sustainable Agriculture,* Gliessman, S. R., Ed., Springer-Verlag, New York, 1990, 47.
Power, A. G., Virus spread and vector dynamics in genetically diverse plant populations, *Ecology,* 72, 232, 1991.
Power, A. G., Host plant dispersion, leafhopper movement and disease transmission, *Ecol. Entomol.,* 17, 63, 1992.
Pratt, R. G. and Knight, W. E., Relationships of planting density and competition to growth characteristics and internal crown breakdown in arrowleaf clover, *Phytopathology,* 73, 980, 1983.
Pratt, B. H., Heather, W. A., and Shepherd, C. J., Recovery of *Phytophthora cinnamomi* from native vegetation in a remote area of New South Wales, *Trans. Br. Mycol. Soc.,* 60, 197, 1973.
Risch, S. J., Andow, D., and Altieri, M. A., Agroecosystems diversity and pest control: data, tentative conclusions, and new research directions, *Environ. Entomol.,* 12, 625, 1983.
Roame, M. K., Griffin, G. J., and Rush, J. R., Chestnut blight, other Endothia diseases, and the genus *Endothia,* American Phytopathological Society, St. Paul, MN, 1986.
Robinson, R. A., *Plant Pathosystems,* Springer-Verlag, Berlin, 1976.
Robinson, R. A., New concepts in breeding for disease resistance, *Annu. Rev. Phytopathol.,* 18, 189, 1980.
Routley, R. and Routley, V., Note on the recovery of *Phytophthora cinnamomi* from the Budawang Ranges, *Trans. Br. Mycol. Soc.,* 63, 413, 1974.
Ruiz, G. M. and Zapata, J. M., *The Variety Colombia: A Variety of Coffee with Resistance to Rust (*Hemileia vastatrix *Berk. & Br.),* CENICAFE, Chinchina-Caldas (Colombia), 1990.
Saxena, K. N. and Saxena, R. C., Patterns of relationships between certain leafhoppers and plants. III. Range and interaction of sensory stimuli, *Entomol. Exp. Appl.,* 18, 194, 1975.
Schmid, B., Effects of genetic diversity in experimental stands of *Solidago altissima* — evidence for the potential role of pathogens as selective agents in plant populations, *J. Ecol.,* 82, 165, 1994.
Segal, A., Manisterski, J., Fischbeck, G., and Wahl, I., How plant populations defend themselves in natural ecosystems, in *Plant Disease: an Advanced Treatise,* Vol. 2, Horsfall, J. G. and Cowling, E. B., Eds., Academic Press, New York, 1980, 287.
Sengooba, T., Comparison of disease development in beans in pure stand and maize intercrop, *Annu. Rep. Bean Improve. Coop.,* 33, 57, 1990.
Shapovalov, M., Blood, H. L., and Christinsen, R. M., Tomato plant populations in relation to curly-top control, *Phytopathology,* 31, 864, 1941 (abstract).

Soria, J., Bazan, R., Pinchinat, A. M., Paez, G., Mateo, N., and Moreno, R., Investigacion sobre sistemas de produccion agricola para el pequeno agricultor del tropico, *Turrialba,* 25, 283, 1975.
Starmer, W. T., An analysis of the fundamental and realized niche of cactophilic yeasts, in *The Fungal Community,* Wicklow, D. T. and Carroll, G. C., Eds., Marcel Dekker, New York, 1981, chap. 8.
Stoetzer, H.A.I. and Omunyin, M.E., Microclimatological observations in bean mono-crops and associations with maize, *Annu. Rep. Bean Improve. Coop.,* 27, 96, 1984.
Thomashow, L. S. and Weller, D. M., Current concepts in the use of introduced bacteria for biological disease control. Mechanisms and antifungal metabolites, in *Plant-Microbe Interactions,* Vol. I, Stacey, G. and Keen, N. T., Eds., Chapman and Hall, New York, 1996, chap. 6.
Thompson, J. N. and Burdon, J. J., Gene-for-gene coevolution between plants and parasites, *Nature,* 360, 121, 1992.
Thrall, P. H., Biere, A., and Antonovics, J., Plant life-history and disease susceptibility — the occurrence of *Ustilago violaceae* on different species of the Caryophyllaceae, *J. Ecol.,* 81, 489, 1993.
Thurston, H. D., *Sustainable Practices for Plant Disease Management in Traditional Farming Systems,* Westview Press, Boulder, CO, 1992.
Tilman, D., Wedin, D., and Knops, J., Productivity and sustainability influenced by biodiversity in grassland ecosystems, *Nature,* 379, 718, 1996.
Toba, H. H., Kishaba, A. N., Bohn, G. W., and Hield, H., Protecting muskmelons against aphid-borne viruses, *Phytopathology,* 67, 1418, 1977.
Tobiesson, P. and Werner, M. B., Hardwood seedling survival under plantations of Scotch pine and red pine in central New York, *Ecology,* 61, 25, 1980.
Trenbath, B. R., Interactions among diverse hosts and diverse parasites, *Ann. N.Y. Acad. Sci.,* 287, 124, 1977.
Trutmann, P., Voss, J., and Fairhead, J., Management of common bean diseases by farmers in the Central African Highlands, *Int. J. Pest Manage.,* 39, 334, 1993.
Vandermeer, J., *The Ecology of Intercropping,* Cambridge University Press, Cambridge, 1989.
Vanderplank, J. E., *Plant Diseases: Epidemics and Control,* Academic Press, New York, 1963.
Van Rheenen, H. A., Hasselbach, O. E., and Muigai, S. G. S., The effect of growing beans together with maize on the incidence of bean diseases and pests, *Neth. J. Plant Pathol.,* 87, 193, 1981.
Uilich, V., Crop rotation with pure stands and mixtures of barley and wheat to control stem and root rot diseases, *Crop Prot.,* 12, 373, 1993.
Vilich-Meller, V., Mixed cropping of cereals to suppress plant diseases and omit pesticide applications, *Biol. Agric. Hortic.,* 8, 299, 1992.
Vyas, S. C., *Nontarget Effects of Agricultural Fungicides,* CRC Press, Boca Raton, FL, 1988.
Weltzien, H. C., Geophytopathology, *Annu. Rev. Phytopathol.,* 10, 277, 1972.
Wennström, A. and Ericson, L., Environmental heterogeneity and disease transmission within clones of *Lactuca sibrica, J. Ecol.,* 80, 71, 1982.
Wicklow, D. T. and Carrol, G. C., Eds., *The Fungal Community,* Marcel Dekker, New York, 1981.
Wicklow, D. T. and Carroll, G. C., Eds., *The Fungal Community,* 2nd ed., Marcel Dekker, New York, 1992.
Williams, C. T., Effects of plant age, leaf age and virus yellow infection on the population dynamics of *Myzus persicae* (Homoptera: Aphididae) on sugarbeet in field plots, *Bull. Entomol. Res.,* 85, 557, 1995.
Windels, C. E. and Lindow, S. E., *Biological Control on the Phylloplane,* APS Press, St. Paul, MN, 1985.
Wolfe, M. S., The current status and prospects of multiline cultivars and variety mixtures for disease resistance, *Annu. Rev. Phytopathol.,* 23, 251, 1985.
Wolfe, M. S., Recent developments in using variety mixtures to control powdery mildew of barley, in *Integrated Control of Cereal Mildews,* Riso National Laboratories, Roskilde, Denmark, 1991, 235.
Wolfe, M. S., Barley diseases: maintaining the value of our varieties, *Barley Genet.,* 6, 1055, 1992.
Wolfe, M. S. and Barrett, J. A., Can we lead the pathogen astray?, *Plant Dis.,* 64, 148, 1980.
Wyant, J. G., Alig, R. J., and Bechtold, W. A., Physiographic position, disturbance and species composition in North Carolina coastal plain forests, *For. Ecol. Manage.,* 41, 1, 1991.
Xie, C. Y., Johnstone, W. D., and Ying, C. C., Spacing and provenance effects on the performance of shore pine *(Pinus contorta* var. *contorta):* 20-year test results, *Can. J. For. Res.,* 25, 567, 1995.
Yang, X. and Madden, L. V., Effects of ground cover, rain intensity and strawberry plants on splash of simulated raindrops, *Agric. For. Meteorol.,* 65, 1, 1993.
Yarwood, C. E., Cross protection with two rust fungi, *Phytopathology,* 46, 540, 1956.
Yoda, K., Kira, T., Ogawa, H., and Hozumi, K., Self thinning in overcrowded pure stands under cultivated and natural conditions, *J. Biol. Osaka City Univ.,* 14, 107, 1963.

Part II

Chemical Control

CHAPTER 3

Use of Chemical Measures

Ron A. Hamlen, Bernard Labit, John A. Bruhn, Mark J. Holliday, Constance M. Smith, and Clayton A. Shillingford

1-8493-2627-3/97/$0.00+$.50

I. INTRODUCTION: FUNGICIDES

As the global standard of living continues to improve and the public becomes increasingly aware of the importance of adequate amounts of fruits and vegetables in their diet, the demand for these foods will expand further. This challenge is but a portion of the greater task of feeding an expected global population of 8 to 10 billion by the year 2020 (Borlaug, 1990). Crop protection chemicals have, and will continue to have, a major role in meeting the challenge of ensuring the continuous global availability of affordable and safe food that is so vital for good health and quality of life (Abelson, 1994). Additionally, the core position that cereals and rice occupy in the global diet requires reliable yields each year — an impossible expectation without disease control. This chapter will discuss the global markets in which the primary plant disease control chemicals, the fungicides, are required and the factors that sustain their use. The discovery and safety testing process, the benefits and risks from use and nonuse, as well as integrated control approaches to sustain and maximize the utility of these important crop production tools also will be reviewed.

Fungicides are chemical products active in controlling pathogens which damage or destroy plants. A fungicide usually contains one or more active ingredients and is formulated with various additives, e.g., solvents, wetting agents, stickers, which allow the active ingredients to reach their optimum activity on the targeted plant pathogens while being safe for the treated plants, applicators, and the environment. In modern agriculture, fungicides are key crop protection tools that assist farmers to reliably and economically grow crops, increase harvest yields and quality, and protect produce from losses in the field and during storage. Fungicides are a core component of crop production systems, as numerous integrated pest management (IPM) programs are designed and successfully implemented based on availability of effective fungicides (Ragsdale et al., 1991).

A. How Fungicides are Used: Contemporary Global Market

In 1994 the global fungicide market was evaluated at \$5.4 billion (U.S.) at the user level, representing around 20% of the total pesticide value (Anon., 1995b). After a period of rapid growth in the 1970s and 1980s, the fungicide market is now progressing at a much slower pace, slightly higher than the rate of inflation, i.e., 4 to 5% per year. The stagnation in growth of the fungicide market is a result of a number of interrelated factors including (1) the General Agreement on Tariffs and Trade (GATT) along with the revised European Common Agricultural Policy (CAP), which aim at reducing agricultural subsidies, forcing farmers to reduce their costs and especially fungicide inputs (North, 1995); and (2) an increased promotion of IPM methods by country authorities which, in some cases, has led to a decrease in pesticide use. Although fungicide use is increasing in some rapidly developing economies like those of India, China, and Central Europe, the total volume of fungicide use is now fairly stable. This tendency is reinforced by the replacement of old active ingredients by new highly active molecules used at lower rates with better environmental characteristics.

All economically important plants are damaged by one or several diseases that can greatly reduce the yield potential and the quality of the harvested produce (Agrios, 1988; Ragsdale et al., 1991). Fungicide use is primarily in areas of intensive agricultural production where farmers integrate all production factors, i.e., high potential yield cultivars, fertilizers, herbicides, insecticides, etc. to realize the high-quality produce that is required by consumers. Fungicides are necessary to protect the following crops: (1) fruits and vegetables including grapevines, pome and stone fruits, potato, tomato, and cucurbits, which represent around 46% of the world fungicide market value; (2) cereals, i.e., wheat, barley, and rye, which represent 23% of the world fungicide market value, mainly located in Western and Central Europe; (3) rice, around 17% of the world fungicide market value, mostly in Japan; and (4) various other crops including cotton, maize, soybean, and sugar beet, as well as ornamental plants and flowers (Anon., 1995b).

B. Major Fungicide Chemical Classes

Today's fungicides offer a diverse array of action sites, or modes of action, but can be grouped into a few very broad classes based on how they interact with host and pathogen (Lyr, 1995; Siegel and Sisler, 1977). Protectant fungicides remain on the surface of the host plant either as surface deposits or as residues embedded in the epicuticular waxes. These chemicals are used with the aim of producing a protectant barrier against pathogen infections and typically affect spore germination or prepenetration stages of infection. Copper and sulfur, in use about 200 years ago, were the first chemical products that demonstrated the possibility to protect plants against diseases. Bordeaux mixture, based on copper salts, was the first fungicide protecting grapevines against downy mildew *(Plasmopara viticola).* Sulfur dusts protected many plants against powdery mildew diseases, e.g., *Uncinula necator* (Lyr, 1995). These two inorganic fungicides are still widely used under various chemical forms and are often mixed with newer active ingredients. Today, they account for around 8% of the fungicide market value (Anon., 1995b).

In the 1950s the first organic fungicides, the dithiocarbamates, also called EBDCs, e.g., mancozeb, maneb, and zineb, as well as the phthalimides, e.g., captan, were discovered and represent the first class of organic protectant fungicides to find broad usage (Lyr, 1995). These fungicides were more active, less phytotoxic, and easier to use than earlier chemicals. Other organic protectants include chlorothalonil, captafol, and dinocap. However, acting only preventively through contact with fungal spores, applications had to be renewed after any significant rainfall that washed off the spray deposits (Bruck et al., 1981). Today, contact fungicides are widely used because of their low cost and account for more than 20% of the fungicide market value (Anon., 1995b). They are also used in mixtures with newer, more specific actives, often referred to as fungicides at-risk, for antiresistance management strategies (Fry and Milgroom, 1990).

In the late 1960s the first systemic and curative fungicides, i.e., benzimidazoles, were discovered (Smith, 1988). The molecules from this new class, i.e., benomyl, carbendazim, and the related thiophanates, were highly active against many pathogens at relatively low rates. The benzimidazoles penetrate quickly into treated plant tissue, are redistributed through the xylem into nontreated areas, including established infections, and eradicate the pathogen from lesions. Because of their ability to cure infections inside plant tissues during the incubation phase, benzimidazoles brought a new flexibility in application timing to farmers. However, after several years of exclusive and intensive use, resistant strains were detected for several pathogens (Smith, 1988). To avoid further loss of activity, it was recommended that benzimidazoles be used in a limited number of treatments per season and usually in mixtures with other fungicides having a different mode of action. The benzimidazole share of the world fungicide market value is now estimated at around 8% (Anon., 1995b). Market share has significantly decreased over the last 10 years, mainly due to resistance, but also because of the availability of new active compounds having new modes of action.

In the 1970s two new chemical classes of fungicides were discovered that control fungi by inhibiting biosynthesis of fungal sterols. These classes are referred to collectively as sterol biosynthesis-inhibiting (SBI) fungicides and inhibit different steps in the sterol pathway (Lyr, 1995). The azole class, consisting mainly of triazoles, but also imidazoles, pyridines, and pyrimidines, represent over 20 different molecules produced by several manufactures and are used widely in cereals, grapes, fruit trees, plus many other crops. Worldwide sales in 1994 exceeded $1 billion (U.S.) (>20% share) (Anon., 1995b). The other new class, the morpholines/piperidines, most commonly fenpropidin and fenpropimorph, is more generally used against cereal powdery mildews *(Erysiphe graminis)* and comprise approximately 5% of the fungicide market value (Anon., 1995b). SBIs are preventive, curative, and systemic fungicides and have contributed greatly to significant increases in production and quality improvements in many crops. Although intensive use of SBIs has led to shifts towards decreased sensitivity in some fungal populations, e.g., *Erysiphe graminis* f. sp. *hordei*, these fungicides remain effective and highly desired by farmers who are using them now in mixtures with other fungicides to broaden the number of pathogens controlled and to delay the build-up of resistant strains (Fletcher and Wolfe, 1981; Hollomon et al., 1984, 1990).

Another important chemical class discovered in the late 1970s, the phenylamides, e.g., metalaxyl and oxadixyl, are systemic and curative fungicides for the control of downy mildews, i.e., the Peronosporales (Staub, 1994). After only a few years of exclusive and intensive use, resistant populations evolved, e.g., *Plasmopara viticola* and *Phytophthora infestans* (Staub and Sozzi, 1983). The phenylamides are now sold only as premixtures, with partner fungicides having different modes of action, and are restricted in the number of sprays per season. Their market share is estimated to be >10% (Anon., 1995b).

Table 1 Examples of Major Chemical Classes of Fungicides

Class	Type of Action	Examples
Inorganic	Protectant	Copper (Bordeaux mixture), sulfur
Organic	Protectant	EBDCs (mancozeb, zineb, etc.), phthalimides (captan)
		Dicarboximides (iprodione, vincolozolin, etc.)
		Chlorothalonil, captafol, etc.
Organic	Systemic	Benzimidazoles (benomyl, carbendazim, thiophanate-methyl)
		Azoles (triadimenol, propiconazole, flusilazole, epoxyconazole, etc.)
		Morpholines/piperidines (tridemorph, fenpropimorph, fenpropidin)
		Phenylamides (metalaxyl, oxadixyl, etc.)
		Cymoxanil, etc.

In addition to these previously mentioned important fungicide classes, many other compounds were discovered, developed, and brought to the market over the last 30 years. Many of the newer molecules, e.g., cymoxanil, and more recently the strobilurin fungicides, are systemic and have a unique mode of action (Kohle et al., 1994). These fungicides provide alternative tools for farmers to build highly effective, integrated programs for disease control and resistance management. The group of fungicides that inhibit melanin biosynthesis, e.g., tricyclazole and pyroquilon, are unique in that they move systemically within a rice plant, but are protectants, because at the site of action infection by *Magnaporthe grisea* is prevented while established infections are not affected. Recently, a class of plant disease control chemicals, referred to as systemic acquired resistance (SAR) agents, has been announced. These chemicals are not directly fungicidal, but instead affect disease development through stimulation of host resistance responses (Kessmann et al., 1994; Neuenschwander et al., 1995). These more recent fungicides and SAR agents are generally more active on the target pathogens than previous types, and thus are used in relatively small dosages. Examples of the major fungicide classes appear in Table 1. Refer to Lyr (1995) and Siegel and Sisler (1977) for an in-depth discussion of fungicide classes and their properties.

C. Importance of Controlling Plant Diseases

Before the widespread use of fungicides, plant diseases destroyed entire harvests as evidenced by the mid-1800s late blight epidemic of potatoes in Ireland that caused the death of an estimated 1.5 million people from starvation and forced the emigration of 1 million others (Dowley and O'Sullivan, 1995; Flynn, 1988; Schwinn, 1992). More recently, famines in Germany in 1916 and 1917 caused by potato late blight and in Bengal in 1943 by rice brown spot have had serious social as well as political repercussions (Schwinn, 1992). In modern intensive agriculture, certain cultural practices used to ensure high output have enhanced disease losses, and with the continuing need for increased harvests, will continue to intensify the potential losses to plant diseases. These practices include the use of genetically uniform susceptible cultivars of crops in continuous monoculture and high nitrogen fertilizers, both of which result in increased susceptibility to disease epiphytotics (de Waard et al., 1993). For example, in European small-grain production, high-density seeding coupled with high nitrogen fertilization levels promotes a dense crop canopy. This succulent canopy favors the development of powdery mildew, i.e., *Erysiphe graminis*, which reduces the photosynthetic activity of the leaves and eventually reduces yield (Jorgensen, 1988).

Plant disease can significantly reduce the agricultural production in harvest volume and quality (Locke, 1995). It is estimated that globally \$90 billion (U.S.) worth of crops per year, or 12% of potential production, is destroyed by diseases (James et al., 1991). This includes losses of 22% for potatoes, 10% for vegetables, 16% for fruits, and 15% for coffee, cocoa, and tea (James et al., 1991; de Waard et al., 1993). Furthermore, depending on weather conditions and other environmental factors, disease epiphytotics can cause serious yield losses in individual years as exemplified by the 23% decrease in Arkansas rice yields in 1984 due to sheath blight infections *(Rhizoctonia oryzae)* (Ragsdale et al., 1991). In some production areas, crops could not be successfully grown without disease control programs dependent on fungicide use (Dowley and O'Sullivan, 1995).

Fungicides used at their recommended rate and timing are frequently the most cost-effective and reliable tools to protect crops against their pathogens (Ragsdale et al., 1991). Modern fungicides are used in two ways: (1) as soil or seed treatments to prevent infections of young seedlings from soil-borne, seed-borne, or foliar pathogens, and (2) as foliar or fruit sprays usually applied in a preventive schedule, but also applied according to disease infection risk forecasts or disease thresholds (de Waard et al., 1993; Martin, 1994). Approximately half of the crop diseases requiring control are treated only once or twice per season; however, under intensive or continuous production conditions, e.g., potatoes, bananas, and coffee, 10 to 20 treatments may be required (Brent, 1995).

Fungicides must be used preventively before infection or, for systemic products, at the early stages of infection before the appearance of disease symptoms if damage is to be minimized and the harvest secured. Review of the disease status for cereals and apples illustrates the impact of fungal diseases in reducing production and the importance of disease intervention through fungicide applications.

Each year, cereals are damaged by a complex of diseases damaging roots, stems, leaves, and ears, e.g., wheat eye spot *(Pseudocercosporella herpotrichoides)*, powdery mildew *(Erysiphe graminis)*, wheat leaf and glume blotch *(Septoria tritici* and *S. nodorum)*, leaf rusts (*Puccinia* spp.), barley scald *(Rhynchosporium secalis)* and net blotch *(Pyrenophora teres)*. In extensive cultivation areas such as the U.S. midwest plains or in Russia, a fungicide seed treatment protecting young seedlings from soil/seed pathogens and early foliar infections is estimated to increase yields by 20 to 30% (Martin, 1994). In the highly intensive production areas of Western Europe, 2 to 3 foliar sprays, recommended by crop specialists to keep plants healthy until harvest, can provide an average of 20 to 35% yield increase, largely outweighing the treatment costs (Oerke et al., 1994; Leprince, 1995). Additionally, the quality of harvested unprotected grains is poor, i.e., empty and shriveled kernels often containing fungal-produced toxins, which translates into a lower price per ton to the farmer (Orson, 1995; Leprince, 1995).

Apple scab *(Venturia inaequalis)* is a very damaging disease for most orchards. In humid and mild springs, severe infections can completely defoliate and provoke fruit drop on the most sensitive varieties. When fruit trees are not adequately protected by fungicides, black spots and cracking can develop on fruit, which become commercially unacceptable according to consumer and export requirements. This can dramatically reduce the final value of the crop.

In modern agriculture, the need for high-quality agricultural produce demands that growers protect their crops against pathogens. Fungicides have proven to be effective tools to help farmers produce high-quality food while ensuring a profitable return. Specific studies addressing the benefits of fungicide use appear later in this chapter.

D. Long-Term Value of Fungicides

It is well acknowledged that the world population will increase significantly in the future decades. It is commonly believed that food production will have to increase 2- to 3-fold to feed 8 to 10 billion people by the year 2020 (Beyer, 1991; Borlaug, 1990). This will not happen by an increase in cultivated land, as this would be at the expense of forests and wetlands, but will need to occur mainly by a significant increase in the harvest per unit area, and at a price affordable to farmers and consumers. Therefore, the intensification of agriculture must occur and the use of crop protection fungicides is and will continue to greatly contribute to meeting the global food demand. Within the agricultural community it is well known that correctly applied fungicides contribute to increased production at an affordable price for the farmers, while meeting the high quality expected by consumers (de Waard et al., 1993; Ragsdale and Sisler, 1994). Even without demand for greater future harvests, plant disease control is needed to maintain the world food supply at its current level of volume and quality.

Crop protection products including fungicides must: (1) be capable of safe use by applicators during treatment and to workers during harvest, (2) leave no unacceptable residues in harvested produce for consumers, (3) have no significant impact on the environment, i.e., wildlife, soil, and aquatic organisms, and (4) be economical to farmers. To ensure public acceptance of fungicides, they must convey visible value to society and be used judiciously. All pesticide regulations around the world are aimed at assessing the risks of each product and defining its application pattern to minimize any potential adverse effects. Thus, the theme that will appear again and again throughout this chapter is that fungicides, as a crop protection tool, must be effective, safe, and economical if their use is to continue in the future.

Technologies will continue to be discovered and introduced to help farmers control diseases and to safely produce quality food at minimum cost. Several additional environmentally acceptable technologies for increased harvest quantities and quality include:

1. Use of more efficacious disease-resistant varieties that will be damaged less by pathogens.
2. Further introduction of disease diagnostics (Smith et al., 1990), along with more precise disease prediction models (Fry and Fohner, 1984), will allow the early detection of infections and the evaluation of epiphytotic risks to help farmers make the best treatment decisions on timing, product choice, and rate.
3. Development of chemicals with greater intrinsic activity able to be used at lower dosages.
4. Introduction of more active and reliable biological control agents.
5. Improvement of product delivery systems, e.g., formulations, packaging, and application equipment, that will further reduce potential worker exposure, better target applications, and minimize wastes.

All of these technologies judiciously combined in IPM programs will ensure that farmers have the right disease control solutions available to use at the right time and in the right place.

II. FACTORS SUSTAINING FUNGICIDE USE

Fungicides are used because they provide effective and reliable disease control, deliver production benefits in the form of crop yield and quality at an economical price, and can be used safely. In this section we consider how fungicides are used to control plant disease, factors that influence the use and effectiveness of fungicide control, and economic aspects of the decision to apply fungicides for disease control.

A. Effective Disease Control

The course of plant disease requires a virulent pathogen, a susceptible host plant, and an environment that is favorable for disease development. Changes at any vertex of this disease triangle influence the level of disease development (Agrios, 1988). Thus, given the same environment and host, a highly virulent pathogen will produce more disease within a period of time than will a less virulent pathogen. Most fungicides affect disease by making the environment component of the triangle less favorable through the introduction of a xenobiotic that disrupts some critical phase of the pathogen's development. Here, "environment" refers to those combinations of factors that immediately affect pathogen development and does not refer to aspects of the broader environment. It will be shown in later sections that modern plant disease controls are developed and used with the aim of having an acceptable affect on the broader environment.

B. Pathogen Epidemiology

Effective disease control is realized through effects on the population dynamics or epidemiology of the pathogen. Van der Plank (1963) recognized two broad strategies for plant disease control based on pathogen population dynamics. The first approach is to reduce the amount of inoculum available to initiate an epiphytotic. This strategy is most effective when a single pathogen life cycle is the aim of control, e.g., *Pseudocercosporella herpotrichoides.* The second approach, reducing the rate of pathogen population increase, is employed when plants must be protected over multiple pathogen life cycles and particularly when the pathogen has a very short generation time and produces a large number of progeny, e.g., *Phytophthora infestans*. Modern plant disease control chemicals are used with each of these strategies (Fry, 1982).

Soil fumigants, soil-applied fungicides, and traditional seed treatment fungicides are used to reduce the amount of initial inoculum. When applied as seed dressings, chemicals may move into the seed to eradicate latent infections, disinfest seed surfaces, or protect the emerging seedling from infections by soil-borne inoculum, e.g., *Pythium*, *Rhizoctonia*, and *Fusarium* spp., all of these actions have the net effect of reducing initial inoculum. Fungicides aimed at reducing initial inoculum represent a relatively small segment (5 to 7%) of the total world fungicide market in value terms (Martin, 1994). While the total sales value is low, this strategy is used extensively. Seed treatments represent the most broadly used chemical control method in that virtually all of the wheat, corn and cotton seed, potato seed pieces, and most of the rice seed planted in the developed world and much of the developing world is treated with fungicides (Martin, 1994).

Plant disease control chemicals applied to protect the foliar portions of a crop are typically directed at reducing the rate of epiphytotic development. These foliar fungicides, representing over 90% of the world fungicide market in terms of value and amount of active ingredient applied (Anon., 1995b), are the focus of the remaining discussion.

C. Fungicide Dynamics

Using foliar-applied fungicides to reduce the rate of epiphytotic development of the pathogen can be viewed as establishing and maintaining a protective barrier on foliage and fruit against the pathogen over the time of crop susceptibility. This period of crop susceptibility varies with crop and pathogen but is often the entire growing season, which may be year-round in areas of continuous production. Commonly, fungicides are applied several times over the season to maintain control over the desired period.

A simple model of fungicide dynamics illustrates the challenge of determining how much chemical to apply and how often to make applications (Figure 1). First, let's simplify the goal of reducing the rate of epiphytotic development to the task of maintaining the foliar deposit of a protectant fungicide above some threshold residue, TR, over a period of time, t. Another fungicide application might be made after this time, or time t might cover the full period of crop susceptibility. When a fungicide is applied to foliage at rate AR, at time 0, a dilution process begins. This dilution process is represented here by a gradual decline of foliar residues over time. The challenge for achieving effective disease control with the chemical is to adjust AR at application time (time = 0) such that the dilution of foliar residues over time t results in the threshold concentration TR. Various fungicidal attributes affect the relationships in this model.

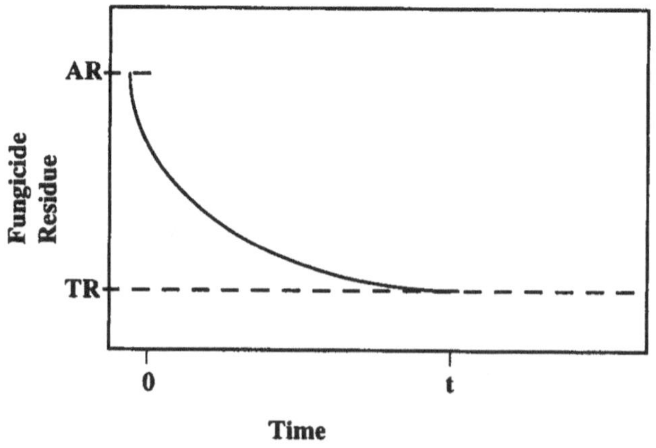

Figure 1 The removal of foliar fungicide residues over time. Maintaining effective disease control over time with a foliarly applied fungicide can be viewed as applying sufficient compound, AR, at time 0, so that residues remain above a threshold, TR, over the time period, t.

Application efficiency (the ratio of amount deposited on the leaf vs. the amount applied) is affected by crop canopy architecture, leaf surface, application equipment, and the weather during application (Matthews, 1979). Aspects of the formulation may improve application efficiency through wetting, spreading, and sticking of deposits to the leaf surface. Systemic fungicides can also improve application efficiency by moving into untreated tissue and creating a uniform deposit throughout the crop canopy.

The reduction of foliar fungicide residues over time is the result of a variety of processes (Nigg and Allen, 1979; Willis and McDowell, 1987). Some of the fungicide deposit degrades into inactive components by photolysis, hydrolysis, plant or microbe metabolism, or other processes. Leaves rubbing in the wind, or rain, may physically remove more of the deposit. Finally, new host growth produces unprotected tissue and further dilutes the effectiveness of the original deposit. Chemicals can be quite different in susceptibility to these processes. Formulation can also affect the rate of degradation. For example, the experimental fungicide, kresoxim-methyl, is very susceptible to metabolism by the host plant (Kohle et al., 1994). Particulate formulations of the compound keep the active ingredient on the plant surface and away from plant metabolic action, thus extending the period of efficacy of the compound (Gold and Leinhos, 1994; Kohle et al., 1994). Microencapsulation and other slow-release technologies also limit exposure of residues to degradation processes. Because systemic fungicides penetrate the plant surface, systemic movement can protect against removal by rainfall and other external factors. As seen with kresoxim-methyl, however, systemic movement also increases exposure to plant metabolism.

The interplay between the time of protection t, and the threshold residue TR, is also influenced by fungicidal attributes. Chemicals differ dramatically in their intrinsic activity. Given similar degradation rates and application efficiency, the greater the intrinsic activity of an active compound the lower the amount needed at application. Host susceptibility also influences the amount of chemical required for effective disease control such that application rates can be reduced or application intervals increased relative to the level of disease resistance of the host plant, i.e., host plant resistance can decrease the fungicide control component. If the pathogen is not present to cause disease, or if the environment does not favor disease development, then t can be increased or TR can be reduced. Plant disease forecasting systems adjust the timing of fungicide applications based on the degree to which the environment favors disease development (Fry and Fohner, 1984; Royle and Shaw, 1988). Some forecasting systems have integrated the effects of host resistance, pathogen development, and fungicide degradation to adjust the timing of fungicide applications. Finally, the ability of a fungicide to eradicate established infections affects the relation between t and TR by allowing foliar residues to drop below TR over the period of its eradicative effect. Thus, a fungicide that eradicates 3-day-old infections can be reapplied 3 days after residues drop below TR and still maintain the desired level of control.

D. Fungicide Economics

Effective and reliable disease control is only one of three necessary attributes of plant disease control chemicals. Economical disease control is a second attribute. At the simplest level, the benefits derived from controlling the plant disease must exceed the cost of using the chemical. Ideally, the amount of chemical applied is adjusted toward an optimum where profits (the difference between benefits and costs) are maximized.

At the farm level, the costs of fungicide use include the cost of the chemical and the cost of applying the chemical, such as fuel, labor, and the costs of owning or renting application and safety equipment. If a forecasting system is used to time fungicide applications, the grower may also incur costs associated with monitoring the crop environment or scouting the crop for disease occurrence. Because most chemicals are applied in anticipation of disease development, i.e., preventively, there are also costs associated with the risk of failing to apply a chemical when needed.

The use of fungicides produces benefit at the farm level by preventing crop damage due to the development of plant diseases. Quantifying the benefit side of this relationship is complicated by two factors (Oerke et al., 1994). First, fungicides, like all crop protection practices, do not create yield or quality improvements directly. Rather, the use of fungicides protects the yield gains from a particular crop production system from being lost to damage from plant disease. Thus, given the same potential for disease damage, a production system with higher yield and quality potential is likely to benefit from fungicide use more than one with lower potential. That is one reason why fungicide use is associated with intensive crop production systems. A second complication is uncertainty about the occurrence and level of disease development. Most fungicides must be applied in anticipation of damaging levels of plant disease and frequently before disease can be observed in the field. If the decision maker, e.g., farmer, crop production manager, or harvest buyer is averse to risk, this uncertainty about the benefit side of the relationship leads to more fungicide use than would be predicted from simply maximizing profits. An in-depth discussion of the benefits of fungicide use and risk of nonuse occurs later in this chapter.

The kind of available information and the population dynamics of the pathogen affect how fungicide use can be optimized (Mumford and Norton, 1984). If disease levels are known at application and if the rate of disease increase is slow, then fungicide use can be based on an economic threshold or can be based on profit maximization. The economic threshold approach (Stern et al., 1959) is a break-even strategy where fungicide applications are made when the level of disease is such that yield and quality loss from plant disease exceeds the cost of using the chemical control. Often disease levels cannot be estimated and/or the pathogen is capable of rapid epiphytotic development. In this case, a theoretic decision approach can be used. Here, the probability of disease development is estimated from historical records or from the judgment of experts and then optimal application decisions are identified using the average and variability of profits associated with the management options under consideration. While optimal solution techniques provide useful insight into the concepts of plant disease control decision making, most practical decisions are based on experience and practice. This is a reflection of the complexities and uncertainties associated with the high relative risk of making and adjusting standard plant disease control programs, and the need of farmers to give time to other aspects of crop production and marketing.

Costs and benefits of using plant disease control chemicals also can be examined at a societal level. The most obvious societal benefit is the assurance of a reliable, safe, abundant, and readily available supply of a diverse array of high-quality, inexpensive fruits, vegetables, and grains (Abelson, 1994). Regulatory agencies around the world can be viewed as society's agents for assuring that these benefits can be realized without compromising society's need for personal and environmental safety. The costs of maintaining this system are passed on to the consumer through food prices and the taxes used to fund the activities of regulatory agencies.

III. BIOLOGICAL IMPLICATIONS IN FUNGICIDE DISCOVERY AND USE

A. Discovery, Commercialization, and Registration

The primary goal in the discovery of a new fungicide is to create a product that provides a solution to a farmer's disease problems. The product must be convenient, easy to use, provide consistent results, and be cost effective. Above all, the product should not present unacceptable risks to the crop, the environment, the user, and the consumer. The focus on safety has never been greater than today, and the challenge to find new products that outperform those already available in the marketplace is exceedingly high. Thus, fungicide discovery has become a long, complex, and costly process. Commercialization of a newly discovered fungicide may take more than 8 years with a cost of $60 to 100 million (U.S.).

In the 1960s and 1970s, when competition in the marketplace and regulatory hurdles were lower, the fungicide discovery and commercialization process was considerably simpler than it is today. The primary consideration was disease control efficacy. Compounds were screened empirically on a battery of plant diseases and those showing desirable activity were moved toward development and commercialization. Bioefficacy was the focus of this process well into the developmental stages, where regulatory issues had only just begun to be considered. Usually, a candidate was likely to pass the less stringent regulatory requirements of that time. If concerns regarding the safety of a product did arise, the candidate product was abandoned in favor of more promising leads. Since this would often occur late on the path to commercialization, the decision was costly. Not unexpectedly, renewed discovery efforts tended to steer away from the chemical class representing the abandoned development candidate.

Today's fungicide discovery process is designed to meet the challenge of a more competitive market, heightened public concern for environmental and safety issues, and more rigorous global governmental regulations. The bygone empirical approach has evolved into a dynamic process where the direction of chemical synthesis is guided not only by important biological parameters, such as preventive or curative disease control, but also by product safety attributes related to human health and the environment. To understand why today's discovery process offers opportunity to yield new products that present no unacceptable risks to people and the environment, it is necessary to take a closer look at this process itself.

The discovery of new fungicides occurs in two sequential processes, "Lead Generation" and "Lead Optimization" (Beyer, 1991). Successful execution of these two processes yields a single molecule that can be developed into a commercial product.

1. Lead Generation

In lead generation, chemists synthesize compounds and biologists test them to identify those that control diseases. This process can be either empirical in nature or it can be variously directed. When test compounds are synthesized strictly through an interest in novel synthesis, detection of leads is mostly an empirical process.

In contrast, the directed process uses some degree of biological forethought to guide the synthesis program, and this can occur with varying degrees of sophistication. Sometimes compounds are synthesized as analogues of known fungicide chemistry published in the literature. Or, natural products with known biological activity may serve to direct a synthesis effort. In its most sophisticated form, approaching that of the pharmaceutical industry, directed lead discovery occurs through biochemical design (Pillmoor and Foster, 1994).

In biochemical design, compounds are initially created to be inhibitors of biochemical targets in fungi. Typically, these targets are enzymes or receptors performing key functions in biochemical pathways believed essential to fungal proliferation. More recently, this approach has been aided by an increasing ability to visualize inhibitor binding sites in enzymes and receptors. Techniques such as X-ray diffraction analysis can reveal critical molecular interactions between proteins and inhibitors to predict inhibitor substructure modifications necessary to create more active inhibitors.

Lead discovery through biochemical design offers several opportunities for developing products with reduced impacts on humans and the environment. First, highly active target site inhibitors offer the promise of reducing chemical load in the environment by providing the required level of disease control at lower field application rates. Use rates might also be lowered by seeking low-abundance targets involved in cascade events, such as signal receptors, where a few inhibitor molecules can have a large inhibitory effect on the development of pathogenic fungi.

Further, the biochemical design approach offers the possibility to develop products that specifically inhibit pathways occurring only in the target pathogen, thus reducing the potential for adverse effects on nontarget organisms including plants, mammals, birds, fish, bacteria, etc. Even more appealing is the possibility to design products that only block pathways critical to causation of disease. These types of products need not be fungitoxic at all, but simply interfere with a pathogen's ability to infect and develop in and on the host (Sisler and Ragsdale, 1995). As discussed earlier, another exciting approach, also not relying on fungitoxicity for effectiveness, is the design of chemicals that specifically turn on host defense mechanisms to protect against pathogen attack (Kessmann et al., 1994; Neuenschwander et al., 1995).

The opportunity to use biochemical design to discover low use rate, pathogen-specific fungicides will rise significantly with an increase in the application of molecular biology techniques to explore fungal biochemistry and disease causation. Many important genes are now being discovered, and with the emergence of efficient DNA sequencing technology and growing gene data bases it will be possible to identify nearly all genes of pathogenic fungi within the next decade. Creative use of antisense mRNA and other techniques to substantiate postulated roles for specific genes in fungal growth and development will undoubtedly yield novel targets for biochemical design of highly selective fungicides. While this technology offers unprecedented opportunity to develop better focused and reduced fungicide use solutions to the farmer's disease problems, specificity and high activity are not the limit of available technology to reduce risk. Safety and environmental concerns are further addressed in the optimization phase of discovery.

Screens for lead generation usually include representative diseases of crops from major fungicide markets such as wheat, rice, grapes, potatoes, and tomatoes. In these screens the threshold for detecting activity is set far below commercial standards so that even weakly active areas of chemistry are identified. Although highly active individual compounds are sometimes discovered in the lead generation process, more often than not, moderate activity on diseases representing one or more major fungicide markets triggers the optimization process.

2. Lead Optimization

Lead optimization begins with the identification of a chemical lead area through the lead generation process just described. Usually, at this point only a handful of active analogues have been synthesized. For intellectual property protection, the area of chemistry has been analyzed for novelty and, if appropriate, patents are filed to protect the invention. Early on, an appreciation for the regions of chemistry available for synthetic modification almost invariably leads to the daunting realization that any potential product lies, in the conceptual realm, among millions or tens of millions of possible combinations of chemical structure. Since it is impractical, if not impossible, to synthesize all conceivable structures in a chemistry lead area, the challenge is to predict which structures will have the desirable competitive biology and product safety attributes. The goal is to minimize the number of analogues that need to be made; typically, several hundred to several thousand analogues are synthesized in a lead area before commercial-level activity is achieved.

To this end, the optimization process relies on an iterative approach where the data generated for each analogue synthesized are evaluated, and any new knowledge gained is applied to the synthesis of additional analogues. Relevant data usually include dose/response curves for disease control, measuring a variety of attributes including preventive and curative control, longevity, rainfastness, systemic movement, etc. Any chemistry lead area undergoing optimization tends to focus on a subset of measures according to the known biological strengths and weaknesses of the area. Along with biological measures, a range of physical properties are evaluated to form the basis for linking biological activity with chemical properties associated with structure. This link is usually supported by computer-generated quantitative structure/activity analyses that identify stearic and electronic features important to physical properties and biological activity (Steffens and Kleier, 1995).

In today's competitive environment, the optimization process is not limited to bioefficacy alone; safety and environmental issues must be considered in the formative stage of a new product. Just as with biological activity, environmental safety parameters can vary greatly with minor chemical modifications. Important parameters might include soil and water dissipation, physical and biological degradation, carcinogenicity, eye or skin sensitivity, oral toxicity, aquatic species toxicity, or fat accumulation. Depending on the chemistry area, any of these parameters may need to be considered during optimization in order to achieve the highest possible biological activity in the most environmentally compatible structure. This process has recently been facilitated by several software packages that can generate predictive models for toxicology and environmental fate (Carsel et al., 1985; Enslein et al., 1994).

3. Development and Commercialization

The development and commercialization of a new fungicide usually exceeds the discovery of that fungicide in cost and complexity of work. Engineering, process design, formulation development, and economic studies related to market opportunity and manufacturing must be conducted.

Several years of field studies are needed to show efficacy in all agronomic conditions under which the formulated product will be used. In the European Community a minimum of two years of official field efficacy and selectivity trials are required for each pathogen before a label is granted. Products must be tested for optimum formulation and for performance in a variety of mixtures with other fungicide products. Irrespective of the innate activity of the new fungicidal molecule, it must be formulated in a way that allows the intrinsic value to be realized in practical use. One of the most important physical properties of a fungicide is the particle size of the active ingredient. Particle size distribution controls a number of important physical as well as biological properties, e.g., volatilization, sublimation, solubilization, and physical wash-off by rain or irrigation (Backman et al., 1976; Feichtmeir, 1949; Hyre, 1942; Kudsk et al., 1991). Inerts are identified and integrated into the formulation to enhance product usability and performance. Surfactants which may act as wetting agents, stickers, and spreaders may aid retention or increase cuticular penetration of the active fungicidal molecule, as well as confer benefits such as resistance to evaporation of spray droplets and drift control. Formulations for seed treatment require considerations as complex as those for foliar fungicides. Formulation design is further complexed as farmers will apply product in water at different pHs, will add various adjuvants, mix with other pesticides, or utilize carriers other than water, such as oil.

Whereas the modern fungicide discovery process offers unprecedented opportunity to consider human and environmental safety in the design of new product molecules, the potential hazards of any product must be investigated through standardized studies which also satisfy governmental regulations (Anon., 1991b, 1995a). A general list of studies aimed at meeting regulatory requirements usually includes physical and chemical analysis, suitable analytical methods development, and residue analysis including the nature and magnitude of residues according to the method of use. These data are developed for the treated crop and for potential rotational crops which may be subsequently planted into the treated areas. A full understanding of the environmental fate is required, including how and where the product breaks down and the identity of breakdown products. Toxicology tests must be conducted to address acute effects (oral, dermal, and inhalation studies, neurotoxicity, and sensitization), subchronic effects (short-term exposure), and chronic effects (long-term exposure) which include oncogenicity and teratogenicity. Reproduction studies and several different types of mutagenicity studies are needed. Furthermore, the potential effects on nontarget organisms including plants, wildlife (birds and mammals), aquatic organisms (fish, invertebrates), and insects must be understood. Also, special studies related to potential worker exposure or spray drift issues may be required. This battery of over 200 individual studies, some of which require more than 4 years to complete, provide the basis on which a product of maximum benefit and acceptable risk is delivered to society. A summary of U.S. Environmental Protection Agency (EPA) data requirements for registration is listed in Table 2. Global requirements differ in data specifics, but in general, regulatory agencies require similar data.

Table 2 Summary of Environmental, Health and Safety Data Requirements (EPA)

Data Requirement Area	Study or Research Example
Environmental fate	Volatility; soil metabolism, phytolysis, mobility, leaching, adsorption/desorption, degradation rate, soil/ground water modeling; water hydrolysis, photolysis, aquatic metabolism, biodegradability, sediment study, surface water modeling, acquatic monitoring.
Toxicology	Acute (technical and formulation) oral, dermal, skin/eye irritation, inhalation, skin sensitization, neurotoxicity; subchronic; chronic rodent lifetime/reproduction, dog; teratology; mutagenicity; metabolism/pharmacokinetics; dermal; aquatic organisms, fish, daphnia, marine organisms, aquatic plants; microcosm/mesocosm studies; terrestrial organisms, avian, beneficials, plants.
Product chemistry	Chemical identity, physical/chemical properties (technical and formulations), manufacturing process description, analytical methods.
Metabolism	Animal and plant studies.
Residue	Field residues, crop rotation, analytical methods, animal feeding studies, worker exposure/reentry studies/modeling, dietary exposure assessment.

4. *Registration*

Once the appropriate studies are conducted and possible hazards are well understood, the data must be reviewed and approved by the appropriate regulatory agencies in the countries where the product will be sold. In effect, registration is the acquiring of a license to sell the product. Acceptable residue tolerances based on toxicity studies and residue analysis must be established for each new crop use for the product. In the regulatory review of relevant data and information, any identified issues are addressed through additional research, with final use authorizations resolved to address health and environmental concerns. The desired approach between the regulatory agencies and the company desiring a product registration is one of cooperation, and not competition (Kohn, 1992). Both parties work together from their individual mandates to find the best joint solution that addresses the considerations of each. When all requirements have been satisfactorily met, the regulatory agency approves registration through the authorized use of an approved product label. The label, generally in the local language, must be attached to every package of product and contains required and pertinent information about the product, including specific directions for use and any cautions. In most countries it is illegal to use an agricultural product in a manner inconsistent with the labeling instructions. The label directions for use and precautions, or contradictions of use, are comparable to information on labels found on pharmaceuticals.

B. Benefits and Risks of Use and Nonuse

1. *Benefits of Fungicide Use*

When used judiciously, fungicides provide manifold benefits with negligible risks. Foremost among these is their critical role in ensuring a reliable, affordable, safe, and abundant food supply available throughout the year (Beyer, 1991; Curtis, 1991; Ragsdale et al., 1991). Fungicides support a burgeoning global population. Through their contributions, world agricultural productivity has more than doubled during the latter half of this century (Beyer, 1991; Borlaug, 1990; de Waard et al., 1993). The importance of plant disease control chemicals is reflected in the situation created by their absence, i.e., serious crop failures, higher food prices, and increased fungal-contaminated produce. Additionally, without fungicides, intensive production for certain export markets would be severely curtailed (Ragsdale et al., 1991), resulting in limited and seasonal availability of foods. Agricultural commodities are an important component of the global economy. Without agrichemicals it is estimated U.S. exports would decrease by $5 billion and further exacerbate the balance-of-trade deficit (Urbanchuk, 1990; Zilberman et al., 1991).

Effective plant disease control technology is needed on a continuing basis in light of the destructive potential of fungal plant diseases on our food supply. All economically important plants worldwide are subject to attack by fungal pathogens, while many plant species are susceptible to multiple diseases by different pathogens (Ragsdale et al., 1991). For example, 18 preharvest and 9 postharvest fungal diseases affect 2.6 million tons of stone fruit annually in the U.S. (Ragsdale et al., 1991).

The benefits of fungicides in assuring an abundant food supply are well documented in several studies assessing fungicide use in the U.S. These investigations have demonstrated that for many crops, particularly fruits and vegetables, fungicide applications are essential. Without their use, production would decline sharply or cease, resulting in severe cost increases to the consumer (Abelson, 1994; Curtis, 1991; Knutson et al., 1990; Ragsdale and Sisler, 1994; Taylor, 1995; Urbanchuk, 1989, 1990; Zilberman et al., 1991). In the absence of disease control chemicals, fruit production would decrease by 32% and vegetables by 21% (Urbanchuk, 1989, 1990). Specifically, losses in production of apples would reach 40%, peaches 49%, strawberries 38%, and tomatoes 30%. The price of fresh produce would increase 34% with price increases in apples of 89%, potatoes 37%, tomatoes 42%, and strawberries 22% (Taylor, 1995; Urbanchuk, 1989). Many imported foodstuffs, such as bananas and other tropical fruits, would be unavailable or uneconomical for consumers (Urbanchuk, 1989). A ban on fungicides/insecticides would result in a 68% decrease in yield of peanuts with as much as a tripling in price; rice production would drop 16% (Anon., 1993; Knutson et al., 1990).

Additionally, the production of many other crops is dependent on fungicides (Curtis, 1991; Ragsdale and Sisler, 1994). Seed treatment in corn, canola, sunflower, alfalfa, cotton, soybeans, and sugar beets is required to prevent severe reductions in germination and to obtain economic stands (Curtis, 1991). Foliar diseases of wheat in the US southeastern and eastern regions as well as eyespot *(Pseudocercosporella herpotrichoides)* in the Pacific Northwest mandate fungicide applications for yields. Fungicides are also essential in disease management in mushroom production and in maintaining the commercial value of nursery crops (Curtis, 1991; Ragsdale and Sisler, 1994).

Beyond documenting the importance of fungicides for yields, these studies also emphasize their role in allowing production of crops year round in different regions, often close to the marketplace (Curtis, 1991; Urbanchuk, 1989). For example, intensive winter production of vegetables (i.e., tomatoes and peppers in Florida) which ensures cold-weather availability of fresh produce, is possible only through fungicide use (Avery, 1995). In similar fashion, fungicides allow fruits, e.g., pome and stone, as well as vegetables, e.g., potatoes, cucumbers, and peppers, to be grown in the humid eastern U.S. where environmental conditions are conducive to disease development, compared to the drier western states (Curtis, 1991). This proximity to large direct markets, e.g., New York City, enhances the freshness and attractiveness of the produce, decreases transportation costs, preserves fossil fuels, and lessens the load of combustion emissions on air quality, all to the benefit of the consumer and the environment. Production of crops in multiple areas is also insurance against natural disasters ruining regional harvests. Additionally, historical experiences of some countries with food shortages and famine have underwritten the nationalist fervor to be as much as possible self-sufficient in food production (Dowley and O'Sullivan, 1995). Thus, the potentially attractive solution of transferring production to areas of the world where diseases are a reduced burden, and where fungicide use is less, is challenged by powerful economic and psychological needs.

The vulnerability of fruits and vegetables to diseases, combined with their importance to a healthful diet, highlights a unique benefit of fungicides, i.e., providing an abundance of these foods at low cost. Extensive studies have determined that diets high in fruits and vegetables can significantly reduce the risk of coronary disease and cancer (Ames et al., 1990b, 1993; Anon., 1996; Giovannucci et al., 1995). For example, consuming five servings of these foodstuffs daily reduces risk of certain types of cancer by half (Ames et al., 1993; Avery, 1995). Fruits and vegetables also contain antioxidants and other chemicals which appear to provide protective effects against other degenerative diseases of aging (Ames et al., 1993). In addition to ensuring the availability of fresh produce, there is evidence that fungicide applications can enhance the nutritional value of food through protecting crops from pest damage or stress. Specifically, Rouchaud et al. (1989) have reported that apples sprayed with fungicides/insecticides have a higher vitamin C content than untreated samples.

Despite the recommendations of health studies, only 9% of Americans eat the suggested servings of fruits and vegetables (Ames et al., 1993; Avery, 1995). In assessing the benefits of fungicides, the scenario created by their absence should be considered. If less than 10% of the population consumes sufficient fruits and vegetables when they are moderately priced, attractive, and readily available, what would be the consumption rate if these foods were highly expensive, limited in supply, and of lower quality? Low-income consumers would be particularly impacted both from an economic and health risk perspective (Abelson, 1994; Ames et al., 1993; Anon., 1992a; Avery, 1995; Graham and Gray, 1995; Urbanchuk, 1989; Zilberman et al., 1991).

Fungicide applications are frequently essential to protect foods from spoilage and contamination by harmful fungal toxins and plant defense chemicals, thus maintaining the highest standards of food quality (Anon., 1992a; Brookes, 1991; Curtis, 1991; de Waard et al., 1993; Graham and Gray, 1995; Ragsdale et al., 1991). At least 45 mycotoxins are known carcinogens or nerve, liver, or kidney poisons, and have been shown to be prevalent in contaminated, moldy foods, i.e., aflatoxin in corn and peanuts; patulin in apples and pears; ergot in grains (Anon., 1989; Ragsdale et al., 1991). For example, the high rates of esophageal cancer in certain African regions have been attributed to the consumption of corn containing aflatoxin (Nelson et al., 1993). Although one quarter of the world's cereals and oilseeds are contaminated with such toxins, primarily in developing countries, the U.S. and other developed countries' food supplies are significantly safer through control of these pathogens with fungicides (Avery, 1995; Ragsdale et al., 1991).

Consumption of elevated concentrations of antifungal compounds called phytoalexins, which are synthesized by plants in response to pathogen challenge, can cause liver and kidney damage in humans and animals (Ragsdale et al., 1991). By controlling fungal infections, fungicide use can reduce phytoalexin production, as the case with benomyl and *Botrytis cinerea* (Ragsdale et al., 1991). Due to the relatively limited information available on the toxicological hazards of mycotoxins and phytoalexins, the USDA/States National Agricultural Pesticide Impact Assessment Program (NAPIP) recommends that the safest course is to prevent fungal infection or contamination of food products whenever possible and emphasizes the important — but not widely recognized — role of fungicides in protecting the U.S. food supply (Ragsdale et al., 1991).

In addition, fungicides can be used to supplement host resistance created through plant breeding which may increase questionable levels of plant-defense chemicals. Plant-defense chemicals, carcinogenic psoralens in certain resistant celery cultivars and glycoalkaloids in certain resistant potatoes, which attained high levels through breeding, are examples of naturally occurring chemicals that have posed serious health risks (Ames et al., 1990b; Ragsdale et al., 1991).

The concept of evaluating risk vs. risk must be rigorously applied in licensing new technologies for use (Graham and Wiener, 1995). The lessons of unintended consequences, and the unintentional shifting of risks, must be well mastered if the creation and use of new technologies are to deliver useful and safe solutions to difficult problems (Senge, 1990). The public must understand that in the quest for pest resistance, and disease control without synthetic pesticides, exposure to naturally occurring substances is not without risk (Anon., 1996). Again, what is important is the judicious integration of all useful control technologies and the avoidance of excessive or exclusive use of single solutions to complex disease control problems.

Beyond the many benefits of fungicides relating to a healthful food supply are their contributions to the economic well-being of farmers and the general public (Taylor, 1995). Disease control chemicals in combination with other pesticides have revolutionized farming practices through reducing labor requirements, facilitating farmwork and harvest, and conserving fossil fuels (Beyer, 1991). By preventing yield and quality losses, fungicides have improved market opportunities as well as the cost/profit ratio of crop production, ensuring the continued economic viability of farming (Knutson et al., 1990; Ragsdale et al., 1991). Specifically, growers realize $3 to 5 (U.S.) in return for every dollar invested in the use of pesticides (Beyer, 1991; Pimetel et al., 1991).

Regarding the general economy, fungicides along with herbicides and insecticides reduce retail food prices and food expenditures by contributing to lower farm prices for crops. Based on impact assessments, reduced crop protection chemical use would trigger a 45% increase in consumer price index for food and higher inflation, accompanied by a decline in consumer economic well-being (Penson et al., 1990; Taylor, 1995; Urbanchuk, 1990). To underscore the current high level of agricultural productivity through the use of crop protection chemicals in the U.S., Americans spend only an average of 10% of their income on food. In contrast, in developing nations in which the use of these chemicals is more limited, food expenditures are significantly higher, with 32% of income spent on food in Mexico and 50% in India and China (Presting, 1993).

Although there is some disagreement over the potential effect of eliminating crop protection chemicals (Pimetel et al., 1991; Gianessi, 1993), the increased cost of organic foods over those grown with synthetic pesticides clearly demonstrates the importance of these products. Studies in 7 European countries found yields of organic fruits and vegetables to average 20 to 40% less than commercially grown produce, with price premiums of 110 to 200% (Avery, 1995). Additionally, the attractiveness and high quality of produce grown with the aid of synthetic fungicides are in contrast to the poorer quality of many organically grown crops (Avery, 1995). These data strongly support the benefits of fungicides in maintaining the availability of quality food at moderate prices.

Fungicides, plus other crop protection chemicals, also allow the intensive, rather than extensive, land use of today's agriculture. Currently, world cropland available per person is about a third of a hectare, down from nearly a half a hectare required in 1961 before the widespread use of modern pesticides (Beyer, 1991). Elimination of pesticides, primarily fungicides used in fruit and vegetable production, would increase the land needed to grow these crops in the U.S. by 44%, or 2.5 million acres (Taylor, 1995). Reduction in cultivation areas has allowed the set-aside of marginal, more highly erosive land for conservation reserve and residential/commercial real estate opportunities (Penson et al., 1990; Urbanchuk, 1990).

Lastly, fungicides are currently the only technology that provides reliable disease control in many situations worldwide, and no alternatives exist at this time that can routinely replace these crop protection chemicals. Coupled with appropriate cultural practices, biological control, and resistant cultivars, these products serve as an integral component of most successful IPM programs. Thus, disease protection chemicals will continue to remain an essential input in many crops in the near term (Beyer, 1991; Ragsdale et al., 1991; Ragsdale and Sisler, 1994).

2. Fungicides and Food Safety

Global health authorities are mandated to assess dietary risks of pesticide exposures and the UN and its agencies act to harmonize rules globally. The World Health Organization (FAO/WHO) through the CODEX ALIMENTARIUS system establishes maximum pesticide residue limits (MRLs). The FAO also developed and implemented the International Code of Conduct on the Use and Distribution of Pesticides. These agencies monitor pesticide use with the goal of mitigating or eliminating risks, especially in developing countries.

Although the U.S. public and environmentalist advocacy organizations have raised questions concerning potential adverse health affects from exposure to pesticide residues in food, strong evidence from many sources documents the safety of our food supply (Anon., 1992a; Flynn, 1988; Graham and Gray, 1995). Pesticides are extensively tested and governmentally regulated commercial products (Anon., 1995a). For example, during the registration process for pesticides the U.S. EPA, and most governmental regulatory agencies, assess the toxicity of the product and potential human exposures in food. Toxicity is predicted by rodent feeding studies. A tolerance or the maximum residue level of a specific pesticide legally permitted on or in food is established (Graham and Gray, 1995; Ragsdale et al., 1991). We have previously discussed additional testing requirements before a product is available for sale.

Concerning exposure to pesticides in food, recent analysis of nearly 9000 food samples by the U.S. Food and Drug Administration (FDA) showed that 60% contained no detectable pesticide residues, 38% had residues within legal tolerances, less than 1% had residues above the EPA tolerance, while less than 1% had residues with no tolerances for that crop. Residues without tolerance generally occurred as an inadvertent exposure through spray drift, and tolerances existed for other crops (Anon., 1992a, 1995c; Avery, 1995). FDA's extensive Total Diet Study demonstrated that pesticide residues in a total diet were 10,000 times lower than the highest dose shown to cause a toxic effect in rodent cancer assessment studies (Brookes, 1991). Moreover, recent studies based on actual measured human exposure to 8 pesticides in foods determined that EPA risk assumptions overstated actual risk by factors of 99,000 to 463,000, while in several cases pesticide residues could not even be detected (Brookes, 1991; Graham and Gray, 1995). Because of the EPA's conservative exposure assumptions, their calculated risks have little relation to actual risks and provide large margins of safety (Ragsdale et al., 1991).

With regard to animal carcinogen testing using maximum tolerated doses, recent theories have questioned their relevancy in predicting human cancer risks from consumption of minute levels of pesticide residues in food (Graham and Gray, 1995). The near-toxic doses tested in the rodent studies can cause mitogenesis, i.e., cell division, which can promote carcinogenesis, whereas the lower, non-cell-killing doses potentially consumed in food would not result in mitogenesis or carcinogenesis (Ames and Gold, 1990; Ames et al., 1990a,b).

Predictions of cancer risk from synthetic pesticides also have been challenged because of the widespread presence of natural pesticides in foods, e.g., psoralens (Anon., 1996). Humans consume between 5000 and 10,000 different natural pesticides which plants produce to defend themselves, and 10,000 times more natural pesticides by weight than synthetic pesticides (Ames et al., 1990a).

Although not extensively tested to date, approximately half of the natural pesticides analyzed are carcinogenic in high-dose rodent tests (Ames and Gold, 1990; Ames et al., 1990a). These results indicate that synthetic pesticide residues are likely to be of minimal concern relative to natural substances (Ames et al., 1990a; Anon., 1996). Moreover, human detoxifying mechanisms, i.e., antioxidant defenses which serve to protect against toxic chemicals, are effective against both synthetic and natural toxins (Ames et al., 1990a,b). Thus, at the low doses of most human exposure to synthetic pesticides, the hazards of these chemicals are negligible (Ames et al., 1990a; Graham and Gray, 1995; Ragsdale et al., 1991; Anon., 1996).

The public is, in general, poorly informed on the many benefits of synthetic pesticides and their negligible risks, particularly in relation to naturally occurring substances in food. This ignorance promotes their desire for chemically free food. The quest for absolute zero chemical residues in food is not only unrealistic in the face of the increasing capability to measure smaller and smaller concentrations, but also appears to be scientifically flawed based on recent information (Anon., 1996).

C. Product Quality Assurance and Stewardship

Fungicides are expected to perform as indicated on the product label, i.e., provide expected disease control and without unacceptable risk to the user or the environment when used according to the label instructions. Companies manufacturing and selling products for disease control do extensive physical, chemical, and biological testing to assure sustainable delivery of acceptable product quality, performance, and safety. All fungicides are formulated to ensure maximum biological effectiveness. Suitable formulations allow for loading, mixing, and delivery from application equipment and aid in enhancing the intrinsic properties of the active fungicidal ingredient. Depending on projected use pattern, fungicides are formulated in various ways, e.g., dusts, wettable powders, suspension concentrates, water-dispersible granules, and emulsions. The use of water-soluble bags for wettable powders and water-dispersible granules has provided advantages in terms of ease of measurement and avoidance of user exposure. Each formulation is manufactured according to a unique set of product specifications that are important for shipment, storage, usability, and effectiveness. A battery of analytical studies are conducted with these formulations to ensure compliance with required physical and chemical specifications.

1. Analytical and Biological Assessments

Physical tests of the formulated product include, where appropriate, appearance, explosivity, flash point, pH, viscosity, bulk density, storage stability, foaming characteristics, suspensibility, emulsion stability, flowability, physical compatibility, wetting ability, dustiness, particle size, dispersion, sedimentation, crystallization, and packaging assessments. Physical tests utilize Collaborative International Pesticides Analytical Counsel (CIPAC) methods where applicable. Physical attributes of the active chemical as well as the inerts and their interaction are assessed. Generally, the manufacturer must supply to the regulatory agencies a confidential statement of formula (CSF) which specifically states the composition of the formulation. Although practices differ according to the country regulatory authorities, usually the use of approved raw materials and inerts is required in formulation. Biological testing in the laboratory, greenhouse, and the field serves to document that the product consistently controls the target disease without detrimental effect to the crop when used according to label directions.

2. Quality Control and Product Stewardship

Quality means offering a product which meets the expectation of the customer. The concept of quality begins with the manufacture of the active chemical, the formulation of the active chemical into a usable product, and the packaging and labeling of the formulation with clear directions for use. There are many manufacturing technologies and processes to assure the production of a quality product, including statistical process control and ISO certification. The ultimate purpose of quality assurance is to continually and reliably meet the customer's needs and to satisfy regulatory standards.

In addition to instructions addressing the proper transportation, storage, use, and disposal of products in the label and directions for use, further education of farmers, applicators, and other agricultural workers is provided. Crop protection chemical manufactures individually, and collectively through the International Group of National Associations of Manufactures of Agrochemical Products (GIFAP), actively participate in training the global farming community to ensure agrichemical products are used appropriately.

IV. MAXIMIZING SUSTAINABLE BENEFITS: FUNGICIDE USE PRACTICES AND RESISTANCE MANAGEMENT

A. What is Resistance?

Since the worldwide use of the benzimidazole fungicides in the early 1970s, farmers, advisors, scientists, and manufacturers of fungicides worldwide have gained considerable knowledge and experience in understanding the utility and limitations of fungicide use (Delp, 1980). We now have an improved understanding of the factors responsible for fungicide resistance development, and the best methods to avoid or delay its appearance (Delp, 1988a; Eckert, 1988). The belief in a "silver bullet" product that and will alleviate all disease problems, and can be used everywhere and without concern for its future utility, has paled under the light of practical experience. Each new fungicide is a tool whose usefulness we must learn to maximize while identifying its limitations. This section provides an overview of fungicide use and how utility can be lessened by the development of resistance. How resistance occurs through natural selection is discussed, as are recommendations for implementing integrated use and resistance management tactics to sustain the full long-term benefits of fungicides.

Fungicide resistance is a complex issue for which there is no single, nor simple solution. However, to optimize disease control and the longevity of fungicide utility, some basic principles must be considered in the development of practical use recommendations and IPM programs. The build-up of pathogen populations resistant to fungicides has challenged all parties involved in food production, e.g., farmers, advisors, scientists, and manufacturers of crop protection chemicals, for the last quarter century (Brent, 1995). Resistance is a serious consideration in the discovery, development, commercialization, sale, and use of fungicides worldwide (Anon., 1979; Sanders, 1991; Urech, 1994; Wade and Delp, 1990).

The ability of fungal pathogens to overcome host plant or varietal resistance is recognized as a natural process of evolution. Resistance to crop protection chemicals likewise is a natural phenomenon occurring in insects, weeds, and fungi (Georghiou, 1990; LeBaron and McFarland, 1990). Fungicide resistance is the inherited ability of a fungal pathogen to survive doses of a fungicide that would normally control it when the product is used according to label recommendations. The capacity of fungi to develop resistance to control agents is not limited to their response to synthetic fungicides, as resistance in *Botrytis cinerea* has been reported to a biological control agent, i.e., *Bacillus subtilis* CL27 (Li and Leifert, 1994).

Fungal populations generally contain very small numbers of individuals which are naturally able to survive exposure to a fungicide. Such individuals result from spontaneous mutations that continually occur during the normal evolutionary process within all populations of living organisms (Georgopoulos and Skylakakis, 1986). Resistance often becomes evident after repeated use of a fungicide allows the naturally occurring resistant individuals within the original population to multiply until they eventually become dominant (Georgopoulos, 1994). As the sensitive individuals are suppressed and/or removed by the action of the fungicide, the rare fungicide-resistant individuals survive, multiply, and occupy the biological niche previously filled by the sensitive population. After repeated applications and/or exposed crop cycles, the fungicide provides less than expected levels of control. Repeated uses of fungicides having identical biochemical modes of action often will lead to resistance.

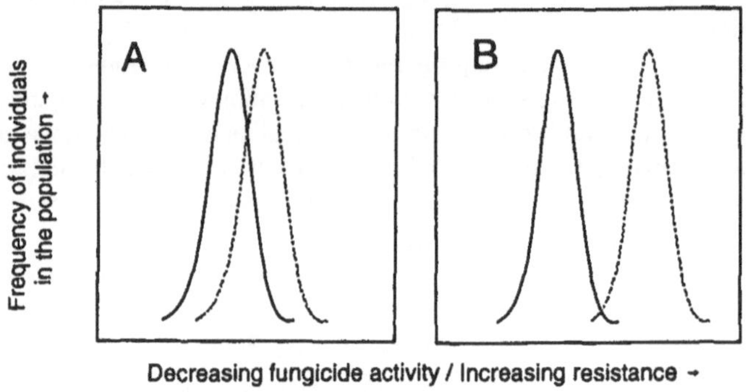

Figure 2 Composition of fungal populations in response to fungicide applications. A progressive type (A) and a disruptive type (B) of resistance evolution. The solid line represents the original, sensitive population, and the dotted line represents the resistant population after the selection process initiated by repeated fungicide applications.

Resistance evolves at different rates and intensities depending on a complex of interacting factors involving the pathogen, the fungicide, and the environment. These factors include pathogen type and population dynamics, fungicide class, mode of action, use patterns, and environmental factors which impact pathogen life and disease cycles and fungicide performance.

B. Resistance Types and Pathogen Fitness

Resistance can occur either through a slow build-up of a few resistant individuals within a sensitive population, or by the rapid appearance of a totally resistant population. The slow build-up type of resistance is commonly referred to as progressive, i.e., a gradual decline or continuous lessening of fungicide effectiveness with a correlated decreasing sensitivity of the pathogen population to the fungicide (Figure 2A). Inheritance of resistance generally involves several genes and is referred to as polygenic resistance (Georgopoulos, 1994). Use of the fungicide, or fungicide class, usually can be continued as long as the level of resistance, or frequency of resistant individuals within the population, is low and field control is satisfactory. This has been the practical field experience with the triazole fungicides (de Waard, 1994; Hollomon et al., 1990; Berg et al., 1990). This also appears to be the case with the morpholine fungicides (Hollomon, 1994).

The rapid type of resistance evolution is often referred to as disruptive as there is a more discrete change in the population as the original sensitivity is replaced with a resistant population (Figure 2B). Inheritance of resistance is by one or a few genes, often referred to as major gene resistance (Georgopoulos, 1994). In this case, resistance usually occurs suddenly and is associated with dramatic loss of control, i.e., the fungicide, or fungicide class, if applied alone no longer provides satisfactory control, even at increased dose levels. Examples are the development of resistance to benomyl and the phenylamides (Davidse and Flach, 1977; Smith, 1988; Staub and Sozzi, 1983).

Whether the resistant individuals dominate and replace the original population, i.e., become stable or fixed as the new resistant population, depends on the fitness or capability of the resistant individuals to compete with the sensitive, original types that have survived, or avoided, fungicide exposure (Fry and Milgroom, 1990). If fitness is high, the resistant individuals are likely to remain dominant or decline only slightly over many years, even when the particular fungicide no longer is applied. Future unrestricted or intensive applications of the particular fungicide are likely to quickly reestablish the resistant population. However, if the resistant individuals possess a fitness deficit, i.e., they are only weakly competitive with sensitive individuals, removal of the fungicide will often result in a decline of resistant types. In this case a controlled or restricted reentry with the fungicide may be possible.

The progressive shifts in sensitivity can be detected through careful monitoring programs (Brent, 1992). Disruptive resistance evolution provides the greatest risk for control failures, as changes in sensitivity may occur so rapidly that the initial detection of resistance may follow disease control failure.

C. Determining Resistance Risk

Most current classes of fungicides have shown potential for encountering resistance difficulties. Understanding the risks and developing management strategies in the early phases of fungicide discovery and commercialization are now critical success factors in product registration, as well as maximizing product life and maintaining customer satisfaction (Furk and Slawson, 1994; Gisi and Staehle-Csech, 1988). Understanding the likelihood for resistance occurrence early in development of a fungicide guides recommendations for product usage based on sound resistance management strategies and monitoring programs (Trivellas, 1988; Brent et al., 1990).

Precommercialization or baseline sensitivity of the target pathogen population must be determined to provide the basis for assessing shifts in sensitivity once the fungicide is commercialized (Brent et al., 1990). Laboratory studies of spontaneous or accelerated mutation rates for resistant individuals and tests for cross resistance to existing products can provide some understanding of the probability for resistance occurrence (Chabane et al., 1993; Shaw, 1995). However, laboratory studies utilizing isolates with limited genetic diversity may not adequately assess the risk potential (Staub et al., 1979). Field studies with fungicide applications sustained full season to large and genetically diverse populations, as occur in commercial fields, can provide additional information on which to base a resistance risk decision. While resistance risk is seldom absolutely predictable, general factors can influence resistance development. Fungicide use and pathogen biological factors known to affect resistance development are

- *Attributes of the fungicide.* Persistence on the plant tissue to be protected, mode of action, and cross resistance to other previously used fungicides.
- *Intensity and use practices.* Fungicide level of control required due to yield or quality demands, number of applications, and timing within the disease cycle, i.e., preventive or curative uses.
- *Pathogen biology.* Reproduction rate (generations within a season or crop cycle), rate of dissemination and epidemic development, genetic variability, ease of producing resistant mutants in the laboratory, fitness of resistant types, and inheritability of resistance.

These factors frequently interact and an understanding of these relationships must be part of the resistance risk potential assessment process.

D. Monitoring to Detect Resistance

Fungicide resistance is monitored by testing field-collected pathogen samples or isolates for their degree of sensitivity to a particular fungicide (Brent, 1992). If resistance is found within field-collected samples, it is important to determine the frequency in which the resistant individuals occur within the samples. The greater the frequency of resistant individuals within a population, the greater the risk for resistance to be expressed in the field.

Monitoring methods, results, and conclusions must be scientifically sound if they are to successfully guide management tactics. Quantitative and reproducible assay methodologies need to be developed and validated to ensure that data accurately represent population sensitivity. Publications of the Fungicide Resistance Action Committee (FRAC), affiliated with GIFAP, contain recommended procedures for specific pathogen and crop complexes (Anon., 1982, 1991a, 1992b). Also important is to understand the relationship between sensitivity data, which are often an *in vitro* responses, and the reaction of the pathogen to the fungicide at the plant-environment interface (Shaw, 1995).

Highly sensitive laboratory methods may detect small variations in pathogen sensitivity that are of no practical significance under field use conditions. Careful monitoring and cautious interpretation of results is necessary to provide useful guidance in resistance assessment. Reports of the appearance of resistant individuals or strains induced in laboratory experiments do not necessarily imply that a loss of control can be expected in the field. Likewise, the appearance of less sensitive strains in the field does not always result in failure of disease control. When the frequency of resistant individuals is low and/or the level of resistance is moderate, fungicide applications in most cases will provide satisfactory control. However, the early detection of a few resistant individuals is a warning that the pathogen population may be shifting toward reduced sensitivity and that appropriate resistance management practices should be implemented quickly.

To avoid the misinterpretation of potential and/or possible resistance occurrences, FRAC developed principles which define actual field resistance (Russell, 1993). The term field resistance is limited to situations where (1) the development of resistance leads to failure of disease control under practical field conditions following application(s) of a fungicide correctly and according to the label, and (2) a demonstration that the loss of control is due to the presence of pathogenic strains with reduced fungicide sensitivity. There must be a clear correlation between the decline in field performance of the fungicide and the existence of resistant individuals at the specific site before field resistance can be correctly identified.

Resistance monitoring also provides a check that management strategies are working, or if changes in fungicide use are called for to manage evolving resistance. Practical experience with benomyl, metalaxyl, and the triazoles have show the importance of tracking resistance status vs. management actions (Cooke, 1995; Delp, 1988b; Hollomon et al., 1990; Smith, 1988). Sensitivity testing of pathogen samples collected from sites where fungicide performance has been unexpectedly low may help to understand performance complaints.

E. Control Failures and Resistance

While resistance may frequently be suggested as the reason for poor disease control, many other factors must be examined before concluding that resistance is the cause. These other factors include use of the correct fungicide for the specific pathogen, correct timing of application(s), recommended application rate and method of application(s), variations in crop susceptibility to the pathogen, and environmental conditions affecting fungicide action and pathogen biology, e.g., temperature, humidity, and rainfall. These important factors, necessary for achieving effective control, are defined in the product label recommendations and must be understood and followed.

F. Management Tactics

With any resistance management program, it is important to prevent the build-up of both sensitive and resistant individuals within the population, i.e., maintain the disease pressure as low as possible (Fry and Milgroom, 1990). Any action, e.g., cultural practices, sanitation, host plant resistance, and fungicides that restricts the size of the pathogen population will be useful in managing resistance. It is also important to minimize fungicide selection pressure by not overusing fungicides with the same mode of action (Delp, 1988b; Staub, 1991). Sound resistance management tactics must be conscientiously applied from the initial use of a new fungicide and carried out over large areas where the fungicide is used due to the wide dispersal of some pathogens.

A core part of resistance management tactics is the recommendation of fungicide mixtures or alternations of fungicides with different modes of action (Delp, 1988b; Fry and Milgroom, 1990; Schwinn and Morton, 1990). A pathogen population challenged by two different modes of action is believed to be less likely to evolve towards resistance (Samoucha and Gisi, 1987). Fungicides having a single mode of action, often referred to as at-risk fungicides, are generally recommended for application with fungicides having a different mode of action either in preformulated mixes or in tank mixes done on the farm, or in alternation with a limited number of consecutive sprays in a season (de Waard, 1988; Urech, 1988).

Fungicides must be used at their recommended label rates to ensure that expected efficacy is achieved under a wide range of conditions. Additionally, reduced rates can allow moderately resistant individuals to survive and increase in frequency within the treated population (Shaw and Pijls, 1994; Zziwa and Burnett, 1994). Prophylactic use of fungicides avoids the high risk of curative or eradicant applications that are generally targeted at nearly epiphytotic pathogen levels. In curative uses, the large pathogen population present at the time of fungicide application affords a greater opportunity for the selection of resistant individuals (Staub, 1994; Delp, 1988a). Any specific disease control programs must carefully integrate all biological and cultural methods along with appropriate fungicide applications. The integration of all sound tactics ensures long-term effectiveness and availability of established and new fungicides (Delp, 1988b; Staub, 1991). The following guidelines are recommended as the most effective means of using fungicides and delaying and/or avoiding the development of resistance.

1. Always understand product attributes of efficacy, persistence, specificity, systemicity, chemical class, mode of action, and compatibility with other products to optimize product fit within local IPM programs.
2. Where feasible do not rely on a single chemical class to control multiple pathogen cycles. Practice fungicide diversity in one of the following ways:
    - Avoid repetitive applications with fungicides having the same mode of action. Select appropriate alternations or rotations of products with different types of action. Follow appropriate FRAC recommendations.
    - Where multiple applications are expected within a season, use mixtures of products that have similar field residual attributes and are not cross resistant, i.e., have a different mode of action. An ideal mixture is one where the efficacy of the mixture is greater than the effect of each product used alone, i.e., synergistic (Grabski and Gisi, 1987).
    - If resistance is strongly suspected as the cause of a control failure, retreat the field with another fungicide having a different mode of action.
3. Always understand the pathogen biology to minimize the numbers of applications.
    - When possible, use cultural and biological control tactics to disrupt timing or relationship between the crop and pathogen, such as early planting and harvesting against mid- to late-season disease occurrence and the use of resistant varieties.
    - Take into account environmental conditions and events that can affect the biology of the pathogen, e.g., temperature and rainfall.
4. Follow label recommendations on specified rates, spray interval, application method and timing, and compatibility for mixing with other fungicides.
    - Stress the importance of timely and adequate coverage from spray applications. Follow well-designed and executed preventive spray programs.
    - Careful use of forecasting and scouting programs can be valuable to optimizing application timings.
5. Emphasize development and use of reliable pathogen sensitivity monitoring programs developed over time and geographic areas to guide the selection and use of appropriate, highly effective fungicides. Monitoring generally needs to be practiced seasonally, especially where the risk of resistance development is high.

V. CONCLUSIONS

In this chapter we have attempted to present a balanced view on the importance and complexity of fungicide use, and to educate the reader as to why fungicides are needed, how they are discovered and assessed for safety, and the use practices that sustain the value they bring to agriculture. Our society is becoming increasingly complex, and at the same time the public is demanding quick and simplistic solutions to complicated issues. It is important to understand that, as with most significant challenges, there is no quick fix nor single exclusive technology currently available for the effective control of plant pathogens. Control of crop diseases demands the judicious integration of all available techniques, i.e., biological, cultural, and chemical, as each has its place in reducing the burden pathogenic fungi place on the quantity and quality of the global harvest. In our quest for abundant and affordable food, choices of options for crop protection approaches are demanded of us. We must carefully and broadly evaluate emerging alternatives for their full environmental impact and do so in a way comparable to those required by regulatory agencies for synthetically derived products. In the development of any technology it is essential that its evaluation be rigorous, with full consideration for the unintended effects or consequences resulting from its use. Crop protection solutions will continually be needed, and it is essential to avoid creating new difficulties while attempting to solve existing problems. Our campaign to reduce risk to our health and environment can come in conflict with itself as one risk is unknowingly exchanged for another. We hope this discussion has shown that fungicides are a safe, sophisticated, reliable, and economical technology that will continue to evolve through the discovery and commercialization of unique molecules providing even additional value to crop production and the global standard of living.

REFERENCES

Abelson, P. H., Adequate supplies of fruits and vegetables, *Science,* 266, 1303, 1994.

Agrios, G. N., *Plant Pathology,* Academic Press, New York, 1988.

Ames, B. N. and Gold, L. S., Chemical carcinogenesis: too many rodent carcinogens, *Proc. Natl. Acad. Sci. U.S.A.,* 87, 7772, 1990.

Ames, B. N., Profet, M., and Gold, L. S., Dietary pesticides (99.99% all natural), *Proc. Natl. Acad. Sci. U.S.A.,* 87, 7777, 1990a.

Ames, B. N., Profet, M., and Gold, L. S., Nature's chemicals and synthetic chemicals: comparative toxicology, *Proc. Natl. Acad. Sci. U.S.A.,* 87, 7782, 1990b.

Ames, B. N., Shigenaga, M. K., and Hagen, T. M., Oxidants, antioxidants, and the degenerative diseases of aging, *Proc. Natl. Acad. Sci. U.S.A.,* 90, 7915, 1993.

Anon., Pest resistance to pesticides and crop loss assessments, in FAO Plant Protection and Production Paper 6/2 AGP: 1979/M/2, Rome, 1979, 41.

Anon., Recommended methods for the detection and measurement of resistance of agricultural pests to pesticides. FAO Plant Protection Bulletin 30, 1982, 36, 141.

Anon., CAST: *Economic and Health Risks Associated with Mycotoxins,* Council for Agricultural Science and Technology, Ames, IA, 1989, 100.

Anon., FRAC methods for monitoring fungicide resistance, *EPPO Bull.,* 21, 291, 1991a.

Anon., European Community Council Directive, Concerning the placing of plant protection products on the market, *Off. J. Eur. Communities,* L 230/1, July 15, 1991b.

Anon., *The Health Significance of Pesticide Residues in Food,* National Food Processors Association, Washington, D.C., 1992a, 37.

Anon., FRAC methods for monitoring the sensitivity of fungal pathogens to phenylamide fungicides, *EPPO Bull.,* 22, 297, 1992b.

Anon., Benomyl benefits assessment on rice in the southern U.S., *La. State Univ. Agric. Center,* 1993, 50.

Anon., U.S. Environmental Protection Agency, *Code of Federal Regulations 40,* parts 150, Protection of Environment, Office of the Federal Register, National Archives and Records Administration, Washington, D.C., July 1, 1995a.

Anon., Wood Mackenzie Report, Wood Mackenzie Agrochemical Service, Products Section, June 1995b.

Anon., Pesticide program: Residue monitoring 1994, FDA, Washington, D.C., *J. Assoc. Off. Anal. Chem.,* 78, Sept./Oct., 1995c.

Anon., Carcinogens and Anticarcinogens in the Human Diet: A Comparison of Naturally Occurring and Synthetic Substances, National Academy Press, 1996, 290.

Avery, D. T., Saving the planet with pesticides and plastic. The environmental triumph of high-yield farming, Hudson Institute, Indiana, 1995, 432.

Backman, P. A., Munger, G. D., and Marks, A. F., The effects of particle size and distribution on performance of the fungicide chlorothalonil, *Phytopathology,* 66, 1242, 1976.

Berg, D., Buchel, K.-H., Holmwood, G., Kramer, W., and Pontzen, R., Sterol biosynthesis inhibitors. Model studies with respect to modes of action and resistance, in *Managing Resistance to Agrochemicals — From Fundamental Research to Practical Strategies,* Green, M. B., LeBaron, H. M. and Moberg, W. K., Eds., American Chemical Society, Washington, DC., 1990, 184.

Beyer, E. M., Crop protection — meeting the challenge, Proc. Brighton Crop Prot. Conf., *Weeds,* 1, 3, 1991.

Borlaug, N. E., The Challenge of feeding 8 billion people, *Farm Chem. Int.* Summer Issue, 10, 1990.

Brent, K. J., Monitoring fungicide resistance: purposes, procedures and progress, in *Resistance '91: Achievements and Developments in Combating Pesticide Resistance,* Denholm, I., Devonshire, A. L., and Hollomon, D. W., Eds., Elsevier, London, 1992, 1.

Brent, K. J., Fungicide resistance in crop pathogens: How can it be managed?, *FRAC Monograph* No. 1, GIFAP, Brussels, 1995.

Brent, K. J., Hollomon, D. W., and Shaw, M. W., Predicting the evolution of fungicide resistance, in *Managing Resistance to Agrochemicals — From Fundamental Research to Practical Strategies,* Green, M. B., LeBaron, H. M. and Moberg, W. K., Eds., American Chemical Society, Washington, D.C., 1990, 303.

Brookes, W. T., EPA's misguided hysteria over pesticide risk, *Farm Chem.,* 28, 1991.

Bruck, R. I., Fry, W. E., Apple, A. E., and Mundt, C. C., Effect of protectant fungicides on the developmental stages of *Phytophthora infestans* in potato foliage, *Phytopathology,* 71, 164, 1981.

Carsel, R. F., Mulkey, L. A., Lorber, M. N., and Baskin, L. B., The pesticide root zone model (PRZM): a procedure for evaluating pesticide leaching threats to groundwater, *Ecol. Modeling,* 30, 49, 1985.

Chabane, K., Leroux, P., and Bompeix, G., Selection and characterization of *Phytophthora parasitica* mutants with ultraviolet-induced resistance to dimethomorph or metalaxyl, *Pestic. Sci.,* 39, 325, 1993.

Cooke, L. R., Monitoring of fungicide resistance and countermeasures: potato late blight, in The International Workshop on Fungicide Resistance: A Factor Limiting Quality Food Production, Fruit Tree Res. Sta., Ministry of Agriculture, Forestry and Fisheries (MAFF), Tsukuba, Japan, 1995, 9.

Curtis, C. R., Fungicide Benefits Assessment Executive Summary, Portion USDA/States Natl. Agri. Pesticide Impact Assessment Program (NAPIP), Fungicide Assessment Project, 1991, 12.

Davidse, L. C. and Flach, W., Differential binding of methyl benzimidazole-yl-carbamate to fungal tubulin as a mechanism of resistance to this antimitotic agent in mutant strains of *Aspergillus nidulans, J. Cell Biol.,* 72 174, 1977.

de Waard, M. A., Interactions of fungicide combinations, in *Fungicide Resistance in North America,* Delp, C. J., Ed., American Phytopathological Society Press, St. Paul, MN, 1988, 98.

de Waard, M. A., Resistance to fungicides which inhibit stereo 14α-demethylation, an historical perspective, *Fungicide Resistance, BCPC Monogr.* 60, 3, 1994.

de Waard, M. A., Georgopoulos, S. G., Hollomon, D. W., Ishii, H., Leroux, P., and Ragsdale, N. N., Chemical control of plant diseases: problems and prospects, *Annu. Rev. Phytopathol.,* 31, 403, 1993.

Delp, C. J., Coping with resistance to plant disease control agents, *Plant Dis.,* 64, 652, 1980.

Delp, C. J., Fungicide resistance problems in perspective, in *Fungicide Resistance in North America,* Delp, C. J., Ed., American Phytopathological Society Press, St. Paul, MN, 1988a, 4.

Delp, C. J., Resistance management strategies for benzimidazoles, in *Fungicide Resistance in North America,* Delp, C. J., Ed., American Phytopathological Society Press, St. Paul, MN, 1988b, 41.

Dowley, L. J. and O'Sullivan, E., A short history of the potato, the famine, late blight and Irish research on *Phytophthora infestans. Teagasc,* Oak Park Research Center, Carlow, Ireland, 1995, 32.

Eckert, J. W., Historical development of fungicide resistance in plant pathogens, in *Fungicide Resistance in North America,* Delp, C. J., Ed., American Phytopathological Society Press, St. Paul, MN, 1988, 1.

Enslein, K., Gombar, V. K., and Blake, B. W., Use of SAR computer-assisted prediction of carcinogenicity and mutagenicity of chemicals by the TOPKAT program, *Mutat. Res.,* 305, 47, 1994.

Feichtmeir, E. F., The effect of particle size and solubility of sulfur in carbon disulfide upon its toxicity to fungi, *Phytopathology,* 39, 605, 1949.

Fletcher, J. T. and Wolfe, M. S., Insensitivity of *Erysiphe graminis* f. sp. *hordei* to triadimefon, triadimenol and other fungicides. Proc. Brighton Crop Prot. Conf., Pests Dis. 2, 633, 1981.

Flynn, L. T., *Pesticides: Helpful or Harmful?,* American Council on Science and Health, 1988, 59.

Fry, W. E., *Principles of Plant Disease Management,* Academic Press, New York, 1982.
Fry, W. E. and Fohner, G. R., Construction of predictive models. 1. Forecasting disease development in *Advances in Plant Pathology,* Vol. 3, Mathematical Modeling of Crop Diseases, Gilligan, C. A. Ed., Academic Press, New York, 1984, 161.
Fry, W. E. and Milgroom, M. G., Population biology and management of fungicide resistance, in *Managing Resistance to Agrochemicals — From Fundamental Research to Practical Strategies,* Green, M. B., LeBaron, H. M., and Moberg, W. K., Eds., American Chemical Society, Washington, D.C., 1990, 275.
Furk, C. and Slawson, D. D., Pesticide resistance and the EC plant protection products directives, *Fungicide Resistance, BCPC Monogr.,* 60, 385, 1994.
Georghiou, G. P., Overview of insecticide resistance, in *Managing Resistance to Agrochemicals — From Fundamental Research to Practical Strategies,* Green, M. B., LeBaron, H. M., and Moberg, W. K., Eds., American Chemical Society, Washington, D.C., 1990, 320.
Georgopoulos, S. G., Early evaluation of fungicide resistance risk, *Fungicide Resistance, BCPC Monogr.,* 60, 389, 1994.
Georgopoulos, S. G. and Skylakakis, G., Genetic variability in the fungi and the problem of fungicide resistance, *Crop Prot.,* 5, 299, 1986.
Gianessi, L. P., Chemicals seen vital to weed, pest control, Forum for Applied Research and Public Policy, Fall, 1993, 87.
Giovannucci, E., Ascherio, A., Rimm, E. B., Stampfer, M. J., Colditz, G. A., and Willett, W. C., Intake of carotenoids and retinol in relation to risk of prostate cancer, *J. Natl. Cancer Inst.,* 87(23), 1767, 1995.
Gisi, U. and Staehle-Csech, U., Resistance risk evaluation of new candidates for disease control, in *Fungicide Resistance in North America,* Delp, C. J., Ed., American Phytopathological Society Press, St. Paul, MN, 1988, 101.
Gold, R. E. and Leinhos, G. M. E., Histological studies on the fungicidal activity of the strobilurin BAS 490F, Proc. Brighton Crop Prot. Conf., *Pests Dis.,* 1, 253, 1994.
Grabski, C. and Gisi, U., Quantification of synergistic interactions of fungicides against *Plasmopara* and *Phytophthora, Crop Prot.,* 6, 64, 1987.
Graham, J. D. and Gray, G. M., How should pesticides be regulated? Risk in Perspective, *Harvard Center for Risk Analysis,* 3(1), 1995.
Graham, J. D. and Wiener, J. B., *Risk Versus Risk: Tradeoffs in Protecting Health and the Environment,* Harvard University Press, Cambridge, 1995, 334.
Hollomon, D. W., Do morpholine fungicides select for resistance?, *Fungicide Resistance, BCPC Monogr.,* 60, 281, 1994.
Hollomon, D. W., Butters, J., and Clark, J., Genetic control of triadimenol resistance in barley powdery mildew, Proc. Brighton Crop Prot. Conf., *Pests Dis.,* 2, 477, 1984.
Hollomon, D. W., Butters, J. A., and Hargreaves, J. A., Resistance to sterol biosynthesis-inhibiting fungicides. Current status and biochemical basis, in *Managing Resistance to Agrochemicals-From Fundamental Research to Practical Strategies,* Green, M. B., LeBaron, H. M., and Moberg, W. K., Eds., American Chemical Society, Washington, D.C., 1990, 199.
Hyre, R. A., Relation of particle size to fungicidal value and tenacity of two insoluble copper fungicides, *Phytopathology,* 32, 388, 1942.
James, W. C., Teng, P. S., and Nutter, F. W., Estimated losses of crops from plant pathogens, in *CRC Handbook of Pest Management in Agriculture,* Vol. 1, Pimentel, D. and Hanson, A. A., Eds., CRC Press, Boca Raton, 1991, 15.
Jorgensen, J. H., *Erysiphe graminis,* powdery mildew of cereals and grasses, *Adv. Plant Pathol.,* 6, 137, 1988.
Kessmann, H., Staub, T., Hofmann, C., Maetzke, T., Herzog, J., Ward, E., Uknes, S., and Ryals, J., Induction of systemic acquired disease resistance in plants by chemicals, *Annu. Rev. Phytopathol.,* 32, 439, 1994.
Knutson, R. D., Taylor, C. R., Penson, J. B., and Smith, E. G., *Economic Impacts of Reduced Chemical Use, Executive Summary,* Texas Agricultural Extension Service, Texas A&M University, College Station, TX, 1990, 44.
Kohle, H. R. E., Gold, R. E., Ammermann, E., Sauter, H., and Rohl, F., Biokinetic properties of BAS 490F and some related compounds: design of mitochondrial transport inhibitors as agrochemicals, *Biochem. Soc. Trans.,* 22, 65S, 1994.

Kohn, A., *No Contest: The Case Against Competition*, Houghton Mifflin, New York, 1992, 324.

Kudsk, P., Mathiassen, S. K., and Kirknel, E., Influence of formulations and adjuvants on the rainfastness of maneb and mancozeb on pea and potato, *Pestic. Sci.*, 33, 57, 1991.

LeBaron, H. M. and McFarland, J., Herbicide resistance in weeds and crops. An overview and prognosis, in *Managing Resistance to Agrochemicals — From Fundamental Research to Practical Strategies*, Green, M. B., LeBaron, H. M. and Moberg, W. K., Eds., American Chemical Society, Washington, D.C., 1990, 336.

Leprince, X., The market for fungicides in European wheat and barley crops, in *A Vital Role for Fungicides in Cereal Production*, Hewitt, H. G., Tyson, D., Hollomon, D. W., Smith, J. M., Davies, W. P., and Dixon, K. R., Eds., BIOS Scientific, Oxford, 1995, 19.

Li, H. and Leifert, C., Development of resistance in *Botryotinia fuckeliana* (de Bary) Whetzel against the biological control agent *Bacillus subtilis* CL27, *Z. Pflanzenkran. Pflanzenschutz*, 101, 414, 1994.

Locke, T., Fungicide resistance and food production, in The International Workshop on Fungicide Resistance: A Factor Limiting Quality Food Production, Fruit Tree Res. Sta., Ministry of Agriculture, Forestry and Fisheries (MAFF), Tsukuba, Japan, 1995, 1.

Lyr, H., *Modern Selective Fungicides: Properties, Applications, Mechanisms of Action*, 2nd ed., Gustav Fischer Verlag, New York, 1995.

Martin, T., Seed treatment: progress and prospects, Proc. symp. University of Kent, Canterbury, British Crop Protection Council, Farnham, U.K., 5-7 January, 1994, 482.

Matthews, G. A., *Pesticide Application Methods*, Longman, London, 1979, 336.

Mumford, J. D. and Norton, G. A., Economics of decision making in pest management, *Annu. Rev. Entomol.*, 29, 157, 1984.

Nelson, P. E., Desjardins, A. E., and Plattner, R. D., Fumonisins, mycotoxins produced by *Fusarium* species: biology, chemistry, and significance, *Annu. Rev. Phytopathol.*, 31, 233, 1993.

North, J. J., Trends in European wheat and barley production, in *A Vital Role for Fungicides in Cereal Production*, Hewitt, H. G., Tyson, D., Hollomon, D. W., Smith, J. M., Davies, W. P., and Dixon, K. R., Eds., BIOS Scientific, Oxford, 1995, 1.

Neuenschwander, U., Friedrich, L., Delaney, T., Vernooij, B., Kessmann, H., and Ryals, J., Activation of plant disease resistance, in *Aspects of Applied Biology 42, Physiological Responses of Plants to Pathogens*, Walters, D. R., Scholes, J. D., Bryson, R. J., Paul, N. D., and McRoberts, N., Eds., Association of Applied Biologists, Wellesbourne, 1995, 217.

Nigg, H. N. and Allen, J. C., A comparison of time and time weather models for predicting parathion disappearance under California conditions, *Environ. Sci. Technol.*, 13, 231, 1979.

Oerke, E.-C., Dehne, H.-W., Schonbeck, F., and Weber, A., *Crop Production and Crop Protection: Estimated Losses in Major Food and Cash Crops*, Elsevier, Amsterdam, 1994, 808.

Orson, J. H., Crop protection practices in wheat and barley crops, in *A Vital Role for Fungicides in Cereal Production*, Hewitt, H. G., Tyson, D., Hollomon, D. W., Smith, J. M., Davies, W. P., and Dixon, K. R., Eds., BIOS Scientific, Oxford, 1995, 9.

Penson, J. B., Taylor, C. R., Smith, E.G., and Knutson, R. D., Restricting chemical use in agriculture. Its paradoxical economic consequences. *AFPC Policy Working Paper 90-12*, Department of Agriculture Economics, Texas Agricultural Exp. Stn., Texas A&M University, College Station, TX, 1990, 18.

Pillmoor, J. B. and Foster, S. G., Molecular approaches to the design of chemical crop protection agents, in *Molecular Biology in Crop Protection*, Marshall, G. and Walters, D., Eds., Chapman & Hall, London, 1994, chap. 2.

Pimetel, D., Mclaughlin, L., Zepp, A., Lakitan, B., Kraus, T., Kleinman, P., Vancini, F., Roach, W. J., Graap, E., Keeton, W. S., and Selig, G., Environmental and economic impacts of reducing U.S. agricultural pesticide use, in *CRC Handbook of Pest Management in Agriculture*, 2nd ed., Pimentel, P. and Hanson, A., Eds., 1991, 679.

Presting, D. B., Respect for the environment — a chemical company's vision for a partnership with nature, *Phytopathology*, 83, 35, 1993.

Ragsdale, N. N. and Sisler, H. D., Social and political implications of managing plant diseases with decreased availability of fungicides in the Unites States, *Annu. Rev. Phytopathol.*, 32, 545, 1994.

Ragsdale, N. N., Hylin, J. W., Sisler, H. D., and Witt, J. M., Health and Environmental Factors Associated with Agricultural Use of Fungicides. Portion USDA/States Natl. Agric. Pesticide Impact Assessment Program (NAPIP), Fungicide Assessment Project, 1991, 117.

Rouchaud, J., Moons, C., Monin, A., and Meyer, J. A., Comparison of the effects of a classical fungicide/insecticide treatment on quality of apples, *Fruit Belge,* 52(408), 319, 1989.
Royle, D. J. and Shaw, M. W., The costs and benefits of disease forecasting in farm practice in *Control of Plant Diseases: Costs and Benefits,* Clifford, B. C. and Lester, E., Eds., Blackwell, Oxford, 1988, 231.
Russell, P. E., Fungicide resistance action committee, *Pestic. Outlook,* 4 (2), 21, 1993.
Samoucha, Y. and Gisi, U., Use of two- and three-way mixtures to prevent buildup of resistance to phenylamide fungicides in *Phytophthora* and *Plasmopara, Phytopathology,* 77, 1405, 1987.
Sanders, P. L., Fungicide resistance in the United States. USDA/States National Agricultural Pesticide Impact Assessment Program (NAPIAP), Fungicide Assessment Project, 1991, 84.
Schwinn, F. J., Significance of fungal pathogens in crop production, *Pestic. Outlook,* 3, 18, 1992.
Schwinn, F. J. and Morton, H. V., Antiresistance strategies, design and implementation in practice, in *Managing Resistance to Agrochemicals — From Fundamental Research to Practical Strategies,* Green, M. B., LeBaron, H. M., and Moberg, W. K., Eds., American Chemical Society, Washington, D.C., 1990, 170.
Senge, P.M., *The Fifth Discipline,* Doubleday, New York, 1990, 424.
Shaw, M. W., Bioassay, in The International Workshop on Fungicide Resistance: A Factor Limiting Quality Food Production, Fruit Tree Res. Stn., Ministry of Agriculture, Forestry and Fisheries (MAFF), Tsukuba, Japan, 1995, 29.
Shaw, M. W. and Pijls, C. F. N., The effect of reduced dose on the evolution of fungicide resistance in *Septoria tritici. Fungicide Resistance, BCPC Monogr.,* 60, 47, 1994.
Siegel, M. and Sisler, H., *Antifungal Compounds,* Vol. 2, interactions in biological and ecological systems, Marcel Dekker, New York, 1977.
Sisler, H. D. and Ragsdale, N. N., Disease control by nonfungitoxic compounds, in *Modern Selective Fungicides,* 2nd ed., Lyr, H., Ed., Gustav Fischer Verlag, New York, 1995, chap. 25.
Smith, C., History of benzimidazole use and resistance, in *Fungicide Resistance in North America,* Delp, C. J., Ed., American Phytopathological Society Press, St. Paul, MN, 1988, 23.
Smith, C. M., Saunders, D. W., Allison, D. A., Johnson, L. E. B., Labit, B., Kendall, S. J., and Hollomon, D. W., Immunodiagnostic assay for cereal eyespot. Novel technology for disease detection, Proc. Brighton Crop Prot. Conf., *Pests Dis.,* 2, 763, 1990.
Staub, T., Fungicide resistance: practical experience with antiresistance strategies and the role of integrated use, *Annu. Rev. Phytopathol.,* 29, 421, 1991.
Staub, T., Early experiences with phenylamide resistance and lessons for continued use. *Fungicide Resistance, BCPC Monogr.,* 60, 131, 1994.
Staub, T. and Sozzi, D., Recent practical experiences with fungicide resistance. Proc. 10th International Congress, *Plant Prot.,* 2, 591, 1983.
Staub, T., Dahmen, H., Urech, P., and Schwinn, F., Failure to select for *in vivo* resistance in *Phytophthora infestans* to acylalanine fungicides, *Plant Dis. Rep.,* 63(5), 385, 1979.
Steffens, J. J. and Kleier, D. A., Computer aided discovery and optimization of fungicides, in *Modern Selective Fungicides,* 2nd ed., Lyr, H., Ed., Gustav Fischer Verlag, New York, 1995, chap. 24.
Stern, V. M., Smith, R. F., van den Bosch, R., and Hagen, K. S., The integrated control concept, *Hilgardia,* 29, 81, 1959.
Taylor, C. R., Economic impacts and environmental and food safety tradeoffs of pesticide use reduction on fruit and vegetables, American Farm Bureau Federation, 1995.
Trivellas, A. F., Benzimidazole resistance monitoring techniques and the use of monitoring studies to guide benomyl marketing, in *Fungicide Resistance in North America,* Delp, C. J., Ed., American Phytopathological Society Press, St. Paul, MN, 1988, 28.
Urbanchuk, J., The Value of Fungicides to the Availability of a Healthy and Affordable food Supply, *GRC Economics,* National Agrichemicals Association, Washington, D.C., Oct. 1989, 33.
Urbanchuk, J., The Value of Crop Protection Chemicals and Fertilizers to American Agriculture and the Consumer, *GRC Economics,* National Agrichemicals Association, Washington, D.C., Feb. 1990, 45.
Urech, P. A., Companion products for use in fungicide resistance strategies, in *Fungicide Resistance in North America,* Delp, C. J., Ed., American Phytopathological Society Press, St. Paul, MN, 1988, 95.
Urech, P. A., Fungicide resistance management: needs and success factors, *Fungicide Resistance, BCPC Monogr.,* 60, 349, 1994.
Van der Plank, J. E., *Plant Diseases: Epidemics and Control,* Academic Press, London, 1963, 349.

Wade, M. and Delp, C. J., The fungicide resistance action committee: an update on goals, strategies, and North American initiatives, in *Managing Resistance to Agrochemicals — From Fundamental Research to Practical Strategies,* Green, M. B., LeBaron, H. M., and Moberg, W. K., Eds., American Chemical Society, Washington, D.C., 1990, 320.

Willis, G. H. and McDowell, L. L., Pesticide persistence on foliage, *Rev. Environ. Contam. Toxicol.,* 100, 23, 1987.

Zilberman, D., Schmitz, A., Casterline, G., Licthenberg, E., and Siebert, J. B., The economics of pesticide use and regulation, *Science,* 253, 518, 1991.

Zziwa, M. C. N. and Burnett, F. J., The effect of reduced doses on the sensitivity of powdery mildew to fenpropimorph in barley field trials, *Fungicide Resistance, BCPC Monogr.,* 60, 303, 1994.

Part III
Biological Control

CHAPTER 4

Bacteria for Biological Control of Plant Diseases

Paul A. Backman, Mark Wilson, and John F. Murphy

I. INTRODUCTION

Bacteria have been added to crops on seed and in furrow for most of this century, though typically their early uses were for enhancement of fertility, and usually consisted of *Rhizobium* spp. intended to improve the nitrogen status of legumes. Development of biological control products has proceeded at a rather slow pace in Western countries due to the availability of and heavy reliance on chemicals for pest control. Early efforts in nonchemical pest control often originated from countries unable to afford the cost of establishing a chemical industry. In this chapter we have chosen to present not only the well-documented Western research, but also the less stringently researched but pioneering efforts centered in Russia and China. Researchers in both of these countries plus the Western countries have played key roles in developing bacteria for plant disease management.

1-8493-2627-3/97/$0.00+$.50

II. RHIZOSPHERE AND SEED-COLONIZING BACTERIA FOR SOIL-BORNE DISEASE CONTROL

A. Early Russian Work

In the 1950s, the Soviet Union introduced what were called "bacterial fertilizers". These pioneering studies largely utilized *Azotobacter chroococcum* and *Bacillus megaterium* (Menkina, 1950); their preparations were referred to as "azotobacterin" and "phosphobacterin", respectively. It was believed that if these bacteria could be introduced and maintained in the rhizosphere, the *Azotobacter* would fix atmospheric nitrogen while the *Bacillus* would mineralize organic phosphorus. This original hypothesis on the modes of action was disproven (Cooper, 1959), but crop growth was promoted nevertheless. By the late 1950s, millions of hectares of Soviet crops were being treated with these bacterial fertilizers with claims of 50 to 70% of total hectarage achieving yield increases of up to 20% (Cooper, 1959). Although many studies did not meet modern standards of statistical design and randomization, yield increases of up to 40% (Mishustin and Naumova, 1962) point to undeniable benefits. Another point to remember is that Soviet industry provided formulations and delivery systems to sustain the yield improvements reported. Soviet scientists suggested several hypotheses regarding the mechanisms responsible for yield increases. The first held that fungistatic as well as other biologically active substances produced by bacteria were responsible for the activity (Mishustin, 1963; Mishustin and Naumova, 1962; Samtsevich, 1962; Voznyakovskaya, 1963); the second again pointed to enhanced phosphorus availability as a mechanism of activity (Kvaratskheliya, 1962; Menkina, 1963); while a third hypothesis involved numerous complex interactions among the introduced bacteria, nitrogen cycle microorganisms, and cellulose decomposers (Samtsevich, 1962). Still later work showed that plants frequently exhibited hormonal effects following bacterial treatment that were indicated by increases in shoot numbers, tillers, plant height, etc. (Sundara et al., 1963). After reviews by Rubenchik (1963) and Brown (1974), increased research activity into root bacterization could be found in several Western countries. However, the level of research effort and adoption of biocontrol practices were delayed, because through the 1960s and 1970s the Western world was consumed with finding the right organic pesticide as a cure for almost any plant disease.

B. Early Progress in the West

Western science was not without some early advances in finding bacteria active in suppression of seed and soil-borne diseases. Dunleavy (1955) found that a *Bacillus subtilis* strain reduced damping-off of sugar beet. While in the early 1970s Broadbent et al. (1971, 1977) identified bacteria (particularly their *B. subtilis* strain A-13) that were quite effective in suppressing seed and seedling diseases under controlled conditions. Additional research in Australia pointed to the ability of *Bacillus* spp. to suppress disease while improving vigor and yield (Merriman et al., 1974, 1975). However, the variability in plant responses to A-13 applied under field conditions prevented its commercial development and research was discontinued in 1976. Research by Kommedahl and Mew (1975) found that *B. subtilis* applied to corn seed could increase yields 8.9% while reducing incidence of stalk rot caused by *Fusarium roseum*.

By the late 1970s, researchers in California launched a major effort to study bacteria as biological control agents, and combined ecological studies of applied bacteria with quantitative estimates of disease to indicate the relationships of successful colonization of the roots (typically with *Pseudomonas* spp.) to the suppression of pathogens and deleterious rhizobacteria (Kloepper et al., 1980; Suslow and Schroth, 1982; Weller, 1988). They also observed that plant growth could be promoted in the absence of disease suppression, and labeled the bacteria producing this response "plant growth-promoting rhizobacteria" or PGPR (Kloepper et al., 1980). Research groups in the USDA found that pseudomonads could be isolated that were extremely effective rhizosphere colonizers, and that could suppress important diseases of cereals such as *Gaeumannomyces tritici* and *Rhizoctonia solani* (Cook, 1994; Weller, 1988). Their research, and also those of others, indicated that antibiosis to pathogens (Fravel, 1988) and competition for iron were key mechanisms for successful biological control (Cook, 1994; Weller, 1988). Typically, the pseudomonads were better root colonizers than most *Bacillus* spp. and also had the ability to grow at lower soil temperatures. However, isolates were quite specific as to the hosts colonized and/or pathogens suppressed (Weller, 1988). An additional key problem was the environmental sensitivity of *Pseudomonas* spp., which do not form endospores like *Bacillus* spp. This latter problem has largely prevented commercialization of superior colonizing biocontrol pseudomonads, because commercial formulations could not guarantee delivery of a viable product to the field or provide a reasonable shelf life (Becker and Schwinn, 1993).

C. Large-Scale Commercialization of *Bacillus*

In addition to the commercialization of *Bacillus* in China and Russia, in the early 1980s Gustafson, Inc., Plano, Texas initiated field trials with the *B. subtilis* A-13 strain of Broadbent's (Broadbent et al., 1977) that had been abandoned some years earlier. Their initial field trials also showed erratic performance when delivered as a seed-applied product, but they had the foresight to attempt combination treatments with seed-treatment fungicides (Backman et al., 1994; Turner and Backman, 1991). The results of these experiments indicated that not only did this endospore-forming *Bacillus* tolerate the seed-treatment fungicides, but it provided enhanced plant growth as well as biological control of root pathogens of peanuts. Though efficient season-long colonization of peanut roots was accomplished only for the upper taproot and nearby branch roots, yields were found to increase on a predictable basis to levels that merited commercial development (Turner and Backman, 1991).

The markets for the A-13 strain and related strains GB03 and GB07 (sold under the trade names Quantum®, Kodiak®, and Epic®, respectively, by Gustafson, Inc.) have continued to increase since their first introduction in 1985. Here the marketing strategy has been to address season-long root health, reducing fungal pathogen damage caused after the seed-treatment fungicides have dissipated (Backman et al., 1994). For the 1996 season, approximately 5 million ha of crops were treated with these strains, targeting diseases of roots caused by *Rhizoctonia, Pythium,* and *Fusarium,* plus promoting increased root mass and plant vigor through hormone-like responses and disease control (Backman et al., 1994; Brannen and Backman, 1994). The ability of these strains to adapt to environmental stresses is evidenced by successful colonization of cotton roots even when applied to seed with acid residues from the delinting process (seed surface pH levels 2.0 to 3.5) (Mahaffee and Backman, 1993). At the present time, 60 to 75% of the cotton crop is treated with *B. subtilis* bacteria, with additional U.S. registrations granted for at least seven other crops including both dicots and monocots. Approximately two-thirds of the cotton that was treated with *B. subtilis* was treated with a combination of two *B. subtilis* strains (GB03 plus GB07, the rest of the acreage receiving GB03 as the sole biocontrol agent), intended to improve cotton response by the addition of the GB07 (personal communication, D. Kenney, Gustafson, Inc.). Marketing is also beginning in Mexico as well, where potatoes are treated with the A-13 Broadbent strain (Bio-Raíz™) in combination with fungicides. Here, recommended application methods are as a tuber immersion dip, tuber spray, or in-furrow spray, resulting in yield increases of 10% or greater above those found in plots receiving the fungicide only (Anonymous, Bio-Raíz™, Industrias Gustafson, Mexico, 1996).

Beginning about 1980 and continuing through the present, China through its Beijing Agricultural University (now China Agricultural University) has developed many yield-increasing bacteria (YIB) that are presently being sold commercially. These bacteria are generally applied on the seed or, often, on seed plus additional sprays over the plant after emergence (Chen et al., 1996). These bacteria largely colonize the roots and affect soil-borne diseases, but there is some evidence that they also colonize above-ground plant parts and provide control of foliar diseases (Mei, 1989). As itemized in the previous citation, there are more than 18 commercial strains or strain mixtures sold, with almost all of these strains derived from the spore-forming genus *Bacillus.* Chinese researchers have reported that YIB efficacy was more variable among *Pseudomonas* than among

Bacillus strains when strains were used in geographical areas outside the area from which they were isolated (Chen et al., 1996). Commercial usage of YIB strains has covered more than 20 million ha, with increases averaging 10% in vegetables but as high as 22.5% for sweet potato (Chen et al., 1996).

D. Further Development of Rhizosphere Colonization

Although the genus *Bacillus* represents the most utilized biocontrol agents delivered on crop seed, or in furrow, this has not occurred because of inherent superiority in biological control when compared to other genera of biocontrol bacteria. Rather, it has occurred because of the resistant endospore that *Bacillus* spp. produce. The endospore provides tolerance to heat and cold, as well as to pH extremes, pesticides, fertilizers, and storage duration. The presence of the endospore has practically assured that if a viable spore leaves the production facility, a viable spore will arrive with the seed when planting finally occurs. Although numerous laboratories have identified superior biocontrol strains from the genera *Enterobacter* (Axelrood et al., 1988; Costa and Loper, 1994; Nelson, 1988), *Pseudomonas* (Cook, 1994; Frey et al., 1994; Leeman et al., 1995; Xu and Gross, 1986a,b), and *Burkholderia* (Hebbar et al., 1992; Huang et al., 1993; Smith, 1994), the ability to deliver viable inocula 6 months to a year later into multiple environments has often been delayed pending the development of formulation technologies that will overcome these organisms' environmental sensitivities. However, there are some new entries into biocontrol markets including a *Burkholderia (Pseudomonas) cepacia* (Deny®, marketed by CCT Corp., Carlsbad, CA) (Wilson and Backman, 1997), a freeze-dried *Streptomyces* sp. primarily targeted to *Fusarium* control (marketed as Mycostop® by Kemira Agro Oy, Finland (Mohammadi, 1994; Mohammadi and Lahdenpera, 1994), and a *Pseudomonas fluorescens* strain WCS374 applied as a seed coating to radish and carnation (Biocoat® by the S&G division of Sandoz Seed [The Netherlands]) for control of *Fusarium* (Leeman et al., 1995a,b).

Presently, the greatest commercial usages of rhizosphere-colonizing bacterial biocontrol agents are in combination with seed-treatment chemical pesticides. Again, these are largely *Bacillus* spp., and the fungicides are still necessary because *Bacillus* spp. are too slow in activating and effectively colonizing the seed and emerging root to deal with fast-growing pathogens such as *Pythium* spp. The bacteria are added to suppress diseases after the fungicides have dissipated, usually occurring more than two weeks after planting. The pathogens addressed by this strategy are largely fungi causing chronic midseason diseases of the roots such as *Fusarium, Rhizoctonia, Pythium, Thielaviopsis*, and others that are very difficult to control with chemical strategies. Additional benefits are likely developing from competition and suppression of deleterious rhizobacteria (Suslow and Schroth, 1982) and from stimulation of root mass and *Rhizobium* nodulation (Turner and Backman, 1991). Since few pesticide applications are presently being replaced, environmental benefits through pesticide use reductions may have to await commercialization of nonspore-forming genera that are more likely to stand alone without the support of fungicides. Research is beginning on the use of combinations of bacteria to improve plant responses and disease control (Pierson and Weller, 1994). The logic here is that combinations of bacteria will provide improved adaptability to cultivars, soils, multiple pathogens, environments, etc. Recent research by Sheng (1996) indicated that two-member mixtures of *B. subtilis* strains provided lower coefficients of variation among soybean cultivars while providing higher levels of colonization of the root systems and improved root mass and seedling stands over those reported for the individual members. However, optimal ratios of the two members were different among cultivars. Combinations join formulation improvements as one of the obvious research areas through which plant pathologists can improve levels of disease control using rhizosphere-colonizing organisms. An additional area for improvement will likely occur through genetic manipulations to add genes, or by up-regulation of beneficial genes, thus enhancing the spectrum of activity or the level of control (Liu et al., 1995). The pest control industry would further benefit because it is easier to patent and maintain proprietary control of microorganisms that have been genetically altered, making them more attractive for commercialization.

E. Biological Control of Crown Gall

The earliest success story in biological control of plant diseases occurred in Australia using *Agrobacterium radiobacter* for suppression of crown gall disease caused by *Agrobacterium tumefaciens* (Campbell, 1994; Clare, 1993; Farrand, 1990; Moore and Warren, 1979). *A. radiobacter* strains were isolated from symptomless plants growing in highly diseased fields. These nonpathogenic bacteria were found to produce bacteriocins called "agrocins" that are encoded by a plasmid and are responsible for antagonism to the pathogenic *A. tumefaciens* that appear to compete for the same niche on the rhizoplane. The commercialization of these strains has been one of the major success stories of biological control. In 1973 *A. radiobacter* strain K84 was commercialized in Australia and others have been sold since under the trade names Galltrol, Norbac 84-C, and Diegall (Clare, 1993; Lumsden et al., 1995). For protection of crops, transplant roots were dipped into a suspension of the antagonist strain, effectively colonizing the root and protecting it from the pathogen. However, there were reported occurences of a breakdown in efficacy of biocontrol that indicated some strains were becoming insensitive to agrocin-based antagonism (Jones and Kerr, 1989). The same plasmid that encodes for agrocin production also contains genes that confer resistance to agrocin and control conjugal transfer (Tra). Jones and Kerr (1989) developed a Traminus derivative of K84, designated K1026, that was unable to conjugate with wild-type *Agrobacterium* species, and thus could not transfer resistance to agrocins. This engineered strain was registered in Australia in 1991, and marketed under the trade name Nogall (Clare, 1993; Lumsden et al., 1995). The marketing of strain K1026 represented the earliest commercial release of a genetically engineered strain of a bacterium used for biological control of a plant disease.

III. BACTERIA FOR SUPPRESSION OF FOLIAGE AND POSTHARVEST DISEASES

A. The Leaf Environment

The microclimate of the leaf is in constant change; particularly variable are relative humidity, temperature, leaf wetness, and ultraviolet radiation. These factors also change depending on leaf position within the plant canopy. Organisms on the leaf are frequently in direct competition both for very limited nutrients and for shelter from UV and infrared radiation which causes heating and drying of the leaf surface (Sutton, 1995). Leaf microflora tend to cluster along anticlinal walls of epidermal cells and around trichomes — probably regions in which leaching of nutrients is higher. An additional environmental stress is the diurnal change from nights of ample moisture produced by dew, which are quickly replaced by desiccating days with resulting osmotic shock as solutes become more concentrated. Addition of a new microorganism to the leaf surface will find that this organism must compete with organisms already occupying the most hospitable locations on the leaf, which magnifies the challenge of it growing to sufficient levels to suppress the invading pathogen (Andrews, 1990, 1992; Blakeman and Fokkema, 1982; Elad, 1993).

B. Biological Control of Foliage Disease

It is not unexpected that introduction of a single antagonistic strain of biocontrol bacteria to this environment has been largely unsuccessful, but there are some recent exceptions. *Pseudomonas fluorescens* A506 is sold as BlightBan™ A506 for the control of *Erwinia amylovora* fire blight of apple and pear (Wilson and Backman, 1997; Lindow, 1992). The formulation is spray-applied early in the bloom period and supported with subsequent applications to colonize newly emerging, largely noncolonized blossoms. The A506 strain preemptively utilizes nutrients and other resources in the blossom environment that are required for the development of pathogenic *E. amylovora* populations (Lindow, 1992; Wilson et al., 1990; Wilson and Lindow, 1993). A novel alternative application method is to force honey bees to walk through a tray of formulated bacteria as they leave their hive, and thus they disseminate the bacterium to the blossoms that they visit (Sutton and Peng, 1993; Thomson et al., 1992). A similar preemptive colonization strategy has been utilized with *Pantoea agglomerans* (syn. *Erwinia herbicola*) for control of fire blight in the same crops (Epton et al., 1994; Johnson et al., 1993). Recently, research began on combinations of A506 and *P. agglomerans* to improve field performance by utilizing bacteria with complementary mechanisms and/or colonization sites (McLaughlin and Roberts, 1992; Stockwell et al., 1993). This latter work was more recently expanded to utilize streptomycin-resistant *P. agglomerans* strains so that biological tactics could be integrated with the antibiotic spray program common in most orchards. Similarly, Cooksey (1988) evaluated a nonpathogenic (Tn5 mutant) copper-resistant strain of *Pseudomonas syringae pv. tomato* in combination with a copper-based spray to control the pathogenic race (cause of tomato speck) of the same pathovar in tomato. Here, the copper product eliminated copper-sensitive competitor organisms from many of the colonization sites on the leaf surface, making them available to the nonpathogenic race, which premptively colonized the open niches to the exclusion of the pathogenic race.

Bacillus has been one of the key genera used for biological control on the leaf surface. As was observed for the rhizosphere-inoculating organisms, and those evaluated for postharvest disease control as well, this genus of bacteria shows great potential since it is much easier to formulate and is relatively promiscuous in its colonization of multiple plant species. For instance, biological control of bean rust has been reported using *B. subtilis* (Baker et al., 1985), and onion rust using *Bacillus cereus* (Doherty and Preece, 1978). *B. cereus* has also been used for biological control of *Alternaria* on tomato (Ploper et al., 1993; Stevens et al., 1996), as well as *Cercospora* on peanut (Kokalis-Burelle et al., 1992).

Since nutrients are a limiting resource in the phyllosphere, researchers have attempted to enhance colonization of the leaf surface by providing applied antagonists with a food source that will sustain high population levels of the beneficial bacteria. Sugars and media products have often been evaluated as likely food sources, but typically led to rapid rises in general populations on the leaf surface (without particular enhancement of the applied bacterium), followed soon after by a rapid fall in the epiphytic populations as the nutrients were exhausted (Andrews, 1992). These added food products were also susceptible to removal by rain or irrigation water, making their effects even more transient.

Recently, Kokalis-Burelle et al. (1992) found that application of a biocontrol bacterium to peanuts, supported by the addition of a selectively available, relatively insoluble polymer (chitin) plus a sticking agent, led to higher populations of the chitinolytic bacterium, and improved control of leafspot disease. Holding capacity of the leaf surface was generally enhanced about one-half log, while the postapplication decline in populations of the applied bacterium were also reduced. One additional contribution of the chitin on the leaf surface was that populations of indigenous chitinolytic bacteria changed from <1% of total microflora to ~40% (added bacterium plus indigenous) of total microflora. In this latter situation, the chitin was found to have supported a chitinolytic population greater than the total population of the nonamended leaves. Recent research has integrated this leaf amendment technology into a biointensive IPM program for vegetables (Ploper et al., 1993; Stevens et al., 1996, 1993), while it has also been successfully adapted to suppression of sugarbeet foliage diseases (B. J. Jacobsen, personal communication). Work is just beginning on evaluation of genetically manipulated organisms for biocontrol of foliage diseases (Fahey et al., 1992).

C. Bacteria for Control of Postharvest Diseases

Fruits and vegetables are highly perishable while still maintaining active metabolism long after harvest. Storage time is reduced by infection with pathogens which cause substantial postharvest losses (Arul, 1994). Postharvest losses are not only important to the producer, but persist through the distribution chain affecting the cost and acceptability of the marketed produce, as well as the storability of the product after the consumer has purchased it. While in the marketing chain, the applied biological control agent may be expected to perform in a cold storage and/or controlled atmosphere environments, and then when purchased by consumers, they may also be expected to function in a warm environment as the ripening process is accelerated. Key infection sites for postharvest diseases are wounds created by harvesting and handling, though some pathogens may directly penetrate. The difficult task in finding bacteria for biological control of postharvest diseases is to find organisms that can function effectively under these rather variable conditions. Present practices of mechanical harvesting and bulk handling have aggravated the need for control measures. Typically, these needs have been met by the use of organic fungicides applied directly onto the stored product. While this strategy is highly successful, it exposes the consumer to potentially toxic or carcinogenic residues, and an additional problem arises where target organisms frequently become resistant. Clearly, the need for biological controls for postharvest diseases has never been greater (Wilson and Wisniewski, 1989).

Over the past 15 years, numerous investigations have been initiated to find biological control agents for suppression and control of postharvest disease. One of the first successes was reported by Pusey et al. (1984, 1986, 1988) who isolated a *B. subtilis* strain B-3 that could suppress brown rot of peach caused by *Monilinia fructicola*. This isolate was found to produce antibiotic substances (Gueldner et al., 1988) that were active in suppression of *M. fructicola*, but inactive against *Rhizopus stolonifer*, which causes an equally important fruit rot of peach. Soon after, Janisiewicz and Roitmann (1988) reported biological control of blue mold and gray mold of apple and pear with an isolate of *Pseudomonas (Burkholderia) cepacia* and later with an isolate of *P. syringae* (Janisiewicz and Marchi, 1992). Smilanick and Denis-Arrue (1992) similarly reported a *Pseudomonas* sp. capable of suppressing green mold of lemons. Strategies for biocontrol here were similar to those for other biological control agents, namely, the preemptive colonization of new wound sites (created during harvesting and packing) with an antagonist population of bacteria that would exclude pathogen populations through both antibiosis and sequestration of resources (Smilanick, 1994). Selection criteria for potential biocontrol agents have been developed from different conceptual bases. While many investigators select from colonists of fruit or vegetable surfaces, particularly from produce where no disease was found, recent research has indicated that the best organisms for biocontrol might be found after several months of storage, or during cool periods of the season (Smilanick, 1994; Stretch, 1989). The logic here is that the conditions of the storage environment are so different from the growth environment that successful biocontrol agents would far less likely be found in this latter environment. It is also obvious that the environment of the wound may be much different from that of the undamaged surface of the produce, and that superior colonists of wounds would more likely be found in those wounds that remain pathogen-free after an extended time in the storage environment.

Wilson and Wisniewski (1994) list nine criteria for the ideal antagonist. These include:

1. Genetic stability
2. High, consistent efficacy
3. Ability to survive under adverse environmental conditions
4. Effectiveness against a wide range of pathogens on a variety of fruits and vegetables
5. Amenability for growth on an inexpensive medium in fermentors
6. Stability of the end product during storage
7. Nonproduction of secondary metabolites that might be toxic to humans
8. Strain resistant to standard fungicides
9. Compatibility with other chemical and physical treatments of the commodity such as heating and waxing.

Needless to say, some of these criteria have been the proverbial needle-in-the-haystack that has prevented the development of numerous biocontrol agents. Particularly, criteria 3 and 6, which have excluded many of the promising nonspore-forming bacteria.

The most problematic infections to control with antagonists are the latent infections (nonvisible) and inactive infections (visible) (Smilanick, 1994). Both types can erupt into destructive phases long after the application of antagonistic bacteria. In order to deal with this problem and to enhance general biological control, Elad and Zimand (1992) demonstrated the utility of an integrated chemical-biological strategy, while El Ghaouth (1994) demonstrated the ability of chitosan used alone and in conjuction with biocontrol bacteria to induce plant resistance mechanisms to complement the more traditional effects of biocontrol bacteria used alone.

IV. ENDOPHYTIC BACTERIA FOR BIOLOGICAL CONTROL OF PLANT DISEASES

Bacteria have been reported in almost all internal plant organs, with a long debate as to whether they were really there or if they were evidence of a poor sterile technique by the investigator. Of late, as the evidence has mounted, this debate has subsided and research has been directed to studies on the role these organisms might play in plant health. Bird et al. (1980, 1981) evaluated multiadversity-resistant cotton cultivars (MAR cultivars) and found that high levels of resistance in these cultivars was related to both endophytic and epiphytic bacteria. Misaghi and Donndelinger (1990) similarly reported endophytes in symptom-free cotton plants, followed by McInroy and Kloepper (1995a,b) who isolated, quantified (finding log 4 to log 6 in tissues), and identified these cotton endophytes.

Kloepper et al. (1992) demonstrated that when endophytes were applied to seeds or roots of cucumbers, five of the six strains evaluated would later be found in internal tissues, and that these would induce systemic resistance to plant diseases. Chen et al. (1995) reported suppression of Fusarium wilt of cotton by five different genera of endophytic bacteria. Work continues to develop a reliable delivery system for endophytes (Musson and Kloepper, 1994). Expanding this technology to develop a delivery system, two reports demonstrate that if pathogens of other hosts were introduced through natural plant openings into substomatal cavities and/or hydothodal pores using a polysilicone adjuvant, the plant would respond by producing high levels of defense proteins that would eventually eliminate the inducing agent, but at the same time protect the plant from true pathogens (Burkett et al., 1996; Jetiyanon et al., 1993). Again, the presence of microorganisms inside plant tissues induced the plant to protect itself. Although most inducing organisms appear to have an endophytic phase, there are numerous examples of endophytes that do not result in ISR. These data point to the potential utility for endophytes and even nonendophytes introduced into internal plant tissues, but research should continue to develop improved selection criteria, delivery systems, and formulations to maximize the plant's responses. Preliminary research reported from China (Chen et al., 1996; Mei, 1989) indicates that some of the seed-applied YIB strains marketed in China may in fact have an endophytic habitat after application, since high populations were recovered from surface-disinfested cotyledons and leaves. This latter research, coupled with the observations of Kloepper's group (1992), indicates the potential for a significant subset of root-inhabiting bacteria to become endophytic.

V. BACTERIA AS INDUCERS OF NATURAL PLANT DEFENSE MECHANISMS

Induced systemic resistance (ISR) of plants to various pathogens has been recognized for over 63 years, when Chester (1933) proposed the term "acquired physiological immunity." Inducers of resistance include pathogens, heat-killed or attenuated pathogens, pathogens of other hosts, wounds, and certain chemicals. Although a large number of plant species have been reported to be induced and thus to be resistant to a broad range of pathogens, there has been great difficulty in reducing this obviously important phenomenon down to agronomic practices that could provide for control of plant diseases in the field. Typically, ISR has been initiated in laboratory settings by either infecting lower leaves with a pathogen and detecting resistance to the same pathogen or other pathogens in the upper leaves (classical ISR), or alternatively, by wounding or injecting the plant with an attenuated pathogen, pathogen of another host, or sometimes a nonpathogen, to initiate the ISR reaction (Tuzun and Kloepper, 1995). Needless to say, to treat lower leaves with pathogens, or to wound inoculate all the plants in a field or greenhouse, is not a viable agronomic practice. Almost simultaneously in 1991 three separate groups reported (Alström, 1991; Van Peer et al., 1991; Wei et al., 1991) that selected rhizobacteria in plants could induce resistance to above-ground diseases, and the door opened to practical applications of this phenomenon. Thereafter, there were three reports by Liu et al. (1992; 1995a,b) and one by Chen et al. (1995) that demonstrated the broad range of pathogens affected by ISR organism, and the agronomic importance of this technology. Soon thereafter the polysilicone adjuvant-mediated delivery system (discussed in the previous section) provided a delivery system for this emerging technology (Burkett et al., 1996; Jetiyanon et al., 1993).

Salicylic acid has previously been found to be produced after an infection by a pathogen, and thereafter this chemical causes ISR products (pathogenesis-related proteins, phytoalexins, etc.) to accumulate (Gaffney et al., 1993), but salicylic acid itself was too rapidly degraded for use in agriculture. Ciba-Geigy Ltd., Basel has developed a nontoxic chemical activator (benzothiadiazole derivative, Actigard® or CGA-245704) of what they have called "systemic activated resistance" that functions to control both fungal and bacterial diseases by using the host plant's natural resistance mechanisms. Particularly, the Actigard product mimics the signal that occurs after a local infection, causing salicylic acid to be produced throughout the plant, which in turn results in generalized induction of ISR (Kessmann et al., 1996). Though not biological control in the strict sense, it shows how biocontrol information is being used to produce commercial chemicals.

Virus diseases of plants can also be suppressed by plant-induced resistance mechanisms. Mann (1968) used cultures of *Bacillus uniflagellatus* and extracts from such cultures to drench tobacco roots in an attempt to induce resistance against TMV. Treatment of plants with *B. uniflagellatus* cultures or with extracts from *B. uniflagellatus* resulted in a significant reduction in lesion numbers upon inoculation with TMV. A reduction in TMV-induced lesion number was observed but effects on TMV in a host that allowed systemic infection were not as obvious (Mann, 1968). Maurhofer et al. (1994) examined *P. fluorescens* strain CHA0 and two mutants derived from CHA0 for their ability to induce resistance against tobacco necrosis virus (TNV) in tobacco. CHA0 induced PR protein accumulation and resistance to TNV to a level similar to that observed using a classical induced systemic resistance approach with local lesion induction by TNV. Based on these and other findings, this group concluded that the mechansims for biological control of root diseases were different from the mechanisms suppressing foliage disease.

Bergstrom et al. (1982) used *Pseudomonas syringae* pv. *lachrymans* (a pathogen of cucumber) in an attempt to induce systemic resistance against cucumber mosaic cucumovirus (CMV) in cucumber. The bacterial preparation was rub-inoculated onto the first two leaves of young cucumber plants, followed seven days later by rub-inoculation of CMV onto the third leaf. Plants treated with *P. lachrymans* showed a reduction in the number of CMV-induced lesions on the inoculated leaf and a 3- to 4-day delay in the appearance of systemic symptoms, relative to controls which consisted of untreated plants or plants rub-inoculated with water. These findings suggest that if the CMV-induced lesions observed on inoculated leaves correlate with infection foci, then the decrease in lesion number observed on the bacteria-treated plants might in turn result in a delay in the onset of systemic symptoms.

Two PGPR strains, 89B-27 *(P. fluorescens)* and 90-166 *(Serratia marcescens),* shown previously to induce systemic resistance (Wei et al., 1991) against bacterial and fungal pathogens, were evaluated for their ability to induce resistance in cucumber against infection and systemic invasion by CMV (Raupach et al., 1996). In the latter study, some plants in each of the PGPR treatments remained symptomless throughout the course of the experiment and were shown by immunoassay to be free of viral antigen.

VI. CONCLUSIONS

As this review has demonstrated, bacteria inhabit all areas of the plant, both internal and external, including flowers, fruits, leaves, and roots. It is probable that many of these play roles in plant growth and survival, even when they occur at natural levels. The research we have reviewed here points directly to the fact that some of these natural plant-inhabiting bacteria provide greater benefits than others. The trick has been to find the most beneficial members of the plant microflora by developing and refining isolation techniques and screening procedures. The evolution of this area of science points to the probability that we may be getting away from single organisms and moving to the use of two- or three-component mixtures and eventually to multimember communities where screening is less important and community modification is more important. Controversies continue as to whether the best criteria for screening are the production of antibiotic products, and/or the ability to colonize, and/or the ability to colonize multiple hosts, and/or the ability to formulate a stable product with prolonged shelf life, etc. The facts definitely point to all of these being important, but there is little doubt that aggressive and consistent colonization is a primary predictor of usefulness. Even in the absence of antibiosis, the ubiquitous colonizer can be modified through genetic engineering (Fahey et al., 1992; Liu et al., 1995) to be a useful tool in suppressing plant diseases. Further improvements can be made by extending screening procedures to find compatible and hopefully synergistic mixtures of organisms that can extend the usefulness of the product to encompass more hosts, target pests, environmental adaptations, and more. What we are presenting here is the early chapter of a book that has only just begun to be written. Present development of this science is constrained by lack of government funds, and by the fact that our agricultural chemical industry does not perceive a need to change from the development and sale of the agricultural chemicals that it has so effectively produced for years. The United States Congress, Office of Technology Assessment (1995) produced a report identifying many of the reasons, mentioned above and below, for the slow progress and adoption of biologically based technologies for pest control.

The typical university or government researcher screens less than 100 isolates to find the organism that demonstrates the principles of his model system. This needs only be compared to the tens of thousands of chemical compounds screened for pest control potential for every one that moves to commercialization to indicate how rudimentary our level of screening efforts really are. The present large-scale commercialization of bacteria for biological control of plant diseases has occurred where there are no chemical products working effectively, or as an adjunct to chemical control where the biological agent is actually applied in combination with pesticides. Alternatively, they have been developed in countries where there has been no resident chemical industry, and little hard money to import pest control chemicals. The reason that the large multinational companies have not moved to find these "best" bacteria is that they are after all largely chemical companies, and their own inertia compels them to do what they do best until something happens to force change. The force most likely to provide this in the future is regulatory changes that might severely restrict chemical use. The first signs are beginning to turn up in Europe, Asia, and the U.S., such as restricting total pesticide use, promoting ecologically based IPM, banning pesticides on crops destined for baby foods, etc. Perhaps the marshaling of these various forces will continue, and the imperative for change to biologically based pest control will become more compelling. If so, the research presented here may be the precursor to a change in the pest control paradigm. If not, scientists will continue to refine their model systems, and adoption of bacteria for control of plant diseases will proceed, but at a slow pace.

REFERENCES

Alström, S., Induction of disease resistance in common bean susceptible to halo blight bacterial pathogen after seed bacterization with rhizosphere pseudomonads, *J. Gen. Appl. Microbiol.*, 37, 495, 1991.

Andrews, J. H., Biological control in the phyllosphere: realistic goal or false hope? *Can. J. Plant Pathol.*, 12, 300, 1990.

Andrews, J. H., Biological control in the phyllosphere, *Annu. Rev. Phytopathol.*, 30, 603, 1992.

Arul, J., Emerging technologies for the control of postharvest diseases of fresh fruits and vegetables, in *Biological Control of Postharvest Diseases: Theory and Practice*, Wilson, C. L. and Wisniewski, M. E., Eds., CRC Press, Boca Raton, FL, 1994.

Axelrood, P. E., Rella, M., and Schroth, M. N., Role of antibiosis in competition of *Erwinia* strains in potato infection courts, *Appl. Environ. Microbiol.*, 54, 1222, 1988.

Backman, P. A., Brannen, P. M., and Mahaffee, W. F., Plant response and disease control following seed inoculation with *Bacillus subtilis*, in *Improving Plant Productivity with Rhizosphere Bacteria*. Ryder, M. H., Stephens, P. M., and Bowen, G. D., Ed., CSIRO, Adelaide, Australia, 1994, 3.

Baker, C. J., Stavely, J. R., and Mock, N., Biocontrol of bean rust by *Bacillus subtilis* under field conditions, *Plant Dis.*, 69, 770, 1985.

Becker, O. J. and Schwinn, F. J., Control of soil-borne pathogens with living bacteria and fungi, Status and outlook, *Pestic. Sci.*, 37, 355, 1993.

Bergstrom, G. C., Johnson, M. C., and Kuc, J., Effects of local infection of cucumber by *Colletotrichum lagenarium, Pseudomonas lachrymans* or tobacco necrosis virus on systemic resistance to cucumber mosaic virus, *Phytopathology*, 72, 922, 1982.

Bird, L. S., Liverman, C., Thaxton, P., and Percy, R. G., Evidence that microorganisms in and on tissures have a role in a mechanism of multi-adversity resistance in cotton: theory and results, Cotton Disease Council 40, 283, Proc. Res. Conf. Natl. Cotton Council Am., Memphis, TN, 1980.

Bird, L. S., Liverman, C., Lazo, G. R, Thaxton, P., Percy, R. G., and Benedict J. H., Do plants have genetics systems controlling selective colonization of tissues by symbiotic organisms which function in mechanisms of resistance to adversities?, Cotton Disease Council 41, 34, Proc. Res Conf., Natl. Cotton Council Am., Memphis, TN, 1981.

Blakeman, J. P. and Fokkema, N. J., Potential for biocontrol of plant diseases on the phylloplane, *Annu. Rev. Phytopathol.*, 20, 167, 1982.

Brannen, P. M. and Backman P. A., Suppression of Fusarium wilt of cotton with *Bacillus subtilis* hopper box formulations, in *Improving Plant Productivity with Rhizosphere Bacteria*. Ryder, M. H., Stephens, P. M., and Bowen, G. D., Eds., CSIRO, Adelaide, Australia, 1994.

Broadbent, P., Baker, K. F., Franks, N., and Holland, J., Effect of *Bacillus* spp. on increased growth of seedlings in steamed and nontreated soil, *Phytopathology*, 67, 1027, 1977.

Broadbent, P., Baker, K. F., and Waterworth, Y., Bacteria and actinomycetes antagonistic to root pathogens in Australian soils, *Aust. J. Biol. Sci.*, 24, 925, 1971.

Brown, M.E., Seed and root bacterization, *Annu. Rev. Phytopathol.*, 12, 181, 1974.

Burkett, J., Jetiyanon K., Backman, P. A., and Tuzun, S., Field evaluation and physiological characterization of induced resistance to control black rot of crucifers, *Proc. Am. Phytopathol. Soc.*, 435A-2, 1996 (Abstr.).

Campbell, R., Biological control of soil-borne diseases: some present problems and different approaches, *Crop Prot.*, 13, 4, 1994.

Chen, C., Bauske E. M., Musson G., Rodriguez-Kabana, R., and Kloepper, J. W., Biological control of Fusarium wilt on cotton by use of endophytic bacteria, *Biol. Control*, 5, 83, 1995.

Chen, Y., Mei, R., Lu, S., Liu, L., and Kloepper, J. W., The use of yield increasing bacteria (YIB) as plant growth-promoting rhizobacteria in Chinese agriculture, in *Management of Soil Borne Diseases*, Utkhede, R. S. and Gupta, V. K., Eds., Kalyani Publishers, New Delhi, India, 1996, 165.

Chester, K., The problem of acquired physiological immunity in plants, *Q. Rev. Biol.*, 8, 129, 1933.

Clare, B. G., *Agrobacterium*: biological plant disease control, in *Advanced Engineered Pesticides*, Kim, L., Ed., Marcel Dekker, New York, 1993, 129.

Cook, R. J., Problems and progress in the biological control of wheat take-all, *Plant Pathol.*, 43, 429, 1994.

Cooksey, D. A., Reduction of infection by *Pseudomonas syringae* pv. *tomato* using a non-pathogenic, copper-resistant strain combined with a copper bactericide, *Phytopathology*, 78, 601, 1988.

Cooper, R., Bacterial fertilizers in the Soviet Union, *Soils Fert*, 22, 327, 1959.

Costa, J. M. and Loper, J. E., Derivation of mutants of *Erwinia carotovora* subsp. *betavasculorum* deficient in export of pectolytic enzymes with potential for biological control of potato soft rot, *Appl. Environ. Microbiol.*, 60, 2278, 1994.

Dimock, M., Turner, J., and Lampel, J., Endophytic microorganisms for delivery of genetically engineered microbial pesticides in plants, in *Advanced Engineered Pesticides*, Kim, L., Ed., Marcel Dekker, New York, 1994, 85.

Doherty, M. A. and Preece T. F., *Bacillus cereus* prevents germination of uredospores of *Puccinia allii* and the development of rust disease of leek, *Allium parium*, in controlled environments, *Physiol. Plant Pathol.*, 12, 123, 1978.

Dunleavy, J., Control of damping-off of sugar beet by *Bacillus subtilis, Phytopathology,* 45, 252, 1955.

El Ghaouth, A., Manipulaion of defense systems with elicitors to control postharvest diseases, in *Biological Control of Postharvest Diseases: Theory and Practice,* Wilson, C. L. and Wisniewski, M. E., Eds., CRC Press, Boca Raton, FL, 1994.

Elad, Y., Microbial suppression of infection by foliar plant pathogens, *IOBC Bull.* 16, 3, 1993.

Elad, Y. and Zimand, G., Integration of biological and chemical control for grey mould, in *Recent Advances in Botrytis Research,* Verhoeff, K., Malathrakis, N. E., and Williamson, B., Eds., Pudoc Scientific, Wageningen, 1992, 272.

Epton, H., Wilson, A. S. M., Nicholson, S., and Sigee, D. C., Biological control of *Erwinia amylovora* with *Erwinia herbicola,* in *Ecology of Plant Pathogens,* Blakeman, J. P. and Williamson, B., Eds., CAB International, Surrey, U.K., 1994, 335.

Fahey, J. W., Dimock, M. B., Tomasino S. F., Taylor, J. M., and Carlson, P., Genetically engineered endophytes as biocontrol agents: a case study from industry, in *Microbial Ecology of Leaves.* Hirano, S. S. and Andrews, J. H., Eds., Springer-Verlag, New York, 1992, 401.

Farrand, S. K., *Agrobacterium radiobacter* strain K84: a model biocontrol system, in *New Directions in Biological Control: Alternatives for Suppressing Agricultural Pests and Diseases,* Baker, R. and Dunn, P. E., Eds., Alan R. Liss, New York, 1990, 679.

Fravel, D. R., The role of antibiosis in the biocontrol of plant diseases, *Annu. Rev. Phytopathol.,* 26, 75, 1988.

Frey, P., Prior, P., Marie, C., Kotoujansky, A., Trigalet-Demery, D., and Trigale, A. T., Hrp mutants of *Pseudomonas solanacearum* as potential biocontrol agents of tomato bacterial wilt, *Appl. Environ. Microsc.,* 60, 3175, 1994.

Gaffney, T., Friedrich, L., Vernooij, B., Negretto, D., Nye, G., Uknes, S., Ward, E., Kessmann, H., and Ryals, J., Requirement of salicylic acid for the induction of systemic acquired resistance, *Science,* 261, 754, 1993.

Gueldner, R. C., Reilly, C. C., Pusey, P. L., Costello, C. E., Arrendale, R. F., Cox, R. H., Himmelsbach, D. S., Crumley F. G., and Cutler, H. G., Isolation and identification of iturins as antifungal peptides in biological control of peach brown rot with *Bacillus subtilis, J. Agric. Food Chem.,* 36, 366, 1988.

Hebbar, K. P., Atkinson, D., Tucker, W., and Dart, P. J., Suppression of *Fusarium moniliforme* by maize root-associated *Pseudomonas cepacia, Soil Biol. Biochem.,* 24, 1009, 1992.

Huang, Y., Deverall, B. J., and Morris, S. C., Effect of *Pseudomonas cepacia* on postharvest biocontrol of infection by *Penicillium digitatum* and on wound responses of citrus fruit, *Australas. Plant Pathol.,* 22, 84, 1993.

Janisiewicz, W. J. and Marchi, A., Control of storage rots on various pear cultivars with saprophytic strain of *Pseudomonas syringae, Plant Dis.,* 76, 555, 1992.

Janisiewicz, W. J. and Roitmann, J., Biological control of blue-mold and grey-mold on apples and pears with *Pseudomonas cepacia, Phytopathology,* 78, 1697, 1988.

Jetiyanon, K., Backman, P. A., Zidack, N. K., Tuzun, S., and Shaw, J., Use of a polysilicone surfactant for delivery of resistance-inducing strains of *Xanthomonas campestris* pvs. for control of black rot of cabbage, *Proc. 6th Int. Cong. Plant Pathol.,* 6, 65, 1993.

Johnson, K. B., Stockwell, V. O., McLaughlin, R. J., Sugar, D., Loper, J. E., and Roberts, R. G., Effect of antagonistic bacteria on establishment of honey bee-dispersed *Erwinia amylovora* in pear blossoms and on fire blight control, *Phytopathology,* 83, 995, 1993.

Jones, D. A. and Kerr, A., *Agrobacterium radiobacter* strain K1026, a genetically engineered derivative of strain K84, for biological control of crown gall, *Plant Dis.,* 73, 15, 1989.

Kessmann, H., Oostendorp, M., Ruess, W., Staub, T., Kunz, W., and Ryals, J., Systemic activated resistance — a new technology for plant disease control, *Pestic. Outlook,* 6, 10 1996.

Kloepper, J. W., Schroth, M. N., and Miller, T. D., Effects of rhizosphere colonization by plant growth-promoting rhizobacteria on potato plant development and yield, *Phytopathology,* 70, 1078, 1980.

Kloepper, J. W., Wei, G., and Tuzun, S., Rhizosphere population dynamics and internal colonization of cucumber by plant growth-promoting rhizobacteria which induce systemic resistance to *Colletotrichum orbiculare,* in *Biological Control of Plant Diseases: Progress and Challenges for the Future,* Tjamos, E. C., Papavizas, G. C., and Cook, R. J., Eds., Plenum Press, New York, 1992, 185.

Kokalis-Burelle, N., Backman, P. A., Rodríguez-Kábana, R., and Ploper, L. D., Potential for biological control of early leafspot of peanut using *Bacillus cereus* and chitin as foliar amendments, *Biol. Control,* 2, 321, 1992.

Kommedahl, T. and Mew, J. C., Biocontrol of corn root infection in the field by seed treatment with antagonists, *Phytopathology,* 65, 296, 1975.

Kvaratskheliya, M. T., The advantages of using bacterial fertilizers, *Mikrobiologiya,* 31, 1102, 1962.

Leeman, M., den Ouden, F. M., vanPelt, J. A., Cornelissen, C., Matamala-Garros, A., Bakker, P. A. H. M., and Schippers, B., Suppression of Fusarium wilt of radish by co-inoculation of fluorescent *Pseudomonas* spp. and root colonizing fungi, *Eur. J. Plant Pathol.,* (in press). 1995a.

Leeman, M., van Pelt, J. A., Hendrickx, M. J., Scheffer, R. J., Bakker, P. A. H. M., and Schippers, B., Biocontrol of Fusarium wilt of radish in commercial greenhouse trials by seed treatment with *Pseudomonas fluorescens* WCS374, *Phytopathology,* 85, 1301, 1995b.

Lindemann, J., Genetic manipulation of microorganisms for biological control, in *Biological Control on the Phylloplane,* Windels, C. E. and Lindow, S. E., Eds., American Phytopathological Society, St. Paul, MN, 1985, 116.

Lindow, S. E., Integrated control of frost injury, fire blight, and fruit russet of pear with a blossom application of an antagonistic bacterium, *Phytopathology,* 82, 1129, 1992 (Abstr.).

Liu, L., Kloepper, J. W., and Tuzun, S., Induction of systemic resistance against cucumber mosaic virus by seed inoculation with select rhizobacteria strains, *Phytopathology,* 82, 1108. 1992.

Liu, L., Kloepper, J. W., and Tuzun, S., Induction of systemic resistance in cucumber against Fusarium wilt by plant growth-promoting rhizobacteria, *Phytopathology,* 85, 695, 1995a.

Liu, L., Kloepper, J. W., and Tuzun, S., Induction of systemic resistance in cucumber against bacterial angular leaf spot by plant growth-promoting rhizobacteria, *Phytopathology,* 85, 843, 1995b.

Lumsden, R. D., Lewis J. A., and Fravel D. R., Formulation and delivery of biocontrol agents for use against soil-borne pathogens, in *Biological Pest Control Agents: Formulation and Delivery,* Hall, F. R. and Barry, J. W., Eds., American Chemical Society, Washington, D.C. 1995.

Mahaffee, W. F. and Backman, P. A., Effects of seed factors on spermosphere and rhizosphere colonization of cotton by *Bacillus subtilis* GB03, *Phytopathology,* 83, 1120, 1993.

Mann, E. W., Inhibition of tobacco mosaic virus by a bacterial extract, *Phytopathology,* 59, 658, 1968.

Maurhofer, M., Hase, C., Meuwly, P., Metraux, J.-P., and Defago, G., Induction of systemic resistance of tobacco to tobacco necrosis virus by the root-colonizing *Pseudomonas fluorescens* strain CHA0. Influence of the *gacA* gene and pyoverdine production, *Phytopathology,* 84, 139, 1994.

McInroy, J. A. and Kloepper, J. W., Population dynamics of endophytic bacteria in field-grown sweet corn and cotton, *Can. J. Microbiol.,* 41, 895, 1995a.

McInroy, J. A. and Kloepper, J. W., Survey of indigenous bacterial endophytes from cotton and sweet corn, *Plant Soil,* 173, 337, 1995b.

McLaughlin, R. J. and Roberts, R. G., Biological control of fire blight with *Pseudomonas fluorescens* strain A506 and two strains of *Erwinia herbicola, Phytopathology,* 82, 1129, 1992 (Abstr.).

Mei, R., *Yield Increasing Bacteria,* Agricultural Press, Beijing, 1989, 128. (In Chinese.)

Menkina, R. A., Bacterial fertilizers and their importance for agricultural plants, *Mikrobiologiya,* 32, 352, 1963.

Menkina, R. A., Bacteria mineralizing combined organic phosphorus, *Mikrobiologiya,* 19, 308, 1950.

Merriman, P. R., Price, R. D., Baker, K. F., Kollmorgen, J. F., Piggott, T., and Ridge, E. H., Effect of *Bacillus* and *Streptomyces* spp. applied to seed, in *Biology and Control of Soilborne Plant Pathogens,* Bruehl, G. W., Ed., American Phytopathological Society, St. Paul, MN, 1975.

Merriman, P. R., Price, R. D., Kollmorgen, J. F., Piggott, T., and Ridge, E. H., Effect of seed inoculation with *Bacillus subtilis* and *Streptomyces griseius* on the growth of cereals and carrots, *Aust. J. Agric. Res.,* 25, 219, 1974.

Misaghi, I. J. and Donndelinger, C. R., Endophytic bacteria in symptom-free cotton plants, *Phytopathology,* 80, 808, 1990.

Mishustin, E. N., Bacterial fertilizers and their effectiveness, *Mikrobiologiya,* 32, 911, 1963.

Mishustin, E. N. and Naumova., A. N., Bacterial fertilizers, their effectiveness and mode of action, *Mikrobiologiya,* 31, 543, 1962.

Mohammadi, O., Commercial development of Mycostop fungicide, in *Improving Plant Productivity with Rhizobacteria,* Ryder, M. H., Stephens, P. M., and Bowen, G. D., Eds., CSIRO, Adelaide, Australia, 1994.

Mohammadi, O. and Lahdenpera, M. L., Impact of application method on efficacy of Mycostop biofungicide, in *Improving Plant Productivity with Rhizosphere Bacteria,* Ryder, M. H., Stephens, P. M., and Bowen, G. D., Eds., CSIRO, Adelaide, Australia, 1994, 279.

Moore, L. W. and Warren G., *Agrobacterium radiobacter* strain 84 and biological control of crown gall, *Annu. Rev. Phytopathol.*, 17, 163, 1979.

Musson, G. and Kloepper, J. W., Development of delivery systems for introducing endophytic bacteria into cotton, *Phytopathology*, 84, 1112, 1994.

Nelson, E. B., Biological control of *Pythium* seed rot and premergence damping-off of cotton with *Enterobacter cloacae* and *Erwinia herbicola* applied as seed treatments, *Plant Dis.*, 72, 140, 1988.

Pierson, E. A. and Weller, D. M., Use of mixtures of fluorescent pseudomonads to suppress take-all and improve the growth of wheat, *Phytopathology*, 84, 940, 1994.

Ploper, L. D., Backman, P. A., Stevens, C., Khan, V. A., and Rodriguez-Kabana, R., Reduction of tomato early blight by combining cultural and biological control strategies, *Proc. 24th Natl. Agric. Plastics Congr.*, 24, 88, 1993.

Pusey, P. L. and Wilson, C. L., Postharvest biological control of stone fruit brown rot by *Bacillus subtilis*, *Plant Dis.*, 68, 753, 1984.

Pusey, P. L., Wilson, C. L., Hotchkiss, M. W., and Franklin, J. D., Compatibility of *Bacillus subtilis* for postharvest control of peach brown rot with commercial fruit waxes, dicloran, and cold-storage conditions, *Plant Dis.*, 70, 587, 1986.

Pusey, P. L., Hotchkiss, M. W., Dulmage, H. T., Baumgardner, R. A., Zehr, E. I., Reilly, C. C., and Wilson, C. L., Pilot tests for commercial production and application of *Bacillus subtilis* (B-3) for postharvest control of peach brown rot, *Plant Dis.*, 72, 622, 1988.

Raupach, G. S., Liu, L., Murphy, J. F., Tuzun, S., and Kloepper, J. W., Induced systemic resistance against cucumber mosaic cucumovirus using plant growth-promoting rhizobacteria (PGPR), *Plant Dis.*, 80, 891, 1996.

Rubenchik, L. I., *Azotobacter and Its Use in Agriculture*, Israel Program for Scientific Translations, Jerusalem, 1963, 278.

Samtsevich, S. A., Preparation, use and effectiveness of bacterial fertilizers in the Ukranian SSR, *Mikrobiologiya*, 31, 923, 1962.

Sheng, H., Colonization and Plant Responses of Soybean Cultivars to *Bacillus subtilis* Inoculants. M.S. Thesis, Auburn University, Auburn, AL, 1996, 114.

Smilanick, J. L. and Denis-Arrue, R., Control of green mold of lemons with *Pseudomonas* species, *Plant Dis.*, 76, 481, 1992.

Smilanick, J. L., Strategies for the isolation and testing of biocontrol agents, in *Biological Control of Postharvest Diseases: Theory and Practice*, Wilson, C. L. and Wisniewski, M. E., Eds., CRC Press, Boca Raton, FL, 1994.

Smith, J., Genetic diversity within *Burkholderia solanacearum* race 3 and the development of a biological control agent, *Mol. Ecol.*, 3, 607, 1994.

Stevens, C., Khan, V. A., Ploper, L. D., Backman, P. A., Rodriguez-Kabana, R., Collins, D. J., Brown, J. E., Wilson, M. A., and Igwegbe, E. C. K., Reduction of tomato early blight by combining soil solarization and biological control strategies, *Proc. 26th Natl. Agric. Plastics Congr.*, 26, 72, 1996.

Stevens, C., Khan, V. A., Wilson, M. A., Ploper, L. D., Backman, P. A., Brown, J. E., Rodriguez-Kabana, R., and Collins, D. J., Evaluation of Black Plastic mulch and row cover on the control of Alternaria foliage diseases of vegetables, *Proc. 24th Natl. Agric. Plastics Congr.*, 24, 88, 1993.

Stockwell, V. O., Johnson, K. B., and Loper, J. E., Compatibility of bacterial antagonists with antibiotics used for management of fire blight, *Phytopathology*, 83, 1383, 1993.

Stretch, A. W., Biological control of blueberry and cranberry fruit rots (*Vaccinium corymbosum* L. and *Vaccinium macrocarpon* Ait.), *Acta Hortic.*, 241, 301, 1989.

Sundara, R., Bajpai, W. V. B., Sharma, J., and Subbiah, B. B., Solubilization of phosphates by phosphorus solubilizing organisms using P^{32} as a tracer, and the inflluence of seed bacterization on the uptake by the drop, *Indian Soc. Soil Sci.*, 11, 209, 1963.

Suslow, T. V. and Schroth, M. N., Role of deleterious rhizobacteria as minor pathogens in reducing crop growth, *Phytopathology*, 72, 111, 1982.

Sutton, J. C., Evaluation of micro-organisms for biocontrol: *Botrytis cinerea* and strawberry, a case study, in *Advances In Plant Pathology*, Vol. 11, Academic Press, 1995, 171.

Sutton, J. C. and Peng, G., Manipulation and vectoring of biocontrol organisms to manage foliage and fruit diseases in cropping systems, *Annu. Rev. Phytopathol.*, 31, 473, 1993.

Thomson, S. V., Hansen, D. R., Flint, K. M., and Vandenberg, J. D., Dissemination of bacteria antagonistic to *Erwinia amylovora* by honey bees, *Plant Dis.*, 76, 1052, 1992.

Turner, J. T. and Backman, P. A., Factors relating to peanut yield increases after seed treatment with *Bacillus subtilis*, *Plant Dis.*, 75, 347, 1991.

Tuzun, S. and Kloepper, J. W., Practical application and implementation of induced resistance, in *Induced Resistance to Disease in Plants*, Hammerschmidt, R. and Kuc, J., Eds., Kluwer Academic, Amsterdam, 1995, 152.

U.S. Congress, Office of Technology Assessment, Biologically Based Technologies for Pest Control, OTA-ENV-636, U.S. Government Printing Office, Washington, D.C., 1995, 204.

Van Peer, R., Niemann, G. J., and Schippers, B., Induced resistance and phytoalexin accumulation in biological control of Fusarium wilt of carnation by *Pseudomonas* sp. strain WCS417r, *Phytopathology*, 81, 728, 1991.

Voznyakovskaya, Y. M., Choice of microorganisms for use in the composition of bacterial fertilizers, *Mikrobiologiya*, 32, 168, 1963.

Wei, G., Kloepper, J. W., and Tuzun, S., Induction of systemic resistance of cucumber to *Colletotrichum orbiculare* by select strains of plant growth-promoting rhizobacteria, *Phytopathology*, 81, 1508, 1991.

Weller, D. M., Biological control of soil-borne plant pathogens in the rhizosphere with bacteria, *Annu. Rev. Phytopathol.*, 26, 379, 1988.

Wilson, C. L. and Wisniewski, M. E., Biological control of postharvest diseases of fruits and vegetables: an emerging technology, *Annu. Rev. Phytopathol.*, 27, 425, 1989.

Wilson, C. L. and Wisniewski, M. E., *Biological Control of Postharvest Diseases: Theory and Practice*, CRC Press, Boca Raton, FL, 1994, 182.

Wilson, M. and Lindow, S. E., Interactions between the biological control agent *Pseudomonas fluorescens* A506 and *Erwinia amylovora* in pear blossoms, *Phytopathology*, 83, 117, 1993.

Wilson, M. and Backman, P. A., Biological control of plant pathogens, in *Handbook of Pest Management*, Ruberson, J. R., Ed., Marcel Dekker, New York, 1997.

Wilson, M., Epton, H. A. S., and Sigee, D. C., Biological control of fire blight of hawthorn *(Crataegus monogyna)* with *Erwinia herbicola* under protected conditions, *Plant Pathol.*, 39, 30, 1990.

Xu, G. W. and Gross D. C., Field evaluation of the interactions among fluorescent pseudomonads, *Erwinia carotovora*, and potato yields, *Phytopathology*, 76, 423, 1986a.

Xu, G. W. and Gross, D. C., Selection of fluorescent pseudomonads antagonistic to *Erwinia carotovora* and suppressive of potato seedpiece decay, *Phytopathology*, 76, 414, 1986b.

CHAPTER 5

Fungal Control Agents

R. S. Mehrotra, K. R. Aneja, and Ashok Aggarwal

I. INTRODUCTION — WHY BIOLOGICAL CONTROL?

Jacobsen and Backman (1993) have stated that biological and cultural control are two alternatives to synthetic chemical pesticides in an integrated pest management program. According to Thompson (1993), there is a worldwide effort to move toward the use of ecologically safe

1-8493-2627-3/97/$0.00+$.50

"environmentally friendly" methods of protecting crops from pests and pathogens. But biocontrol agents will not replace all fungicides in the foreseeable future. There are questions being asked today as to why we should go forward with biological control. The answer to this is very clear. Pesticides are necessary at present but are not a long-term solution to crop and human beings (Mukhopadhyay, 1987, 1996). Besides their nontarget effects and hazardous nature, pesticides are becoming more expensive and often lose their effectiveness due to development of resistant strains. Even the public perceives that residues of pesticides create food safety problems. This perception has caused major changes in pesticide use in the food production and processing industries and has resulted in government-mandated reductions of pesticide use in several countries. Finally, it is now recognized that the use of biological methods is a distinct possibility for the future and can be successfully implemented in modern agriculture.

Moreover, wider use of biological control of plant pathogens is compatible with sustainable agriculture and maintains a biological balance in nature.

II. HISTORICAL ACCOUNT

A. Some Important Landmarks

That activity of a plant pathogen could be inhibited by an accumulation of its own metabolic products was demonstrated as early as 1908 by Potter at Durham University, England. However, neither he nor anyone else followed up on the implications of this discovery. The first important landmark was a paper from Sanford (1926), the first person in charge of the Dominion Laboratory of Plant Pathology at Edmonton, Canada, in 1926 in *Phytopathology,* in which he suggested that the control of potato scab by green manure was a biological control due to the increase in the population of certain saprophytic bacteria that multiplied on the organic material, and he proposed two concepts: (1) saprophytic organisms can control the activity of plant pathogens; and (2) the microbiological balance of the soil can be changed by altering soil conditions, in particular the addition of fresh organic material. It was around this time that Millard and Taylor (1927) reported control of scab in potatoes grown in sterilized soil and inoculated with *Streptomyces scabies,* through simultaneous inoculation of the soil with *S. proaecox,* a vigorous saprophytic species.

Sanford and Broadfoot (1931) provided experimental evidence for Sanford's original hypothesis that infection of wheat seedlings by *Ophiobolus graminis* (now called *Gaeumannomyces graminis*) could be completely suppressed by the antagonistic action of the various individually coinoculated species of bacteria and fungi. According to Garrett (1965), the contribution of a Canadian worker, A. W. Henery of the University of Alberta, has been one of the greatest simple feats of intellectual and experimental analysis yet to have been performed in the field of root disease investigation. Henery (1932) demonstrated that the influence of soil temperature upon the take-all disease in greenhouse trials could be strongly affected by interactions between the pathogens and other soil organisms. Prior to Henery's work, it was Fawcett (1931) who emphasized the importance of investigations on the known mixtures of the organisms. A research student of Fawcett, R. Weindling, published a brilliant series of papers on the parasitism of *Trichoderma viride* on other soil fungi (Weindling, 1932, 1934, 1937, 1941; Weindling and Emerson, 1936; Weindling and Fawcett, 1936), and it can aptly be said that Weindling's work laid the firm foundation toward biological control of plant pathogens.

Another important landmark is the work of Bliss (1951) in California, who studied the control of *Armillaria mellea* in citrus orchards by soil fumigation. The pathogen was killed in soil in which *T. viride* was present. This work of Bliss was examined in detail by Garrett (1957, 1958) to find an answer to the question as to what contribution was made by the dominant population of *T. viride* developing after soil fumigation toward the elimination at *A. mellea* from woody inocula. Garrett showed that there is a direct killing effect of the fumigant on the fungus, and another indirect killing by the antagonist *T. viride,* whose population dramatically increases in the fumigated soils.

Rishbeth's (1963) work is a classical example of controlling Heterobasidium root rot of pine. Rishbeth, at Cambridge in Great Britain, found an ingenious method of controlling *Heterobasidium annosum* (*Fomes annosus*) by replacing the fungus with *Peniophora gigantea* which was difficult to control. Biocontrol of chestnut blight caused by *Endothia parasitica,* which some call *Cryphonectria parasitica,* by introducing into the canker a hypovirulent strain of the same vegetative compatible group as that of the virulent strain has been investigated by Biraghi (1951), Grente and Sauret (1969), and Day et al. (1977). An important observation by Day and associates was that the dsRNA virus-like agent contained in the hypovirulent strain is transmitted to the virulent strain when their hyphae anastomose. Consequently, the virulent strain becomes hypovirulent and the canker heals. MacDonald and Fulbright (1991) have discussed in detail the use and limitations of transmissible hypovirulence. According to them, nature has provided a novel approach for control of this devastating disease, yet biological control has not been achieved by simply adding a hypovirulent strain to the normal disease situation. To utilize hypovirulence we must first understand hypovirulence, and to achieve successful control researchers in the area of plant pathology, ecology, epidemiology, genetics, and molecular biology have to coordinate their energies in ameliorating this disease. Some naturally occurring stands are showing signs of recovery; there are indications that success is possible in this area.

Although there are several other examples where biocontrol of plant pathogens has been achieved by fungal antagonists, the control of plant pathogens by vesicular arbuscular mycorrhizal fungi is an important area which is gaining ground and much momentum. The first important landmark in the control of plant pathogens by VAM fungi came from the work of Marx (1970, 1972) who demonstrated that ectomycorrhizal fungi increased the resistance of pine roots to *Phytophthora cinnamomi.*

Stirling et al. (1976) showed that *Dactylella oviparasitica* gave good field control of root rot nematodes *(Meloidogyne)* on peach in California soils. Another classical work on nematophagous fungi in plant disease control was initiated in Hawaii by Linford (1937) and Linford and Yap (1938). These workers added chopped, green, pineapple tops to eelworm-infested soil in pots, and estimations were made of the nematode populations and the activity of the predatory fungi. The predators rapidly destroyed the nematodes, and there was a marked reduction in the nematode populations below the original level. Some other important discoveries identified in the area of biological control are biological control of *P. cinnamomi* by a continuous cover-crop and rotation of poultry manure to maintain high organic matter (Broadbent and Baker, 1974; Broadbent et al., 1971); *G. graminis* var. *tritici* by transfer of suppressive soil and by monoculture (Shipton et al., 1973; Vojinovic, 1972; Gerlagh, 1968); *Rhizoctonia solani* by monoculture (Henis et al., 1979); *Sclerotinia trifoliorum* by the use of mycoparasite (Tribe, 1957); *Helminthosporium victoriae* by *Chaetomium* spp. (Wood and Tveit, 1965); and *R. solani* and *Verticillium albo-atrum* by *Streptomyces* spp. (Yin et al., 1957, 1965).

B. Some Important Symposia and Conferences

Some of the important international symposia and conferences on biocontrol organized during the last 40 years are: (1) in 1963 at Berkeley, California, U.S. (Baker and Snyder, 1965); (2) in 1968 at London, U.K. (Toussoun et al., 1970); (3) in 1973 at Minneapolis, Minnesota, U.S. (Bruel, 1975); (4) in 1978 at Munich, Germany (Schippers and Gams, 1979); (5) in 1983 at Melbourne, Australia (Parker et al., 1985); and (6) in 1988 at Kyoto, Japan (Hornby, 1990).

C. Some Important Books on Biocontrol

During the last few years, biocontrol research in plant pathology has been one of the most rapidly expanding areas as is evident from the number of recent books on the subject: Hornby, 1990; Baker and Cook, 1974; Burges, 1981; Chet, 1987; Cook and Baker, 1983; Deacon, 1983; Hoy and Herzog, 1985; Mukerji and Garg, 1988; Papavizas, 1981; Schneider, 1982; and Windels and Lindow, 1985, and reviews by Baker and Cook, 1974; Papavizas, 1981; Baker, 1968, 1973, 1980; Baker and Snyder, 1965; Corke and Rishbeth, 1981; Monkau, 1980; Papavizas and Lumsden, 1980; Schroth and Hancock, 1981; Schner and Castagno, 1986; Nelson, 1991; Mehrota et al., 1980, 1988; Mukhopadhyaya, 1994, 1996; Dwivedi, 1993; Aggarwal, 1991; and Bagayraj and Govindan, 1996.

III. MECHANISM OF BIOLOGICAL CONTROL

A. Antibiosis

Biocontrol agents most likely interact with pathogens by the mechanism of antibiosis and competition. Antibiosis includes all antagonistic chemical products produced and released into the environment by different species of antagonists. This includes antibiotic compounds and some extracellular enzyme systems that damage plant pathogens, as well as compounds that may enhance an antagonist, such as *Trichoderma* in the soil. Weindling (1934) and Weindling and Emerson (1936) isolated for the first time a toxic substance from the cultural filtrate of *Trichoderma.* Dennis and Webster (1971) and Webster and Lomas (1964) isolated gliotoxin which was later identified to be produced by *Gliocladium.* Also, the isolates that produced gliotoxin and viridin were later shown to be *Gliocladium* (Brian et al., 1946; Brian and Hemming, 1945). However, no existing cultures of *Trichoderma* have been shown to produce gliotoxin and viridin, but Webster and Lomas (1964) reported that some isolates recovered from soil had been shown to produce these two compounds, however it cannot be said that these were not *Trichoderma.* Even the recent work does not report gliotoxin and viridin as a product of *Trichoderma.* Godtfredsen and Vangedal (1965) reported trichodermin, a sesquiterpene, was an antibiotic produced by *Trichoderma.* However, Pyke and Dietz (1966) reported dermadine was a new antibiotic produced by *Trichoderma* which is active against a wide range of fungi and bacteria. Suzukacillin and alamethicine are peptides produced by *Trichoderma* having both antifungal and antibacterial properties (Moubasher, 1963; Ooka et al., 1965; Reusser, 1967).

The production of antibiotics by *Trichoderma* spp. cannot be correlated with mycoparasitism because some enzymes like β1-3 glucanase, chitinase, and cellulase are important factors facilitating the ability of *Trichoderma* to compete for substrate as well as to directly attack a wide range of plant pathogens. Lytic enzymes are most often implicated as major features in nonantibiotic antibiosis of plant pathogens. Rodriguez-Kabana (1969) and Rodriguez-Kabana et al. (1978) have demonstrated that even proteolytic enzymes play a significant role in the destruction of the enzymes activity of *Sclerotium rolfsii.* Some of the recent work has focused on enzymes, particularly chitinase, produced by *Trichoderma* that attack the cell walls of various fungi except in oomycetous fungi. Elad et al. (1983) were able to demonstrate increased chitinase enzymatic activity at the point where *T. harzianum* attacked *S. rolfsii* or *R. solani* with clamps or by coiling. Harman et al. (1981) were able to increase the efficacy of *T. hamatum* by the addition of chitin, presumably by enhancing the chitinase activity of *Trichoderma* for biocontrol of *Pythium* spp. and *R. solani.* But Siven et al. (1984) contended that not all the biocontrol activity of *Trichoderma* sp. can be accounted for by presently known enzymatic activity. Thus both antibiotic and enzymatic activity might be combined for biocontrol.

Antibiotic properties for *Trichoderma* have been suggested many times, but this phenomenon is not properly utilized as a biocontrol procedure (Wells, 1988).

Several researchers have reported significant benefits from the use of Trichodermin-1, Trichodermin-2, Trichodermin-3, and Trichodermin-4 for the control of many diseases (Fedorinck and Chardymskaya, 1968; Fedorinck et al., 1975; Kozlova et al., 1966; Ponamareva, 1965). These Trichodermin materials are crude preparations and should not be confused with the sesquiterpene antibiotic and Trichodermin reported by Godtfredsen and Vangedal (1965). The dosage rates of these Trichodermin-1, -2, -3, and -4 may be too high for practical application in agriculture.

B. Competition

Competition refers to the interaction of two organisms striving for the same thing, e.g., space and nutrients. Although antibiosis and competition are treated as separate entities, they are considered together on the basis that they function together. It is known that intra- and interspecific competition among microbes is one of the most important factors determining density of population in nature. Odum (1953) reported that related species or species that prefer the same ecological niche usually avoid competition and occupy different geographical areas or different habitats in the same area.

The omnipresence of *Trichoderma* in natural soils throughout the world shows that it must be an excellent competitor for space and nutritional resources. Here we explain with examples how the competitive ability of various *Trichoderma* spp. effected suppression of pathogen activity or plant disease expression. Toole (1971) reported that *T. viride* protected southern pine wood from rot by *Poria monticola* without affecting the quality of the wood. There was no indication that factors other than competition were involved.

Pottel and Shigo (1975) and Pottel et al. (1977) used *T. harzianum* to control rots caused by hymenomycetous fungi for up to 2 years. Smith et al. (1981) described that by using *T. harzianum* to treat red maple wounds *Trichoderma* preempted *Phialophora melinii,* a pioneer colonizer of red maple wounds, which prevented colonization by the rotting fungus *Fomes connatus.* A part of this control was attributed to the removal of phenols by *P. melinii,* whereas the *Trichoderma* isolates allowed the normal accumulation of plant phenols which were toxic to *F. connatus.* Here the competition was not directly against the pathogen but was against the primary colonizer, which in turn detoxified the site for invasion by the rotting organism.

In Europe, the use of carbon disulfide has been practiced for the control of *A. mellea;* but Bliss (1951) explained that sublethal doses of carbon disulfide gave *Trichoderma* a competitive advantage in the environment as an antagonist to *A. mellea.* It was hypothesized that the additional competitive ability of *Trichoderma* in fumigated soil allowed it to overcome the protective mechanisms of the pseudosclerotial phase of *A. mellea,* thus effecting disease control. Garrett (1957) stated that in citrus orchards fumigated with CS_2 there is a direct kill due to the fumigant and an indirect killing of *A. mellea* root tissues due to the preponderance of *T. viride* in fumigated soils. Saksena (1960) attributed the dominance of *T. viride* in fumigated soils to its moderate tolerance to the fumigants and high growth rate of the fungus. Lily (1961) and Saksena and Lily (1967) stated that in Indian soils with alkaline pH, the place of *T. viride* is taken up by *Penicillium nigricans.* Mughogho (1968) tried seven fumigants which promoted the development of *T. viride* populations in such soils. Tiwari and Mehrotra (1973) showed that when soils of a *Piper betle* orchard were fumigated with CS_2 and formalin, levels of *Aspergillus terreus, A. niger, A. flavus, Curvularia lunata,* and *Penicillium nigricans* increased. Kumar (1995) showed that in alkaline soils fumigated with CS_2, formalin, or allyl alcohol, the population levels of fungi other than *T. viride* increase in such soils. *Trichoderma* was much lower in the list of dominant soil fungi. In alkaline soils its population does not increase substantially because the initial population of *Trichoderma* is much less in comparison to other tolerant fungi. If such soils are acidified and then fumigated, the population of the antagonist tends to increase.

Munnecke et al. (1973), while using carbon disulfide and methyl bromide as soil fumigants, reported that *A. mellea* was weakened by the fumigant and it was during this period that *T. viride* exerts its competitive and antagonistic ability. The competitive ability of *Trichoderma* as a biocontrol agent has been shown to be affected by soil properties. Weindling (1934) reported that pH affected the growth of *Trichoderma* as an antagonist to *R. solani.* Similarly, the availability of iron is also important for the growth and competitive ability of *Trichoderma.* Warcup (1951) reported that *Trichoderma* preferred moist soils over arid soils.

C. Parasitism/Mycoparasitism

Parasitism is an antagonistic symbiosis between organisms. Parasitism among fungi is generally called mycoparasitism. When the host fungus itself is parasitizing a plant and an animal host, the parasite is a hyperparasite. Recently there has been a considerable interest in exploiting hyperparasitic relationships in the biological control of plant diseases.

Organisms that parasitize and destroy the pathogen should be applied before planting the crop, whereas the antagonist that competes with the pathogen or inhibits its growth by antibiosis should be applied at the time of planting. There have been a number of investigations on hyperparasites, but information on the actual biocontrol is still limited. There are a number of reports about promising results from field trials and some hyperparasites are being developed for commercial utilization which we shall discuss in brief in this chapter.

In most plant diseases caused by the *Aphanomyces* spp. *Phytophthora* spp., and *Pythium* spp., the pathogens survive as oospores between crops. Sneh et al. (1977) reported a number of microorganisms parasitizing oospores of *Phytophthora* spp. Hsu and Lockwood (1984) studied the possibility of using *Hypocytrium catenoides* in the biological control of *P. megasperma* f. sp. *glycinae* in greenhouse tests. Regarding hyperparasitism on sclerotia-forming fungi, only four fungi (*Coniothyrium minitans, Gliocladium* sp., *Sporidesmium sclerotivorum,* and some *Trichoderma* spp.) were destructive to the *Sclerotinia* spp. in soil (Adams and Ayers, 1979). Adams (1990) has reviewed the work on mycoparasites and biological control of plant diseases. *Trichoderma* spp. and *Gliocladium* spp. are common soil inhabitants, especially in organic soils. Dennis and Webster (1971) investigated 80 *Trichoderma* isolates in dual cultures against six test fungi (*Rhizoctonia, Heterobasidion, Fusarium, Pyronema, Mucor, Pythium*) and only 10 isolates did not show coiling; even though high interactions are quite frequent among isolates of *Trichoderma* it was earlier doubted whether *Trichoderma* really act as hyperparasites. But studies have shown that *Trichoderma* species produce enzymes capable of degrading fungal walls composed of chitin, cellulose, etc., and now it is well established that *Trichoderma* spp. do parasitize the mycelium of various fungal pathogens (Wells, 1987; Kumar, 1995; Puri, 1995). Similarly, *Gliocladium* spp. have been shown to be parasitic on a wide variety of plant pathogens including *Botrytis, Fusarium, Eutypa, Phomopsis, Rhizoctonia,* and *Sclerotinia* spp. (Walker and Maude, 1975; Huang, 1978; Ricard et al., 1974; Moody and Gindrat, 1977; Howell, 1982).

Pythium nunn is a very interesting mycoparasite of a number of soil-borne plant pathogens. When attacking *Pythium ultimum* and *P. vexans* the hyphae of *P. nunn* coil around the hyphae of the host (Lifshitz et al., 1984). While attacking several soil-borne pathogens such as *Pythium aphanidermatum, Rhizoctonia solani, Phytophthora parasitica,* or *P. cinnamomi, P. nunn* formed appressoria-like structures and then parasitized the hyphase. Biological control of *Pythium* damping-off of cucumbers with *P. nunn* has been reported by Paulitz and Baker (1987).

Coniothyrium minitans is a mycoparasite that appears to have great potential for biological control of a number of diseases caused by *Sclerotinia* species.

Laetisaria arvalis is another mycoparasite which has attracted the attention of plant pathologists (Hoch and Abawi, 1979; Martin et al., 1986). However, in all these experiments it became clear that the natural soil microflora had an adverse effect on biocontrol agents and pathogen interactions.

Sporidesmium sclerotivorum, a demataceous hyphomycetous fungus, has been much utilized in biocontrol of plant pathogens (Marois and Adams, 1985; Adams et al., 1985).

The most common hyperparasite on powdery mildews is the coelomycete *Ampelomyces quisqualis.* Mathur and Mukerji (1981) sprayed a spore suspension of *Cladosporium spongiosum* onto leaves of *Morus alba* infected with *Phyllactinia guttata* and obtained some disease control. Recently, Urquhart et al. (1994) demonstrated that *Tilletiopsis* species (ballistospore-forming yeast) can be exploited for biocontrol of *Sphaerotheca fuliginea* on greenhouse cucumber.

Eudarluca caicis with the conidial stage *Sphaerellopsis filum* parasitizes many species of rust fungi. Lim and Nik (1983) reported *Verticillium psalliotae* growing on the uredinia of the coffee rust *Hemileia vastatrix* in Malaysia. Some uredinia were completely covered by the white mycelium of the hyperparasite, which prevented further development of the rust. Sundaram (1962) found parasitism of hyphomycetous fungus *Tuberculina costaricana* on uredinia of *Puccinia penniseti* and other *Puccinia* and *Uromyces* species in India.

D. Pathogen-Suppressive Soils

Baker and Cook (1974) defined suppressive soils as soils in which the pathogen does not establish or persist, establishes but causes little or no change, or establishes and cause disease for a while but thereafter the disease is less important, although the pathogen may persist in the soil.

Suppressive soils are known for the following plant pathogens, such as *G. graminis* var. *tritici* where take-all decline occurred (Shipton, 1975), pathotypes of *Fusarium oxysporum* that cause vascular wilts (Toussoun, 1975), *F. solani* f. sp. *phaseioli* (Burke, 1965; Furuya and Ui, 1981), root-knot nematode (Mankau and Prasad, 1977; Stirling and Mankau, 1978), *P. cinnamomi* (Broadbent and Baker, 1975), *Pythium* spp. (Hancock, 1977), and *S. scabies* (Menzies, 1959).

Two different types of disease suppressiveness are known: natural and induced. Natural suppressiveness is relatively indifferent of crop history and is frequently associated with the physical characteristics of soil, while induced is wholly dependent on agricultural practices. In induced suppressiveness, the disease increases for a few crop generations and then it declines. This decline phase of disease is the expression of induced suppressiveness.

The mechanism of suppressiveness is not fully understood. However, explanations given for the suppressive phenomenon includes antibiosis, competition, parasitism, and predation. Soil fungistasis in some instances appears to be involved in soil that is suppressive to disease, but Lockwood (1986) points out that there are some other factors in addition to fungistasis involved in the determination of disease suppressiveness. Alabouvette et al. (1979) suggested that suppressiveness is linked with microbial activity while others believe that some soils are rich in montmorillonite clay which favors the development of antagonistic microflora, which in turn suppresses the activity of pathogens. Transmission of suppressive soils to conducive soil for practical exploitation in biological control of soil-borne pathogens has now proved to be feasible.

Biological suppression of *P. cinnamomi*, the pathogen responsible for the destruction of eucalyptus, has been reported by Malajczuk and McComb (1979) in Australia and New Zealand. They reported massive colonization and breakdown of the hyphae of the pathogen in soil. Lysis of the sporangia due to antagonists of the suppressive soil was also reported. A possible explanation for the suppression was attributed to the presence of antagonistic bacteria in soil. Similar results have been reported by many workers.

E. Cross Protection and Induced Resistance

This is another unique approach to the control of plant pathogens that take place mainly inside the plant. The cross protection and induced resistance can be explained by the case in which the first organism in the plant acts directly (as an antagonist) against the second (challenger) organism, and those cases in which the first organism initiates (induces) a process whereby the host plant itself inhibits the challenger.

Use of induced resistance shows that the applied inoculum itself will not damage the crop. The resistance induced by mild strains of rust fungi on bean and sunflower leaves was effective only near the inoculation point, but the extent of control achieved was comparable to that observed with foliar fungicides (Yarwood, 1956). Similarly, cucumber, watermelon, and muskmelon plants were protected systematically against a challenge inoculation with *Colletotrichum lagenarium* by prior inoculation with mildly pathogenic strains of this same pathogen. Caruso and Kuc (1977) demonstrated this phenomenon both in the field as well as in the greenhouse. Very little information is available on practical aspects of systemic and local resistance induced by inoculation with mild or avirulent strains of fungi.

Cross protection can be very effective in controlling fusarium wilt (Davis, 1967, 1968; Langton, 1969; Molot and Mass, 1975; Baker et al., 1987). Another interesting example of cross protection by a nonpathogenic *Fusarium oxysporum* against fusarium wilt of sweet potatoes has been detected and explained (Ogawa and Komada, 1985). This fungus was seen and isolated from the vessels of healthy plants of sweet potato. When fresh-cut ends of sweet potato sprouts were dipped in the bud-cell suspension of the nonpathogenic *F. oxysporum* for preinoculation and planted in soil infested with the pathogen, satisfactory control of the wilt disease was seen not only in pots but also in commercial tests in fields. Mandeel and Baker (1991) have shown that suppression of wilt of cucumber by *F. oxysporum* is due to the fact that the nonpathogenic strain competed well for infection sites, as well as by enhanced induced resistance in the host.

IV. SOME IMPORTANT ANTAGONISTS WITH SPECIAL REFERENCE TO SOIL-BORNE, SEED-BORNE, AND AERIAL PATHOGENS

A. Biological Control of Soil-Borne Diseases

One particularly novel biological control of a soil-borne plant pathogen achieved by introducing the agent directly into natural soil is that of *Sclerotinia sclerotivorum* by the mycoparasite *S. sclerotivorum* (Ayers and Adams, 1979). In a later work of Adams and Fravel (1990), biocontrol of lettuce drop was achieved with *S. sclerotivorum* at rates as low as 0.2 kg/ha. In a previous study 2300 kg/ha was required for disease control. Research on the growth of *S. sclerotivorum* in culture, activity of the mycoparasite in natural soil, production of sclerotia of *S. minor* on lettuce, and the subsequent spatial arrangement of these sclerotia in the field, made possible strategies for increasing the efficacy of the biocontrol agent which resulted in a substantial reduction in the application rate of the antagonist. Another good example of a biocontrol agent registered by U.S. Environmental Plant Protection Agency (USEPPA) for commercial use is *P. gigentea* for control of *H. annosus* of pine stumps (Rishbeth, 1963, 1979). This antagonist is now widely used in the U.K. and the U.S. for protection of pines. *T. viride* was observed to be an antagonist, and the development of *Phytophthora parasitica* var. *piperina,* responsible for foot rot and leaf rot of *Piper betle,* was checked provided it was given a chance to establish in the rhizosphere flora (Mehrotra, 1961). For this purpose, several trials were planned: applying a conidial suspension of *T. viride* to the sterilized soil, submerging the cuttings used for plantation in a conidial suspension of *T. viride,* acidification of soils, and the adding of a conidial suspension of *T. viride.* In all these conditions, it was found that *T. viride* became established in the rhizosphere and was successful in warding off the attack of *P. parasitica* even in high doses.

Extensive work in soil application of *Trichoderma* and *Gliocladium* preparations have been done by scientists of the USDA, Beltsville, U.S. for control of damping-off and blight of snap bean *(Sclerotium rolfsii),* damping-off of cotton, and Rhizocotonia disease of potatoes (Papavizas and Lewis, 1989; Lewis and Papavizas, 1991). Upadhyaya and Mukhopadhyay (1986) reported biocontrol of Sclerotium root rot of sugarbeet by soil application of *T. harzianum* preparations. Similarly, control of *Pythium* spp. in tobacco, sugar beet, and cauliflower could be achieved by *T. harzianum* through soil applications (Mukhopadhyay et al., 1986; Mukherjee et al., 1989) at G. B. Pant University of Agriculture and Technology, Pantnagar. Diseases of several ornamental plants caused by *Pythium and Rhizoctonia* could be controlled by soil application of *G. virens* (Lumsden and Locke, 1989; Lumsden et al., 1990).

Biological seed treatments provide an economical and relatively nonpolluting delivery system compared to other field application systems. Conidial suspension of *Verticillium bigutattum* applied to potato tuber either as a dip or spray has been used on a field scale for the control of collar rot and black scurf caused by *R. solani* (Jager and Velvis, 1985). Similarly, *Pythium*-induced damping-off and seed rot of pea have been controlled when the seeds were treated with *T. harzianum* or *T. koningii* (Harder et al., 1984). Harman and Taylor (1988) and Harman et al. (1989) reported that damping-off of cucumber and tomato caused by *Pythium* spp. has been effectively controlled by incorporating *Trichoderma* and *Enterobacter* spp. as a biological seed treatment.

B. Biological Control of Seed-Borne Diseases

Antagonists generally applied to seeds are *Chaetomium* sp., *Penicillium oxalicum,* and *Trichoderma* spp. Their application to seed reduces seed-borne fungi and results in vigorous seedling growth. Windels (1981) found that plants grown from pea seeds treated with the spores of *P. oxalicum* were significantly better than those produced from untreated seeds. Similarly, Tveit and Moore (1954) reported that a *Chaetomium* sp. isolated from oat seeds protected oat seedlings from infection by seed-borne *Drechslera victoriae.* Certain isolates of *Chaetomium* controlled fusarium blight in oats as effectively as an organic mercury seed treatment. Wiley and Kommedahl (1981) reported that seed treatment of sweet corn and wheat with *T. harzianum* is very effective in crop management. Magie (1980) has demonstrated that gladiolus corms could be protected by using nonpathogenic fusarias corm treatment with spores against corm rot caused by *F. oxysporum* f. sp. *gladioli.* Certain isolates of *F. moniliforme* 'subglutinans' give biological control even after the pathogen is established in the corm. Odvody et al. (1980) demonstrated that *Laetisaria arvalis* when applied as seed treatment protected sugarbeet against damping-off. Biological control of *Pythium* damping-off of cotton with seed coating preparations of *G. virens* has been obtained by Howell (1991).

C. Biological Control of Airborne Pathogens

According to Hudson (1968), "The first colonisers of leaf are usually parasites but their development may be affected by the saprophytic microbial population in the phyllosphere." These populations increase in number as leaves mature and die and may affect disease development directly by competing with the parasites for space and/or nutrients, by stimulating or inhibiting spore germination, by exciting some form of antagonism, or even by changing the surface of the leaf.

Fokkema (1973) detected 50 isolates of phyllosphere fungi against *Cochliobolus sativus* and found that *Cladosporium* sp., *Sporobolomyces* sp., and *Aureotasidium pullulans* inhibited the growth of *Cochliobolus sativus* in *in vitro* conditions. On rye leaves with pollen grains, these fungi (saprophytes) were equally effective in reducing superficial mycelial development and infection of *C. sativus.* Fokkema (1976) also reported that pathogens like *Botrytis cinerea, Phoma betae,* and *Alternaria* spp. of noncereal hosts could be antagonized by similar phyllosphere fungi. Antagonism between phyllosphere fungi and *C. sativus* is mainly due to competition for nutrients. The artificial enhancement in the population of saprophytes on leaf, when such densities do not naturally exist, help in the biological control of disease in fields. Fokkema (1976) found 50% less infection on flag leaves with an artificially enhanced mycoflora on the water-sprayed control leaves. Competition for nutrition was considered to be the mechanism of antagonism rather than antibiotic production. Andrews (1990) has reviewed the work on biological control in the phyllosphere.

There are many reports of the elimination of pathogenic microbes by saprophytes, and application of these saprophytes as foliar spray are known to control diseases caused by both facultative as well as obligate parasites (Dickinson and Preece, 1976; Laben, 1965; Skidmore, 1976). Weindling and Emerson (1936) reported control of smut of barley, bunt of wheat, and leaf spot of oats due to *Helminthosporium victoriae* by spraying the plants with 'Gliotoxin', a compound isolated from *Trichoderma lignorum.* Wood and Tveit (1955) concluded that there was a possibility of using antagonists against the aerial pathogens and further stated that such antagonists should have a high rate of sporulation, fast growth, survival under unfavorable environmental conditions, and a very aggressive or antagonistic nature. There are also some reports on the antagonistic role of phylloplane fungi against biotrophic fungi. Mishra and Tewari (1976) sprayed fungi isolated from the phylloplane of wheat such as *Cladosporium herbarum. Myrothecium roridum, Nigrospora sphaerica,* and *Penicillium notatum* and achieved control of rust on wheat.

Hyperparasites or mycoparasites are also common on members of Erysiphales and Uredinales. In nature, these hyperparasites or mycoparasites sometimes control or suppress rust and powdery mildew diseases. Boosalis (1964) has done some work to biologically control a rust using hyperparasites. Blakeman and Fokkema (1982) have reviewed the work on the biological control of plant diseases on the phylloplane and its potential in controlling diseases.

V. BIOLOGICAL CONTROL OF POSTHARVEST FRUIT DISEASES

In recent years, different approaches to the microbial control of postharvest diseases of fruits have been made by various workers (Mehrota et al., 1982; Wilson and Pusey, 1985; Janisiewicz, 1988a,b; Wilson and Wisniewski, 1989). Although there are several methods of controlling postharvest diseases of fruits and vegetables, control of postharvest diseases by biological control agents is currently gaining more importance. Wilson and Wisniewski (1989) have reviewed the work done till then on this aspect. Fungicides are a primary means of controlling postharvest

Table 1 Fungal Biocontrol Agents for Control of Some Fruit Pathogens

Host	Pathogen/Disease	Biocontrol Agent	Ref.
1. Apple	Gray mold caused by *Botrytis cinerea*	*Acremonium breve*	Janisiewicz, 1986
	Dry-eye rot caused by *B. cinerea*	*T. harzianum* *T. pseudokoningii*	Tronsmo and Raa, 1977; Tronsmo and Ystaas, 1980;
	Apple scab caused by *Venturia inaequalis*	*Anthelia bombacina*	Heye and Andrews, 1983; and Cullen et al., 1984)
		Chaetomium globosum	Cullen et al., 1984
2. Grape	Blue mold and green mold caused by *Penicillium italicum*, *P. digitatum*	*Debaryomyces hensenii*	Chalutz and Wilson, 1990
	Sour rot caused by *Geotrichum candidum*	*T. harzianum*	Dubos, 1984
	Gray mold caused by *B. cinerea*		
3. Oranges	Blue mold and green mold caused by *P. italicum and P. digitatum*	*Debaryomyces hensenii* *Myrothecium roridum*	Janisiewicz, 1988b Chalutz and Wilson, 1990
4. Strawberries	Gray mold caused by *B. cinerea*	*Cladosporium herbarum*, *Pullularia pullulans*,	Bhatt and Vaughan, 1963
		T. harzianum, *T. viride*	Tronsmo and Dennis, 1977
5. Pineapple	Fruit rot caused by *P. funiculosum*	*P. funiculosum* (nonpathogenic strain)	Lim and Rohrback, 1980

diseases and they have recently come under special scrutiny as posing potential oncogenic risks. Blakeman and Fokkema (1982) found that there are natural antagonists in the phyllosphere and rhizosphere of plants that can suppress disease development. Tronsmo and Dennis (1977) were able to control *Botrytis* rot of strawberry with *Trichoderma.* Since then, a variety of antagonists have been found to control different rot pathogens of apple (Janisiewicz, 1988a). Janisiewicz (1988a) has reviewed the work done on postharvest control of fruit diseases. Table 1 presents in a summarized manner some effective biocontrol agents of some important postharvest fruit diseases.

VI. VESICULAR ARBUSCULAR MYCORRHIZAL (VAM) FUNGI IN CONTROLLING SOIL-BORNE DISEASES

It is known that VAM fungi may protect plants against several pathogens. The mode of action involved in the inhibition of soil-borne pathogens by VAM is (1) as an antagonist to the pathogen; (2) it increases the nutritional status in roots, thereby giving resistance; (3) it brings about biochemical and structural changes in the hot tissues; and (4) it may act as a mechanical barrier to check the entry of pathogens into the host. The work indicating the effect of VAM fungi on root diseases has been summarized in several reviews (Hussey and Roncadori, 1981; Schenck, 1981; Dehne, 1982; Upadhaya and Rai, 1985; Jalali and Jalali, 1991). Chandra (1992) has listed several soil-borne plant pathogens which are prevented by VAM-fungi from causing root diseases. Some important soil-borne pathogens controlled by the VAM fungi are given in Table 2.

According to Chet (1987), the potential to increase plant growth and vigor under a number of stress conditions, including the presence of some root pathogens, by utilizing mycorrhizal fungi are very great. Mycorrhizal fungi provide two advantages: firstly, they increase the growth by providing more phosphorus and nitrogen to the plants, and secondly they give a protective mechanism to the roots against fungal and other soil-borne infections.

Table 2 Inhibitory Effect of VAM Fungi on Soil-Borne Fungal Pathogens

Crop	Pathogen	Disease	Ref.
Tobacco	*Phytophthora cinnamomi*	Root rot	Bartcshi et al., 1981
	Thielaviopsis basicola	Root rot	Baltruschat and Schonbeck, 1975
	Pythium spp.	Root rot	Subhashini, 1990
Onion	*Pyrenochaeta terrestris*	Wilt	
	Fusarium oxysporum f. sp. *cepi*	Wilt	Dehne, 1982; Becker, 1976
	Pyrenochaeta terrestris	Root rot	Safir, 1968
Tomato	*F. oxysporum* f. sp. *lycopersici*	Wilt	Dehne and Schonbeck, 1979
	F. oxysporum f. sp. *radicis lycopersici*	Root rot	Caron et al., 1966
Citrus	*P. parasitica*	Root rot	Davis and Menge, 1981
Chickpea	*E. oxysporum* f. sp. *ciceri*	Wilt	Jalali and Thareja, 1981; Reddy et al., 1988
Brinjal	*Verticillium albo-atrum*	Wilt	Melo et al., 1985
Soybean	*Macrophomina phaseolina*	Root rot	Zambolin and Schenck, 1983
	Rhizoctonia solani		
	Fusarium solani		
Wheat	*Fusarium* spp.	Root rot	Khsuscheheva and Milenina, 1978
	Pythium spp.		
	Gaeumannomyces graminis var. *tritici*	Take-all disease	Graham and Menge, 1982
Rice	*Sclerotium oryzae*	Stem rot	Gangopadhyay and Das, 1987
Brassica	*Olpidium brassicae*	Root rot	Schonbeck and Dehne, 1979
Cotton	*Thielaviopsis basicola*	Seedling rot	Schonbeck and Dehne, 1977
Pea	*Aphanomyces euteiches*	Root rot	Vidyasekaran, 1990
Lettuce	*Olpidium brassicae*	Root rot and damping-off	Schonbeck and Dehne, 1979
Chestnut	*Phytophthora cinnamomi*	Pink disease	Vrot and Grante, 1985
Black gram Green gram Chickpea	*Macrophomina phaseolina*	Dry root rot	Kehri and Chandra 1990; Jalali et al., 1990

VII. SOIL SOLARIZATION AND BIOLOGICAL CONTROL

Disease control by soil solarization of mulching soil with transparent plastic covers during summer is a recent innovation. It was first used in Israel and Japan in the 1970s and soon afterwards in the U.S. (Katan et al., 1976; Kodama and Fukui, 1979). This method is quite effective for ornamental and horticultural crops in warmer regions of the world. Mahrer (1979) reports that during solarization the plastic covers trap solar energy and prevent heat losses caused by evaporation and conduction, thus creating a greenhouse effect. Many reports show that fungal pathogens are killed by temperatures within the range of 45 to 55°C (Kodama and Fukui, 1979; Porter and Merriman, 1983, 1985; Pullman et al., 1981). The principal effect of soil solarization is to raise soil temperature to levels which are lethal or harmful to soil-borne pathogens.

Porter and Merriman (1983) have isolated *Penicillium* spp., *Trichoderma* spp., *Gliocladium* spp., and *Fusarium* spp. from dead sclerotia in solarized soil from zones where temperatures were sublethal than in untreated soils. These fungi have been shown to be antagonistic to sclerotia (Merriman, 1976; Papavizas, 1985; Tu, 1989). Bliss (1951) showed that sublethal heating of soil infested with *A. mellea* stimulated the more heat-resistant resident *Trichoderma* spp. to proliferate and parasitize, and weakened *A. mellea*. Ashworth and Gaona (1982) also reported a reduction in the inoculum density of *Verticillium dahliae* in pistachio nut groves as a result of soil solarization. They have suggested that microsclerotia under plastic covers were induced to germinate suicidally in the absence of host roots and that germination was stimulated by an unknown substances that accumulates under mulches. This substance was either directly toxic to the germinating microsclerotia or acted indirectly by partially nullifying fungistasis. This result is supported by those of Katan et al. (1976) who found increased germination of conidia of *F. oxysporum* in soils preheated for either 1 or 3 h at 45 to 50°C compared with soil not preheated. Another possibility was the direct parasitism by antagonistic microorganisms, which was favored by a shift in the microbial population during solarization. The most outstanding example of long-term benefits of soil solarization was the control of *F. oxysporum* on cotton in Israel for at least three years after treatment (Katan et al., 1983).

Porter (1992) has discussed several possible mechanisms of biocontrol by soil solarization and these can be summarized with the work of others as follows:

1. Microbial attack and subsequent death of the pathogen already weakened by either heat or a reduction in soil redox potential during solarization.
2. Partial or complete nullification of fungistasis and subsequent lysis of the germinating propagule.
3. Parasitism of the pathogen by the changed microflora initiated by solarization.
4. Competitive saprophytic ability and reduced inoculum potential of the pathogen due to increased antibiosis and competition for nutrients by soil saprophytes.
5. Cross protection on the host roots by colonizations with either saprophytic microflora or avirulent strains of the pathogen.
6. Accumulation of toxic substances such as ammonia and ethylene during solarization have been implicated in biological control of soil-borne pathogens.

Thus it can be said that the principal use of soil solarization is to create a layer of pasteurized or disinfested soil where root growth is relatively unaffected by disease.

VIII. NEMATOPHAGOUS FUNGI AS BIOCONTROL AGENTS

Nematophagous fungi are natural enemies of nematodes and play a vital role in causing a decline in their populations in soils. Amongst the different nematode-trapping fungi, some of the species of *Dactylaria, Arthobotrys, Dactylella, Trichothecium, Verticillium, Cystopage,* etc. are abundantly found in soil, and amongst these *Dactylaria* spp. are greatly predacious on nematodes. These fungi capture nematodes by different methods such as adhesive hyphae, adhesive branches, adhesive nets, adhesive knobs, nonconstricting rings, and constricting rings. Zopf (1980) pioneered the discovery of fungi that parasitize nematodes. *A. robusta* can control *Dictylenchus myceliophagus* and *Meloidogyne* (Cayrol and Frankowski, 1979; Cayrol et al., 1978). There are some fungi known to parasitize nematode eggs and prominent among egg parasites are *Paecilomyces lilacinus* and *Dactylella oviparasitica* (Dunn et al., 1982). Jatala (1986) has reviewed the information on biocontrol of plant pathogenic nematodes till 1986. Swarup (1990) has summarized the information till 1990 in this area of biocontrol.

IX. BIOCONTROL OF INSECT PESTS BY ENTOMOPATHOGENIC FUNGI

Insects occupy virtually every ecological niche with both a boon and a curse to the biological entity it is affecting. There are some friendly insects like the honeybee and silkworm, but the majority such as beetles, aphids, hoppers, moths, borers, scale insects, etc. are inimical to us and cause enormous damage to crop plants and forest trees. However, insects too, like other living things, have their enemies in nature. There are diverse kinds of microorganisms such as bacteria, fungi, and viruses, which parasitize and kill insects. There are more fungal parasites of insects than bacteria and viruses (Agarwal, 1990). Entomopathogenic fungi range from ectoparasites, which do not seriously affect their insect hosts, to pathogens that are lethal (Hawsworth et al., 1983).

The association of fungi with insects was noted long ago. Bassi (1835) was perhaps the first to report infection of silkworm by *Beauveria bassiana,* the muscardine fungus, and suggested that an infectious microorganism might be used to control insect pests. This work was followed by isolation of several entomopathogenic fungi from several insects, such as *Empusa muscae* from houseflies (Cohn, 1855), *Beauveria bassiana* from silkworm (Pasteur, 1870), and *Metarhizium anisopliae* from beetles (Metschnikoff, 1879). Thaxter (1888) and Petch (1921, 1931) did pioneering work on entomopathogenic fungi in the U.S. and Ceylon (now known as Sri Lanka), respectively. A literature search reveals that presently entomopathogenic fungi include representatives of all the major groups of fungi: the Mastigomycotina, Zygomycotina, Ascomycotina, Deuteromycotina, and Basidiomycotina (Hawsworth et al., 1983; Aizawa, 1982; Evans and Prior, 1990).

There is a wide range of naturally occurring entomopathogenic fungi. Of these, several have a broad host range (e.g., *M. anisopliae* with more than 200 known hosts and *Beauveria* spp. with about 500 host species). However, different individual strains of the some species can have different host ranges and pathogenicities, and this can be of use in tailoring the biocontrol strain to particular pests (Dart, 1990).

Fungi for biological control of insects should ideally possess four attributes: (1) a good infection capability; (2) high postinfection virulence; (3) adequate survival potential; and (4) an efficient dispersal mechanism (Cooke, 1977). The fungal conidia/spore can be ingested or breathed in, or more usually develop into a hypha which invades the insect cuticle. Once the fungus enters the insect hemocoel, growth spreads throughout the insect, usually killing it after about 4 days, sometimes by toxin production. Once a fungus proliferates in the insect's body it is covered with mycelia and conidia. The newly formed conidia are dispersed and cause subsequent infection. This property of fungi proves to be very advantageous in the microbial control of insect pests. One of the main limitations to their use is the requirement of a high relative humidity (about 90%) for germination of spores, and some species require a film of water on the host surface (Aizawa, 1982; Dart, 1990).

The entomogenous fungi can be used in four different ways for insect control. The method of choice will be determined by the characteristics of the biocontrol agent (fungus) and the target insect (Agarwal, 1990):

1. *Introduction:* The fungus is introduced into an area where it becomes established in insect populations and maintains the density of that population below the economic threshold;
2. *Augmentation:* The density of a native entomogenous fungus within an insect population is increased to maintain the pest density at an acceptable level;
3. *Inundation:* The fungus is propagated in the laboratory and applied against that developmental stage of the insect which is most suitable for infection, colonization, and sporulation, in the manner of an insecticide to a pest population for short-term suspension of pest numbers;
4. *Integration:* Use of entomogenous fungus with other control methods which may be biological, chemical, and cultural.

According to Aizawa (1982), Pasteur and Metchnikoff were the originators of the idea of microbial control of insect pests in the 1870s. Current awareness about environmental pollution and the misuse of chemical pesticides has focused attention on biological control, especially on the microbial pathogens of pest insects. This has led to the evaluation of several fungal pathogens for their biocontrol potential against several insect pests. There are a number of important reviews which deal with the use of microorganisms as biological control agents of insect pests: Agarwal, 1990; Aizawa, 1976, 1982; Dart, 1990; Fawcett, 1944; Steinhaus, 1963, 1967; Madelin, 1966, 1968; Narasimhan, 1970; Kamat and Rao, 1975; Ferron, 1978; Samways, 1980; and Agarwal and Rajak, 1985. A literature search reveals that during the last 50 years several notable successes have been obtained in the control of insect pests with entomopathogenic fungi in the developed countries such as the U.S., Russia, Japan, England, Australia, and France (Table 3). The need of the hour is that both entomologists and mycologists should collaborate, especially in the developing

Table 3 Examples of Entomopathogenic Fungi Being Used as Biocontrol Agents on Pests

Fungi	Target	Trade	Country
Beauveria bassiana	Colorado potato beatle and coding moth	Boverin or beauverin	U.S.S.R.
Metarhizium anisdpliae	European corn borer	—	China
	Citrus root weevil	—	U.S.
	Spittle bug	Metaquino	Brazil
	Frog hopper	—	Brazil
	Cockroaches	—	U.S.
Normuraea rileyi	Tobacco budworm	—	U.S.
Hirustella thomsonii	Citrus rust mite	Mycar	U.S.
Verticillium lecanii	Greenhouse aphids	Vertalac	U.K.
Aschersonia plancenta	Whiteflies	—	U.S.S.R.
Aschersonia sp.	Whiteflies and scale insects	—	U.S.S.R., Japan
Culicinomyces clavisporus	Mosquito, dipteran larvae	—	Australia

countries of the world, in comprehensive collections of diseased insects and compile an inventory of entomopathogenic flora for later testing of these pathogens for their host specificity and biocontrol efficacy. If found host specific and efficacious, they should be developed as biopesticides.

X. MASS PRODUCTION AND DELIVERY SYSTEMS OF BIOCONTROL AGENTS

Several advancements have been made during the last few years with regard to the field application of biocontrol agents. Conway (1986) used a seeding gel, laponite 508 (Laportekned Ltd., U.K.), to deliver *T. harzianum* and *T. hamatum* conidia concomitantly with germinated vegetable seeds. Hader et al. (1984) used cellulosic gels for delivering the conidia of *Trichoderma* species and sclerotia of *Laetisaria arvalis* to control *R. solani* on pepper, and sclerotia of *L. arvalis* or chalmydospores of *T. harzianum* to control *S. rolfsii* in apple (Conway, 1986). In order to improve the efficiency of several bioprotectants, seed priming has been applied. Solid matrix priming (SMP) was developed by John Eastin Inc. (after Harman and Lumsden, 1990). Seed pelleting has been done with bioprotectants. Backman and Rodrigues-Kabana (1975) developed diatomaceous earth granules impregnated with molasses, etc. on which *Trichoderma* was grown. Jones et al. (1989) developed a liquid carrier system while Fravel et al. (1985) encapsulated the biocontrol agents in an alginate clay matrix. Recently, in this laboratory margosa seeds *(Azadirachta indica)* were utilized for growing *Trichoderma* species and incorporating them in the soil. The antagonist grew luxuriantly on autoclaved seeds and remained active in soil for a sufficiently long period (Puri, 1995).

The basic requirement for mass production of an antagonist is to standardize its nutritional requirements and cultural conditions, which differ for every organism, so mass production techniques cannot be generalized. The formulation of a biocontrol agent is an equally important aspect to consider for mass production. First, the adjuvent should not be toxic to the biocontrol agent. Second, the physical manipulation done during formulation should have no adverse effect on the viability of the organism. Last but not the least, a food base may be required to be added which helps in the initial establishment of some of the biocontrol agents.

Supplemented with a suitable nitrogen source for growth, submerged fermentation is used for mass production of various fungi. Currently, there are many types of formulations that are in use. Dry formulations include dusts and granules. Dust formulations contain 5 to 10% spores by weight and the rest is diluent, but they are mixed dry, while granular formulations are obtained either by coating previously harvested spores, or through growth and sporulation of the fungus on the surface of a nutritional granular substrate. Granular formulations contain about 5 to 20% spores, 80 to 95% carriers, and 1.5% binder. Wettable powders can be stored dry until use. Liquid formulations are essentially a suspension of insoluble spores in a liquid. Here the active ingredient is around 10 to 40%, suspending agent 1 to 3%, surfactant 3 to 8%, dispersant 1 to 5%, and the solvent vehicle, normally water or oil, 35 to 65%.

XI. APPLICATION OF BIOCONTROL AGENTS

Here we will discuss an approach to the development of application procedures based upon how the biocontrol agent integrates with a total biocontrol system.

The economics and mechanism of application will determine which system is utilized. For application procedures to elicit specific agent responses, it may be important to consider the substrate and carrier individually. In some cases, the substrate and carrier are the same, i.e., wheat straw and beet pulp. The straw is both the physical unit upon which the agent is delivered and the chemical source of energy. But in some cases they are different things, for example lignite granules and their liquid silage. Here the lignite acts as a carrier and the silage is the energy source. For some plant pathogens virulence is a function of pathogen nutrition, so the form and concentration of nitrogen and carbon is important to the kinetics of antibiotic synthesis and also to the excretion of the compounds as well. If the biocontrol agent is tied to the synthesis of antibiotics then the substrate upon which the biocontrol agent is grown is an important step in the biocontrol system, therefore, it is important not only to provide a substrate that will enhance the synthesis of the diseased enzymes, but also to provide sufficient substrate so as not to limit the synthesis of the enzymes at the time they are required.

Lewis and Papavizas (1984) have reported that mycelial preparations of germinated conidia of *T. harzianum* and *G. virens* had significantly greater biocontrol than nongerminated conidia when added to soil. So it is clear that different propagules have different capabilities and just their presence will not help in biocontrol.

There are several methods of delivery of biocontrol agents to the soil. The best and most efficient approach is to drench the soil with a suspension of agent propagules (Abd-EL-Moity and Shatla, 1981; Marois, 1981). Some attempts have been made to control several soil-borne pathogens by incorporating natural substrates colonized by antagonists of the pathogen into soil. Embedding propagules of the biocontrol agent in a matrix formed by sodium alginate and clay appears to be another effective method for delivery of *G. virens, Penicillium oxalicum, Talaromyces flavus,* and *T. viride* (Fravel et al., 1985). But the survival of applied biocontrol agents has generally been poor, so more attention should be given to the role of the delivery system onto serial parts. Time of application of biocontrol agents is equally important and should be considered with respect to crop development and the effect of the environment on the survival and activity of the agent and the target. With this information, appropriate decisions can be made for when and how to delivery the biocontrol agent to the plant.

A. Commercial Products for Biocontrol Agents

There are some biocontrol products registered for use for control of plant diseases:

1. BINAB-T — a Trichoderma-based mycofungicide for protection of pruning wounds.
2. F-stop — a Trichoderma-based product for control of damping-off of beans, corn, and peas.
3. GL-21 — a *Gliocladium virens*-based product that will be marketed for control of Pythium and Rhizoctonia root rot of some ornamental plants.
4. Soilgard — a *Gliocladium*-based product that has high potential for controlling a number of soil-borne pathogens.
5. Binah-T — has been on the market for many years.

One of the limiting factors of biological control products has been their efficacy, reliability, and activity when compared to chemical pesticides.

XII. AN OVERVIEW OF CURRENT AND FUTURE STRATEGIES

Recent developments in the field of molecular biology offers scope for producing better strains by genetic modification. Such techniques include mutations or sexual or asexual recombinations of the genomes of the two parents. With the help of these techniques, addition or deletion of genes can be introduced into a single organism by recombination, which will produce a superior biocontrol agent.

The phenomenon of mutation has been used by many workers to produce improved strains of biocontrol agents (Abd-EL-Moity et al., 1982; Papavizas and Lewis, 1983; Papavizas et al., 1982). Papavizas et al. (1982) reported that strains produced by UV radiation were more effective than the wild strain in disease suppression. Similarly, Ahmad and Baker (1986) reported strains of *T. harzianum* resistant to benomyl and having strong rhizosphere competance were produced by mutation.

Katan et al. (1984), by hybridization, developed three mutants resistant to benomyl and two resistant to dicarboximides from a mild strain of *Talaromyces flavus*. Protoplast fusion has been shown in *Trichoderma* and *Gliocladium* (Sivan et al., 1990; Stasz et al., 1988). Sivan et al. (1990) isolated nuclei from protoplasts of an auxotrophic mutant of *T. harzianum* and transferred them into protoplasts obtained from another auxotroph of the same strain. This type of transfer may lead to better methods of transferring genetic information into an organism.

G. virens cannot be used for biocontrol where benomyl is used because this fungicide binds to the fungal B-tubulin protein and prevents microtubule polymerization, and thus acts as an antimitotic agent. Production of benomyl resistance in *G. virens* by UV mutagenesis is a difficult task but recently Ossanna an Mischke (1990) isolated and regenerated protoplasts from *G. virens* and transformed them to benomyl resistance with a *Neurospora crassa* B-tubulin gene.

XIII. CONCLUSIONS

Bagyaraj and Govindan (1996) state that the greatest success in biological control has been achieved in indirect manipulation of soil microflora by techniques like green manuring, crop rotation, or other agricultural practices. The direct use of microbes for biocontrol of fungal root pathogens has achieved only limited success at the field level, despite voluminous scientific reports on potential biocontrol agents. There is a very complex nature of interaction between the microbe, the plant, and the environment. In fact, the environment is the referee which will ultimately decide the outcome of the host and pathogen interaction.

Finding a potential antagonist in the laboratory is one thing, but its successful utilization at the field level is also a difficult task. There are several problems. One of the most important is the amount of antagonist inoculum needed to be incorporated in the soil, or sprayed on the aerial parts, or applied to the seeds, seedlings, cuttings, or other propagating material.

Mass production of the biopesticide is another area where difficulties arise, as is the period for which these products can be stored and remain viable or useful for biocontrol. Another area that needs immediate attention is that the raw materials required for the production of biopesticides should be made easily affordable by the farmers and should be based on available farm products as far as possible.

Chemical pesticide industries should also patronize the production of biopesticides. A very concerted effort is needed by government agencies as well as the agricultural institutes where work on biocontrol is centered. We cannot afford to attempt biocontrol of each and every pathogen, but we can definitely suggest methods of indirectly encouraging antagonists in the field conditions so that these become useful as biocontrol agents.

Biotechnology is making its impact everywhere. In controlling plant pathogens better antagonists are needed, and the recent techniques of production of better biocontrol agents will include mutation, hybridization, parasexual recombination, protoplast fusion, and genetic engineering (of both antagonists as well as host plants).

It may be emphasized that much greater advances than have been made heretofore will be required in research to significantly replace some of the lost chemicals with biological controls. Nevertheless, replacement is a very realistic hope. For the next generation of astronomers, the stars will still be there, but for the next generation of people the conservation of our ecosystem is in peril. The biological control of pests, therefore, becomes all the more important in the next century.

REFERENCES

Abd-El Moity, T. H., Papavizas, G. C., and Shatla, M. N., Introduction of new isolates of *Trichoderma harzianum* tolerant to fungicides and their experimental use for control of white rot of onion, *Phytopathology,* 72, 396, 1982.

Ab-El Moity, T. H. and Shatla, M. N., Biological control of white rot disease of onion *(Sclerotium cepivorum)* by *Trichoderma harzianum, Phytopathol. Z.,* 100, 29, 1981.

Adams, P. B., The potential of mycoparasites for biological control of plant diseases, *Annu. Rev. Phytopathol.,* 28, 59, 1990.

Adams, P. B. and Ayers, W. A., Ecology of *Sclerotinia* species, *Phytopathology,* 69, 896, 1979.

Adams, P. B., Ayers, W. A., and Marois, J. J., Energy efficiency of the mycoparasite *Sporidesmium sclerotivorum in vitro* and soil, *Soil Boil. Biochem.,* 17, 155, 1985.

Adams, P. B. and Fravel, D. R., Economical biological control of Sclerotinia lettuce drop by *Sporidesmium sclerotivorum, Phytopathology,* 80, 1120, 1990.

Agarwal, G. P., Entomogenous fungi in India and management of insect pests, *Indian Phytopathol.,* 43, 131, 1990.

Agarwal, G. P., Biological Plant Protection: Recent Development, Presidential Address, Bot, Sec. 78th India Sci. Cong., 1, 1991.

Agarwal, G. P. and Rajak, R. C., Entomogenous fungi in biological control of insect pests, in, *Trends in Plant Research,* Govil, C. M. and Kumar, V., Eds., Bishen Singh Mahendra Pal Siungh, Dehradun, 1985, 485.

Ahmad, J. S. and Baker, R., Competitive saprophytic ability and cellulolytic activity of rhizosphere competent mutants of *Trichoderma harzianum, Phytopathology,* 77, 358, 1986.

Aizawa, K., Microbial insecticides in plant protection in Japan, *Agric. Asia,* 10, 387, 1976.

Aizawa, K., Microbial control of insect pests, in *Advances in Agricultural Microbiology,* Vol. IV, Subba Rao, N.S., Ed., Oxford and IBH Publishing, New Delhi, 1982, 397.

Alabouvette, C., Rouxel, F., and Louvet, J., Characteristics of fusarium wilt suppressive soils and prospects for their utilization in biological control, in *Soil-Borne Plant Pathogens,* Schippers, B. and Gams, W., Eds., Academic Press, London, 1979, 165.

Andrews, J. H., Biological control in the phyllosphere: realistic goal or false hope, *Can. J. Plant Pathol.,* 12, 300, 1990.

Ashworth, L. J., Jr. and Gaona, S. A., Evaluation of clear polyethylene mulch for controlling Verticillium wilt in established pistachio nut groves, *Phytopathology,* 72, 243, 1982.

Auld, B. A., Mass production of fungi for biopesticides, *Plant Prot. Q.,* 8, 1992.

Ayers, W. A. and Adams, P. B., Mycoparasitism of sclerotia of *Sclerotina* and *Sclerotium* species by *Sporidesmium sclerotivorum, Can J. Microbiol.,* 25, 17, 1979.

Backman, P. A. and Rodrigez-Kabana, R., A system for the growth and delivery of biological control agents to the soil, *Phytopathology,* 61, 819, 1975.

Bagyaraj, D. J. and Govindan, M., Microbial control of fungal root pathogens, in *Advances in Botany,* Mukerji, K. G., Ed., APH Publishing, New Delhi, 1996, 293.

Baker, K. F., Biological control of soil-borne plant pathogens (First A.W. Dimock Lecture, Cornell Univ.), Cornell University, Ithaca, NY, 1973, 14.

Baker, K. F., Developments in Plant Pathology and Mycology, 1930–1980, in, *Commonwealth Agric. Bureau. Prespectives in World Agric.,* Commonwealth Agricultural Bureaux, Slough, 1980, 207.

Baker, K. F. and Cook, R. J., *Biological Control of Plant Pathogens,* American Phytopathological Society, St. Paul, MN, 1974, 433.

Baker, K. F. and Snyder, W. C., *Ecology of Soil-Borne Plant Pathogens. Prelude to Biological Control,* University of California Press, Berkeley, 1965, 571.

Baker, R., Mechanism of biological control of soil-borne pathogens, *Annu. Rev. Phytopathol.*, 6, 263, 1968.

Baker, R., Hanchey, P., and Dottatra, S. D., Protection of carnation against Fusarium stem rot by fungi, *Phytopathology*, 68, 1495, 1987.

Baltruschat, H. and Schonbeck, F., The influence of endotrophic mycorrhiza on the infestation of tobacco by *Thielaviopsis basicola, Phytopathol. Z.*, 84, 172, 1975.

Bartschi, H. V., Gianinazzi-Pearson, V., and Vegh, I., Vesicular arbuscular mycorrhizae formation and root rot disease *(Phytophthora cinnamomi)* development in *Chamaecypris lawsaniana, Phytopath. Z.*, 102, 212, 1981.

Bassi, A., Del-maldelsegno calcinacctor moscardino malaffia che offigebachi de seta. Part 1. *Teronica Tip Ocesi, Lodi*, 1835.

Becker, W. N., Quantification on Onion Vesicular-Arbuscular Mycorrhizae and Their Resistance to *Pyrenochaeta terrestris*, Ph.D. Dissertation, University of Illinois, Urbana, 1976, 72.

Bhatt, D. D. and Vaughan, E. K., Interrelationships among fungi associated with strawberries in Oregon, *Phytopathology*, 53, 217, 1963.

Biraghi, A., Sul proposto reaggruppamento di *Endothia fluens* (Sow.) S. et S.e di *Endothia parasitica* (Murr.) P. J. et H. W. And. in un uniea specie, *Boll. Staz. Patol. Veg. (Florence)*, 9 (3), 133, 1951.

Blakeman, J. P. and Fokkema, N. J., Potential for biological control of plant diseases on the phylloplane, *Annu. Rev. Phytopathol.*, 20, 167, 1982.

Bliss, D. E., The destruction of *Armillaria mellea* in citrus soils, *Phytopathology*, 41, 665, 1951.

Boosalis, M. G., Hyperparasitism, *Annu. Rev. Phytopathol.*, 2, 363, 1964.

Brian, P. W., Curtis, P. J., Hemming, H. G., and McGowan, J. C., The production of viridin by pigment-forming strains of *Trichoderma viride, Ann. Appl. Biol.*, 33, 190, 1946.

Brian, P. W. and Hemming, H. G., Gliotoxin, a fungastatic metabolic product of *Trichoderma viride, Ann. Appl. Biol.*, 32, 214, 1945.

Broadbent, P. and Baker, K. F., Behaviour of *Phytophthora cinnamomi* in soil suppressive and conducive to root rot, *Aust. J. Agric. Res.*, 25, 121, 1974.

Broadbent, P. and Baker, K. F., Soil suppressive to Phytophthora root rot in Eastern Australia, in *Biology and Control of Soil-Borne Plant Pathogens*, Bruehl, G. W., Ed., American Phytopathological Society, St. Paul, MN, 1975, 152.

Broadbent, P., Baker, K. F., and Waterworth, Y., Bacteria and actinomycetes antagonistic to fungal root pathogens in Australian soils, *Aust. J. Biol. Sci.*, 24, 925, 1971.

Bruehl, G. W., *Biology and Control of Soil-Borne Plant Pathogens*, American Phytopathology Society, St. Paul, MN, 1975, 216.

Burke, D. W., Fusarium root rot of beans and behavior of the pathogen in different soils, *Phytophthology*, 55, 1122, 1965.

Burges, H. D., *Microbial Control of Pests and Plant Diseases, 1970–1980*, Academic Press, New York, 1981, 914.

Caron, M., Fortin, J. A., and Richards, C., Effect of inoculation sequence on the interaction between *Glomus intraradices* and *Fusarium oxysporum* f. sp. *radicis lycopersici* in tomatoes, *Can. J. Plant Pathol.*, 8, 12, 1966.

Caruso, F. L. and Kuc, J., Field protection of cucumber, watermelon and muskmelon against *Colletotrichum lagenarium* by *Colletotrichum lagenarium, Phytopathology*, 67, 1290, 1977.

Cayrol, J. C. and Frankowski, J. P., Und methode de lutte biologuique centre les nematodes a gallos des racines appartenante au genre *Meloidogyne, Rev. Hortic.*, 193, 15, 1979.

Cayrol, J. C., Frankowski, J. P., Laniece, A., d'Hordemere, G., and Talon, J. P., Centre les nematodes en champignonniere, Mise au point d'une methode de lutte biologique a l'aide d'un hyphomycete predateur *Arthobotrys robusta* souche *antipolis, Rev. Hortic.*, 184, 23, 1978.

Chalutz, E. and Wilson, C. L., Postharvest biocontrol of green and blue mold and sour sot of citrus fruit by *Debaryomyces hansenii, Plant Dis.*, 74, 134, 1990.

Chandra, S., VA-Mycorrhiza — Dimensions of its application, *Indian Phytopathol.*, 45(4), 391, 1992.

Chet, I., *Innovative Approaches to Plant Disease Control*, John Wiley & Sons, New York, 1987.

Chet, I., Trichoderma — application, mode of action and potential as a biocontrol agent of soil-borne plant pathogenic fungi in, *Innovative Approaches to Plant Disease Control*, Chet, I., Ed., John Wiley & Sons, New York, 1987, 137.

Cohn, F., *Nova Acta K. Acad. Caes. Leop. Carol Carm. Nat.*, 25, 301, 1855.

Conway, K. E., Use of fluid drilling gets to deliver biological control agents to soil, *Plant Dis.*, 70, 835, 1986.

Cooke, R., *Biology of Symbiotic Fungi,* John Wiley & Sons, London, 1977.

Cook, R. J. and Baker, K. F., *The Nature and Practice of Biological Control of Plant Pathogens,* American Phytopathological Society, St. Paul, MN, 1983, 539.

Corke, A. T. K. and Rishbeth, J., Use of microorganisms to control plant diseases, in *Microbial Control Pests and Plant Diseases,* Burges, H. D., Ed., Academic Press, London, 1981.

Cullen, D., Berbee, F. M., and Andrews, J. H., *Chaetomium globosum* antagonizes the apple scab pathogen, *Venturia inaequalis,* under field conditions, *Can. J. Bot.*, 62, 1814, 1984.

Dart, P. J., Agricultural microbiology: introduction, in *Agricultural Biotechnology — Opportunities for International Development,* Pevsely, G. J., Ed., C.A.B. International, Kew, 1990.

Davis, D., Cross protection in Fusarium wilt disease, *Phytopathology,* 57, 311, 1967.

Davis, D., Partial control of Fusarium wilt in tomato by formae of *Fusarium oxysporum, Phytopathology,* 58, 121, 1968.

Davis, R. M. and Menge, J. A., *Phytophthora parasitica* inoculation technique and intensity of vesicular arbuscular mycorrhizae in citrus, *New Phytol.*, 87, 705, 1981.

Day, P. R., Dodds, J. A., Elliston, J. E., Jaynes, R. A., and Anagnostakis, S. L., Double-stranded RNA in *Endothia parasitica, Phytopathology,* 67, 1393, 1977.

Deacon, J. W., *Microbial Control of Plant Pests and Diseases,* Van Nostrand Reinhold, Workingam, U.K., 1983.

Dehne, H. W., Interaction between vesicular arbuscular mycorrhizal fungi and plant pathogens, *Phytopathology,* 72, 1115, 1982.

Dehne, H. W. and Schonbeck, F., The influence of endotrophic mycorrhiza on Fusarium wilt of tomato, *Z. Pflkrankh.*, 82, 630, 1979.

Dennis, D. and Webster, J., Antagonistic properties of species-groups of *Trichoderma.* I. Production of non-volatile antibiotics, *Trans. Br. Mycol. Soc.*, 57, 25, 1971.

Dickinson, C. H. and Preece, T. F., *Microbiology of Aerial Plant Surfaces,* Academic Press, New York, 1976, 669.

Dubos, B., Biocontrol of *Botrytis cinerea* on grapevines by antagonistic strain at *Trichoderma harzianum,* in *Current Prespectives in Microbial Ecology,* Klub, M. J. and Reddy, C. A., Eds., American Society of Microbiology, Washington, D.C., 1984, 370.

Dunn, M. T., Syre, R. M., Carrell, A., and Wergin, W. P., Colonization of nematode eggs by *Paecilomyces lilacinum* (Thom.) Samson, as observed with scanning electron microscopy, *Scanning Electron Microsc.*, 3, 1351, 1982.

Dwivedi, R. S., Biopesticidal Plant Protection for sustainable crop productivity, Presidential Address, Bot. Sec. 80th Indian Sci. Congr., Goa, 1, 1993.

Elad, Y., Chet, I., Boyle, P., and Hennis, Y., Parasitism of *Trichoderma* spp. on *Rhizoctonia solani* and *Sclerotium rolfsii* scanning electron microscopy and fluorescence microscopy, *Phytopathology,* 73, 85, 1983.

Evans, H. C. and Prior, C., Entomopathogenic fungi, in *The Armored Scale Insects, Their Biology, Natural Enemies and Control,* Vol. B, Elsevier, New York, 1990.

Fawcett, H. S., Fungus and bacterial diseases of insects as factors in biological control, *Bot. Rev.*, 10, 327, 1944.

Fawcett, H. S., The importance of investigations on the effect of known mixtures of microorganisms, *Phytopathology,* 21, 545, 1931.

Fedorinck, N. S. and Chardymskaya, A. V., I Spytamie biopreparta Trikhodermin-3v bor'bes Komeedom Uskhodov Suekly, *Byull. Vses. Mauchno-isled. Inst. Zasch. Rast.*, 1, 3, 1968.

Fedorinck, N. S., Tarunina, T. A., Tyutynnkov, M. G., and Kudryavstreva, K. I., Trichodermin 4, a new biological preparation for plant disease control, *8th Int. Congr. Plant Prot.*, 3, 67, 1975.

Ferron, R., Biological control of insect pests by entomogenous fungi, *Annu. Rev. Entomol.*, 23, 409, 1978.

Fokkema, N. J., The role of saprophytic fungi in antagonism against *Drechslera sorokiniana (Helminthosporium sativum)* on agar plates and on rye leaves with pollen, *Physiol. Plant Pathol.*, 3, 195, 1973.

Fokkema, N. J., Antagonism between fungal saprophytes and pathogens on aerial plant surfaces, in *Microbiology of Aerial Plant Surfaces,* Dickinson, C. H. and Preece, T. F., Eds., Academic Press, London, 1976, 487.

Fravel, D. R., Maroris, J. J., Lumsden, R. D., and Connick, W. J., Jr., Encapsulation of potential biocontrol agents in an alternate-clay matrix, *Phytopathology,* 75, 774, 1985.

Furuya, H. and Ui, T., The significance of soil microorganisms on the inhibition of the macroconidial germination of *Fusarium solani* f. sp. *phaseoli* in a soil suppressive to common bean root rot, *Ann. Phytopathol. Soc. Jpn.,* 47, 42, 1981.

Gangopadhyay, S. and Das, K. M., Control of soil borne diseases of rice through VAM, in *Mycorrhiza Round Table, Proc. Natl. Workshop,* JNU, New Delhi, India, 1987, 560.

Garrett, S. D., Effect of soil microflora selected by carbon disulphide fumigation on survival of *Armillaria mellea* in woody host tissues, *Can. J. Microbiol.,* 3, 135, 1957.

Garrett, S. D., Inoculum potential as a factor limiting lethal action of *Trichoderma viride* Fr. on *Armillaria mellea* (Fr.) Quet, *Trans. Br. Mycol. Soc.,* 41, 157, 1958.

Garrett, S. D., Towards biological control of soil-borne plant pathogens, in *Ecology of Soil-Borne Plant Pathogens,* Baker, K. P. and Snyder, W. C., Eds., University of California Press, Berkeley, 1965, 4.

Gerlagh, M., Introduction of *Ophiobolus graminis* into new polders and its decline, *Neth. J. Plant Pathol.,* 74 (2), 1, 1968.

Godtfredsen, W. O. and Vangedal, S., Trichodermin, a new sesquiterpene antibiotic, *Acta Chem. Scand.,* 19, 1088, 1965.

Graham, J. H. and Menge, J. A., Influence of VAM and soil phosphorus on take all disease of wheat, *Phytopathology,* 72, 95, 1982.

Grente, J. and Sauret, S., L'hypovirulence exclusive phenome'ne original en pathologie vegetale, *C.R. Acad. Sci. Paris,* 268, 2347, 1969.

Hader, Y., Harman, G. E., and Taylor, A. G., Evaluation of *Trichoderma koningii* and *T. harzianum* from New York soils for biological control of seed rot caused by *Pythium* spp., *Phytopathology,* 74, 489, 1984.

Hancock, J. G., Factors affecting soil population of *Pythium altimum* in the San Joaquin Valley of California, *Hilgardia,* 45, 107, 1977.

Harman, G. E., Chet, I., and Baker, R., Factors affecting *Trichoderma hamatum* applied to seeds as a biocontrol agent, *Phytopathology,* 71, 569, 1981.

Harman, G. E. and Lumsden, R. D., Biological disease control, in *The Rhizosphere,* Lynch, J. M., Eds., John Wiley & Sons, Chichester, England, 1990.

Harman, G. E. and Taylor, A. G., Improved seedling performance by integration of biological control agents at favourable pH levels with soil matrix priming, *Phytopathology,* 78, 520, 1988.

Harman, G. E., Taylor, A. G., and Stasz, T. E., Combining effective strains of *Trichoderma harzianum* and solid matrix priming to improve biological seed treatment, *Plant Dis.,* 73, 631, 1989.

Hawsworth, D. L., Sutton, B. C., and Ainsworth, G. C., *Ainsworth & Bisby's Dictionary of the Fungi,* 7th ed., Commonwealth Mycological Institute, Kew, 1983.

Henis, Y., Ghaffar, A., and Baker, R., Factor affecting suppressiveness to *Rhizoctonia solani* in soil, *Phytopathology,* 69, 1164, 1979.

Henery, A. W., Influence of soil temperature and soil sterilization on the reaction of wheat seedlings to *Ophiobolus graminis* Sacc., *Can. J. Res.,* 7, 198, 1932.

Heye, C. C. and Andrews, J. H., Antagonism of *Athelia bombacina* and *Chaetomium globosum* to the apple scap pathogen, *Venturia inaequalis, Phytopathology,* 73, 650, 1983.

Hoagland, R. E., *Microbes and Microbial Products as Herbicides,* American Chemical Society, Washington, D.C., 1990.

Hoch, H. C. and Abawi, G. S., Biological control of *Pythium* root rot of table beet with *Corticum* sp., *Phytopathology,* 69, 417, 1979.

Hornby, D., *Biological Control of Soil-Borne Plant Pathogens,* C.A.B. International, Wallingford, U.K., 1990, 465.

Howell, C. R., Effect of *Gliocladium virens* on *Pythium ultimatum, Rhizoctonia solani,* and damping-off of cotton seedlings, *Phytopathology,* 72, 496, 1982.

Howell, C. R., Biological control of *Pythium* damping off cotton with seed coating preparation of *Gliocladium virens, Phytopathology,* 81, 738, 1991.

Hoy, M. A. and Herzog, D. C., *Biological Control in Agricultural IPM Systems,* Academic Press, London, 1985.

Hsu, S. C. and Lockwood, J. L., Biological control of *Phytophthora* root rot of soybean by *Hypochytrium catenoides* in greenhouse tests, *Phytopathol. Z.,* 109, 139, 1984.

Huang, H. C., *Gliocladium catenulatum:* hyperparasite of Sclerotinia sclerotiorum and *Fusarium* species, *Can. J. Bot.,* 56, 2243, 1978.

Hudson, H. J., The ecology of fungi on plant remains above the soil, *New Phytol.,* 67, 837, 1968.

Hussey, R. S. and Roncadori, R. W., Influence of *Aphelenchus avanae* on vesicular-arbuscular endomycorrhizal growth response in cotton, *J. Nematol.,* 13(1), 52, 1981.

Jacobsen, B. L. and Backman, A., Biological and cultural plant disease control: Alternatives and supplements to chemicals in IPM systems, *Plant Dis.,* 77(3), 311, 1993.

Jager, G. and Velvis, H., Biological control of *Rhizoctonia solani* on potatoes by antagonists. 4. Inoculation of seed tubers with *Verticillium biguttatum* and other antagonists in field experiments, *Neth. J. Plant Pathol.,* 91, 49, 1985.

Jalali, B. L., Chabra, M. L., and Singh, R. P., Interaction between vesicular-arbuscular mycorrhizal endophyte and *Macrophomina phaseolina* in mung bean, *Indian Phytopathol.,* 43, 527, 1990.

Jalali, B. L. and Jalali, I., Mycorrhiza in plant disease control, in *Handbook of Applied Mycology,* Arora, D. K., Elander, R. P., and Mukerji, K. G., Eds., Marcel Dekker, New York, 1991.

Jalali, B. L. and Thareja, M. L., Recent trends in plant growth responses to vesicular arbuscular mycorrhiza, *Forage Res.,* 7, 119, 1981.

Janisiewicz, W. J., Post-harvest biological control of blue mold of apples, *Phytopathology,* 77, 481, 1986.

Janisiewicz, W. J., Biological control of diseases of fruits, in *Biocontrol of Plant Diseases,* Vol. II, Mukerji, K. G. and Garg, K. L., Eds., CRC Press, Boca Raton, FL, 1988a, 153.

Janisiewicz, W. J., Biocontrol of post-harvest diseases of apples with antagonist mixtures, *Phytopathology,* 78, 194, 1988b.

Jatala, P., Biological control of plant parasitic nematodes, *Annu. Rev. Phytopathol.,* 24, 453, 1986.

Jones, J. K., Grady, T. S., and Bedbrook, K. J., Isolation and characterization of genes encoding two chitinase enzymes from *Serratia mercescens, EMBO J.,* 5, 467, 1989.

Kamat, M. N. and Rao, V. G., Entomogenous fungi and their taxonomy, in *Advances in Mycology and Plant Pathology,* Raychaudhuri, S. P., Verma, A., Bhargava, K. S., and Mehrotra, R. S., Eds., Published by Prof. R. N. Tandon's Birthday Celebration Committee, New Delhi, 1975, 57.

Katan, T., Dann, M. T., and Papavizas, G. C., Genetics of fungicides resistance in *Talaromyces flavus, Can. J. Microbiol.,* 30, 1079, 1984.

Katan, J., Fisler, G., and Grinstein, A., Short and long-term effects of soil solarization and crop sequence on Fusarium wilt and yield at cotton in Israel, *Phytopathology,* 73, 1215, 1983.

Katan, J., Greenberger, A., Alon, H., and Grinstein, A., Solar heating by polyethylene mulching for the control of diseases caused by soil borne pathogens, *Phytopathology,* 76, 683, 1976.

Kehri, H. K. and Chandra, S., Mycorrhizal association in crops under sewage farming, *J. Indian Bot. Soc.,* 69, 267, 1990.

Khrucheheva, A. P. and Milenina, N. P., Mycorrhiza and root rot of spring wheat, *Mycol. Fitopathol.,* 12(i), 43, 1978.

Kodama, T. and Fukui, T., Solar heating sterilization in closed vinyl house against soil borne disease. I. The movements of soil temperature and determination of thermal lethal condition for some soil-borne pathogens, *Bull. Nara. Pref. Agric. Exp. Sta.,* 10, 71, 1979.

Kozlova, L. N., Zlothina, G. D., and Bashmakova, V. I., Rezultaty izucheniya metodor borby uvydanium khlopchatnika V tadzhikistane, *Bvull. Nauchno. Tekh, Inf. Tadzh. Nauchno. Issled. Inst. Selsk. Khoz.,* 4, 95, 1966.

Kumar, J., Studies on Indian Species of *Trichoderma* and *Gliocladium* with Special Reference to the Production of Metabolites and Biocontrol of Some Important Soil Borne and Aerial Pathogen, Ph.D. Thesis, Kurukshetra University, Kuruksherta, India, 1995, 158.

Laben, C., Epiphytic microorganisms in relation to plant disease, *Annu. Rev. Phytopathol.,* 3, 209, 1965.

Langton, F. A., Interaction of the tomato with two formae speciales of *Fusarium oxysporum, Ann. Appl. Biol.,* 62, 413, 1969.

Leeman, M., Den Ovden, F. M., Van Pelt, J. A., Cornelissen, C., Matamala-Garros, A., Bakker, P. A. H. M., and Schippers, B., Suppression of fusarium wilt of radish by coinoculation of fluorescent *Pseudomonas* spp. and root-colonizing fungi, *Eur. J. Plant Pathol.,* 102, 21, 1996.

Lewis, J. A. and Papavizas, G. C., Chlamydospores formation by *Trichoderma* spp. in natural substrates, *Can. J. Microbiol.,* 30, 1, 1984.

Lewis, J. A. and Papavizas, G. C., Biocontrol at cotton damping-off caused by *Rhizoctonia solani* in the field with formulations of *Trichoderma* spp. and *Gliocladium virens, Crop Prot.,* 10, 95, 1991.

Lifshitz, R., Dapler, M., Elad, Y., and Baker, R., Hyphal interaction between a mycoparasite, *Pythium nunn*, and several soil fungi, *Can. J. Microbiol.*, 30, 1482, 1984.

Lily, K., Ecological Studies on Soil Fungi, Ph.D. Thesis, University of Saugar, Sagar, M.P. (India), 1961, 268.

Lim, T. K. and Nik, W. Z., Mycoparasitism of the coffee rust pathogen, *Hemileia vastatrix* by *Verticillium psalliotae* in Malaysia, *Pertanika*, 6, 23, 1983.

Lim, T. and Rohrbach, K. G., Role of *Penicillium funiculosum* strains in the development of pine apple fruit diseases, *Phytopathology*, 70, 663, 1980.

Linford, M. B., The feeding of root-knot nematodes in root tissue and nutrient solution, *Phytopathology*, 27, 824, 1937.

Linford, M. B. and Yap, F., Root-knot nematode injury restricted by a fungus, *Phytopathology*, 29, 596, 1938.

Lockwood, J. L., Soil-borne plant pathogens: concepts and connections, *Phytopathology*, 76, 20, 1986.

Lumsden, R. D. and Locke, J. C., Biological control of damping off caused by *Pythium ultimum* and *Rhizoctonia solani* wilt *Gliocladium virens* in soilless mix, *Phytopathology*, 79, 361, 1989.

Lumsden, R. D., Locke, J. C., and Lewis, J. A., Evaluation of *Gliocladium virens* for biocontrol of *Pythium* and *Rhizoctonia* damping-off of bedding plants, *Biol. Cultural Tests*, 5, 90, 1990.

MacDonald, W. L. and Fulbright, D. V., Biological control of chestnut blight: use and limitations of transmissible hypovirulence, *Plant Dis.*, 75, 656, 1991.

Madelin, M. F., Fungal parasites of insects, *Annu. Rev. Entomol.*, 11, 423, 1966.

Madelin, M. F., Fungal parasites of invertebrates. 1. Entomogenous fungi, in *The Fungi — An Advanced Treatise*, Vol. 3, Ainsworth, G. C. and Sussman, J. S., Eds., Academic Press, New York, 1968, 227.

Magie, R. O., Fusarium diseases of gladioli controlled by inoculation of corms with non-pathogenic fusaria, *Proc. Fla. State Hortic. Soc.*, 93, 172, 1980.

Mahrer, Y., Prediction of soil temperature of a soil mulched with transparent polyethylene, *J. Appl. Meteorol.*, 18, 1263, 1979.

Malajczuk, N. and McComb, A. J., The microflora of unsuberized roots of *Eucalyptus calaphylla* R. Br. and *E. marginata* Donn ex. Sm. seedlings grown in soil suppressive and conducive to *Phytophthora cinnamomi* Rands. I. Rhizoplane bacteria, actinomycetes and fungi. II. Mycorrhizal roots and associated microflora, *Aust. J. Bot.*, 27, 235, 1979.

Mandeel, Q. and Baker, R., Mechanisms involved in biological control of Fusarium wilt of cucumber with strains of non-pathogenic *Fusarium oxysporum*, *Phytopathology*, 81, 462, 1991.

Mankau, R., Biological control of nematode pests by natural enemies, *Annu. Rev. Phytopathol.*, 18, 415, 1980.

Mankau, R. and Prasad, N., Infectivity of *Bacillus penetrans* in plant parasitic nematodes, *J. Nematol.*, 9, 40, 1977.

Marois, J. J. and Adams, P. B., Frequency distribution analysis of lettuce drop caused by *Sclerotinia minor*, *Phytopathology*, 75, 957, 1985.

Marois, J. J., Mitchell, D. J., and Sonada, R. M., Biological control of fusarium crown rot of tomato under field conditions, *Phytopathology*, 71, 1257, 1981.

Martin, S. B., Abawi, G. S., and Hoch, H. C., The relationship of population densities of the antagonist, *Laetisaria arvails*, the seedling diseases of table beet incited by *Pythium ultimum*, *Can. J. Microbiol.*, 32, 156, 1986.

Marx, D. H., The influence of ectotrophic mycorrhizal fungi on the resistance of pine roots to pathogenic infections. V. Resistance of mycorrhizae to infection by vegetative mycelium of *Phytophthora cinnamomi*, *Phytopathology*, 60, 1471, 1970.

Marx, D. H., Ectomycorrhizae as biological deterrents to pathogenic root infections, *Annu. Rev. Phytopathol.*, 10, 429, 1972.

Mathur, M. and Mukerji, K. G., Antagonistic behaviour of *Cladosporium spongiosum* against *Phyllactinia dalbergiae sissoo*, *Angew. Bot.*, 55, 75, 1981.

Mehrotra, R. S., Studies on Soil Fungi from *Piper betle* Linn. Orchards with Special Reference to the Diseases Caused by *Phytophthora parasitica* var. *piperina* Dastur and Their Control, Ph.D. Thesis, University of Saugar, Sagar, India, 1961.

Mehrota, R. S., Aneja, K. R., and Goel, M. K., Approaches to biological control with antagonistic microorganisms, in *Annual Review of Plant Science*, Vol. II, Malik, C. P., Ed., Kalyani Publ. New Delhi, 1980, 595.

Mehrotra, R. S., Aneja, K. R., Gupta, A. K., and Aggarwal, A., Fungal-Agents of Biological Control, in *Biocontrol of Plant Diseases,* Vol. I, Mukerji, K. G. and Garg, K. L., Eds., CRC Press, Boca Raton, FL, 1988, 37.

Mehrotra, R. S., Sumbali, G., and Aneja, K. R., Control of post-harvest diseases of fruits and vegetables, *Bot. Progr.,* 5, 1, 1982.

Melo, I. S., Costa, C. P., and Sulveria, A. P. D., Effect of vesicular arbuscular mycorrhizae on arbergina wilt caused by *Verticillium albo-attrum, Phytopathologia,* 17(3-4), 173, 1985.

Menzies, J. D., Occurrence and transfer of a biological factor in that suppresses potato scab, *Phytopathology,* 49, 648, 1959.

Merriman, P. R., Survival of sclerotia of *Sclerotinia sclerotiorum, Soil Biol. Biochem.,* 8, 385, 1976.

Metschnikoff, E., *Odessa* (In Russian), 1879.

Millard, W. A. and Taylor, C. B., Antagonism of microorganism as the controlling factor in the inhibition of scab by green-manuring, *Ann. Appl. Biol.,* 14, 202, 1927.

Mishra, R. R. and Tewari, R. P., Studies on biological control of *Puccinia graminis tritici,* in *Microbiology of Aerial Plant Surfaces,* Dickinson, C. H. and Preece, T. F., Eds., Academic Press, London, 1976, 559.

Molot, P. M. and Mass, P., Influence of temperature on cross protection against *Fusarium oxysporum* f. sp. *melonis* induced in muskmelon by an incompatible race, *Ann. Phytopathol.,* 7, 175, 1975.

Moody, A. R. and Gindrat, D., Biological control of cucumber black root rot by *Gliocladium roseum, Phytopathology,* 67, 1159, 1977.

Moubasher, A. H., Selective effects of fumigation with carbon disulfide o the soil fungus flora, *Trans. Br. Mycol. Soc.,* 46, 338, 1963.

Mughogho, L. K., The fungus flora of fumigated soils, *Trans. Br. Mycol. Soc.,* 51, 441, 1968.

Mukerji, K. G. and Garg, K. L., *Biocontrol of Plant Diseases,* Vol. I and II, CRC Press, Boca Raton, FL, 1988.

Mukherjee, P. K., Upadhyay, J. P., and Mukhopadhyay, A. N., Biological control of Pythium damping-off of cauliflower by *Trichoderma harzianum, J. Biol. Control,* 3, 119, 1989.

Mukhopadhyay, A. N., Biological control of soil-borne plant pathogens by *Trichoderma* spp., *Indian J. Mycol. Plant Pathol.,* 17, 1, 1987.

Mukhopadhyay, A. N., Biocontrol of soil-borne Fungal plant pathogens: current status, future prospects and potential limitations, *Indian Phytopathol.,* 45(25), 119, 1994.

Mukhopadhyay, A. N., Recent innovations in plant disease control by ecofriendly biopesticides, *Presidential Address, Agric. Sci. 88th Indian Sci. Cong.,* Patiala, 1, 1996.

Mukhopadhyay, A. N., Brahmbhatt, A., and Patel, G. J., *Trichoderma harzianum* — a potential biocontrol agent for tobacco damping off, *Tobacco Res.,* 12, 26, 1986.

Munnecke, D. E., Kolbezen, M. J., and Wilber, W. D., Effect of methyl bromide or carbon disulfide on *Armillaria* and *Trichoderma* growing on agar medium and relation to survival of *Armillaria* in soil following fumigation, *Phytopathology,* 63, 1352, 1973.

Narasimhan, N. J., Entomogenous fungi and possibility of their use for biological control of insect pests in India, *Indian Phytopathol.,* 23, 16, 1970.

Nelson, E. B., Current limit to biological control of fungal phytopathogens, in *Handbook of Applied Mycology,* Vol. I, Soil and Plant, Arora, D. K., Rai, B., and Mukerji, K. G., Eds., Marcel Dekker, New York, 1991, 327.

Odum, E. P., *Fundamentals of Ecology,* W.B. Saunders, Philadelphia, 1953, 384.

Odvody, G. N., Boosalis, M. G., and Kerr, K. D., *Biological control of* Rhizoctonia solani *with a soil-inhabiting Basidiomycete, Phytopathology,* 70, 655, 1980.

Ogawa, K. and Komada, H., Biological control of Fusarium wilt of sweet potato by non-pathogenic *Fusarium oxysporum,* in *Ecology and Management of Soil-Borne Plant Pathogens,* Rovira, C. A., Moore, A. D., Wong, K. J., and Kollmoregen, J. F., Eds., American Phytopathology Society, St. Paul, MN, 1985, 121.

Ooka, T., Shimojima, Y., Akimoto, T., Senoh, S., and Abe, J., A new antibacterial peptide "Suzukacillin", *Agric. Biol. Chem.,* 30, 700, 1965.

Ossanna, N. and Mischke, S., Genetic transformations of the biocontrol fungus *Gliocladium virens* to benomyl resistance, *Appl. Environ. Microbiol.,* 56, 3056, 1990.

Papavizas, G. C., *Biological Control of Crop Production,* Beltsville Symp. Agric. Res. Allanheld, Osmum Co., London, 461, 1981.

Papavizas, G. C., *Trichoderma* and *Gliocladium:* biological control of soil-borne fungal propagules, *Annu. Rev. Phytopathol.,* 18, 389, 1985.

Papavizas, G. C., Lewis, J. A., and Abd-El Moity, T. H., Evaluation of new biotypes of *Trichoderma harzianum* for tolerance to benomyl and enhanced biocontrol capabilities, *Phytopathology,* 72, 126, 1982.

Papavizas, G. C. and Lewis, J. A., Physiological and biocontrol characteristics of stable mutants of *Trichoderma viride* resistant to MBC fungicides, *Phytopathology,* 73, 407, 1983.

Papavizas, G. C. and Lewis, J. A., Effect of *Gliocladium* and *Trichoderma* on damping off and blight of snapbean caused by *Sclerotium rolfsii* in the green house, *Plant Pathol.,* 38, 277, 1989.

Papavizas, G. C. and Lumsden, R. D., Biological control of soil-borne fungal propagules, *Annu. Rev. Phytopathol.,* 18, 389, 1980.

Parker, C. A., Rovira, A. D., Moore, K. J., Wong, P. T. W., and Kollmorgem, J. F., *Ecology and Management of Soil-Borne Plant Pathogens,* American Phytopathological Society, St. Paul, MN, 1985, 358.

Pasteur, L., *Studies Sur la Maladie des Vers a Sole,* Vol. I, Gauthien Villars, Paris, 1870.

Paulitz, T. C. and Baker, R., Biological control of Pythium damping-off of cucumbers with *Pythium nunn:* population dynamics and diseases suppression, *Phytopathology,* 77, 335, 1987.

Petch, T., Studies on entomogenous fungi, *Trans. Br. Mycol. Soc.,* 7, 167, 1921.

Petch, T., Notes on entomogenous fungi, *Trans. Br. Mycol. Soc.,* 16, 55, 1931.

Ponamareva, G. Ya., Khimicheskie i biologicheskie metady zaschchityrostenii, *Tr. Vses. Inst. Rast.,* 24, 182, 1965.

Porter, J. J., Soil solarization and biological control, in *Recent Developments in Biocontrol of Plant Diseases,* Mukerji, K. G., Tewari, J. P., Arora, P. K., and Saxena, G., Eds., Aditya Books, New Delhi, India, 1992, 164.

Porter, J. J. and Merriman, P. R., Effects of solarization of soil on nematode and fungal pathogens at two sites in Victoria, *Soil Biol. Biochem.,* 15, 39, 1983.

Porter, J. J. and Merriman, P. R., Evaluation of soil solarization for control of root diseases of row crops in Victoria, *Plant Pathol.,* 34, 108, 1985.

Pottel, H. W., Shigo, A. L., and Blanchard, R. O., Biological control of wood hymenomycetes by *Trichoderma, Plant Dis. Rep.,* 61, 687, 1977.

Pottel, H. W. and Shigo, A. L., Treatment of wounds of *Acer rubrum* with *Trichoderma viride, Eur. J. For. Pathol.,* 5, 274, 1975.

Potter, M. C., On a method of checking parasitic diseases in plant, *J. Agric. Sci.,* 3, 102, 1908.

Pullman, G. S., DeVay, J. E., Garber, R. H., and Weinhold, A. R., Soil solarization: effect of Verticillium wilt of cotton and soil-borne populations of *Verticillium dahliae, Pythium* spp., *Rhizoctonia solani* and *Thielaviopsis basicola, Phytopathology,* 71, 954, 1981.

Puri, A., Studies on *Trichoderma* and *Gliocladium* Species with Special Reference to Enzyme Production and Biocontrol, Ph.D. Thesis, Kurukshetra University, Kurukshetra, India, 1995.

Pyke, T. R. and Dietz, A., U-21, 1963, a new antibiotic. I. Discovery and biological activity, *Appl. Microbiol.,* 14, 506, 1966.

Reddy, M. V., Rao, J. N., and Krishna, K. R., Influence of mycorrhiza on chickpea, News letters No. 19, *ICRISAT,* 1988.

Reusser, F., Biosynthesis of Antibiotic U-22, 324, a cyclic polypeptide, *J. Biol. Chem.,* 242, 243, 1967.

Ricard, J. L., Grosclaude, C., and Ale-Agha, N., Antagonism between *Eutypa armeniacae* and *Gliocladium roseum, Plant Dis. Rep.,* 58, 983, 1974.

Rishbeth, J., Stump protection against *Fomes annosus.* III. Inoculation with *Peniophora gigantea, Ann. Appl. Biol.,* 52, 63, 1963.

Rishbeth, J., Modern aspects of biological control of *Fomes and Armillaria, Eur. J. For. Pathol.,* 2, 331, 1979.

Rodriguez-Kabana, R., Enzymatic interaction of *Sclerotium rolfsii* and *Trichoderma viride* in mixed soil culture, *Phytopathology,* 59, 910, 1969.

Rodriguez-Kabana, R., Kelley, W. D., and Curl, E. A., Proteolytic activity of *Trichoderma viride* in mixed culture of *Sclerotium rolfsii* in soil, *Can. J. Microbiol.,* 24, 487, 1978.

Safir, G., The Influence of Vesicular-Arbuscular Mycorrhiza on the Resistance of Onion to *Pyrenochaeta terrestris,* M.S. Thesis, University of Illinois, Urbana, 1968, 62.

Saksena, S. B., Effect of carbon disulphide fumigation on *Trichoderma viride* and other soil fungi, *Trans. Br. Mycol. Soc.,* 43, 111, 1960.

Saksena, S. B. and Lily, K., Studies on he interaction between soil microorganisms with special reference to *Penicillium nigricans*, *J. Indian Bot. Soc.*, 46, 185, 1967.

Samways, M. J., *Biological Control of Pests and Weeds*, Edward Arnold, London, 1980.

Sanford, G. B., Some factors affecting the pathogenicity of *Actinomyces scabies*, *Phytopathology*, 16, 525, 1926.

Sanford, G. B. and Broadfoot, W. C., A note on the biological control of root rots of cereals. Studies of the effects of other soil-inhabiting microorganisms on the virulence of *Ophiobolus graminis* Sacc., *Sci. Agric.*, 11, 460, 1931.

Schenck, N. C., Can mycorrhiza control root diseases, *Plant Dis.*, 65, 230, 1981.

Schner, F. M. and Castagno, J. R., Biocontrol, a view from industry, *Can. J. Plant Pathol.*, 8, 222, 1986.

Schippers, B. and Gams, W., *Soil-borne Plant Pathogens*, Academic Press, New York, 1979, 686.

Schneider, R. W., *Suppressive Soils and Plant Disease*, American Phytopathological Society, St. Paul, MN, 1982.

Schonbeck, F. and Dehne, H. W., Damage to mycorrhizal cotton seedlings by *Thielaviopsis basicola*, *Plant Dis. Res.*, 61, 266, 1977.

Schonbeck, F. and Dehne, H. W., The influence of endotrophic mycorrhiza on plant diseases. Fungal parasites on aerial plant parts, *Olpidium brassicae*, *TMV*, *Z. Pflanzenkr. Pflanzensch.*, 86, 103, 1979.

Schroth, M. N. and Hancock, J. G., Selected topics in biological control, *Annu. Rev. Microbiol.*, 35, 453, 1981.

Shipton, P. J., Cook, P. J., and Sitton, J. W., Occurrence and transfer of a biological factor in soil that suppresses take-all of wheat in eastern Washington, *Phytopathology*, 63, 511, 1973.

Shipton, P. J., Take-all decline during cereal monoculture, in *Biology and Control of Soil-Borne Plant Pathogens*, Bruehl, G. W., Ed., American Phytopathological Society, St. Paul, MN, 1975, 137.

Sivan, A., Elad, Y., and Chet, I., Biological control effects of a new isolate of *Trichoderma harzianum* on *Pythium aphanidermatum*, *Phytopathology*, 74, 498, 1984.

Sivan, A., Harman, G. E., and Stasz, T. E., Transfer of isolated nuclei into protoplasts of *Trichoderma harzianum*, *Appl. Environ. Microbiol.*, 56, 2404, 1990.

Skidmore, A. M., Interactions in relation to biological control of plant pathogens, in *Microbiology of Aerial Plant Surfaces*, Dickinson, C. H. and Preece, T. F., Eds., Academic Press, London, 1978, 507.

Smith, K. T., Blanchard, R. O., and Schortle, W. C., Postulated mechanism of biocontrol of decay fungi in red maple wounds treated with *Trichoderma harzianum*, *Phytopathology*, 71, 494, 1981.

Sneh, B., Humbel, S. J., and Lockwood, L. J., Parasitism of oospores of *Phytophthora megasperma* var. *sojae*, *P. cactorum*, *Pythium* sp., and *Aphanomyces eutieiches* in soil by oomycetes, chytridiomycetes, hyphomycetes, actinomycetes, and bacteria, *Phytopathology*, 67, 622, 1977.

Stasz, T. E., Harman, G. E., and Weeden, N. F., Protoplast preparation and fusion in two biocontrol strains of *Trichoderma harzianum*, *Mycologia*, 80, 141, 1988.

Steinhaus, E. A., *Insect Pathology*, Academic Press, New York, 1963, 689.

Steinhaus, E. A., *Principles of Insect Pathology*, Hafner, New York, 1967.

Stirling, G. R. and Mankau, R., *Dactylella oviparasitica*, a new fungal parasite of Meloidogyne eggs, *Mycologia*, 80, 774, 1978.

Stirling, G. R., McKenry, M. V., and Mankau, R., Biological control of root-knot nematodes (*Meloidogyne* spp.) on peach, *Phytopathology*, 69, 806, 1976.

Subhashini, D. V., The role of VAM in controlling certain root diseases of tobacco, in *Current Trends in Mycorrhizal Research*, Jalali, B. L. and Chand, H., Eds., Proc. Natl. Conf. Mycorrhiza, Haryana Agric. Univ., Hisar, India, 1990, 121.

Sundaram, N. V., Studies in parasites of the rusts, *Indian J. Agric. Sci.*, 32, 266, 1962.

Swarup, G., Nematode-microorganism interaction — an appraisal, *Indian Phytopathol.*, 43, 1, 1990.

Thaxter, R., Entomophthorae of United States, *Mem. Boston Soc. Nat. Hist.*, 4, 133, 1888.

Thompson, J. A., Biological control of plant pests and pathogens: alternative approaches, in *Biotechnology in Plant Disease Control*, Chet, I., Ed., Wiley-Liss, New York, 1993, 275.

Tiwari, D. P. and Mehrotra, R. S., Survival and control of *Phytophthora parasitica* in fumigated soils, *J. Indian Bot. Soc.*, 52, 138, 1973.

Toole, E. R., Interaction of mold and decay fungi on wood in laboratory tests, *Phytopathology*, 61, 124, 1971.

Toussoun, T. A., Fusarium-suppressive soils, in *Biology and Control of Soil-Borne Plant Pathogens*, Bruehl, G. W., Ed., American Pathological Society, St. Paul, MN, 1975, 145.

Toussoun, T. A., Bega, R. V., and Nelson, P. E., *Root Diseases and Soil-borne Pathogens*, University of California Press, Berkeley, 1970, 252.

Tribe, H. T., On the parasitism of *Sclerotina trifoliorum* by *Coniothyrium minitans*, *Trans. Br. Mycol. Soc.*, 40, 487, 1957.

Tronsmo, A. and Dennis, C., The use of *Trichoderma* species to control strawberry fruit rots, *Neth. J. Plant Pathol.*, 83, 449, 1977.

Tronsmo, A. and Raa, J., Antagonistic action of *Trichoderma pseudokonigii* against the apple pathogen *Botrytis cinerea*, *Phytopathol. Z.*, 89, 216, 1977.

Tronsmo, A. and Ystaas, J., Biological control of *Botrytis cinerea* on apple, *Plant Dis.*, 64, 1009, 1980.

Tu, J. C., *Gliocladium virens*, a destructive mycoparasite of *Sclerotina sclerotivorum*, *Phytopathology*, 70, 690, 1989.

Tveit, M. and Moore, M. B., Isolates of *Chaetomium* that protects oats with *Helminthosporium victoriae*, *Phytopathology*, 44, 686, 1954.

Upadhyaya, J. P. and Mukhopadhyay, A. N., Biological control of *Sclerotium rolfsii* by *Trichoderma harzianum* in sugarbeet, *Trop. Pest Manage.*, 32, 215, 1986.

Upadhyaya, R. S. and Rai, B., Biological control of soil-borne plant pathogens, in *Frontiers in Applied Microbiology*, Mukerji, K. G., Pathak, N. C., and Singh, V. P., Eds., Print House, Lucknow, India, 1985, 261.

Urquhart, E. J., Menzies, J. G., and Punja, Z. K., Growth and biological control activity of *Tilletiopsis* species against powdery mildew *(Sphaerotheca fuliginea)* on green house cucumber, *Phytopathology*, 84, 341, 1994.

Vidyasekaran, P., Mycorrhiza-induced resistance, a mechanism for management of crop diseases, in *Current Trends in Mycorrhizal Research*, Jalali, B. L. and Chand, H., Eds., Proc. Natl. Mycorrhiza, H.A.U., Hisar, India, 1990, 14.

Vojinovic, Z. D., Antagonists from soil rhizosphere to phytopathogens, *Final Tech. Rep. Inst. Soil. Sci., Beograd. Yugoslavia*, 1972, 130.

Vrot, S. and Grante, J., Research for a method of biological control of hik disease by using mycorrhizal symbiosis, *Eur. J. For. Pathol.*, 15(506), 379, 1985.

Walker, J. A. and Maude, R. B., Natural occurrence and growth of *Gliocladium roseum* on the mycelium and sclerotia of *Botrytis allii*, *Trans. Br. Mycol. Soc.*, 65, 335, 1975.

Warcup, J. H., Soil steaming: a selective method for the isolation of ascomycetes from soil, *Trans. Br. Mycol. Soc.*, 34, 515, 1951.

Webster, J. and Lomas, N., Does *Trichoderma viride* produce gliotoxin and viridin?, *Trans. Br. Mycol. Soc.*, 47, 535, 1964.

Weindling, R., *Trichoderma lignorum* as a parasite of other soil fungi, *Phytopathology*, 22, 837, 1932.

Weindling, R., Studies on a lethal principle effective in the parasitic action of *Trichoderma lignorum* on *Rhizoctonia solani* and other soil fungi, *Phytopathology*, 24, 1153, 1934.

Weindling, R., Isolation of toxic substances from the culture filtrates of *Trichoderma* and *Gliocladium*, *Phytopathology*, 27, 1175, 1937.

Weindling, R., Experimental consideration of the mold toxins of *Gliocladium* and *Trichoderma*, *Phytopathology*, 31, 991, 1941.

Weindling, R. and Emerson, O. H., The isolation of toxic substance from the culture filtrate of *Trichoderma*, *Phytopathology*, 26, 1068, 1936.

Weindling, R. and Fawcett, H. S., Experiments in the control of Rhizoctonia damping-off of citrus seedlings, *Hilgardia*, 10, 1, 1936.

Wells, H. D., *Trichoderma* as a biocontrol agent, in *Biocontrol of Plant Diseases*, Vol. I, Mukerji, K. G. and Grag, K. L., Eds., CRC Press, Boca Raton, FL, 1988, 71.

Wiley, H. B. and Kommedahl, T., Biological seed treatment in sweet corn and wheat as a component of crop management, *Phytopathology*, 91, 265, 1981.

Wilson, C. L. and Pusey, P. L., Potential for biological control of postharvest plant diseases, *Plant Dis.*, 69, 375, 1985.

Wilson, C. L. and Wisniewski, M., Biological control of post harvest disease, *Annu. Rev. Phytopathol.*, 27, 425, 1989.

Windels, C. E., Growth of *Penicillium oxalicum* as a biological seed treatment on pea seed in soil, *Phytopathology*, 71, 929, 1981.

Windels, C. E. and Lindow, S. C., *Biological Control on the Phylloplane,* American Phytopathological Society, St. Paul, MN, 1985.

Wood, R. K. S. and Tveit, M., Control of plant diseases by use of antagonist organisms, *Bot. Rev.,* 21, 441, 1955.

Wood, R. K. S. and Way, M. J., *Biological Control of Pests, Pathogens, and Weeds,* The Royal Society, London, 1988.

Yarwood, C. E., Cross protection with two fungi, *Phytopathology,* 46, 540, 1956.

Yin, S. Y., Chang, J. K., and Xun, P. C., Studies in the mechanism of antagonistic fertilizer "5406". IV. The distribution of the antagonist in soil and its influence on the rhizosphere, *Acta Microbiol. Sin.,* 11, 259, 1965.

Yin, S. Y., Keng, D. C., Yang, K. Y., and Cheu, D., A further study on the biological control of Verticillium wilt of cotton, *Acta Phytopathol. Sin.,* 3, 55, 1957.

Zambolin, L. and Schenck, N. C., Reduction of the effect of pathogenic root infecting fungi in soyabean by the mycorrhizal fungus *G. mossae, Phytopathology,* 73, 1402, 1983.

Zopf, W., Die Pilze in Morphologischer, *Physiol. System. Beziehung, Breslau,* 1980.

CHAPTER 6

Viral Control Agents

Charles M. Rush and John L. Sherwood

I. INTRODUCTION

A. Traditional Approaches to Virus Disease Control

Plant virus diseases cost producers billions of dollars each year by reducing crop yields and quality. For example, rice tungro causes $1.5 billion worth of damage to the rice crop in Southeast Asia annually, and rhizomania, caused by beet necrotic yellow vein virus (BNYVV), has been largely responsible for a 33% reduction in acreage planted to sugar beets in California since 1983 (Hull and Davies, 1992; Rush and Heidel, 1995a). Although producers minimize the impact of many virus diseases through a variety of management practices, introduction of new viruses and mutation of existing ones has continually challenged those involved in virus disease control.

The primary means of managing virus diseases has been through the use of resistant varieties, specific cultural practices, and avoidance. These techniques can be highly effective but each has limitations. Use of resistant varieties is generally considered the best option for virus disease control, but genes for resistance are often unavailable or difficult to incorporate into agronomically acceptable cultivars. Furthermore, resistance genes can be so specific that they only confer resistance

1-8493-2627-3/97/$0.00+$.50

to a single strain of a given virus. Cultural practices, such as elimination of alternate hosts or adjusting time of planting, may be effective in reducing incidence and severity of certain virus diseases but unacceptable for economic reasons. For instance, disease losses to wheat streak mosaic virus can be minimized in the Southern Great Plains by planting winter wheat in mid to late October instead of late August or early September. However, wheat in this region is typically used as winter pasture for grazing cattle, and maximum forage production can only be achieved by early planting. Therefore, depending on the price of cattle, disease control may or may not be a primary consideration. The third commonly used disease control strategy, i.e., avoidance, may be the best control option for some types of virus diseases such as those caused by soil-borne viruses. When previously uncultivated ground is available for farmers to rotate to, disease control can be immediate and absolute. Still, for a variety of agronomic, economic, or logistical reasons, avoidance may not be a realistic or practical option for disease control. In such situations where disease loss is excessive, resistant varieties are unavailable, cultural practices are ineffective, and avoidance is impractical, biological control may be a fourth option for virus disease control.

B. Biological Control of Plant Virus Diseases

Biological control is a term not generally associated with control of plant virus diseases. However, since the late 1920s, it has been recognized that infection of a plant by one virus may interfere in various ways with infection by a second one. Cross protection is the term generally used in the literature to describe this interaction. McKinney (1929) was one of the first to report on the phenomenon of cross protection in describing the ability of a light-green mosaic strain of tobacco mosaic virus (TMV) to prevent symptom expression of another strain which induced a yellow mosaic. This stimulated considerable interest within the research community and, over the years, there have been numerous studies to explore the mechanisms and practical applications of cross protection. A number of review articles providing in-depth coverage of specific aspects of cross protection have been published (Beachy, 1988; Dodds, 1990; Fulton, 1986; Gonsalves and Garnsey, 1989; Hamilton, 1980; Palukaitis and Zaitlin, 1984; Ponz and Bruening, 1986; Sherwood, 1987b; Urban et al., 1990; White and Antoniw, 1991; Wilson, 1989; Zaitlin and Hull, 1987).

Our goal in this article is to review the potential of using cross protection as a biological control for plant virus diseases. We use the term cross protection, not suggesting a specific type of viral interaction, but in its broadest general context to describe the interference of one virus with another in causing disease. Although the term cross protection has traditionally implied interactions between strains of a given virus, we will include in our discussion interactions between viruses and their satellites, interactions between unrelated viruses, and interactions between viroids.

II. TRADITIONAL CROSS PROTECTION

A. History of Concepts

Cross protection has been known by many names: acquired immunity, acquired tolerance, cross immunization, and premunity. Wingard (1928) was the first to report that recovered leaves of tobacco systemically infected with tobacco ringspot virus (TobRSV) were "immune" from further symptom development when reinoculated with TobRSV. McKinney (1929) observed this phenomenon with isolates of TMV collected in the Canary Islands that caused different mosaics on various *Nicotiana* spp. When plants with a mild dark-green mosaic were inoculated with an isolate causing a yellow mosaic, symptoms characteristic of the yellow strain resulted. However, if plants infected with a light-green mosaic strain were reinoculated with an isolate causing a yellow mosaic, no change in symptoms occurred, i.e., the plants were protected. McKinney did not apply specific terminology to these events. Thung (1931) observed that when a plant with a white mosaic (thought to have been similar to McKinney's yellow mosaic) was inoculated with the common type TMV, it did not develop symptoms of the common type. However, in contrast to McKinney's results, plants affected by the common type failed to develop white type mosaic upon inoculation with the white type, or produced a mosaic suggestive of infection by both isolates. Thung's inability to isolate the common type strain from plants with a white mosaic that had been challenged with the common type strain suggested that the common type had not been able to multiply. According to Fulton (1986), Thung used the terms "acquired virus immunity," "antagonistic action," and "domination" to describe his observations. Salaman (1933, 1938), apparently independent of Thung, discovered five days after infection that a mild strain (G) of potato virus X (PVX) protected tobacco or a *Datura* sp. from infection by a severe strain, L, even though symptoms of strain G did not appear until nine days after inoculation. As Thung had found, only the mild strain could be reisolated from the doubly inoculated plants. Salaman also showed that protection was virus specific because protection did not occur when the second inoculation was with potato virus Y (PVY) or TMV. Salaman (1937) discovered a mild strain of PVY that protected against a severe strain and was among the first to suggest that this phenomenon could have practical use. He used the terms "protective inoculation" and "acquired immunity" to describe the results.

Price (1932) applied the terms "recovery" and "acquired immunity" to the symptom sequence in tobacco infected with TobRSV. In a systemically infected plant, the new growth had mild or no symptoms, contained less virus than directly inoculated leaves, and was protected from reinfection by most, but not all, strains of TobRSV (Price 1936a,b). He demonstrated, as Salaman had done, that protection was specific for TobRSV and did not occur for other viruses (Price 1936b). Smith (1939) suggested "acquired tolerance" for this phenotype of recovery because the plant remained infected. Valleau (1935, 1941) indicated that the symptoms of "recovered" leaves of a TobRSV systemically infected plant were not mild enough to warrant the term "recovered," nor was it appropriate to name the syndrome "acquired immunity." Valleau (1935) pointed out the similarity in the results of many of the current and previous investigators in this area and concluded that the phenomenon could be explained by competition between virus strains.

Kunkel (1934) studied acquired immunity *sensu* Salaman, using "cross immunization" tests involving *Nicotiana sylvestris* and TMV and aucuba mosaic virus. His thorough study of this system demonstrated some fundamental principles of cross protection. He showed that plants systemically infected with TMV or attenuated strains of aucuba mosaic virus were protected from a necrotic strain of aucuba mosaic virus, but were not protected against cucumber mosaic virus (CMV) or TobRSV. The immunity resulting from systemic infection was closely correlated to the tissue invaded. By using the symptom of vein clearing as an indication of when the tissue was systemically invaded, he showed that only the tissue in a leaf that was infected was protected from the necrotic strain. Protection was induced on a local level by using a high concentration of a systemically infecting strain prior to the challenge inoculation. Protection began in two days and was complete four or five days after the first inoculation. The protected cells did not have to be directly inoculated. Inoculation of the abaxial side of the leaf with a systemic strain resulted in protection of the adaxial side against infection by a necrotic strain.

The term "cross protection" seems to have been introduced into the literature by Price (1940) as a synonym for the cross immunization tests of Kunkel which, according to Price, demonstrated acquired immunity. The term cross protection is now used to describe demonstration of the protection developed in an infected plant against related strains of a virus and for the events that occur in this phenomenon as well. Quanjer (1946) suggested that "premunity," from the word "premunition," be used to refer to the protection resulting from prior infection. This word denotes that insusceptibility persists only as long as the first pathogen is present. This term is not widely used.

Since Price (1940) introduced the term, many papers have appeared on cross protection, but this review will focus on those examining the possible mechanism(s) of cross protection or implementation of cross protection as a control measure. Although the result of cross protection is sought, i.e., the protective effect, instances in which protection has not occurred or only partially occurred, i.e., superinfection, have provided the most useful experimental results in providing information on how cross protection may work.

B. Mechanisms of Cross Protection

Because morphological changes in a virus-infected plant would probably not account for specificity in cross protection, the area that has received the most interest is that of physiological changes in the protected plant. These studies have resulted in many hypotheses concerning the nature of cross protection. Hamilton (1985) indicated that the condition of insusceptibility of the protected plant to superinfection by the challenge virus results from a limitation on the completion of the life cycle of the challenger. The limitations on the challenge has been measured as a (1) a decrease in the size of lesions, (2) a decrease in the number of lesions, (3) a decrease in the concentration of virus finally attained, and (4) a prevention (or delay) of virus movement or symptoms. Thus, cross protection is generally measured as a relative effect.

1. Dark Green Islands and Protoplasts

Most mechanisms proposed to account for specificity in cross protection assume that the events that control the outcome of the interaction of virus strains are relatively independent of the host and that viruses are uniformly distributed in the infected plant. However, the dark green areas of a mosaic leaf contain much less virus than the light green areas and are insusceptible, though not immune, to superinfection (Fulton 1951). The question arises as to what is responsible for the reduced virus content and the resistance to superinfection? In order to try to protect dark green areas from superinfection by a necrotic strain, Fulton (1951) inoculated mosaic leaf halves with a mosaic strain 4 to 21 days prior to inoculation by the necrotic strain. Upon inoculation with the necrotic strain, however, dark green areas of inoculated and noninoculated half leaves were equally susceptible to superinfection. Thus, the mosaic strain had been unable to manifest itself in the same manner as Kunkel found when a healthy leaf was inoculated and subsequently challenged. Atkinson and Matthews (1970) examined dark green islands in TMV-infected tobacco and found that these areas maintained a low concentration of virus over a period of weeks, even though they were in contact with light green areas with high concentrations of TMV. The dark green areas were cytologically normal and maintained normal cytological connections with light green areas. The TMV concentration in these areas could not be increased by mechanical inoculation. They proposed that dark green areas did not arise as escapes from infection during the ontogeny of the leaf, but developed and was maintained by a "dark green agent."

Murakishi and Carlson (1976) regenerated virus-free plants from cells excised from dark green islands, demonstrating that virus-free cells do exist in these areas. They also observed that the regenerated plants were insusceptible to reinfection for the first 2 months after their initiation in tissue culture, but the resistance was lost within an additional 1.5 months. Sap from insusceptible plants was inhibitory to infection by TMV on bean, as measured by reduction in lesion counts. By using other viruses, it would have been interesting to see if the factor(s) present in the sap were specific for TMV. They concluded that the inhibitory agent possibly was that proposed by Atkinson and Matthews, and that loss of insusceptibility of regenerated plants could indicate that contact with light green areas was essential for maintaining the insusceptible state. Using *N. tabacum* var. John Williams Broadleaf with a specific genetic marker, Carlson and Murakishi (1978) followed the ontogeny of clonally derived cells in the leaf. By regenerating virus-free plants from nonclonally derived cells in the same dark green areas, they concluded that the dark green areas did not arise from a single cell and that an "intercellularly communicable agent was involved in producing green islands." Using TMV, Shalla and Peterson (1978) studied the development of insusceptibility of directly inoculated leaves to superinfection. After inoculation with a systemically infecting strain (U1), they could only detect virus serologically in a small percentage of the leaf cells, yet the whole leaf was immune to superinfection by the necrotic strain (U5). They postulated that out of 4 million potentially infectible cells, over 75% remained noninfected by the protective virus, but the leaf was insusceptible to the challenger. Thus, they concluded, "there is some substance formed in response to infection which moves into cells surrounding the initial U1 infection centers which somehow inhibits replication of either U1 or U5 when either virus enters those cells, but does not interfere with the replication of other unrelated viruses."

In reciprocal cross protection tests in *N. tabacum* cvs. Samsun and Xanthi with two TMV isolates, Rezende and Sherwood (1991) found that the susceptibility to superinfection was associated with the dark green islands. The challenge virus readily spread in the plant after successful inoculation of light green areas with viral RNA and dark green areas with viral RNA or virus. Protoplasts from light and dark green areas could be superinfected as well, but less challenge virus accumulated in protoplasts from light green areas than dark green areas. As the uneven distribution of the protecting strain can influence the outcome of challenge inoculations, examination of cross protection in transgenic tobacco expressing a cDNA copy of TMV, as Yamaya et al. (1988) suggested, would be interesting if the virus produced in transgenic plants is more evenly distributed than in a natural infection.

Loebenstein et al. (1977) found that the dark green islands of CMV-infected tobacco contained significantly less virus than light green areas and had virus-free cells as indicated by lack of serological reaction with CMV antiserum. Dark green areas were resistant to superinfection by three strains of CMV, but not TMV, thus indicating the cells were capable of supporting viral replication. When infected plants were incubated for 2 to 3 weeks at 30°C, there was an increase in extractable infectivity in dark green areas. Yet, protoplasts from dark green areas incubated at 25°C did not increase in infectivity, indicating a continuously resistant state. This was in contrast to the increase in viral RNA synthesis when protoplasts from dark green islands of TMV-infected tobacco were incubated at 25°C (Folgein et al., 1975). Perhaps the lack of increase of infectivity of incubated protoplasts and inability to superinfect in the CMV system is due to a different mechanism of cross protection than that in the TMV system.

Other work with plant protoplast systems has suggested that the protective virus must be present to inhibit infection by the challenger. One must be careful in evaluating these systems in that the nature of the individual protoplast may not reflect the physiological state of the intact plant. Otsuki and Takebe (1976) demonstrated that two related strains of TMV could infect the same protoplast but, in successive inoculations, there was a greater exclusion of the challenger the longer the period between the two inoculations. Though protoplasts lose their susceptibility in culture, the loss of susceptibility of protoplasts infected with one strain of TMV was much greater to infection by another strain of TMV than to CMV.

Barker and Harrison (1978) found similar results with two strains of raspberry ringspot virus (RRV). When tobacco protoplasts were inoculated with a mixture of RRV-S and RRV-E, both strains could be detected in over 50% of the protoplasts by fluorescent antibody staining. As Otsuki and Takebe (1976) had found, sequential inoculation resulted in the exclusion of the challenger, and the longer the interval between inoculations the greater the effect. The effect was also virus specific with RRV because the exclusion of tobacco rattle virus (TRV) did not occur to any significant degree. One interesting finding was that recovered leaves on a RRS-S-infected plant could not be superinfected with the RRV-E isolate, but 11% of the protoplasts from recovered leaves of RRV-S-infected plants could be superinfected by RRV-E. This could indicate that either RRV-E infects the intact leaves but no necrosis results, or that the protoplasts from an RRV-S-infected plant behave in a manner different from the intact plant. This may further support the notion that cellular communication is required for complete expression of cross protection.

In viewing the information concerning the physiological basis of cross protection, one must note observations concerning superinfection, too. McKinney (1929) described the first reported incidence of superinfection. A mild, dark green mosaic on tobacco did not protect the plant from infection by a yellow mosaic. Fulton (1950) found that there was not complete protection between some strains of CMV in cowpea and *Zinnia* sp. Price (1932) found "recovered" tobacco was not immune to infection by all strains of TobRSV. Reports of superinfection between isolates of PVX, maize streak virus, and sugar beet curly-top virus have been summarized by Bennett (1951). Fulton (1951) challenged *N. sylvestris* systemically infected with TMV with isolates of TMV causing local lesions. He was able to produce lesions only in the dark green areas, and found that the isolates that most readily superinfected were those that caused necrosis in healthy *N. sylvestris* with the least multiplication. Cassels and Herrick (1977) studied cross protection in tomato with systemically infecting mild and severe strains of TMV. They found that the presence of a mild strain in tomato delayed symptom expression of the severe strain, yet the severe strain was serologically detectable. Upon increasing the concentration of the severe strain used to challenge plants, symptoms of the severe strain became evident earlier. Dominance of the severe strain in the plant did not result in a reduction of the mild strain, i.e., protection was unilateral. The authors suggest that "in the interaction between the mild cross protection strain and the severe strain, the mild strain, entering the cell before the severe strain, suppresses the production of the severe strain inoculum." The severe strain may become dominant, however, because "it moves faster in the plant" and "it is produced in greater quantity." It invades the cell first prior to the mild strain, thus "eliminating the interaction." The question remains, however, why does the severe strain have no effect on the mild strain? Fulton (1978) studied superinfection using a number of isolates of tobacco streak virus (TSV). As had been found with other viruses, the degree of protection varied between strains. Of the 28 pairs of strains tested, 8 reciprocally protected, 8 unilaterally protected, and 12 did not protect in either direction. When nonsuperinfecting strains were combined with the challenger, they interfered with infection by the challenger. Thus, specificity probably does not lie in the infection site. He also assigned the function of superinfection to the bottom particle by making heterologous mixtures between strains. The bottom component contains the RNA that encodes the coat protein.

2. Protection between Unrelated Viruses

The question of specificity is crucial to cross protection. This is an area of particular confusion considering the number of examples in the literature where protection may occur between unrelated viruses. Price (1932) found that TMV produced milder symptoms when inoculated onto tobacco systemically infected with TobRSV or if TMV and TobRSV were mixed prior to inoculation onto healthy tobacco. McKinney (1941) noted that *Nicotiana* spp., systemically infected by different viruses, had reduced susceptibility to infection by a lesion-forming strain of TMV. The extent of reduction was not as great, however, as when TMV was the systemically infecting strain. Thomson (1958), as well, made similar observations with a number of different viruses. Procházhová (1970) found that systemic infection with CMV or PVY reduced susceptibility of tobacco to superinfection by a necrotic lesion strain of TMV. Kassanis et al. (1974) found that leaves either previously inoculated or systemically infected with viruses unrelated to TMV were less susceptible to infection by TMV. Nitzany and Cohen (1960) found that alfalfa mosaic virus (AMV) would protect against CMV but not TMV. Fulton (1975) noted that tomato ringspot virus (TmRSV) protected against unrelated cherry leafroll virus (CLRV), but protection did not occur in the other direction. The inference one can make from these observations is that protection between viruses entails many aspects. Viral infection may reduce susceptibility of the plant to infection from another virus regardless of the relationship of the two viruses. In addition, there is a mechanism (or mechanisms) that accounts for specificity. This has resulted in many hypotheses to account for specificity in cross protection.

3. Mechanisms of Specificity in Cross Protection

a. Utilization and Depletion of Host Metabolites or Structures

Ross (1974) proposed the available ribosomes would be utilized by a rapid increase in RNA of the initial virus and that would leave the challenge virus RNA unbound and, therefore, susceptible to degradation. A protein that is required as part of the RNA-dependent RNA polymerase of an RNA virus possibly could also be depleted to prevent challenge virus replication (Ponz and Bruening, 1986).

b. Specific Sequestering of Nucleic Acid

Palukaitis and Zaitlin (1984) suggested that for (+)RNA viruses, infection by the challenging virus would be reduced or prevented by inhibiting its ability to synthesize (+)RNA. This could occur when the nascent (–)RNA of the challenge virus became hybridized to the excess (+)RNA of the protecting strain. If a mechanism is to account for cross protection that occurs between strains of viroids as well as between strains of viruses, it will most likely involve the regulation of replication of nucleic acid.

c. Involvement of Coat Protein

de Zoeten and Fulton (1975) suggested that the coat protein present in the cell would encapsidate the viral RNA of the challenge strain before it could be expressed. Zaitlin (1976) tested this hypothesis using a coat protein mutant of TMV (PM1) that produced insoluble coat protein that did not encapsidate TMV RNA. When plants inoculated with the coat protein mutant were challenge-inoculated with TMV (U1), less of the U1 strain was recovered than in the controls. This suggests protection was achieved by the coat protein mutant, and encapsidation of the challenge RNA was not involved in protection in this system. Although coat protein may not encapsidate the RNA of the challenge virus, Horikoshi et al. (1987) found that the coat protein of brome mosaic virus, *in vitro*, blocked the binding site of the replicase, thereby interfering with the RNA synthesis. They suggested that the regulation of replication by coat protein may be the basis of cross protection.

The inability of the challenge virus to uncoat when inoculated onto plants containing a protecting strain was proposed to be the basis for cross protection with TMV in *Nicotiana sylvestris* (Sherwood and Fulton, 1982). When a leaf with mosaic is inoculated with a strain of TMV that produces local lesions in *N. sylvestris*, the lesions are confined to the dark green areas of the mosaic (Fulton, 1951; Sherwood and Fulton, 1982). When a leaf with mosaic symptoms is inoculated with RNA from a strain of TMV that produces local lesions, the lesions are produced in both the light green and the dark green areas (Sherwood and Fulton, 1982). This suggests that the necrotic lesion-producing strains are unable to uncoat in the light green areas where the concentration of TMV is the greatest in the mosaic leaf. de Zoeten and Gaard (1984) reported that there is 2 to 7.5 times more TMV antigen detected in cell walls of light green areas than in dark green areas. The first encounter between the challenge virus and the protected host may be at the cell wall. Other supporting evidence for the involvement of coat protein in cross protection was found by Dodds et al. (1985) in tomato protected with CMV. As with TMV in *N. sylvestris*, superinfection occurred when the leaves were challenged with viral RNA but not with intact virions.

Results from experiments with coat protein-free mutants are equivocal. Sherwood (1987a,b) inoculated leaves of *N. sylvestris* with the DT-1G coat protein-free strain of TMV (Sarkar and Smitamana, 1981) or TMV-C. The leaves were then challenged with a necrotizing strain of TMV or turnip mosaic virus (TuMV) to test nonspecific protection. The leaves inoculated with strain DT-1G had equal susceptibility to TuMV and the necrotizing strain of TMV. Infection was less with the necrotizing strain than with TuMV when TMV-C was the protecting virus. Gerber and Sarkar (1989) found that the DT-1G mutant of TMV induced up to a 90% degree of cross protection against the U1 strain of TMV in 'Samsun' tobacco. These conflicting reports may indicate that the mechanism of cross protection may be different with different hosts.

Coat protein is likely not the only factor involved in cross protection. Zinnen and Fulton (1986) compared the cross protection effects of two tobamoviruses: sunn-hemp mosaic virus (SHMV) and TMV-C. SHMV-infected cowpeas were challenged with either virions or RNA of a necrotic lesion-producing strain of SHMV produced by nitrous acid. In both cases, protection occurred. However, when SHMV-infected plants were challenged by TMV-C RNA encapsidated in SHMV coat protein or TMV-C intact virions, the SHMV-encapsidated TMV-C-inoculated plants produced 5- to 27-fold fewer lesions on the SHMV-infected cowpeas.

d. Prevention of Systemic Spread of the Challenge Virus

The initial virus infection establishes a mechanism for virus spread throughout the plant. The challenge virus may be limited in its ability to spread from the site of initial inoculation by saturation or changes in the putative receptors involved in cell-to-cell movement of the virus. Dodds (1982) and Dodds et al. (1985), using tomato as the host and two strains of CMV, showed that the challenge strain was able to increase only in the inoculated leaves. When the protecting strain was not systemically well established in the plant, an increase in the challenge strain also occurred in the upper leaves.

Using serologically distinct strains of TMV that reciprocally cross protect in *Arabidopsis thaliana*, it was found that the challenge strain multiplies in inoculated leaves to concentrations detectable by ELISA, but it did not move systemically in the plant (Rezende et al., 1992). This suggests an impairment of systemic movement of the challenge virus and a possible interaction of the 30-kDa movement protein with a host component. Blum et al. (1989) suggested a viral-induced, host-specific inhibitor of viral transport. They hypothesized that the challenge virus cannot synthesize its movement protein because the protecting strain induces the host to synthesize an inhibitor of 30 kDa synthesis. They further suggested that if the challenge virus cannot use the movement system of the protecting strain, or the protein has already dissipated, the challenge virus will be unable to spread. Moser et al. (1988) found that in a TMV local lesion host, 'Samsun NN,' the amount of movement protein in cell wall fractions decreased when necrosis became visible and the production of coat protein ceased.

C. Cross Protection and Virus Disease Control

Although Salaman (1937) and Johnson (1937) suggested the use of cross protection for disease control over 50 years ago, there are few examples of the use of cross protection for the control of plant virus diseases. Filling all the requirements to implement cross protection in the field can be difficult. These include selection of suitable protecting isolates, preliminary greenhouse and field trials, deployment of the protective isolate or material inoculated with the protective isolate, and integration of cross protection into grower practices. Although the literature is replete with references identifying protective isolates and success in small trials, the complete stepwise process leading to control has only been successful for a few virus diseases.

Some approaches to obtain potential protective isolates are the selection of naturally occurring mild strains, using nitrous acid to induce mutation followed by selection, use of low temperature to select attenuated isolates (Kosaka and Fukunishi, 1993), and passage through a selective host (Yeh and Cheng, 1989). Some studies have indicated that the relatedness of the coat protein between isolates might be a predictor of the effectiveness of cross protection between virus isolates (Bitterlin and Gonsalves, 1988; Krstic et al., 1995; Wen et al., 1991). Culver (1996) has developed a PVX vector that may permit testing the protective ability of coat protein constructs without having to go through the transformation process. TMV coat protein was expressed in *Nicotiana benthamiana* plants inoculated with the PVX vector containing the ORF for TMV coat protein, and plants were protected from subsequent inoculation with TMV, but not TMV RNA or a TMV construct expressing a distantly related tobamovirus coat protein. As the protective ability of the isolate can vary with host, e.g., PVY isolates in potato cultivars (Singh and Singh, 1995), the potential candidates must be thoroughly evaluated in the host of interest as well.

Cacao swollen shoot in Africa (Posnette and Todd, 1955) and passion fruit woodiness in Queensland, Australia (Simmonds, 1959) were the first virus diseases in which plants protected with mild strains showed good development and yield increase under field conditions. Programs for the control of these diseases by cross protection have not continued, and the diseases are still of economic importance. In Ghana, the potential for using cross protection to control cacao swollen shoot virus is considered positive (Hughes and Ollennu, 1994). Citrus tristeza, tomato mosaic, and papaya ringspot diseases appear to be the only extensive examples in which cross protection has been earnestly investigated for disease control.

Grant and Costa (1951) demonstrated that cross protection could be used to control citrus tristeza in Brazil. In 1961, Müller and Costa (1977) initiated a research program to utilize cross protection to control citrus tristeza in Sáo Paulo State, Brazil. The isolation of suitable mild strains required an immense effort. The identification of isolates to be used, based on mild symptom expression, was not sufficient. Each isolate had to be checked in each type of citrus to be protected. From some 45 mild isolates, each checked on Pêra sweet orange, Galego lime, and Ruby Red grapefruit, which had each been budded on three different rootstocks, only six isolates were satisfactory. Three isolates could be used for Pêra sweet orange, two isolates for Galego lime, and one isolate for Ruby Red grapefruit. An isolate that worked in one type of citrus did not work in another. By 1987, a total of 50,000,000 Pêra sweet orange trees protected with a mild strain had been planted in Brazil (Salibe, 1987). Control of citrus tristeza with cross protection has been successful in other countries (Hamilton, 1985), but the long-term utility of this control measure in South Africa and Florida is not conclusive (Powell et al., 1992; van Vuuren et al., 1993).

Cross protection was also widely applied for control of tomato mosaic caused by tomato mosaic virus (Broadbent, 1976; Channon et al., 1978; Fletcher and Rowe, 1975; Rast, 1975). However, difficulties have developed in trying to control tomato mosaic using cross protection. Mild strains generally protect less effectively at 25 to 30°C than at lower temperatures. The mild strain MII-16, produced by nitrous acid mutation of Strain I (Rast, 1975), does not protect tomatoes against all other strains of the tomato mosaic virus. In some areas where the MII-16 strain has been used, there has been an increase in the prevalence of Strain I. Today, growers primarily rely on resistant varieties to control the disease (Fulton, 1986).

The use of cross protection to control papaya ringspot disease has been investigated in Taiwan and Hawaii (Wang et al., 1987; Yeh and Gonsalves, 1984; Yeh et al., 1988) and Brazil (Rezende et al., 1981; Rezende, 1985). A mild strain of papaya ringspot virus (PRV), produced by nitrous acid treatment of PRV, was used in field experiments in Taiwan. Papaya plants protected by the mild strain of PRV had 82% greater fruit yield than unprotected trees, resulting in a 111% increase in grower income. Protection was effective when protected plants were planted in solid blocks, and disease pressure within the test orchard was minimized by rouging severely-infected plants once every 10 days up to flowering stage. However, protection waned under high disease pressure by other strains from areas near the protected trees. Due to the success of these experiments, more than 1 million papaya seedlings inoculated with the mild strain of PRV were planted in the field in 1986 (Yeh et al., 1988). In Brazil, in spite of very satisfactory results of protection under greenhouse and field experiments, protected papaya plants in the field showed mild symptoms for only 6 to 8 months. Afterwards, the symptoms increased in severity. Symptom intensification occurred in a synchronized manner in all plants. The change of symptoms was not considered a breakdown of protection, but rather a change in the mild strains selected from the field due to mutation and selective competition (Rezende and Costa, 1987). The success of cross protection as a method of controlling papaya ringspot in Brazil is dependent upon finding stable mild strains.

Although cross protection has been commercially applied to control a few diseases, it may be an effective option for control of many other diseases. A zucchini yellow mosaic virus variant (ZYMV-WK) was shown in France (Lecoq et al., 1991), Taiwan (Wang et al., 1991) and California (Perring et al., 1995) to control ZYMV in a number of cucurbits. A unique use of cross protection in Japan was to prevent seed transmission of soybean mosaic virus (Kosaka and Fukunishi, 1994). Cross protection has been proposed for control of cauliflower mosaic virus (CaMV) in brussels sprouts (Tomlinson and Shepherd, 1978), CMV in tomato (Dodds, 1982; Yoshida et al., 1985) and pepper (Tien and Chiang, 1983), TMV in sweet pepper (Goto et al., 1984), PVY in tobacco (Latorre and Flores, 1985), tomato aspermy virus in tomato (Kuti and Moline, 1986), and grapevine fanleaf virus in grapevines (Huss et al., 1989). Fulton (1986) suggested that avocado sun blotch, concave gum and psoriasis of citrus, and some stone and pome fruit viruses may be effectively controlled by cross protection. Greater attention to the practical application of cross protection could result in control of more virus diseases by this method.

D. Potential Problems in Use of Cross Protection

The decision to use cross protection to control diseases must be well thought out and precautions should be taken to prevent additional problems (Fulton, 1986). One should be concerned about the degree of protection the mild strain affords since cross protection is not necessarily complete. Other concerns are that a strain that is mild in the crop of interest may not be mild in other crops in the area, the strain may interact with other viruses in a synergistic fashion, or the mild strain may mutate to a more severe strain. Finally, the cost vs. potential success of using cross protection should be determined. Of particular importance is the feasibility of inoculating plants. The anxiety of potential problems could be the basis for the scarcity of examples where cross protection has been used for disease control in the field.

Of some concern has been the possibility of virus interaction resulting in new viruses. A number of mechanisms generate variability in viral genomes. Some of the mechanisms that may lead to variability in RNA viruses, the largest group of viruses infecting plants, are transcapsidation, template switching (de Zoetin, 1991), RNA-RNA recombination (Simon and Bujarski, 1994), and replication slippage (Hancock et al., 1995).

III. POSSIBILITIES FOR NONTRADITIONAL CROSS PROTECTION

A. Viroids

Viroids are low molecular weight RNAs which are capable of causing disease in a variety of host species. Viroids differ from the majority of conventional plant viruses in five respects:

1. The RNA molecule is unencapsidated *in vivo*;
2. Virion-like particles are not readily detectable in infected tissue;
3. The infectious RNAs are of low molecular weight;
4. Unlike satellite RNAs or satellite viruses, viroids replicate autonomously in susceptible cells, i.e., without a helper virus; and
5. The infectious RNA consists of only one molecular species (Diener, 1987c).

Since none of the presently recognized viroid diseases were known to exist before the twentieth century, questions as to their origin have naturally arisen. Although it is possible that pathogenic viroids have developed through mutation of normal cellular RNAs, it is more likely that viroids have existed for years in wild host species and only by chance have been transferred to cultivated species (Diener, 1987a). Because the recognition of viroid diseases has generally paralleled the development of modern agriculture, the lack of genetic diversity in many of today's intensively cultivated crops possibly is associated with the occurrence of these diseases. Attempts to identify and incorporate genes for resistance to viroid diseases have generally met with failure, and use of cultural practices to reduce disease incidence and severity have been only partially successful (Broadbent and Garnsey, 1987; Diener, 1987b; Randles, 1987). Because the potential and actual losses from several viroid diseases are great, cross protection has been considered and evaluated as an option for disease control.

Several of the earlier studies of cross protection with viroids demonstrated that inoculation of plants with a protecting mild strain could delay symptom appearance following inoculation with a severe strain (Fernow, 1967; Niblett et al., 1978). However, unlike many examples of classical cross protection with plant viruses, the challenge strain was able to replicate in protected plants, and symptoms of the severe strain eventually appeared. Similar results have been reported in a series of more recent studies from Canada using potato spindle tuber viroid (PSTV) as a model (Khoury et al., 1988; Singh et al., 1989, 1990). Khoury et al. (1988) reported that a mild strain of PSTV delayed the onset of symptoms but the severe challenge strain not only replicated in protected tomato plants, but eventually replaced the protecting strain. Later, Singh et al. (1989) demonstrated that as the interval increased between inoculation of the protecting and challenge strains, the number of doubly infected plants decreased. A similar result was achieved by increasing the inoculum concentration of the protecting strain, although the effect was less pronounced. The authors concluded that both the concentration of the protecting strain and the establishment of the protecting strain in most tissues of the plant was important in maximizing the duration of cross protection. This conclusion was supported by the finding that second-generation potato plants infected with a mild strain were completely protected from subsequent infection and the severe strain could not be detected (Singh et al., 1990). However, this protection could be overcome when the severe challenge strain was introduced via graft inoculation. Singh et al. (1990) suggested that the use of cross protection for control of PSTV might not be useful if protection was not complete, and that the extent of cross protection of PSTV was highly dependent on the host plant.

Cross protection for control of citrus exocortis viroid (CEV) is also unlikely (Garnsey et al., 1988; Duran-Vila and Semancik, 1990). Garnsey et al. (1988) reported that mild isolates of CEV did not confer cross protection against severe challenge strains. However, Duran-Vila and Semancik (1990) showed that the "strains" used by Garnsey were actually mixtures of the small Group II and III citrus viroids which are not true strains of CEV. When true strains of CEV were used in cross protection experiments, the degree and duration of protection was dependent on the interval between inoculation with the protecting and challenge strains. As this interval was lengthened, the appearance of severe symptoms was delayed. Although symptom appearance of the severe strain was delayed, the severe strain still multiplied to substantial levels in symptomless plants. Furthermore, when extracts from these plants were used as inoculum, severe symptoms developed in inoculated plants. Based on these results and the definition of cross protection put forth by Palukaitis and Zaitlin (1984), Duran-Vila and Semancik (1990) suggested that the term cross protection was inappropriate to describe the interaction between mild and severe strains of viroids. Dodds (1990) disagreed and argued that the term cross protection was applicable to interactions between viroids, and Singh et al. (1990), following the suggestion of Hamilton (1980), used the terms interference and immunity to describe the interactions they observed.

Regardless of the terms used to describe the interaction between mild and severe strains of viroids, the effective use of mild strains for biological control of viroid diseases is unlikely. This conclusion is based on the following observations from several studies: (1) protection is variable and highly dependent on host species, (2) symptom expression is delayed but viroid replication continues, (3) symptom suppression is dependent on the interval between protection and challenge inoculation, and (4) in most cases, severe symptoms are eventually expressed.

Even though the potential for biological control of viroid diseases through cross protection is unlikely, study of the interactions between mild and severe strains may help elucidate the mechanisms of cross protection. Studies to date have probably revealed more about mechanisms that are not active in cross protection with viroids than those that are. Obviously, any theory associating cross protection with coat protein activity is inapplicable to interactions between viroids. The theory of RNA sequestration put forth by Palukaitis and Zaitlan (1984) as a possible mechanism for cross protection in virus and viroid systems is also questionable. In most studies of cross protection with viroid systems, even though symptom expression is suppressed or delayed, viroid replication continues at near-normal levels (Singh et al., 1989, 1990; Duran-Vila and Semancik, 1990). The fact that symptom attenuation is unrelated to viroid titer suggests that sequestration is not occurring at any appreciable level. Furthermore, the different responses observed in various hosts protected and challenged with the same isolates suggests that, with viroids, the protective ability of a particular strain is strongly influenced by factors from the host.

B. Unrelated Viruses

As indicated above, viral infection may reduce the susceptibility of a plant to infection from another virus regardless of the relationship of the two viruses. This fact has provided the impetus to investigate the possibility of using cross protection as a biological control of the rhizomania disease of sugar beet caused by beet necrotic yellow vein virus (BNYVV).

BNYVV is a furovirus (fungal-transmitted, rod-shaped virus) vectored by *Polymyxa betae* (Richards and Tamada, 1992). Recently, Kruse et al. (1994) reported the existence of two major strain groups of BNYVV. The two groups were defined by minor but highly stable differences in nucleotide sequence. The nucleotide variation, however, did not result in serological or biological differentiation of the two strains. Although it was suggested that observed differences in disease resistance of new sugar beet genotypes grown in various geographical regions could be related to strain distribution, there was no indication that any sort of cross protection was involved. Conversely, Prillwitz and Schlosser (1993) demonstrated that beet soil-borne virus (BSBV) could reduce titer and attenuate symptom development in sugar beets challenged with BNYVV. Like BNYVV, BSBV is a furovirus but is serologically unrelated and has a considerably different genomic organization (Koenig et al., 1996). Even though BSBV was able to reduce damage from infection by BNYVV, it would not make a suitable biocontrol agent because BSBV also reduced root yield 50% compared to the virus-free check. However, BSBV might have important epidemiological effects on the development of rhizomania. Prillwitz and Schlosser (1993) suggested that because BSBV has a lower temperature optimum than BNYVV, it might infect plants earlier in the growing season and reduce damage from subsequent infection by BNYVV.

Beet soil-borne mosaic virus (BSBMV) is another virus which has been evaluated for its affect on BNYVV (Lovic and Rush, 1995; Rush and Heidel, 1995a,b). BSBMV has a polyadenylated, quadripartite genome which possesses considerable nucleotide sequence homology with BNYVV (Heidel and Rush, 1993; Rush et al., 1993, 1996). These features, plus similarities in host range, capsid size, and a weak serological relation, make BSBMV more similar to BNYVV than any other virus and prompted speculation that BSBMV was possibly a mild strain of BNYVV (Rush et al., 1993; Rush and Heidel, 1995a). However, recent studies of sequence homology between BSBMV and BNYVV suggest that BSBMV is probably a distinct virus (Rush et al., 1996).

In preliminary studies, when BSBMV and BNYVV were coinoculated to various hosts, symptom expression by BNYVV was completely suppressed (Rush et al., 1995). This led to controlled field experiments to evaluate BSBMV as a possible biocontrol agent (Lovic and Rush, 1995). Using a seed treatment technique previously shown to be an effective means of infecting sugar beet seedlings with BNYVV (Harveson and Rush, 1993), specific isolates of BSBMV were tested for their ability to protect sugar beets from BNYVV. In the first year of testing, one BSBMV isolate significantly reduced disease severity, increased sucrose content, and increased root yield 67% compared to unprotected control plants (unpublished data). Two other isolates had no effect on rhizomania development. These experiments are noteworthy because they are the first to evaluate cross protection with furoviruses in field studies using the natural vector to infect plants with the protecting virus. The use of seed treated with viruliferous resting spores of the natural vector as a means of infecting plants with the protecting virus is also a novel approach to cross protection. It reduces two of the major problems associated with the use of cross protection as a means of disease control, i.e., an economically feasible method of introducing the protecting virus and spread to other cultivated hosts.

C. Satellites

Satellites are small RNAs typically ranging in size from just under 200 nucleotides (nt) to approximately 1500 nt. They are totally dependent on a helper virus for replication and normally lack significant sequence homology with their helper (Roossinck et al., 1992). However, there are examples of chimeric RNA molecules, which have been referred to as satellites, that have sequence homology with satellite and genomic RNA (Li and Simon, 1990; Zhang et al., 1991). Some satellites encode their own capsid (satellite viruses), while others rely on their helper for encapsidation (satellite RNAs). Small RNAs from multipartite viruses, such as RNAs 3 and 4 of BNYVV and RNA 3 of GRV, possess all of the above-mentioned characteristics of satellite RNAs but, because they are required for survival in nature (Murant, 1990; Murant and Kumar, 1990; Murant et al., 1988; Richards and Tamada, 1992), they should not be considered as true satellites. Since their discovery in 1962 (Kassanis, 1962), satellite RNAs (satRNA) have received considerable attention because of the effects they can have on symptom expression and replication of the helper virus in infected plants. Complete coverage of these unique molecules and the interactions they have with their helper and host is beyond the scope of this chapter, but several excellent reviews addressing various aspects of satellite research have been published (Collmer and Howell, 1992; Fritsch and Mayo, 1989; Kaper, 1992; Kaper and Collmer, 1988; Murant and Mayo, 1982; Roossinck et al., 1992; Simon, 1988; Tien and Wu, 1991). Because of the extensive amount of research conducted on CMV and its satRNA, the focus here will be to provide an overview of CMV satRNAs as potential biocontrol agents and the potential problems associated with their use.

CMV, the type member of the cucumovirus group, is a small icosahedral tripartite virus that includes a subgenomic RNA (RNA 4) and may also support satRNA (RNA 5). Satellites of CMV have been isolated from around the world and have a wide range of effects on symptom expression and virus accumulation in infected plants (Collmer and Howell, 1992; Francki, 1985; Fritsch and Mayo, 1989; Jorda et al., 1992; Kaper and Collmer, 1988; Wu et al., 1989; Yoshida et al., 1985). Depending on its CMV helper strain and host species, an individual CMV satRNA may attenuate, exacerbate, or not affect disease severity. Furthermore, small mutations involving only a few nt can alter a satellite's ability to modulate symptom expression of its helper and affect other important biological functions (Gal-on et al., 1995; Jaegle et al., 1990; Kurath and Palukaitis, 1990; Masuta and Takanami, 1989; Moriones et al., 1991; Moriones et al., 1992; Palukaitis, 1988; Roossinck et al., 1992; Sehgal et al., 1993; Sleat and Palukaitis, 1990, 1992).

In recent years, diseases caused by CMV and associated satRNAs have become increasingly important in several crops, but especially tomato. For instance, beginning in 1988, tomato production in southern Italy was plagued by outbreaks of lethal tomato necrosis which is caused by CMV plus a necrogenic satRNA (Crescenzi et al., 1993; Gallitelli et al., 1991; Kaper et al., 1990). Measures to control aphid vectors typically failed to reduce disease incidence and cultural practices were also ineffective in disease control. The recognition that some CMV satRNAs were able to attenuate disease symptoms along with the lack of other options, prompted studies to evaluate satellite-mediated cross protection as a means of biological control (Gallitelli et al., 1991; Montasser et al., 1991). Gallitelli et al. (1991) mechanically inoculated several hundred tomato plants with a mild strain of CMV plus a nonnecrogenic satRNA and, after 34 days, transplanted the "protected" plants to the field. At the time of transplanting none of the inoculated plants showed obvious disease symptoms, but 65 of 66 plants tested positive for both CMV and its satRNA, and 55 days after transplanting, lethal necrosis appeared in the field. Incidence of disease in protected plants was less than 5%, and fruit yields were double those from unprotected plants. Fruit quality seemed to be unaffected by the mild strain plus its satRNA. Disease incidence among unprotected plants in the field was approximately 40%, which was considerably lower than surrounding fields where disease losses reached 80 to 100%. It was suggested that the lower disease incidence in the test field could have resulted from natural aphid transmission of the protecting strain. However, Tien and Wu (1991) reported that mild CMV strains used in cross protection studies were inefficiently transmitted by aphids.

In a companion study using the same mild CMV and nonnecrogenic satRNA as Gallitelli et al. (1991), Montasser et al. (1991) showed that protected plants were able to resist the pathogenic effects of a severe strain of CMV when challenged three weeks after the protective inoculation. They concluded that satellite-mediated protection was superior to traditional cross protection using mild CMV alone because it was more protective and established sooner. This conclusion is in agreement with other studies (Piazzolla et al., 1982; Sayama et al., 1993; Wu et al., 1989).

Sayama et al. (1993) tested the efficacy of a mild strain of CMV, designated KO2, for protecting against lethal necrosis of tomato. Plants were preinoculated with CMV-KO2 with or without an attenuating nonnecrogenic satellite, which had a unique sequence compared to previously reported satRNAs. Plants protected with CMV-KO2 and subsequently challenged with a severe CMV strain exhibited minimal disease development, but when a necrogenic satRNA was added to the challenge inoculum, severe disease developed. Conversely, when the nonnecrogenic satRNA was included with CMV-KO2 in the protecting inoculum, plants were protected from severe disease even when the necrogenic satRNA was added to the challenge inoculum, and yields were increased up to 200% over unprotected controls. These results demonstrated the importance of including a nonnecrogenic satRNA in the protecting inoculum and indicated that a mild satRNA can protect against one which exacerbates symptoms of its helper. In support of this concept, Kurath and Dodds (1994) demonstrated that satRNAs could interfere with each other in a cross protection-like phenomenon even when they differed in sequence homology by as little as five nucleotides. Furthermore, Rubino et al. (1992) reported that satRNA of cymbidium ringspot virus in transgenic plants could suppress accumulation of DI RNA.

The mechanism(s) by which a satRNA modulates symptom expression of its helper virus is uncertain. However, one theory which fits much of the existing data rather well proposes that symptom attenuation, which is often accompanied by a reduction in virus accumulation, is a result of competition between the helper virus and the satRNA for replicase-related factors (Collmer and Howell, 1992; Collmer et al., 1992; Gal-on et al., 1995; Montasser et al., 1991; Piazzolla et al., 1982; Tien et al., 1987). According to this theory, satRNA is able to out-compete the helper virus for the limited but unknown factors required for viral replication. As these factors are depleted, viral concentrations drop and disease symptoms are modulated. However, since satRNA is dependent on the helper for replication, as titer of the helper drops so does that of the satRNA. This results in a cycle of increase and decline for both the helper and the satRNA. In such a scenario, any new virus, such as a severe CMV strain which attempts to infect the host, will trigger an increase in satRNA and be prevented from replicating to damaging levels in the satRNA-protected plant (Montasser et al., 1991). As stated, this theory fits much of the existing data and observations of disease development in nature. However, there are examples in which symptom attenuation is not accompanied by a concomitant reduction in virus accumulation and other instances in which virus accumulation is reduced but symptom expression is not (Gal-on et al., 1995; Wilson, 1993). This suggests that, as with traditional cross protection, more than one mechanism of disease attenuation is involved.

Many of the fundamental questions concerning mechanisms involved with satRNA-mediated protection are unanswered. Likewise, just as with traditional cross protection, there are many important questions concerning the wisdom and safety of using satRNA-mediated protection that should be answered. Although there are examples where satRNA-mediated protection has been successfully used in the field (Tien and Wu, 1991), there are definitely risks that need to be addressed before such technology is deployed on a larger scale (Collmer and Howell, 1992; Roossinck et al., 1992; Sayama et al., 1993; Wilson, 1989). Because of the inherent dangers in releasing highly mutable satRNA into the environment, many have investigated the feasibility of using satRNA sequences in transgenic plants (Baulcombe et al., 1986; Harrison et al., 1987; Jacquemond et al., 1988; McGarvey et al., 1994; Quemada et al., 1991; Scholthof et al., 1993; Wilson, 1993; Yie et al., 1992). The advantages of disease control through development of transgenic plants are numerous, but this too has potential dangers. For more complete coverage of the use of genetic engineering in achieving disease control, see Chapter 14 in this volume.

IV. CONCLUSIONS

Perhaps an appropriate approach to concluding this chapter is to borrow the titles from two comments to the Letter to the Editor section of *Phytopathology:* "Understanding Generates Possibilities" by de Zoeten and Fulton (1975) and "Viral Cross-protection: More Understanding is Needed" by Zaitlin (1976). Testable hypotheses were generated indicating an emerging understanding on how cross protection may work about the time the possibilities for plant transformation came to fruition. This resulted in a number of comparisons between pathogen-derived resistance (*sensu* Sanford and Johnston, 1985) and cross protection. Although the phenomenon of pathogen-derived resistance with tobamovirus coat protein and cross protection with TMV have many similarities (reviewed in Register and Nelson, 1992), this may not have been the case with potyviruses (Cassidy and Nelson, 1995; Goodwin et al., 1996). That more understanding is needed is clear from the paucity of examples where cross protection has been used and the lack of conclusions on how cross protection works. The mechanism of cross protection likely involves events in addition to the ones discussed here. The use of transgenic plants to identify the function of specific viral and plant genes and their products brings a new dimension to this on-going research. The role the host plays in the events that occur in the cross protection phenomenon must not be overlooked. A concurrent development of an understanding of the biology of plant viruses, of hosts of plant viruses, and of the interaction that occurs between the two will be necessary to fully utilize cross protection as a control measure for virus-induced disease.

REFERENCES

Atkinson, P. H. and Matthews, R. E. F., On the origin of dark green tissue in tobacco leaves infected with tobacco mosaic virus, *Virology*, 40, 344, 1970.

Barker, H. and Harrison, B. D., Double infection, interference and superinfection in protoplasts exposed to two strains of raspberry ringspot virus, *J. Gen. Virol.*, 40, 647, 1978.

Baulcombe, D. C., Saunders, G. R., Bevan, M. W., Mayo, M. A., and Harrison, B. D., Expression of biologically active viral satellite RNA from the nuclear genome of transformed plants, *Nature*, 321, 446, 1986.

Beachy, R. N., Virus cross-protection in transgenic plants, in *Temporal and Spatial Regulation of Plant Genes*, Verma, D. P. S. and Goldberg, R. B., Eds., Springer-Verlag, New York, 1988, 313.

Bennett, C. W., Interference phenomena between plant viruses, *Annu. Rev. Microbiol.*, 5, 295, 1951.

Bitterlin, M. W. and Gonsalves, D., Serological grouping of tomato ringspot virus isolates: Implications for diagnosis and cross-protection, *Phytopathology*, 78, 278, 1988.

Blum, H., Gross, H. J., and Beier, H., The expression of the TMV-specific 30-kDa protein in tobacco protoplasts is strongly and selectively enhanced by actinomycin, *Virology*, 169, 51, 1989.

Broadbent, L., Epidemiology and control of tomato mosaic virus, *Annu. Rev. Phytopathol.*, 14, 75, 1976.

Broadbent, P. and Garnsey, S. M., Citrus exocortis, in *The Viroids*, Plenum Press, New York, 1987, 235.

Carlson, P. S. and Murakishi, H. H., Evidence on the clonal vs. non-clonal origin of dark green islands in virus infected tobacco leaves, *Plant Sci. Lett.*, 13, 377, 1978.

Cassels, A. C. and Herrick, C. C., Cross protection between mild and severe strains of tobacco mosaic virus in doubly inoculated tomato plants, *Virology*, 78, 253, 1977.

Cassidy, B. G. and Nelson R. S., Differences in protection phenotypes in tobacco plants expressing coat protein genes from peanut stripe potyvirus with or without an engineered ATG, *Mol. Plant Microb. Interact.*, 8, 357, 1995.

Channon, A. G., Cheffins, N. J., Hitchon, G. M., and Barker, J., The effect of inoculation with an attenuated mutant strain of tobacco mosaic virus on the growth and yield of early glasshouse tomato crops, *Ann. Appl. Biol.*, 88, 121, 1978.

Collmer, C. W. and Howell, S. H., Role of satellite RNA in the expression of symptoms caused by plant viruses, *Annu. Rev. Phytopathol.*, 30, 419, 1992.

Collmer, C. W., Stenzler, L., Chen, X., Fay, N., Hacker, D., and Howell, S. H., Single amino acid change in the helicase domain of the putative RNA replicase of turnip crinkle virus alters symptom intensification by virulent satellites, *Proc. Natl. Acad. Sci.*, 89, 309, 1992.

Crescenzi, A., Barbarossa, L., Gallitelli, D., and Martelli, G. P., Cucumber mosaic cucumovirus populations in Italy under natural epidemic conditions and after a satellite-mediated protection test, *Plant Dis.*, 77, 28, 1993.

Culver, J. N., Tobamovirus cross protection using a potexvirus vector, *Virology*, 226, 228, 1996.

de Zoeten, G. A., Risk assessment: Do we let history repeat itself?, *Phytopathology*, 81, 585, 1991.

de Zoeten, G. A. and Fulton, R. W., Understanding generates possibilities, *Phytopathology*, 65, 221, 1975.

de Zoeten, G. A. and Gaard, G., The presence of viral antigen in the apoplast of systemically virus-infected plants, *Virus Res.*, 1, 713, 1984.

Diener, T. O., Introduction, in *The Viroids*, Plenum Press, New York, 1987a, 2.

Diener, T. O., Potato spindle tuber, in *The Viroids*, Plenum Press, New York, 1987b, 221.

Diener, T. O., *The Viroids*, Plenum Press, New York, 1987c.

Dodds, J. A., Approaches to studying viral cross protection, in *New Directions in Biological Control: Alternatives for Suppressing Agricultural Pests and Diseases*, Alan R. Liss, New York, 1990, 213.

Dodds, J. A., Cross-protection and interference between electrophoretically distinct strains of cucumber mosaic virus in tomato, *Virology*, 118, 235, 1982.

Dodds, J. A., Lee, S. Q., and Tiffany, M., Cross protection between strains of cucumber mosaic virus: Effect of host and type of inoculum on accumulation of virions and double-stranded RNA of the challenge strain, *Virology*, 144, 301, 1985.

Duran-Vila, N. and Semancik, J. S., Variations in the "cross protection" effect between two strains of citrus exocortis viroid, *Ann. Appl. Biol.*, 117, 367, 1990.

Fernow, K. H., Tomato as a test plant for detecting mild strains of potato spindle tuber virus, *Phytopathology*, 57, 1347, 1967.

Fletcher, J. T. and Rowe, J. M., Observations and experiments on the use of an avirulent mutant strain of tobacco mosaic virus as a means of controlling tomato mosaic, *Ann. Appl. Biol.*, 81, 171, 1975.
Folgein, F. J., Kalpagam, C., Bates, D. C., Premecz, G., Nyitrai, A., and Farkas, G. L., Viral RNA synthesis is renewed in protoplasts isolated form TMV-infected Xanthi tobacco leaves in an advanced stage of virus infection, *Virology*, 67, 74, 1975.
Francki, R. I. B., Plant virus satellites, *Annu. Rev. Microbiol.*, 39, 151, 1985.
Fritsch, C. and Mayo, M. A., Satellites of plant viruses, in *Plant Viruses, Structure and Replication*, Mandahar, C. L., Ed., CRC Press, Boca Raton, FL, 1989, 290.
Fulton, J. P., Studies on strains of cucumber virus 1 from spinach, *Phytopathology*, 40, 729, 1950.
Fulton, R. W., Superinfection by strains of tobacco mosaic virus, *Phytopathology*, 41, 579, 1951.
Fulton, R. W., Unilateral cross-protection among some NEPO viruses, *Acta Hortic.*, 44, 29, 1975.
Fulton, R. W., Superinfection by strains of tobacco streak virus, *Virology*, 85, 1, 1978.
Fulton, R. W., Practices and precautions in the use of cross protection for plant virus disease control, *Annu. Rev. Phytopathol.*, 24, 67, 1986.
Gallitelli, D., Vovlas, C., Martelli, G., Montasser, M. S., Tousignant, M. E., and Kaper, J. M., Satellite-mediated protection of tomato against cucumber mosaic virus. II. Field test under natural epidemic conditions in southern Italy, *Plant Dis.*, 75, 93, 1991.
Gal-on, A., Kaplan, I., and Palukaitis, P., Differential effects of satellite RNA on the accumulation of cucumber mosaic virus RNAs and their encoded proteins in tobacco vs. zucchini squash with two strains of CMV helper virus, *Virology*, 208, 58, 1995.
Garnsey, S. M., Lee, R. F., and Semancik, J. S., Lack of cross protection between citrus exocortis viroid and citrus viroids associated with mild symptoms Etrog citron, 5th Int. Congr. Plant Pathology, Kyoto, Japan, 1988, 69 (Abstr.).
Gerber, M. and Sarkar, S., The coat protein of tobacco mosaic virus does not play a significant role for cross-protection, *J. Phytopathol.*, 124, 323, 1989.
Gonsalves, D. and Garnsey, S. M., Cross-protection techniques for control of plant virus diseases in the tropics, *Plant Dis.*, 73, 592, 1989.
Goodwin, J., Chapman, K., Swaney, S., Parks, T. D., Wernsman, E. A., and Dougherty, W. G., Genetic and biochemical dissection of transgenic RNA-mediated virus resistance, *Plant Cell*, 8, 95, 1996.
Goto, T., Iizuka, N., and Komochi, S., Selection and utilization of an attenuated isolate of pepper strain of tobacco mosaic virus, *Annu. Phytopathol. Soc. Jpn.*, 50, 221, 1984.
Grant, T. J. and Costa, A. S., A mild strain of the tristeza virus of citrus, *Phytopathology*, 41, 114, 1951.
Hamilton, R. I., Defenses triggered by previous invaders. Viruses, in *Plant Disease: An Advanced Treatise*, Vol. V., Horsfall, J. G. and Cowling, E. B., Eds., Academic Press, New York, 1980, 279.
Hamilton, R. I., Using plant viruses for disease control, *HortScience*, 20, 848, 1985.
Hancock, J. M., Chaleeprom, W., Chaleeprom, W., Dale, J., and Gibbs, A., Replication slippage in the evolution of potyviruses, *J. Gen. Virol.*, 76, 3229, 1995.
Harrison, B. D., Mayo, M. A., and Baulcombe, D. C., Virus resistance in transgenic plants that express cucumber mosaic virus satellite RNA, *Nature*, 328, 799, 1987.
Harveson, R. M. and Rush, C. M., A simple method for field and greenhouse inoculation of *Polymyxa betae* and beet necrotic yellow vein virus, *Phytopathology*, 83, 1216, 1993.
Heidel, G. B. and Rush, C. M., Incidence of beet necrotic yellow vein virus, beet distortion mosaic virus, and an unnamed soil-borne sugar beet virus in Texas, *J. Sugar Beet Res.*, 30, 98, 1993.
Horikoshi, M., Nakayama, M., Yamaoka, N., Furosawa, I., and Shishiyama, J., Brome mosaic virus coat protein inhibits viral RNA synthesis in vitro, *Virology*, 158, 15, 1987.
Hughes, J., d'A. and Ollennu, L. A. A., Mild strain protection of cocoa in Ghana against cocoa swollen shoot virus — a review, *Plant Pathol.*, 43, 442, 1994.
Hull, R. and Davies, J. W., Approaches to nonconventional control of plant virus diseases, *Crit. Rev. Plant Sci.*, 11, 17, 1992.
Huss, B., Walter, B., and Fuchs, M., Cross-protection between arabis mosaic virus and grapevine fanleaf virus isolates in *Chenopodium quinoa*, *Ann. Appl. Biol.*, 114, 45, 1989.
Jaegle, M., Devic, M., Longstaff, M., and Baulcombe, D., Cucumber mosaic virus satellite RNA (Y strain). Analysis of sequences which affect yellow mosaic symptoms on tobacco, *J. Gen. Virol.*, 71, 1905, 1990.
Jacquemond, M., Amselem, J., and Tepfer, M., A gene coding for a monomeric form of cucumber mosaic virus satellite RNA confers tolerance to CMV, *Mol. Plant-Microbe Interact.*, 1, 311, 1988.

Johnson, J., An acquired partial immunity to the tobacco streak disease, *Trans. Wisconsin Acad. Sci. Arts Lett., Arts and Letters,* 30, 27, 1937.

Jorda, C., Alfaro, A., Aranda, M. A., Moriones, E., and Garcia-Arenal, F., Epidemic of cucumber mosaic virus plus satellite RNA in tomatoes in eastern Spain, *Plant Dis.,* 76, 363, 1992.

Kaper, J. M., Satellite-induced viral symptom modulation in plants. A case of nested parasitic nucleic acids competing for genetic expression, *Res. Virol.,* 143, 5, 1992.

Kaper, J. M. and Collmer, C. W., Modulation of viral plant diseases by secondary RNA agents, in *RNA Genetics. Variability of RNA Genomes,* Domingo, E., Holland, J. J., and Ahlquist, P., Eds., CRC Press, Boca Raton, FL, 1988, 171.

Kaper, J. M., Gallitelli, D., and Tousignant, M. E., Identification of a 334-ribonucleotide viral satellite as principal aetiological agent in a tomato necrosis epidemic, *Res. Virol.,* 141, 81, 1990.

Kassanis, B., Properties and behaviour of a virus depending for its multiplication on another, *J. Gen. Microbiol.,* 27, 477, 1962.

Kassanis, B., Gianinazzi, S., and White, R. F., A possible explanation of the resistance of virus-infected tobacco plants to second infection, *J. Gen. Virol.,* 23, 11, 1974.

Khoury, J., Singh, R. P., Boucher, A., and Combs, D. H., Concentration and distribution of mild and severe strains on potato spindle tuber viroid in cross-protected tomato plants, *Phytopathology,* 78, 1331, 1988.

Koenig, R., Beier, C., Commandeur, U., Luth, U., Kaufmann, A., and Luddecke, P., Beet soil-borne virus RNA 3. A further example of the heterogeneity of the gene content of furovirus genomes and of triple gene block-carrying RNAs, *Virology,* 216, 202, 1996.

Kosaka, Y. and Fukunishi, T., Attenuated isolates of soybean mosaic virus derived at a low temperature, *Plant Dis.,* 77, 882, 1993.

Kosaka, Y. and Fukunishi, T., Application of cross protection to the control of black soybean mosaic disease, *Plant Dis.,* 78, 339, 1994.

Krstic, B., Ford, R. E., Shukla, D. D., and Tosic, M., Cross-protection studies between strains of sugarcane mosaic, maize dwarf mosaic, Johnsongrass mosaic, and sorghum mosaic potyviruses, *Plant Dis.,* 79, 135, 1995.

Kruse, M., Koenig, R., Hoffmann, A., Commandeur, U., Solovyev, A. G., Savenkov, I., and Burgermeister, W., Restriction fragment length polymorphism analysis of reverse transcription — PCR products reveals the existence of two major strain groups of beet necrotic yellow vein virus, *J. Gen. Virol.,* 75, 1835, 1994.

Kunkel, L. O., Studies on acquired immunity with tobacco and aucuba mosaics, *Phytopathology,* 24, 437, 1934.

Kurath, G. and Dodds, J. A., Satellite tobacco mosaic virus sequence variants with only five nucleotide differences can interfere with each other in a cross protection-like phenomenon in plants, *Virology,* 202, 1065, 1994.

Kurath, G. and Palukaitis, P., Serial passage of infectious transcripts of a cucumber mosaic virus satellite RNA clone results in sequence heterogeneity, *Virology,* 176, 8, 1990.

Kuti, J. O. and Moline, H. E., Effects of inoculation with a mild strain of tomato aspermy virus on the growth and yield of tomatoes and the potential for cross-protection, *J. Phytopathol.,* 115, 56, 1986.

Latorre, B. A. and Flores, V., Strain identification and cross protection of potato virus Y affecting tobacco in Chile, *Plant Dis.,* 69, 930, 1985.

Lecoq, H., Lemaire, J. M., and Wipf-Scheibel, C., Control of zucchini yellow mosaic virus in squash by cross protection, *Plant Dis.,* 75, 208, 1991.

Li, X. H. and Simon, A. E., Symptom intensification on cruciferous hosts by the virulent satellite RNA of turnip crinkle virus, *Phytopathology,* 80, 238, 1990.

Loebenstein, G., Cohen, J., Shabtai, S., Coutts, R. H. A., and Wood, K. R., Distribution of cucumber mosaic virus in systematically infected tobacco leaves, *Virology,* 81, 117, 1977.

Lovic, B. R. and Rush, C. M., BYVV-related indigenous mild viral strains for biocontrol of rhizomania. Characterization of candidate isolates and production of inoculum for field testing, *Phytopathology,* 85, 1136, 1995.

Masuta, C. and Takanami, Y., Determination of sequence and structural requirements for pathogenicity of a cucumber mosaic virus satellite RNA (Y-satRNA), *Plant Cell,* 1, 1165, 1989.

Matthews, R. E. F., *Plant Virology,* 3rd ed., Academic Press, New York, 1991, 835.

McGarvey, P. B., Montasser, M. S., and Kaper, J. M., Transgenic tomato plants expressing satellite RNA are tolerant to some strains of cucumber mosaic virus, *J. Am. Soc. Hortic. Sci.*, 119, 642, 1994.

McKinney, H. H., Mosaic diseases in the Canary Islands, West Africa, and Gibraltar, *J. Agric. Res.*, 39, 557, 1929.

McKinney, H. H., Virus-antagonism tests and their limitations for established relationship between mutants and non-relationship between distance viruses, *Am. J. Bot.*, 28, 770, 1941.

Montasser, M. S., Tousignant, M. E., and Kaper, J. M., Satellite-mediated protection of tomato against cucumber mosaic virus. I. Greenhouse experiments and simulated epidemic conditions in the field, *Plant Dis.*, 75, 86, 1991.

Moriones, E., Diaz, I., Rodriguez-Cerezo, E., Fraile, A., and Garcia-Arenal, F., Differential interactions among strains of tomato aspermy virus and satellite RNAs of cucumber mosaic virus, *Virology*, 186, 475, 1992.

Moriones, E., Fraile, A., and Garcia-Arenal, F., Host-associated selection of sequence variants from a satellite RNA of cucumber mosaic virus, *Virology*, 184, 465, 1991.

Moser, O., Gagey, M.-J., Godefroy-Colburn, T., Stussi-Garaud, C., Ellwart-Tshurtz, M., Nitschko, H., and Mundry, K.-W., The fate of the transport protein of tobacco mosaic virus in systemic and hypersensitive tobacco hosts, *J. Gen. Virol.*, 69, 1367, 1988.

Müller, G. W. and Costa, A. S., Tristeza control in Brazil by preimmunization with mild strains, *Proc. Int. Soc. Citriculture*, 3, 868, 1977.

Murakishi, H. H. and Carlson, P. S., Regeneration of virus-free plants from dark-green islands of tobacco mosaic virus-infected tobacco leaves, *Phytopathology*, 66, 931, 1976.

Murant, A. F., Dependence of groundnut rosette virus on its satellite RNA as well as on groundnut rosette assistor luteovirus for transmission by *Aphis craccivora*, *J. Gen. Virol.*, 71, 2163, 1990.

Murant, A. F. and Kumar, I. K., Different variants of the satellite RNA of groundnut rosette virus are responsible for the chlorotic and green forms of groundnut rosette disease, *Ann. Appl. Biol.*, 117, 85, 1990.

Murant A. F. and Mayo, M. A., Satellites of plant viruses, *Annu. Rev. Phytopathol.*, 20, 49, 1982.

Murant, A. F., Rajeshwari, R., Robinson, D. J., and Raschke, J. H., A satellite RNA of groundnut rosette virus that is largely responsible for symptoms of groundnut rosette disease, *J. Gen. Virol.*, 69, 1479, 1988.

Niblett, C. L., Dickson, E., Fernow, K. H., Horst, R. K., and Zaitlin, M., Cross protection among four viroids, *Virology*, 91, 198, 1978.

Nitzany, F. E. and Cohen, S., A case of interference between alfalfa mosaic virus and cucumber mosaic virus, *Virology*, 11, 771, 1960.

Otsuki, Y. and Takebe, I., Double infection of isolated tobacco leaf protoplasts by two strains of tobacco mosaic, in *Biochemistry and Cytology of Plant Parasite Interaction*, Tomiyama, K., Daly, J. M., Uranti, I., Oka, H., and Ouchi, S., Eds., Kodanaha Ltd., Tokyo, 1976.

Palukaitis, P., Pathogenicity regulation by satellite RNAs of cucumber mosaic virus. Minor nucleotide sequence changes alter host responses, *Mol. Plant-Microbe Interact.*, 1, 175, 1988.

Palukaitis, P. and Zaitlin, M., A model to explain the "cross-protection" phenomenon shown by plant viruses and viroids, in *Plant-Microbe Interactions: Molecular and Genetic Perspectives*, Kosuge, T. and Nester, E. W., Eds., Macmillan, New York, 1984.

Perring, T. M., Farrar, C. A., Blua, M. J., Wang, H. L., and Gonsalves, D., Cross protection of cantaloupe with a mild strain of zucchini yellow mosaic virus. Effectiveness and application, *Crop Prot.*, 14, 601, 1995.

Piazzolla, P., Tousignant, M. E., and Kaper, J. M., Cucumber mosaic virus-associated RNA 5. IX. The overtaking of viral RNA synthesis by CARNA 5 and dsCARNA 5 in tobacco, *Virology*, 122, 147, 1982.

Ponz, F. and Bruening, G., Mechanisms of resistance to plant viruses, *Annu. Rev. Phytopathol.*, 24, 355, 1986.

Posnette, A. F. and Todd, J., Mc. A., Virus diseases of cacao in West Africa. IX. Strain variation and interference in virus 1A, *Ann. Appl. Biol.*, 43, 433, 1955.

Powell, C. A., Pelosi, R. R., and Cohen, M., Superinfection of orange trees containing mild isolates of citrus tristeza virus with severe Florida isolates of citrus tristeza virus, *Plant Dis.*, 76, 141, 1992.

Price, W. C., Acquired immunity to ring-spot in Nicotiana, *Contrib. Boyce Thompson Inst.*, 4, 359, 1932.

Price, W. C., Virus concentration in relation to acquired immunity from tobacco ring spot, *Phytopathology,* 26, 503, 1936a.
Price, W. C., Specificity of acquired immunity from tobacco ring-spot diseases, *Phytopathology,* 26, 665, 1936b.
Price, W. C., Acquired immunity form plant virus diseases, *Q. Rev. Biol.,* 15, 338, 1940.
Prillwitz, H. and Schlosser, E., Interactions between beet soil-borne virus (BSBV-2) and beet necrotic yellow vein virus (BNYVV), in *Proc. Second Symp. Int. Working Group Plant Viruses with Fungal Vectors,* Hiruki, C., Ed., American Society of Sugar Beet Technologists, Denver, CO, 1993, 71.
Procházhová, Z., Interaction of cucumber mosaic virus and potato virus Y with tobacco mosaic virus, *Biol. Plant.,* 12, 297, 1970.
Quanjer, H. M., "Premunity," *Phytopathology,* 36, 892, 1946.
Quemada, H. D., Gonsalves, D., and Slightom, J. L., Expression of coat protein gene from cucumber mosaic virus strain C in tobacco. Protection against infections by CMV strains transmitted mechanically or by aphids, *Phytopathology,* 81, 794, 1991.
Randles, J. W., Coconut cadang-cadang, in *The Viroids,* Plenum Press, New York, 1987, 265.
Rast, A. T. B., Variability of tobacco mosaic virus in relation to control of tomato mosaic in glasshouse tomato crops by resistance breeding and cross protection, Agricultural Research Report No. 834, Institute of Phytopathological Research, Wageningen, Netherlands, 1975, 76.
Register, J. C., III and Nelson, R. S., Early events in plant virus infection. Relationships with genetically engineered protection and host gene resistance, *Semin. Virol.,* 3, 441, 1992.
Rezende, J. A. M., Tentativas de Premunizacao Para o Controle do Mosaico de Mamoeiro, University Sao Paulo, Piracicaba, SP, Brazil, MS Thesis, 1985, 64.
Rezende, J. A. M. and Costa, A. S., Intensificao dos sintomas dificulta o controle do mosaico do mamoeiro por premunizao, *Fitopatol. Bras.,* 12, 100, 1987.
Rezende, J. A. M., Costa, A. S., and Soares, N. B., Ocorrencia de um isolado fraco do virus do mosaico do mamoeiro *Carica papaya* L., *Fitopatol. Bras.,* 6, 534, 1981.
Rezende, J. A. M. and Sherwood, J. L., Breakdown of cross protection between strains of tobacco mosaic virus due to susceptibility of dark green areas to superinfection, *Phytopathology,* 81, 1490, 1991.
Rezende, J. A. M., Urban, L., Sherwood, J. L., and Melcher, U., Host effect on cross protection between two strains of tobacco mosaic virus, *J. Phytopathol.,* 136, 147, 1992.
Richards, K. E. and Tamada, T., Mapping functions on the multipartite genome of beet necrotic yellow vein virus, *Annu. Rev. Phytopathol.,* 30, 291, 1992.
Roossinck, M. J., Sleat, D., and Palukaitis, P., Satellite RNAs of plant viruses. Structures and biological effects, *Microbiol. Rev.,* 56, 265, 1992.
Ross, A. F., Interaction of viruses in the host, *Acta Hortic.,* 36, 247, 1974.
Rubino, L., Carrington, J. C., and Russo, M., Biologically active cymbidium ringspot virus satellite RNA in transgenic plants suppresses accumulation of DI RNA, *Virology,* 188, 429, 1992.
Rush, C. M., French, R. C., and Heidel, G. B., Texas 7 a possible strain of beet necrotic yellow vein virus, in *Proc. Second Symp. Int. Working Group Plant Viruses with Fungal Vectors,* Hiruki, C., Ed., American Society of Sugar Beet Technologists, Denver, CO, 1993, 59.
Rush, C. M. and Heidel, G. B., Furovirus diseases of sugar beets in the United States, *Plant Dis.,* 79, 868, 1995a.
Rush, C. M. and Heidel, G. B., Variation in symptomatology and serotype among furoviruses infecting sugar beet, *J. Sugar Beet Res.,* 32, 157, 1995b.
Rush, C. M., Heidel, G. B., and Manohar, S. K., Variation among isolates of beet soil-borne mosaic virus, *Phytopathology,* 85, 1041, 1995.
Rush, C. M., Scholthof, K.-B. G., Manohar, S. K., and Heidel, G. B., Similarities between beet soil-borne mosaic virus and beet necrotic yellow vein virus RNA2 nucleotide sequence and genomic organization, *Proc. Third Symp. Int. Work. Group Plant Viruses with Fungal Vectors,* Scottish Crop Research Institute, Dundee, Scotland, 1996.
Salaman, R. N., Protective inoculation against a plant virus, *Nature,* 131, 468, 1933.
Salaman, R. N., Acquired immunity against the 'Y' potato virus, *Nature,* 139, 924, 1937.
Salaman, R. N., The potato virus "X". Its strains and reactions, *Philos. Trans. R. Soc. (B),* 229, 137, 1938.
Salibe, A. A., Clones nucelares de citros no Estado de So Paulo, *Laranja,* 2(8), 443, 1987.

Sanford, J. C. and Johnston, S. A., The concept of parasite derived resistance — deriving genes from the parasite's own genome, *J. Theor. Biol.*, 113, 395, 1985.

Sarkar, S. and Smitamana, P., A proteinless mutant of tobacco mosaic virus. Evidence against the role of a viral coat protein for interference, *Mol. Gen. Genet.*, 184, 158, 1981.

Sayama, H., Sato, T., Kominato, M., Natsuaki, T., and Kaper, J. M., Field testing of a satellite-containing attenuated strain of cucumber mosaic virus for tomato protection in Japan, *Phytopathology*, 83, 405, 1993.

Scholthof, K.-B. G., Scholthof, H. B., and Jackson, A. O., Control of plant virus diseases by pathogen-derived resistance in transgenic plants, *Plant Physiol.*, 102, 7, 1993.

Sehgal, O. P., Sinha, R. C., Gellatly, D. L., Ivanov, I., and AbouHaidar, M. G., Replication and encapsidation of the viroid-like satellite RNA of lucerne transient streak virus are supported in divergent hosts by cocksfoot mottle virus and turnip rosette virus, *J. Gen. Virol.*, 74, 785, 1993.

Shalla, T. A. and Peterson, L. J., Studies on the mechanism of viral cross protection, *Phytopathology*, 68, 1681, 1978.

Sherwood, J. L., Demonstration of the specific involvement of coat protein in tobacco mosaic virus (TMV) cross protection using a TMV coat protein mutant, *J. Phytopathol.*, 118, 358, 1987a.

Sherwood, J. L., Mechanisms of cross-protection between plant virus strains, in *Plant Resistance to Viruses*, Evered, D. and Harnett, S., Eds., John Wiley & Sons, Chichester, U.K., 1987b.

Sherwood, J. L. and Fulton, W., The specific involvement of coat protein in tobacco mosaic virus cross protection, *Virology*, 119, 150, 1982.

Simmonds, J. H., Mild strain protection as a means of reducing losses from the Queensland woodiness virus in the passion vine, *Queensl. J. Agric. Sci.*, 16, 371, 1959.

Simon, A. E., Satellite RNAs of plant viruses, *Plant Mol. Biol. Rep.*, 6, 240, 1988.

Simon, A. E. and Bujarski, J. J., RNA-RNA recombination and evolution in virus-infected plants,*Annu. Rev. Phytopathol.*, 32, 337, 1994.

Singh, R. P., Boucher, A., and Somerville, T. H., Cross-protection with strains of potato spindle tuber viroid in the potato plant and other solanaceous hosts, *Phytopathology*, 80, 246, 1990.

Singh, R. P., Khoury, J., Boucher, A., and Somerville, T. H., Characteristics of cross-protection with potato spindle tuber viroid strains in tomato plants, *Can. J. Plant Pathol.*, 11, 263, 1989.

Singh, M. and Singh, R. P., Host dependent cross-protection between PVYn, PVYo, and PVA in potato cultivars and *Solanum brachycarpum*, *Can. J. Plant Pathol.*, 17, 82, 1995.

Sleat, D. E. and Palukaitis, P., Site-directed mutagenesis of a plant viral satellite RNA changes its phenotype from ameliorative to necrogenic, *Proc. Natl. Acad. Sci.*, 87, 2946, 1990.

Sleat, D. E. and Palukaitis, P., A single nucleotide change within a plant virus satellite RNA alters the host specificity of disease induction, *Plant*, 2, 43, 1992.

Smith, K. M., The present status of plant virus research, *Biol. Rev.*, 8, 136, 1939.

Thomson, A. D., Interference between plant viruses, *Nature*, 181, 1547, 1958.

Thung, T. H., Smetstof en plantencel by enkele virusziekten van de tabaksplant, Handel, Ned.-Ind., Naturwetensch, Congr. 6th, 1931, pp. 450, 1931; abstr. in *Rev. Appl. Mycol.*, 11, 750, 1932.

Tien, P. and Chiang, X. H., Control of two seed-borne virus diseases in China by the use of protective inoculation, *Seed Sci. Technol.*, 11, 969, 1983.

Tien, P. and Wu, G., Satellite RNA for the biocontrol of plant disease, *Adv. Virus Res.*, 39, 321, 1991.

Tien, P., Zhang, X., Qiu, B., Qin, B., and Wu, G., Satellite RNA for the control of plant diseases caused by cucumber mosaic virus, *Ann. Appl. Biol.*, 111, 143, 1987.

Tomlinson, J. A. and Shepherd, R. J., Studies on mutagenesis and cross-protection of cauliflower mosaic virus, *Ann. Appl. Biol.*, 90, 223, 1978.

Urban, L. A., Sherwood, J. L., Rezende, J. A. M., and Melcher, U., Examination of mechanisms of cross protection with non-transgenic plants, in *Recognition and Response in Plant-Virus Interactions*, Fraser, R. S. S., Ed., Advanced Study Institute Series, Springer-Verlag, Berlin, Vol. H41, 467 pp., 1990.

Valleau, W. D., Do tobacco plants recover and develop an immunity from ring spot?, *Phytopathology*, 25, 37, 1935.

Valleau, W. D., Experimental production of symptoms in so-called recovered ring-spot tobacco plants and its bearing on acquired immunity, *Phytopathology*, 31, 522, 1941.

van Vuuren, S. P., Collins, R. P., and da Graça, J. V., Evaluation of citrus tristeza virus isolates for cross protection of grapefruit in South Africa, *Plant Dis.*, 77, 24, 1993.

Wang, H. L., Gonsalves, D., Provvidenti, R., and Lecoq, H. L., Effectiveness of cross protection by a mild strain of zucchini mosaic virus in cucumber, melon, and squash, *Plant Dis.*, 75, 203, 1991.

Wang, H. L., Yeh, S.-D., Chiu, R.-J., and Gonsalves, D., Effectiveness of cross-protection by mild mutants of papaya ringspot virus for control of ringspot disease of papaya in Taiwan, *Plant Dis.*, 71, 491, 1987.

Wen, F., Lister, R. M., and Fattouh, F. A., Cross protection among strains of barley yellow dwarf virus, *J. Gen. Virol.*, 72, 791, 1991.

White, R. F. and Antoniw, J. E., Virus-induced resistance responses in plants, *Crit. Rev. Plant Sci.*, 9, 443, 1991.

Wilson, T. M. A., Plant viruses: a tool-box for genetic engineering and crop protection, *BioEssays*, 10, 179, 1989.

Wilson, T. M. A., Strategies to protect crop plants against viruses. Pathogen-derived resistance blossoms, *Proc. Natl. Acad. Sci.*, 90, 3134, 1993.

Wingard, S. A., Hosts and symptoms of ringspot, a virus disease of plants, *J. Agric. Res.*, 37, 127, 1928.

Wu, G., Kang, L., and Tien, P., The effect of satellite RNA on cross-protection among cucumber mosaic virus strains, *Ann. Appl. Biol.*, 114, 489, 1989.

Yamaya, J., Yoshioka, M., Mehsi, T., Okada, Y., and Ohno, T., Cross protection in transgenic tobacco plants expressing a mild strain of tobacco mosaic virus, *Mol. Gen. Genet.*, 215, 173, 1988.

Yeh, S.-D. and Cheng, Y.-H., Use of resistant *Cucumis metuliferus* for selection of nitrous acid induced attenuated strains of papaya ringspot virus, *Phytopathology*, 79, 1257, 1989.

Yeh, S.-D. and Gonsalves, D., Evaluation of induced mutants of papaya ringspot virus for control by cross protection, *Phytopathology*, 74, 1086, 1984.

Yeh, S.-D., Gonsalves, D., Wang, H.-L., Namba, R., and Chiu, R.-J., Control of papaya ringspot virus by cross protection, *Plant Dis.*, 72, 375, 1988.

Yie, Y., Zhao, F., Zhao, S. Z., Liu, Y. Z., Liu, Y. L., and Tien, P., High resistance to cucumber mosaic virus conferred by satellite RNA and coat protein in transgenic commercial tobacco cultivar G-140, *Mol. Plant-Microbe Interact.*, 5, 460, 1992.

Yoshida, K., Goto, T., and Iizuka, N., Attenuated isolates of cucumber mosaic virus produced by satellite RNA and cross protection between attenuated isolates and virulent ones, *Ann. Phytopathol. Soc. Jpn.*, 51, 238, 1985.

Zaitlin, M., Viral cross-protection: more understanding is needed, *Phytopathology*, 66, 382, 1976.

Zaitlin, M. and Hull, R., Plant virus-host interactions, *Annu. Rev. Plant Physiol.*, 38, 291, 1987.

Zhang, C., Cascone, P. J., and Simon, A. E., Recombination between satellite and genomic RNAs of turnip crinkle virus, *Virology*, 184, 791, 1991.

Zinnen, T. M. and Fulton, R. W., Cross protection between sunn-hemp mosaic and tobacco mosaic viruses, *J. Gen. Virol.*, 67, 1679, 1986.

Part IV
Physiological Aspects

CHAPTER 7

Constitutive Barriers and Plant Disease Control

Dov Prusky

I. INTRODUCTION

Fruit and vegetable crops are exposed to a variety of microorganisms, the vast majority of which, however, do not invade the plant tissue and cause disease development. During their life cycle, fungal pathogens pass through several stages — from spores landing on the host surface, through infection, and symptom development. Spores have to attach to the surface, germinate, produce penetration structures, penetrate directly or through wounds, and activate pathogenicity factor(s) in order to induce symptoms. A successful pathogen must also have the ability to overcome host defenses and initiate attack under prevalent physiological and environmental conditions.

1-8493-2627-3/97/$0.00+$.50

Plants are equipped with a wide array of constitutive and inducible defenses. The inducible defense may be nonspecifically induced by factors such as mechanical injury, or specifically induced as a result of the presence of a potential pathogen. Recognition of the pathogen results in a series of localized responses, the main event being the rapid death of cells in the area of infection, known as the hypersensitive response (HR), the accumulation of toxic metabolites, and phytoalexins in the necrotic area (Kuc, 1994). In the inducible defense response, a single elicitor signal sets in motion a complex cascade of events comprising many metabolic changes, including changes induced in the expression of many different genes. This dynamic defense reaction may include the release of endoglycosidase aimed at inhibiting the growth of pathogenic fungi; reinforcement of the cell walls with callose, lignin, or other phenolic polymers aimed at hindering their degradation by the pathogen's hydrolytic enzymes; and the secretion of antimicrobial peroxidase(s). Since inducible defense mechanisms operate only after infection has been attempted, their products are normally either absent from healthy plant tissues or are present in amounts below detection.

In contrast to the active defense reaction, the constitutive or preformed resistance includes the constitutive accumulation of low molecular weight secondary metabolites. The preformed resistance does not involve temporally differential gene activation; it involves the presence of biologically active compounds in healthy plants whose activity affords protection during most of the lifetime of the plant (Ingham, 1973; VanEtten et al., 1994). Preformed antifungal compounds may directly affect the pathogen itself and/or the pathogenicity factors produced by the pathogen to enable it to attack the host. In some cases, such as cyanogenic glycosides and glucosinolates, compounds occur as inactive precursors and are activated by plant enzymes released in response to tissue damage or pathogen attack (Osbourn, 1996b). These compounds are still regarded as constitutive because they are derived directly from preexisting precursors (VanEtten et al., 1994). Preformed compounds are only part of what are generally considered as preformed barriers, which may also include the presence of thickened cuticle and high molecular weight proteins with antimicrobial activity.

A number of factors may affect the concentrations of preformed antimicrobial compounds in the host. The nature and level of these compounds in a plant may vary, depending on such factors as host genotype, age, and environmental conditions. Within the plant, distribution is often tissue specific (Bennett and Wallsgrove, 1994; Grayer and Harborne, 1994; Osbourn, 1996b). It is common for antifungal compounds to be concentrated in the outer cell layers of plant organs, such as the peel of fruits (Droby et al., 1986). Some compounds, such as catechol and protocatechuic acid may even diffuse to the plant surface (Schlosser,1994; Schonbeck and Schlosser, 1976). In other cases, the compounds are sequestered in vacuoles or organelles in healthy plants, and the active concentration depends on the extent of tissue damage caused by the fungus. Biotrophs may prevent the release of preformed inhibitors by minimizing damage to the host, while necrotrophs may stimulate the release of substantial amounts of these compounds.

Since preformed antimicrobial barriers include fungitoxic secondary compounds and proteins with specific inhibitory activity or enzymes involved in restricting fungal development, there are possiblities for strengthening or creating defense mechanisms by manipulating their component factors. This could be done (1) by preventing the reduction of or inducing the enhancement of the level of preformed barriers; or (2) by selecting cultivars which naturally contain higher concentrations of preformed barriers. The possibility of modulating these systems exogenously so as to alter the level of the preformed inhibitory barriers and, at the same time, to affect host resistance, may possibly represent a new tool for disease control. The purpose of this chapter is to assess the importance of some examples in which the modulation of preformed barriers could be used as a tool for disease control.

II. INHIBITION OF FUNGAL DEVELOPMENT BY PREFORMED CONSTITUTIVE BARRIERS

Few conclusive studies on the physiological importance of preformed resistance factors have been described. This is mainly due to the difficulty of obtaining clear results that can test the importance of this kind of resistance. Problems are encountered in accurately assessing the quantities of inhibitory compounds that contact an invading pathogen, in defining their *in vivo* biological activity, and in relating changes in the concentration to those in resistance. In several systems studied, changes in concentration of inhibitory compounds did not coincide with changes in susceptibility (Schulz, 1978) and infusion of a host with antifungal substances did not always improve its resistance to disease (Sitterly and Shay, 1960; Brown and Swinburne, 1980). Dopamine, for example, was isolated from the peel of unripe banana in concentrations which inhibited *Colletotrichum musae in vitro* and it was considered a possible preformed antifungal compound. However, changes in its concentration were not synchronized with changes in decay development (Muirhead and Deverall, 1984). Only a few examples of mechanisms accounting for changes in the *in vivo* concentration of preformed inhibitory compounds are available.

A. Modulation of Preformed Barriers by Enhancement, Prevention of Reduction, or Selecting Cultivars with High Levels of Preformed Barriers

1. Mango System

In mango fruits, a mixture of antifungal compounds consisting of 5-12-*cis*-heptadecenyl resorcinol and 5-pentadecenyl resorcinol was found at fungitoxic concentration in the peel of unripe mango fruits that were resistant to *Alternaria alternata* (Droby et al., 1986, 1987). A new substituted resorcinol, not reported in previous studies, has also been detected recently (R. Reved, unpublished data). The evidence suggests that the preformed substituted resorcinols might be involved in the resistance of mango fruits to fungal development. The 5-substituted resorcinols occurred at fungitoxic levels in the peel of unripe fruits of several mango cultivars, and decreased to nontoxic levels in ripening fruits at the same time as symptoms of decay appeared in inoculated fruits. When two mango cultivars were compared during ripening the decrease in the levels of the 5-substituted resorcinols was faster in the susceptible cv. Tommy Atkins than in the more resistant cv. Haden. These observations suggest that the resistance of unripe mango fruits results from fungitoxic concentrations of the antifungal resorcinols present in the peel and that the modulation of the level of the compounds can determine the activation of mango pathogens *A. alternata* and *Colletotrichum gloeosporioides*. Modulation of resistance could be achieved by deferring the reduction in levels of the antifungal compound (Droby et al., 1987). Exposure of peeled fruits to air resulted in a rapid increase from subfungitoxic to fungitoxic levels of the antifungal resorcinols in the flesh tissue; this was accompanied by an increase in disease resistance to *A. alternata*. Elicitation of disease resistance was also observed in the peel of mangos, cv. Haden, which had been exposed to 30% CO_2 concentrations (Prusky and Gat, 1992). In some cases, exposure to even higher CO_2 concentrations (80 to 90%) also affected the development of decay caused by *C. gloeosporioides* in cv. Nam Dok Mai (N. Visarathanonth, personal communication).

2. Avocado System

The resistance of unripe avocado fruits to the quiescent germinated appressoria of *C. gloeosporioides* is correlated with the presence of high concentrations of preformed antifungal compounds. The major antifungal compound has been shown to be 1-acetoxy-2-hydroxy-4-oxo-heneicosa-12,15-diene (Prusky et al., 1982); the amount of this compound decreased tenfold to subfungitoxic concentrations during fruit ripening. A second antifungal compound was subsequently purified from unripe avocado fruits and identified as 1-acetoxy-2,4-dihydroxy-*n*-heptadeca-16-ene (Prusky et al., 1991). This monoene was less fungitoxic than the diene and its concentration in the peel of unripe fruits was about half that of the diene, but it decreased in concentration as decay symptoms appeared. It therefore appears that the monoene accounts for some activity, but the diene accounts for most of the antifungal activity. All these observations are consistent with the view that activation of fungus development depends on the reduction below a threshold concentration of the antifungal diene that is present in unripe fruits.

Because disease susceptibility in avocado fruits is correlated with decreased concentrations of the preformed antifungal compound, the mechanism controlling this reduction may predispose fruit to activation of quiescent *Colletotrichum* infection. The antifungal diene serves as a substrate for oxidation by lipoxygenase. Several experimental findings (Prusky, 1988; Prusky et al., 1983) indicated that lipoxygenase is capable of degrading the antifungal diene and of inducing decay susceptibility in ripening fruits: (1) the apparent specific activity of the enzyme increased by 80% during fruit ripening, (2) avocado lipoxygenase could oxidize the antifungal diene *in vitro*, and (3) treatments with α-tocopherol acetate — a nonspecific inhibitor, or with 5,8,11,14-eicosatetraynoic acid — a specific inhibitor of lipoxygenase, inhibited enzyme activity *in vitro* and delayed decay development by *C. gloeosporioides*.

Lipoxygenase activity in unripe and ripening avocado fruits has been reported to be affected by an endogenous inhibitor, epicatechin, present in the avocado peel (Prusky and Keen, 1993; Prusky et al., 1988). This flavan-3-ol competitively inhibited lipoxygenase activity, but it decreased in concentration during fruit ripening, thus permitting lipoxygenase activity to increase. On the basis of these results, modulation of the levels of antifungal compounds in order to regulate fruit resistance appears to be possible. The antifungal diene could be maintained at fungitoxic levels by exogenously supplying inhibitors of lipoxygenase. A commercial formulation of butylated hydroxy anisole, Xedaphen-20, applied to freshly harvested fruits, provided significant decay control (Prusky, 1988). Inhibition of lipoxygenase activity in these fruits was accompanied by a delayed decrease of the antifungal diene and reduced fungal colonization. The possibility that the preformed diene could be induced to higher levels offers an additional possibility to modulate resistance. Freshly harvested avocado fruits exposed to an atmosphere of 30% CO_2 produced higher levels of the preformed compound and showed reduced decay. However, the concentration of CO_2 and maturity of the fruit were critical to enhancing the levels of the preformed compound and achieving a significant delay in symptom development. Interestingly, inoculation of unripe avocado fruits with spores of a nonpathogenic mutant of *C. magna* also induced a higher and persistent increase in levels of the antifungal diene, without development of symptom expression (Prusky et al., 1994). This effect of a nonpathogenic strain was used to delay the activation period of wild-type *C. gloeosporioides* by coinoculation of fruit with both strains (Prusky et al., 1994). The induction of preformed antifungal compounds by nonpathogenic *Colletotrichum* spp. suggests that nonspecific eliciting factors are probably present in the hyphae of the pathogen species and could be used to induce plant resistance (Prusky et al., 1994).

In all cases where the levels of the preformed compound were enhanced or its diminution delayed, the key factor affected by the modulating factors was the level of epicatechin. On the basis of these facts the differential susceptibility of avocado cultivars to *C. gloeosporioides* was tested. The concentration of epicatechin in fruits of the highly susceptible cv. 'Fuerte' decreased concomitantly with the decrease in fruit firmness during ripening, and symptom expression occurred when the epicatechin concentration was at its lowest. In resistant cultivars, like Hass, infection by *C. gloeosporioides* was activated and symptoms expressed several days after the fruit had become ripe. The epicatechin concentration was initially high, and a considerable amount was still present in the peel when the fruits were soft and ready to be consumed. Host resistance was, therefore, related to the period required for complete reduction of epicatechin in softening fruits. This indicates that resistance of *C. gloeosporioides* in avocado might be initially regulated by epicatechin acting as an inhibitor of lipoxygenase activity and, consequently, delaying degradation of the antifungal diene (Prusky and Keen, 1993).

3. *Vitis* System

In response to *B. cinerea* infection, the leaves of *Vitis* spp. produce the stilbene, resveratrol, which is converted into the antifungal trimer, ε-viniferin (Langcake and McCarthy, 1979; Pryce and Langcake, 1977). The viniferins are constitutive compounds in the lignified tissue but are inducible in leaves. Young leaves are highly susceptible to *B. cinerea* and synthesize only a small amount of resveratrol; as they develop, their resistance and their ability to synthesize resveratrol increase concomitantly. With the fruits of grapes, the situation is reversed: immature grapes can synthesize large amount of resveratrol but, as they mature, the ability to synthesize it steadily decreases and the fruit become more susceptible.

In grapevine leaves, an efficient elicitation factor may alter the activity of rate-limiting enzyme, leading to stilbene formation. Stilbene synthase activity is the rate-limiting step and various biotic and UV-light treatments were able to activate stilbene synthase gene to a different extents (Schroder et al., 1988; Lanz et al., 1990). In grape, more than seven genes encode for stilbene synthase, but only two of them are expressed at higher rates upon elicitation (Melchior and Kindl, 1991; Weise et al., 1994). A simple way to modulate resistance is the use of ozone (Sarig et al., 1996), which was found to induce resistance to *Rhizopus stolonifer* in treated grape. Several *Vitis vinifera* cultivars exposed to ozone for 10 min showed induction of resveratrol similar to those induced by UV-c irradiation (Sarig et al., 1996). Ozone induction of plant defense has been attributed to the induction of the phenylpropanoid pathway (Eckney-Kaltenbach et al., 1994) and other systems (Kangasjarvi et al., 1994). Experiments with transgenic tobacco (Hahn et al., 1993) and tomato (Kindl, 1994) clearly proved that the transfer of the stilbene synthase gene to a plant not capable of forming stilbene results in disease resistance.

4. Tomato System

a. Fungal Inhibitors

The involvement of preformed inhibitors in resistance to infection by *B. cinerea* in tomato fruits has been suggested (Verhoeff, 1970, 1974; Verhoeff and Liem, 1975). The saponin, α-tomatine, present in high concentrations in the peel of green tomatoes, is inhibitory to the fungus at the concentrations present in the peel, but although there was no further development of the lesions after ripening, tomatine could not be detected in mature fruits. However, because of the potent antifungal properties associated with many saponins, there has been considerable speculation about their possible function as preformed determinants of resistance of plants to attack by saponin-sensitive fungi (Osbourn et al., 1994a, 1995; Sandrock et al., 1995). α-Tomatin can be detoxified by terminal deglucosylation by tomatinase produced by specific pathogens (Osbourn et al., 1994a, 1995). Significant evidence for a functional role of preformed tomatine in disease resistance has been produced by using genetically related pathogen strains that differ in their ability to tolerate or degrade plant antifungal compounds (Arneson and Durbin, 1967; DeFago et al., 1983). Induced mutants of *Nectria haematococca* were more tolerant to α-tomatine and became pathogenic to green tomato fruits, which contain high levels of α-tomatine (DeFago et al., 1983); the wild-type strains were pathogenic only on ripe fruits, which contain little or no α-tomatine (Verhoeff and Liem, 1975).

The recent cloning of the gene encoding for tomatinase should clarify whether degradation of α-tomatine by the enzyme is necessary and sufficient for *N. haematococca* to parasitize tomato (Osbourn et al., 1995; Sandrock et al.,1995); it should also provide further evidence on the role of detoxifying enzymes in pathogenesis (Bowyer et al.,1995). Attempts have been made to adddress the contribution of α-tomatine to the specific resistance of near-isogenic tomato lines to the vascular wilt fungi, *Fusarium oxysporum* f. sp. *lycopersici* (Smith and MacHardy, 1982) and *Verticillium albo-atrum* (Pegg and Woodward, 1986). No significant accumulation of α-tomatine was seen in resistant interactions of incompatible lines, in comparison with compatible ones, suggesting that α-tomatine does not have a role in variety-specific resistance (Osbourn, 1996b). However, since saponins have been implicated as preformed resistance determinants which protect fruits against attack by fungi that could not detoxify saponin (Verhoeff and Liem, 1975; Rick et al., 1994), the inactivation of fungal-specific enzymes involved in saponin detoxification could provide an efficient mean to obtain a more resistant host (Osbourn, 1996).

b. Fungal Restrictors

A highly anionic peroxidase isolated from tomato fruits was found in the exocarp of unripe fruit (Sherf and Kolattukudy, 1993); it is encoded by a single gene *(tap)*. The constitutive expression of *tap* in green fruits undergoes a twofold increase during the development from immature to mature green fruits. Once fruits reach their climacteric stage, however, *tap* transcriptions diminish and remain undetectable in the ripe fruit that is susceptible to decay development. The anionic peroxidase is a key enzyme in polymerization of the phenolic polymers which act as barriers in cell walls. Lignification and suberization of the plant cell walls have been suggested as being part of the elaborate defense strategies of the plant. Like lignin, suberin comprises an aromatic polymer that is covalenty attached to carbohydrates of the plant cell wall (Kolattukudy, 1984) and is possibly attached to proteins such as hydroxyproline-rich glycoproteins. Suberin, however, also contains aliphatic components and associated waxes that are not present in lignin, and these, together with the aromatic domain, form a hydrophobic barrier which waterproofs the wounded plant tissues (Kolattukudy, 1987). The deposition of these phenolic polymers as barriers on the cell walls of the wound-healing or infected tissue has been postulated as preventing the entry or spread of phytopathogens (Kolattukudy, 1980). The polymerization of the phenolic moieties of the aromatic domain are thought to be catalyzed by a cell wall-associated anionic peroxidase, rendering the cell walls highly resistant to mechanical and enzymatic disruption (Kolattukudy, 1987; Pearce and Rutherford, 1981). Detaching green fruits from the vine, however, leads to the complete loss of the *tap* transcripts. The rapid decline in the peroxidase mRNA levels observed in detached green fruits may possibly result from the loss of a positive regulatory molecule which originates in the parent plant and functions to sustain the constitutive expression of the gene in maturing green tomatoes. Administration of cytokinins through the pedicels of detached green fruits revealed that this growth regulator was able to sustain a level of *tap* expression in fruits which had been detached for 72 h, and was comparable with that found in fruits attached to the vine. It is, therefore, possible that the constitutive expression of defense-related genes such as the anionic peroxidase in green tomato fruits reflects an evolutionary strategy that ensures elevated pest resistance (Sherf and Kolattukudy, 1993).

5. Citrus System

Citrus fruits are resistant to wound pathogens during fruit growth, even though a considerable amount of inoculum of *Penicillium* is present and considerable wounding occurs naturally in the orchard. It has been suggested that the resistance of young mature green lemons is related to the presence of a preformed monoterpene aldehyde, citral, that decreases in older, yellow fruits, enabling decay to develop rapidly (Rodov et al., 1995). Cautious correlations between the levels of resistance of fruits after harvest and the *in vivo* concentrations of preformed inhibitory compounds have suggested the possible involvement of citral in the resistance of lemon fruits to postharvest decay (Rodov et al., 1995). Resistance to *P. digitatum* in lemon fruits could be modulated by growth regulators and physical treatments. Exposure of fruits to heat treatment at 36°C for 72 h following storage at 17°C and 85% RH significantly reduced decay development. The incidence of decay was further reduced by film wrapping, thus increasing the humidity around the heated fruit, suggesting that antifungal compounds and wound curing could curtail fungal development (Ben Yehosua et al.,1995). Postharvest applications of growth regulators comprising 50 μg/ml giberellic acid (GA) and 200 μg/ml 2,4-D were also effective in delaying the decline in antifungal activity in the fruit flavedo and in preserving the resistance of the fruits. The marked effects of heat, GA, and 2,4-D on the levels of preformed antifungal activity suggest the importance of preformed compounds in the modulation of citrus resistance.

6. Celery System

Psoralen phytoalexins (furanocoumarins) including psoralen, bergapten, xanthotoxin, and isopimpinellin induced in celery plants by pathogen infection have been hypothesized not to be involved in its resistance (Afek et al., 1993). The resistance of celery stalks could result from the presence of the preformed antifungal psoralen precursor, mermesin, since this is a very active preformed compound that is almost ten times as fungicidal as the inducible psoralens. Decreased concentration of mermesin during storage, and not the induction of psoralens, is the main factor affecting the resistance of celery to *B. cinerea* (Afek et al., 1995). Treatment of celery stalks with GA delayed the decrease in mermesin and also reduced the incidence of *Botrytis* (Afek et al., 1994). One of the possiblities to modulate susceptiblity of celery stalks to *Botrytis* is by storage at proper temperatures. When celery stalks were stored at optimal storage conditions of 2°C a slower reduction in the rate of mermesin occurred compared to storage at 0°C. Incidence of decay occurred earlier when mermesin concentration decreased faster (Afek et al., 1995).

7. Cotton System

The preformed compound, gossypol, is present in the pigment gland of cotton leaves and is responsible for resistance to insects (Bell, 1967,1969). On the other hand, tissues of cotton, when stressed by a pathogen like *Verticillium* or following chemical injury, accumulate antifungal substances in the xylem. Schnathorst et al. (1966) observed inhibition of fungus development in the presence of xylem fluids from inoculated tolerant plants. Bell and co-workers (1978) presented evidence that this biological activity is due to *de novo* production of gossypol and/or gossypol-related compounds in the cotton stem after fungal infection. The main antifungal compound seems to be a gossypol precursor, hemigossypol, a sesquiterpenoid aldehyde. It was found that there is a direct relationship between the rates of production of these compounds by different cotton cultivars and their resistance to *Verticillium* wilt. When three cultivars, susceptible, tolerant, and resistant to *Verticillium albo-atrum* were compared, the index of Verticillium development was generally inversely related to gossypol concentration. The more resistant cultivars enhanced gossypol accumulation more rapidly than the susceptible ones. These results suggest that the modulation of the enhancing process in selected cultivars represents a potential method for wilt disease prevention.

III. INHIBITION OF PATHOGENICITY FACTORS BY CONSTITUTIVE BARRIERS

The inactivation of pectic enzymes by inhibitors has been hypothesized to be a means of modulating fungal pathogenicity. According to Byrde and co-workers (Byrde et al.,1960), polyphenols and tannins may account for their inactivation. A problem encountered with this hypothesis was that several of the compounds are localized within vessels and are not, therefore, generally available to the attacking pathogen. Another class of inhibitors of cell wall-degrading enzymes comprises the polygalacturonase-inhibiting proteins (PGIP) which are present in infected and uninfected plant tissues (Abu-Goukh and Labavitch, 1983; Abu-Goukh et al.,1983a,b; Brown and Adikaram, 1983).

A. Preformed Phenols Inhibiting Pathogenicity Factors and Their Possible Modulation

1. Strawberry System

Botrytis infects strawberry fruits via the floral parts, but the fungus remains in a quiescent form at the bottom of the receptacle until the fruit matures. Infection by direct penetration through the epidermis is relatively rare. No preformed or postinfectionally elicited compounds were found in immature, resistant green fruits (Jersch et al., 1989). The inhibition of fungal development could be entirely attributed to the presence of proanthocyanidins, which are present in a solid layer beneath the epidermis and surrounding the receptacle. These compounds are oligomers, composed of pelargonidin, cyanidin, or delphinidin subunits, with cyanidin being the most common. Since extracellular hydrolases have been shown to be essential for pathogenesis of *B. cinerea* (Huth and Schlosser, 1982), their binding to proanthocyanidins and consequent inactivation would explain the plant resistance to this pathogen in unripe fruits. When fruits mature the proanthocyanidins are still present, but have lost their inhibitory power (Jersch et al., 1989) because of a high degree of polymerization. Pathogen enzymes produced at this stage will permit a fast decay development. This model offers the possiblity of resistance modulation by means of the development of new strawberry cultivars. Newly developed cultivars in which mature, red-colored fruits have a white collar of immature tissue around the fruit stem, indicate that in that tissue the proanthocyanidins are as active as in resistant white-green fruits (Schlosser, 1994).

2. *Vitis* System

Proanthocyanidins have also been found in the grape. These proanthocyanidins are located mainly in the skin of the grape, and they inhibit macerating enzymes produced by *B. cinerea.* Their concentration in young berries is sufficient to completely inhibit the extracellular hydrolases of *Botrytis.* With berry development the inhibitory power declines steadily (Hill et al., 1981). Modulation of this process in grapes has so far not been shown.

3. Avocado System

It was recently shown that there is a close relationship between changes in the level of the flavan-3-ol, epicatechin, present in the peel of avocado fruit, and the inhibition of pectolytic enzymes by *C. gloeosporioides*. Purified polygalacturonase and pectate lyase produced by *C. gloeosporioides* were also inhibited *in vitro* by epicatechin (Prusky et al., 1989; Wattad et al., 1994). At 20 μg/ml, epicatechin inhibited the enzyme maceration capability by 64%. Since the flavan is present in unripe fruit at much higher concentrations (about 350 μg/g fresh weight or approximately 270 μg/ml) than the inhibitory concentrations, epicatechin may contribute to the resistance of avocado fruits by inhibiting the activity of pectate lyase of *Colletotrichum*. The possibility of selecting cultivars with high epicatechin concentrations or prevent its decrease could affect the activity of pectate lyase activity, and thereby inhibit development of the anthracnose symptom.

B. Preformed Proteins Inhibiting Pathogenicity Factors

1. PGIPs in Different Hosts

Plant cell walls contain proteins that can specifically and effectively inhibit polygalacturonases of fungal origin. These polyglacturonase-inhibiting proteins (PGIPs) have been reported in numerous plant species including pear fruit (Abu-Goukh et al., 1983a), apple fruit (Yao et al., 1995), pepper (Brown and Adikaram, 1983), bean (Cervone et al., 1987), alfalfa (Degra et al., 1988), soybean (Favaron et al., 1994), raspberry fruit (Johnson et al., 1993), pea leaf (Hoffman and Turner, 1984), and tomato (Stotz et al., 1994). PGIPs are relatively heat-stable glycoproteins that may inhibit fungal PG by both competitive and noncompetitive mechanisms. Purified pear PGIPs inhibited various fungal PGs including that of *B. cinerea*, but did not affect endogenous pear fruit PG activity (Abu-Goukh et al., 1983b). In pears, PGIP activity was observed throughout development (4 to 14 weeks after anthesis), and changes in susceptibility to decay development after fruit ripening were accompanied by a decrease in the PG inhibitor content of Bartlett pear fruit. PGIPs from different plant species are likely to differ in their inhibition kinetics and target specificity (Abu-Goukh et al., 1983a,b). The molecular characterization of PGIP from pear was recently reported (Stotz et al., 1993). The expression of polygalacturonase inhibitor is regulated in a tissue-specific manner: PGIP mRNA was approximately 100 times more abundant in fruit than in flowers and was not detectable in pear leaves. PGIP specific activity in fruit was 200 times higher than in flowers and 1400 times higher than in leaves. The Labavitch group has suggested that the pear PGIP promotor may effect a fruit-specific expression of the gene, resulting in the inhibition of *B. cinerea* (Stotz et al., 1993). The characterization of the protein inhibitor from pear (Stotz et al.,

1993) and from other plants (Fielding, 1981) should lead to expression of PGIPs in transgenic plants and possibly to inhibit decay development. In fact, Powell and co-workers (1994) reported that transgenic tomato fruits expressing the pear PGIP gene were more resistant to *B. cinerea* than control fruits. By constitutively expressing the PGIP gene, or by retarding the dissociation of PGIP from the cell wall by means of calcium and heat treatment, as has been reported for apple fruits, it is feasible to reduce disease development (Conway et al., 1988; Yao et al., 1995).

2. Cabbage System

Recently, a serine proteinase inhibitor that may be involved in the inhibition of the enzymes necessary for microbial development was isolated from cabbage (Lorito et al., 1994). The trypsin and chemotrypsin inhibitory activity in the foliage of cabbage was found to originate from four proteins, stable at high temperatures and over a broad pH range, that are developmentally regulated (Broadway, 1993) This inhibitor significantly inhibited *B. cinerea* and *Fusarium solani* f. sp. *pisi* growth by blocking chitin synthesis and hence causing cytoplasmic leakage. Interestingly, there was no apparent effect of the cabbage protein inhibitors on a host-specific pathogen, *Alternaria brassicicola*, suggesting that there is specificity between the inhibitor and the fungus. The possibility of constitutively expressing these proteins in other systems could lead to an efficient method for delaying disease development.

IV. THE RELEVANCE OF PREFORMED BARRIERS IN PLANT DISEASE CONTROL

A logical application of the information concerning defense compounds would be the possible modulation of these compounds to protect plants against disease. The possible scenarios would involve prevention of compound reduction, enhancement of its concentration, or selection of cultivars rich in such antifungal compounds. The assumptions underlying such action are that the compounds are safe because they occur naturally, and that they would be effective since they have evolved in nature for specific plant protection and have enabled existing plants to survive the selection pressure of evolution. However, naturally occurring compounds are not necessarily safe (Osbourn, 1996a); many of the world's most potent poisons are derived from plants, and some plant tissues are extremely toxic to animals because they contain protective compounds, e.g., potato, tomato, and tobacco foliage. Furthermore, the levels of preformed compounds can change during the host's lifetime, usually they decrease (Prusky and Keen, 1993), thus necessitating induction during specific periods in the lifetime. The exogenic application of naturally occurring defense compounds to protect plants would be uneconomical: their synthesis would be difficult and the cost of their isolation would probably be high.

In spite of the possible flaws in this scenario, the protection of the host during specific periods of susceptibility by modulation of the levels of the natural compounds seems to be a safe way to preserve host resistance. The possibility of maintaining a high level of the antifungal diene, for example, in avocado fruits during fruit storage but to allow decrease of the compound before consumption of the fruit could be a safe way to use this natural compound.

An alternative to the external application of defense compounds is their elicitation, specifically or not specifically within the plant, by regulation of gene expression, and hence of gene product, leading to the synthesis and further accumulation of the desired compounds. Preformed compounds have been defined as "static phenomenon involving fungistatic or fungitoxic chemical inhibitors or preformed mechanical structures within the tissue of the host plant" (Ingham, 1973). VanEtten et al. (1994) further defined preformed antifungal compounds as "phytoanticipins" that "are present in plants before challenge by microorganisms or are produced after infection solely from preexisting constituents." In spite of the fact that phytoanticipins do not consider the presence of inducible preformed compounds, several examples of preformed antifungal compounds have been shown to be enhanced to higher levels by biotic and abiotic factors (Prusky and Keen, 1993). The possibility of induction of preformed compounds by nonpathogenic strains of *Colletotrichum* suggests that this process could be done by biological means during fungal colonization. However, the use of such a biological elicitor or other physical elicitors (CO_2, γ-radiation) has not always proved effective because the host, e.g., avocado, is receptive to the signals only during very specific periods after harvest (Prusky and Keen,1995).

Suggestions to enhance the resistance barriers of plants by incorporation of specific inhibitors of fungal enzymes that detoxify preformed antifungal compounds have been made. If ongoing studies on the preformed α-tomatine suggest that saponin-degrading enzymes may have a more general role in pathogenicity, then inhibitors of these enzymes could become attractive as a basis for disease control strategies (Osbourn et al., 1995; Sandrock et al., 1995). The extracellular location of the detoxyfing enzymes should facilitate approaches involving the use of chemicals or the expression of saponinase inhibitors in genetically engineered plants, since there should be no requirements for the inhibitors to penetrate the fungal hyphae (Osbourn et al., 1995; Osbourn, 1996a).

The possibility of overexpressing compounds that have preformed activity in a susceptible host which does not express those compounds has also been tested. The transfer of stilbene synthase from grape to a resveratrol-nonproducing plant such as tobacco, leading to the synthesis and accumulation of resveratrol in tobacco, might be an exciting example (Hahn et al., 1993): the transgenic tobacco-synthesizing and rapidly accumulating resveratrol had enhanced resistance to disease caused by *B. cinerea*. However, since there are several preformed antifungal barriers, and since secondary metabolites and their biosynthetic pathways are complex and not always known, the transgenic approach might be difficult. Another possible way to modulate the basic constitutive barriers may be to engineer hosts with constitutive barriers by overexpressing host constitutive inhibitors such as PGIP, or inducible host inhibitors that had not shown any significant preformed activity. Transgenic tomato plants expressing PGIP were found more resistant to fungal attack (Powell et al., 1994). Also, transgenic tobacco and rape plants containing an inducible bean chitinase gene with a constitutive promotor have been shown to exhibit higher basal levels of chitinase and concomitant increased resistance to *Rhizoctonia solani* compared with control plants (Broglie et al., 1991). In these cases resistance to fungal attack was obtained by "creating a host with a preformed barrier" as a result of engineering the plants by continuously overexpressing an inducible chitinase gene. In another study (Neuhaus et al., 1991) tobacco plants which had been transformed with a basic chitinase from tobacco, under the regulation of a constitutive promotor, accumulated up to 120 times more active chitinase than nontransformed plants. However, the transformed plants were found to be no more resistant to the fungus *Cercospora nicotiana* than the controls. It is possible that the location of the additional chitinase — intracellularly in the transgenic plant — is compartmentalized in such a way that it does not come into contact with the penetrating fungus. This is a very common problem in preformed barriers. It may also be that combination with other genes encoding antifungal products is necessary for providing resistance.

The general understanding of the plant biosynthetic pathways which lead to the formation of preformed barriers, and of their contribution to disease resistance, is of primary importance for the manipulation of these compounds to enhance resistance. Such understanding offers the possibility that the potential level of preformed barriers might be modulated by genetic, biological, physical, and chemical means to provide the basis for novel crop protection strategies that could lead to the reduction of pesticide use in the future.

ACKNOWLEDGMENTS

I would like to thank Ilana Kobiler and Dr. R. Ben-Arie for their useful remarks while reviewing the manuscript, and P. Sarig, and R. Ben-Arie, and A. Osbourn for supplying unpublished results. The author is supported by funds from BARD, GIARA, and CDR-AID.

REFERENCES

Abu-Goukh, A. A. and Labavitch, J. M., The *in vivo* role of 'Bartlett' pear fruit polygalacturonase inhibitors, *Physiol. Plant Pathol.*, 23, 123, 1983.

Abu-Goukh, A. A., Greve, L. C., and Labavitch, J. M., Purification and partial characterization of 'Bartlett' pear polygalacturonase inhibitors, *Physiol. Plant Pathol.*, 23, 111, 1983a.

Abu-Goukh, A. A., Strand, L. L., and Labavitch, J. M., Development-related changes in decay susceptibility and polygalacturonase inhibitor content of 'Bartlett' pear fruit, *Physiol. Plant Pathol.*, 23, 101, 1983b.

Afek, U., Aharoni, N., and Carmeli, S., A possible involvement of gibberellic acid in celery resistance to pathogens during storage, Int. Symp. Natural Phenols in Plant Resistance, Geible, M., Treutter, D., Feucht, W., Eds., Vol I, *Acta Hortic.*, 381, 583, 1994.

Afek, U., Aharoni, N., and Carmeli, S., The involvement of mermesin in celery resistance to pathogens during storage and the effect of temperature on its concentration, *Phytopathology*, 85, 1033, 1995.

Arneson, P. A. and Durbin, R. D., Hydrolysis of tomatin by *Septoria lycopersici:* a detoxification mechanism, *Phytopathology*, 57, 1358, 1967.

Bell, A. A., Formation of gossypol in infected or chemically irritated tissues of *Gossypium* species, *Phytopathology*, 57, 759, 1967.

Bell, A. A., Phytoalexin production and *Verticillium* wilt resistance in cotton, *Phytopathology*, 59, 1119, 1969.

Bell, A. A., Stipanovic, R. D., O'Brien, D. H., and Fryxell, P. A., Sesquiterpenoid aldehyde quinones and derivatives in pigment glands of *Gossypium*, *Phytochemistry*, 17, 1297, 1978.

Ben Yehosua, S., Rodov, V., Fang, D. Q., and Kim, J. J., Preformed antifungal compounds of citrus fruits: effect of postharvest treatments with heat and growth regulators, *J. Agric. Food Chem.*, 43, 1062, 1995.

Bennet, R. N. and Wallsgrove, R. M., Secondary metabolites in plant defence mechanism, *New Phytol.*, 127, 617, 1994.

Bowyer, P., Clarke, B. R., Lunness, P., Daniels, M. J., and Osbourne, A. E., Host range of a plant pathogenic fungus determined by a saponin detoxification enzyme, *Science*, 267, 371, 1995.

Broadway, R. M., Purification and partical characterization of trypsin/chemotrypsin inhibitors from cabbage foliage, *Phytochemistry*, 33, 21, 1993.

Broglie, K., Chet, I., Holliday, M., Cressman, R., Biddle, P., Knowlton, S., Mauvais, C. J., and Broglie, R., Transgenic plants with enhanced resistance to the fungal pathogen *Rhizoctonia solani*, *Science*, 254, 1194, 1991.

Brown, A. E. and Adikaram, N. K. B., The differrential inhibition of pectic enzymes from *Glomerella cingulata* and *Botrytis cinerea* by a cell wall protein from *Capsicum annuum* fruit, *Phytopathol. Z.*, 105, 277, 1982.

Brown, A. E. and Adikaram, N. K. B., A role for pectinase and protease inhibitors in fungal rot development in tomato fruits, *Phytopathol. Z.*, 106, 239, 1983.

Brown, A. E. and Swinburne, T. R., The resistance of immature banana fruits to anthracnose [*Colletotrichum musae* (Berk. & Curt.) Arx], *Phytopathol. Z.*, 99, 70, 1980.

Byrde, R. J. W., Fielding, A. H., and Williams, A. H., The role of oxidized polyphenols in the varietal resistance of apples to brown rot, in *Phenolics in Plants Health and Disease*, Pridham, J. B., Ed., Pergamon Press, Oxford, 1960, 95.

Cervone, F., De Lorenzo, G., Degra, L., Salvi, G., and Bergami, M., Purification and characterization of a polyglacturonase-inhibiting protein from *Phaseolus vulgaris* L., *Plant Physiol.*, 855, 6531, 1987.

Conway, W. S., Gross, K. C., Boyer, C. D., and Sams, C. E., Inhibition of *Penicillium expansum* polygalacturonase activity by increased apple cell wall calcium, *Phytopathology*, 78, 1052, 1988.

DeFago, G., Kern, H., and Sedlar, L., Genetic analysis of tomatin insensitivity, sterol content and pathogenicity for green tomato fruits in mutants of *Fusarium solani*, *Physiol. Plant Pathol.*, 22, 39, 1983.

Degra, L., Salvi, G., Mariotti, D., De Lorenzo, G., and Cervone, F., A polygalacturonase-inhibiting protein in alfalfa callus cultures, *J. Plant Physiol.*, 133, 364, 1988.

Droby, S., Prusky, D., Jacoby, B., and Goldman, A., Presence of an antifungal compound in the peel of mango fruits and their relation to latent infections of *Alternaria alternata, Physiol. Mol. Plant Pathol.*, 29, 173, 1986.

Droby, S., Prusky, D., Jacoby, B., and Goldman, A., Induction of antifungal resorcinols in flesh of unripe mango fruits and its relation to latent infection by *Alternaria alternata, Physiol. Mol. Plant Pathol.*, 30, 285, 1987.

Eckney-Kaltenbach, H., Ernst, D., Heller, W., and Sandermann, H., Biochemical plant responses to ozone. IV. Cross-induction of defensive pathways in parsely *(Petroselinum crispum)* plants, *Plant Physiol.*, 104, 67, 1994.

Favaron, F., D'Ovidio, R., Porceddu, E., and Alghisi, P., Purification and molecular characterization of a soybean polygalacturonase-inhibiting protein, *Planta,* 195, 80, 1994.

Fielding, A. H., Natural inhibitors of fungal polygalacturonases in infected fruit tissues, *J. Gen. Microbiol.*, 123, 377, 1981.

Gazit, S. asnd Adato, I., Effect of carbon dioxide atmosphere on the course of astringency disappearance of persimmon (*Diospyrus kaki* Linn.) fruits, *J. Food Sci.*, 37, 125, 1972.

Grayer, R. J. and Harborne, J. J., A survey of antifungal compounds from higher plants 1982–1993, *Phytochemistry,* 37, 1942, 1994.

Hahn, R., Reif, H., Krause, E., Langebarteis, R., Kindl, H., Varnam, B., Weise, W., Scheizer, E., and Stenzel, K., Disease resistance results from foreign phytoalexin expression in a novel plant, *Nature,* 361, 153, 1993.

Hill, G. K., Stellwaag-Kittler, F., Huth, G., and Schlosser, E., Resistance of grapes in different development stages to *Botrytis cinerea, Phytopathol. Z.*, 102, 328, 1981.

Hoffman, R. and Heale, J. B., Cell death, 6-methoxymellin accumulation, and induced resistance to Botrytis cinerea in carrot root slices, *Physiol. Mol. Plant Pathol.*, 30, 67, 1987.

Hoffman, R. M. and Turner, J. G., Occurrence and specificity of a eudopolygalacturonase inhibitor in *Pisum sativum, Physiol. Plant Pathol.*, 27, 49, 1984.

Huth, G. and Schlosser, E., Role of extracellular microbial hydrolases in pathogenesis, *Med. Fac. Landbounww. Rijksuniv. Gen.*, 47, 885, 1982.

Ingham, J. L., Disease resistance in higher plants. The concept of pre-infection and post-infectional resistance, *Phytopathol. Z.*, 78, 314, 1973.

Jersch, S., Scherer, C., Huth, G., and Schlosser, E., Proanthocinaidins as a basis for quiescence of *Botrytis cinerea* in immature strawberry fruits, *Z. Pflanzenkr. Pflanzenschutz.*, 96, 365, 1989.

Johnson, D. J., Raamanathan, V., and Williamson, B., A protein from immature raspberry fruits which inhibits endopolyglacturonases from *Botrytis cinerea* and other micro-organisms, *J. Exp. Bot.*, 44, 971, 1993.

Kangasjarvi, J., Talvinen, J., Utriainen, M., and Karjalainen, R., Plant defence systems induced by ozone, *Plant Cell Environ.*, 17, 783, 1994.

Kindl, H., Biochemical mechanism controlling the formation of phenols in plants, Int. Symp. Natural Phenols in Plant Resistance, Geible, M., Treutter, D., and Feucht, W., Eds., Vol. I, *Acta Hortic.*, 381, 176, 1994.

Kolattukudy, P. E., Biopolyester membranes of plants: cutin and suberin, *Science,* 208, 990, 1980.

Kolattukudy, P. E., Biochemistry and function of cutin and suberin, *Can. J. Bot.*, 62, 2918, 1984.

Kolattukudy, P. E., Lipid-derived defensive polymers and waxes and their role in plant-microbe interaction, In *The Biochemistry of Plants,* Vol. 9, Stumpf, P. K. and Conn, E. E., Eds., Academic Press, New York, 1987, 291.

Kuc, J., Relevance of phytoalexins — a critical review. Int. Symp. Natural Phenols in Plant Resistance, Vol. II, Geible, M., Treutter, D., and Feucht, W., Eds., *Acta Hortic.*, 381, 526, 1994.

Langcake, P. and McCarthy, W. V., The relationship of resveratrol production to infection of grapevine leaves by *Botrytis cinerea, Vitis,* 18, 244, 1979.

Lanz, T., Schroder, G., and Schroder, J., Differential regulation of genes for resveratrol synthase in cell cultures of *Arachis hypogaea* L., *Planta,* 181, 169, 1990.

Lorito, M., Broadway, R. M., Hayes, C. K., Woo, S. L., Noviello, C., Williams, D. L., and Harman, G. E., Proteinase inhibitors from plants as a novel class of fungicides, *Mol. Plant Microbe Interac.*, 7, 525, 1994.

Melchior, F. and Kindl, H., Coordinated and elicitor-dependent expression of stilbene synthase and phenylalanine ammonia-lyase gene in *Vitis* cv. Optima, *Arch. Biochem. Biophys.*, 288, 552, 1991.

Muirhead, I. F. and Deverall, B. J., Evaluation of 3,4-dihydroxybenzaldehyde, dopamine and its oxidation products as inhibitors of *Colletotrichum musae* (Berk. & Curt.) Arx in green banana fruits, *Aust. J. Bot.*, 32, 575, 1984.

Neuhaus, J. M., Ahl-Goy, P., Hinz, U., Flores, S., and Mains, F., High-level expression of a tobacco chitinase gene in *Nicotiana sylvestris.* Susceptiblity of transgenic plants to *Cercospora nicotianae* infection, *Plant Mol. Biol.*, 16, 141, 1991.

Osbourn, A., Saponins and plant defence — a soap story, *Trends Plant Sci.*, 1, 4, 1996a.

Osbourn, A., Pre-formed antimicrobial compounds and plant defence against fungal attack, *The Plant Cell*, 8, 1821, 1996b.

Osbourn, A., Bowyer, P., Bryan, G., Lunness, P., Clarke, B., and Daniels, M., Detoxification of plant saponins by fungi, in *Advances in Molecular Genetics of Plant-Microbe Interactions*, Daniels, M. J., Downie, J. A., and Osbourn, A. E., Eds., Kluwer Academic, Amsterdam, 1994a, 215.

Osbourn, A. E., Clarke, B. R., Lunness, P., Scott, P. R., and Daniels, M., An oat species lacking avenacin is susceptible to infection by *Gaeumannomyces graminis* var. *tritici*, *Physiol. Mol. Plant Pathol.*, 45, 457, 1994b.

Osbourn, A., Bowyer, P., Lunness, P., Clarke, B., and Daniels, M., Fungal pathogens of oat roots and tomato leaves employ closely related enzymes to detoxify different host plant saponins, *Mol. Plant Microsc. Int.*, 8, 671, 1995.

Pearce, R. B. and Rutherford, J., A wound-associated suberized barrier to the spread of decay in the sapwood of oak (*Quercus robur* L.), *Physiol. Plant Pathol.*, 19, 359, 1981.

Pegg, G. F. and Woodward, S., Synthesis and metabolism of α-tomatine in tomato isolines in relation to resistance to *Verticillium albo--atrum*, *Physiol. Mol. Plant Pathol.*, 28, 187, 1986.

Prusky, D., The use of antioxidants to delay the onset of anthracnose and stem end rot in avocado fruits after harvest, *Plant Dis.*, 72, 381, 1988.

Prusky, D. and Keen, N. T., Involvement of preformed antifungal compounds in the resistance of subtropical fruits to fungal decay, *Plant Dis.*, 77, 114, 1993.

Prusky, D. and Keen, N. T., Inducible preformed compounds and their involvement in the resistance of plants to pathogens, in *Novel Approaches to Integrated Pest Management*, Reuveni, R., Ed., Lewis Publishers, Boca Raton, FL, 1995, 151.

Prusky, D., Freeman, S., Rodriguez, R. J., and Keen, N. T., A nonpathogenic mutant strain of *Colletotrichum magna* induces resistance in avocado fruits to *C. gloeosporioides*, *Mol. Plant Microsc. Int.*, 7, 326, 1994.

Prusky, D. and Gat, Z., Preharvest conditions and postharvest treatments affecting the incidence of decay in mango fruits during storage, *Acta Hortic.*, 1992.

Prusky, D., Gold, S., and Keen, N. T., Purification and characterization of an endopolygalacturonase produced by *Colletotrichum gloeosporioides*, *Physiol. Mol. Plant Pathol.*, 35, 121, 1989.

Prusky, D., Keen, N. T., and Eaks, I., Further evidence for the involvement of a preformed antifungal compound in the latency of *Colletotrichum gloeosporioides* on unripe avocado fruits, *Physiol. Plant Pathol.*, 22, 189, 1983.

Prusky, D., Keen, N. T., Sims, J. J., and Midland, S. L., Possible involvement of an antifungal diene in the latency of *Colletotrichum gloeosporioides* on unripe avocado fruits, *Phytopathology*, 72, 1578, 1982.

Prusky, D., Kobiler, I., Fishman, Y., Sims, J. J., Midland, S. L., and Keen, N. T., Identification of an antifungal compound in unripe avocado fruits and its possible involvement in the quiescent infection of *Colletotrichum gloeosporioides*, *J. Phytopathol.*, 132, 319, 1991.

Prusky, D., Kobiler, I., and Jacoby, B., Involvement of epicatechin in cultivar susceptibility of avocado fruits to *Colletotrichum gloeosporoides* after harvest, *J. Phytopathol.*, 123, 140, 1988.

Prusky, D., Plumbley, R. A., and Kobiler, I., Modulation of natural resistance of avocado fruits to *Colletotrichum gloeosporioides* by CO_2 treatment, *Physiol. Mol. Plant Pathol.*, 39, 325, 1991.

Pryce, R. J. and Langcake, P., ε-viniferin: an antifungal resveratrol trimer from grapevines, *Phytochemistry*, 16, 1452, 1977.

Powell, A. L. T., D'Hallewin, G., Hall, B. D., Stotz, H., Labavitch, J. M., and Bennett, A. B., Glycoprotein inhibitors of fungal poygalacturonases: expression of pear PGIP improves resistance in transgenic tomatoes, *Plant Physiol.*, 105, 159, 1994.

Rick, C. M., Uhlig, J. W., and Jones, A. D., High α-tomatine content in ripe fruit of Andean *Lycopersicon esculentum* var. *carasiforme*; developmental and genetic aspects, *Proc. Natl. Acad. Sci. U.S.A.*, 81, 12877, 1994.

Rodov, V., Ben-Yehoshua, S., Fang, D., Kim, J. J., and Ashkenazi, R., Preformed antifungal compounds of lemon fruit: citral and its relation to disease resistance, *J. Agric. Food Chem.*, 43, 1057, 1995.

Sarig, P., Zahavi, T., Zutkhi, Y., Yannai, S., Lisker, N., and Ben-Arie, R., Ozone for control of postharvest decay of table grapes caused by Rhizopus stolonifer, *Physiol. Mol. Plant Pathol.*, In press.

Sandrock, R. W., DellaPenna, D., and VanEtten, H. D., Purification and characterization of β_2-tomatinase, and enzyme involvement in the degradation of α-tomatine and isolation of the gene encoding β_2-tomatinase from *Septoria lycopersici*, *Mol. Plant Microsc. Int.*, 8, 960, 1995.

Schlosser, E., Preformed phenols as resistance factors, Int. Symp. Natural Phenols in Plant Resistance, Geible, M., Treutter, D., and Feucht, W., Eds., Vol. II, *Acta Hortic.*, 381, 615, 1994.

Schnathorst, W. C., Coplin, D., Presley, J. T., and Carns, H. R., Inhibition of germination of conidia of *Verticillium albo-atrum* in tracheal fluids from infected tolerant cotton plants, *Phytopathology*, 56, 899, 1966 (Abstr).

Schonbeck, F. and Schlosser, E., Preformed substances as potential protectants, in *Physiological Plant Pathology*, Heitefuss, R. and Williams, P. H., Springer-Verlag, Berlin, 1976, 653.

Schroder, G., Brown, J. W. S., and Schroder, J., Molecular analysis of resveratrol synthase: cDNA clones and relationship with chalcone synthase, *Eur. J. Biochem.*, 172, 161, 1988.

Schulz, F. A., Some physiological and biochemical aspects of the action mechanism of fungal parasites during fruit storage, *Fruits*, 33, 15, 1978.

Sherf, B. A. and Kolattukudy, P. E., Developmentally regulated expression of the wound- and pathogen-responsive tomato anionic peroxidase in green fruits, *Plant J.*, 3, 829, 1993.

Sitterly, W. R. and Shay, J. R., Physiological factors affecting the onset of susceptibility of apple fruit to rotting by fungus pathogens, *Phytopathology*, 50, 91, 1960.

Smith, C. A. and MacHardy, W. E., The significance of tomatine in the host response of susceptible and resistant tomato isolines infected with two races of *Fusarium oxysporum* f. sp. *lycopersici*, *Phytopathology*, 72, 415, 1982.

Stotz, H. U., Powell, A. L. T., Damon, S. E., Greve, C. L., Bennett, A. B., and Labavitch, J. M., Molecular characterization of a polyglacturonase inhibitor from *Pyrus communis* L. cv. Bartlett, *Plant Physiol.*, 102, 133, 1993.

Stotz, H. U., Contos, J. A., Powell, A. L. T., Bennett, A. B., and Labavitch, J. M., Structure and expression of an inhibitor of fungal poyglacturonases from tomato, *Plant Mol. Biol.*, 25, 607, 1994.

VanEtten, H. D., Mansfield, J. W., Bailey, J. A., and Farmer, E. E., Two classes of plant antibiotics: phytoalexins vs. "phytoanticipins", *Plant Cell*, 9, 1191, 1994.

Verhoeff, K., Spotting of tomato fruits caused by *Botrytis cinerea*, *Neth. J. Plant Pathol.*, 76, 219, 1970.

Verhoeff, K., Latent infections by fungi, *Annu. Rev. Phytopathol.*, 12, 99, 1974.

Verhoeff, K. and Liem, J. I., Toxicity of tomatin to *Botrytis cinerea* in relation to latency, *Phytopathol. Z.*, 82, 333, 1975.

Wattad, C., Dinoor, A., and Prusky, D., Purification of pectate lyase produced by *Colletotrichum gloeosporioides* and its inhibition by epicatechin: a possible factor involved in the resistance of unripe avocado fruits to anthracnose, *Mol. Plant Microsc., Int.*, 7, 293, 1994.

Wiese, W., Vornam, B., Krause, E., and Kindl, H., Structural organization and differential expression of three stilbenes synthase genes located on a 13-kb grapvine DNA fragment, *Plant Mol. Biol.*, 26, 667, 1994.

Yao, C., Conway, W. S., and Sams, C. E., Purification and characterization of a polyglacturonase-inhibiting protein from apple fruit, *Phytopathology*, 85, 1373, 1995.

CHAPTER 8

Induced Resistance to Disease

Ray Hammerschmidt and Elizabeth K. Dann

1-8493-2627-3/97/$0.00+$.50

I. OVERVIEW AND DISTRIBUTION OF INDUCED RESISTANCE

A. Terminology

Infection of one leaf of a plant with virulent or avirulent pathogens that cause localized necrotic lesions often results in physiological and biochemical changes in the plant that enhance the resistance of the plant to further infection. This type of response, which can be either very localized or manifested throughout the plant (systemic), is commonly referred to as induced resistance, acquired resistance, or immunization. We will be using the term "induced resistance" to describe the phenomenon of enhanced resistance following a specific biological or chemical treatment, as defense processes are induced in response to such treatments.

The induced resistance response is characterized by nonspecificity with regard to the inducing agent and the pathogens against which resistance develops. Several different pathogens (including fungi, bacteria, and viruses) are often able to induce resistance in a particular plant, and the resistance that develops is effective against the inducing pathogen as well as a diverse array of pathogens from the major pathogen groups. This lack of specificity distinguishes induced resistance from the phenomenon of viral cross protection, in which a mild strain of a virus will protect plants against a more virulent strain.

B. Distribution of Induced Resistance

Induced resistance has been studied in a number of plant families. The level of understanding, however, varies greatly from initial observations to detailed mechanistic studies on the processes involved in the induction and expression of induced resistance. In this section we will provide a brief overview of the distribution of induced resistance in the plant kingdom with emphasis on recent reports. Several thorough reviews have recently been published, and these should be consulted for further details (e.g., Hammerschmidt and Kuc, 1995; Deverall, 1995).

1. Cucurbits

Induced resistance in cucumber and other cucurbits has recently been reviewed extensively by Hammerschmidt and Yang-Cashman (1995) and by Hammerschmidt and Smith Becker (1996). Therefore, this section will provide only a brief overview of the phenomenon in this plant family. Systemic induced resistance in cucumber was first reported by Kuc et al. (1975) and Kuc and Richmond (1977). Inoculating one leaf of anthracnose-susceptible plants with *Colletotrichum lagenarium*, induced systemic resistance of the plant to anthracnose. The spectrum of biological activity of the induced resistance was subsequently expanded to include expression of resistance to several fungi, bacteria, and viruses that attack above-ground parts of the plant (Hammerschmidt and Yang-Chasman, 1995). In addition, a number of other pathogens that caused necrotic lesions on leaves were also shown to be effective inducers of resistance (Hammerschmidt and Yang-Chasman, 1995). Resistance to anthracnose was also reported to be induced by *C. lagenarium* in muskmelon and watermelon (Caruso and Kuc, 1977).

Resistance to Fusarium wilt in cucurbits has also been reported. Inoculation with avirulent forms of *Fusarium oxysporum* induced resistance in cucumber (Mandeel and Baker, 1991) and watermelon (Biles and Martyn, 1989) to pathogenic *formae speciales*. In the study by Mandeel and Baker, the use of a "split root" system demonstrated that the induced resistance was indeed systemic.

Induced resistance to disease in cucumber has also been demonstrated under field conditions. Inoculation of one leaf of cucumber plants in the field with *C. lagenarium* protected the plants against a subsequent challenge inoculation with the same pathogen (Caruso and Kuc, 1977). Over a four year period, we have found that cucumber plants grown under simulated commercial conditions in the field could be protected against angular leaf spot *(Pseudomonas syringae* pv. *lachrymans)* by prior foliar treatment with the resistance-activating chemical 2,6-dichloroisonicotinic acid (Metraux et al., 1991). The induced resistance was effective in all cultivars tested and and was expressed under different environmental conditions (Hammerschmidt and Widders, unpublished results).

Numerous reports on physiological and biochemical changes associated with induced resistance in cucumber have been reported. These will be discussed later in the chapter.

2. Solanaceous Species

Induced resistance has also been reported for tobacco, potato, and tomato. Much of the biology of induced resistance in this family was recently reviewed (Ozeretskovskaya, 1995), and the reader is directed to this review for more details. Tobacco has also served as a model for many biochemical and molecular studies on induced resistance. An overview of these studies will be discussed later in this review.

Ross (1961a,b) reported that inoculation of tobacco containing the N gene for hypersensitive resistance with tobacco mosaic virus (TMV) induced local and systemic resistance to TMV. The induced resistance was seen as a decrease in the size of the lesions. Injecting sporangia of the blue mold pathogen, *Peronospora tabacina*, into the stem tissues of tobacco was also shown to induce systemic resistance to this pathogen (Cruickshank and Mandryk, 1960). Extensive research by Kuc and co-workers demonstrated that TMV inoculation could also induce resistance to blue mold infection. The resistance was characterized by a more rapid restriction of lesion development and lower rates of sporulation by *Peronospora* sp. (Tuzun and Kuc, 1989).

Doke et al. (1987) first reported on the systemic induction of resistance in potato plants to *Phytophthora infestans*, the late blight pathogen. In these experiments, an elicitor preparation from the pathogen was used to induce resistance which was characterized by fewer lesions. The lesions that did appear on the induced plant resembled those that typically appeared on genetically resistant plants. Stromberg and Brishammer (1991, 1993) demonstrated that inoculating lower leaves with either *P. infestans* or the nonpathogen *P. cryptogaea* also induced systemic resistance in potato. Treatment of potato with arachidonic acid, an elicitor from *P. infestans,* induced resistance to *P. infestans* and *Alternaria solani*, the cause of early blight (Coquoz et al., 1995). Systemic resistance to *P. infestans* is also induced in tomato by a prior inoculation with the same fungus (Enkerli et al., 1993; Heller and Gessler, 1986; Kovats et al., 1991b).

3. Crucifers

Uknes et al. (1992) demonstrated that resistance to *Pseudomonas syringae* pv. *tomato* was induced in Arabidopsis following treatment with the resistance-inducing compound 2,6-dichloroisonicotinic acid (INA). Resistance to *Pseudomonas* sp. was also elicited by prior infection with turnip crinkle virus (Cameron et al., 1994) and *Pseudomonas* sp. (Uknes et al., 1993).

Resistance to downy mildew caused by *Peronospora parasitica* was induced by treatment with salicylic acid, an endogenous inducer of resistance (Uknes et al., 1993), or by inoculation with *Fusarium oxysporum* isolated from roots of Arabidopsis (Mauch-Mani and Slusarenko, 1994).

Inoculation of lower leaves of Arabidopsis with a hypersensitive response-inducing *Pseudomonas syringae* pathovar induced systemic resistance to the appearance of necrosis in leaves challenged with the same bacterium (Summermatter et al., 1995). Even though there were less severe symptoms in the induced leaves compared with controls, the population of bacteria after challenge was the same in both control and induced plants.

Preliminary studies in cabbage *(Brassica olercea)* have shown protection to pathogenic *Xanthomonas campestris* pv. *campestris* following an inoculation three weeks earlier with a weakly pathogenic strain of the same bacterium or a nonpathogen, *X. campestris* pv. *malvacearum* (Jetiyanon et al., 1994). Treatment of cabbage seedlings with 2,6-dichloroisonicotinic acid (INA), a resistance-inducing chemical, induced resistance to *P. parasitica* (Hijwegen and Verhaar, 1993) as shown by decreased sporulation of the pathogen.

4. Legumes

Induced resistance in green bean and other leguminous plants has been recently reviewed (Deverall and Dann, 1995; Hammerschmidt and Smith Becker, 1996). Green bean, *Phaseolus vulgaris*, has been the model host for induced resistance studies and reports of the phenomenon in other legumes remains limited.

Initial studies in green bean demonstrated that treatment of etiolated hypocotyl, leaf, or pod tissue with fungal nonpathogens could induce localized resistance to the anthracnose disease caused by *Colletotrichum lindemuthianum* (Rahe et al., 1969; Skipp and Deverall, 1973). Reports of systemic protection against anthracnose were first provided by Sutton (1979) and later confirmed by Cloud and Deverall (1987). It was shown that droplet inoculation of first leaves with *C. lindemuthianum* resulted in reduced symptoms in upper leaves when they were inoculated with the same pathogen at least a week later.

In expanded studies, INA treatment to unifoliate leaves was equally or more effective than *C. lindemuthianum* inoculation in inducing resistance to *C. lindemuthianum* and *Uromyces appendiculatus*, causing rust disease, in first, second, and third trifoliate leaves (Dann and Deverall, 1995). INA treatment also induced systemic resistance to the halo blight bacterium *Pseudomonas syringae* pv. *phaseolicola*, but not to two soil-borne pathogens, *Rhizoctonia* sp. and *Fusarium solani* f. sp. *phaseoli*.

Protection of green bean under field conditions was also induced by *C. lindemuthianum* or INA to anthracnose or rust diseases, respectively (Sutton, 1982; Dann and Deverall, 1996). Additionally, lesion formation by tobacco necrosis virus or alfalfa mosaic virus in untreated leaves was reduced if lower leaves had received a prior treatment with *U. appendiculatus* or the oligosaccharide chitosan, respectively (Kutzner et al., 1993; Pospieszny et al., 1991).

Physiological changes within the host have been observed in association with local and systemic induced resistance in bean. These include the accumulation of phytoalexins (Elliston et al., 1977), cell wall appositions and encroachment of nuclei in host cells at the sites of attempted challenge infection (Cloud and Deverall, 1987), and the increased activities of chitinase and β-1,3-glucanase in treated and systemic leaves (Dann et al., 1996). These and other studies on induced resistance mechanisms in bean will be discussed in greater detail later.

Reports of induced resistance in red clover *(Trifolium pratense)*, soybean *(Glycine max)*, pea *(Pisum sativum)*, and broad bean have been presented in the review by Deverall and Dann (1995).

Induced resistance has also been investigated in alfalfa *(Medicago sativa)*, where inoculation with three nonpathogenic species of *Colletotrichum* resulted in increased survival of seedlings after challenge inoculation with pathogenic *C. trifolii* (O'Neill et al., 1989). A three-year field trial conducted recently by Hammerschmidt and co-workers demonstrated that white mold disease in soybean caused by *Sclerotinia sclerotiorum* was significantly less severe if plants had received three or four sprays of INA (Hammerschmidt et al., unpublished). Four applications of another resistance activator, benzo (1,2,3) thiadiazole-7-carbothioic acid S-methyl ester (BTH, CGA 245704), also resulted in decreased disease symptoms in the only year it has been tested to date.

There is one preliminary study of apparent induced resistance in mung bean, *Vigna radiata* (Arora and Bajaj, 1985). Very young (24-h-old) hypocotyls were exposed to ethephon, an ethylene-releasing compound, for 24 h then inoculated with *Rhizoctonia solani*. Five days later significantly fewer ethephon-treated hypocotyls bore macroscopic symptoms of infection compared with controls. The increased resistance was associated with enhanced activities of peroxidase and polyphenol oxidase enzymes. Further studies with more mature plants are required to confirm the induced resistance response in mung bean.

5. Other Dicots

There have been few studies examining induced resistance in dicotyledonous plants other than those mentioned already. INA treatment of sugar beet *(Beta vulgaris)* and pear induced resistance to diseases caused by *Cercospora beticola* (Nielsen et al., 1994) and *Erwinia amylovora* (Métraux et al., 1991), respectively.

Fewer lesions developed on upper leaves of cotton following inoculation with *Alternaria macrospora* if the cotyledons had previously received treatments with *A. macrospora* or INA (Brock et al., 1994). Studies of the induced resistance response in cotton are continuing.

Prior treatment of sweet potato *(Ipomea batata)* cuttings with an avirulent isolate of *Fusarium oxysporum* resulted in less infection by the wilt pathogen *F. oxysporum* f. sp. *batatas* (described in Cook and Baker, 1983). This report demonstrated the potential application of induced resistance because the experiments were carried out under field conditions. No report was made on the nature of the host response, thus it is not totally clear if this was a case of induced resistance.

More recently, Simms and Vision (1995) demonstrated that systemic resistance could be induced in wild morning glory *(Ipomea purpurea)* to *Colletotrichum dematium* by prior infection of cotyledons with the pathogen. In this study, several lines of the host were both induced and challenged with several isolates of the pathogen. All isolates of the pathogen were capable of inducing systemic resistance, although some were more effective than others. Some differences in the level of resistance were also observed among the plant lines. Interpretation of these results, however, is made more complex by different levels of age-related resistance that developed in the various plant lines. This work is significant because it describes induced resistance in a wild species of plant.

6. Graminaceous Monocots

The literature concerning the induced disease resistance response in the graminaceous monocots has been extensively reviewed by Steiner and Schonbeck (1995). The systems most thoroughly investigated to date are the barley/powdery mildew and rice/blast interactions. Far less is known of the phenomenon in other important species such as wheat, corn, and pasture grasses. This section provides a brief account of some more recent reports of induced resistance in barley, wheat, corn (maize), and pearl millet. The reader is referred to the above-mentioned review for greater detail of earlier reports on the occurrence and mechanisms of the induced resistance response in monocots.

Kogel and co-workers (1994) have made detailed and comparative investigations of the resistance mechanisms in barley to the powdery mildew fungus *Erysiphe graminis* f. sp. *hordei*. Significantly less disease developed on susceptible primary leaves compared with controls after inoculation if INA had been applied as a soil drench 2, 4, or 10 days earlier. Cytological examinations of epidermal cells revealed hypersensitive cell collapse, cell autofluorescence, and appearance of papillae within 21 h of inoculation. The observed HR was associated with the cessation of haustorial development. This response was almost identical to that observed in plants which had not been treated with INA but which carried the race-specific *Mlg* gene for resistance to the fungus. In addition, INA-induced resistance and histological changes in primary leaves were associated with enhanced mRNA transcript accumulation of PR-1, peroxidase and chitinase, but not β-1,3-glucanase. In support of earlier research (Wasternack et al., 1994), Kogel and co-workers (1995) also reported results of careful experimentation in the barley/powdery mildew system which provided strong evidence that jasmonates were not involved as mediators or signals of chemically induced or genetically based resistance, as previously suggested (e.g., Farmer and Ryan, 1992).

Pretreating lower leaves (leaves 1, 2, and 3) of maize with solutions of phosphates induced resistance in leaves 4, 5, and 6 to the Northern leaf blight disease caused by *Exserohilum turcicum* (Reuveni et al., 1994). Protection, observed as decreased numbers and size of lesions, could be achieved when the upper leaves were challenge-inoculated with the fungus as early as 2 to 4 h after treatment with phosphates. The treatments also stimulated growth of the plants; thus a dual role for foliar application of phosphates was proposed.

Recent studies in the wheat/leaf rust interaction demonstrated that treatment of first leaves with 4-chlorosalicylic acid or inoculation with an avirulent leaf rust pathotype resulted in decreased leaf rust infection when second leaves were subsequently challenge-inoculated with virulent *Puccinia recondita* f. sp. *tritici* (Owen, 1995). Interestingly, INA, salicylic acid, or other chlorinated salicylic acid solutions did not induce systemic resistance across a range of tested concentrations.

An observation of apparent "self-induced" or natural resistance of pearl millet *(Pennisetum glaucum)* to the downy mildew disease caused by *Sclerospora graminicola*, prompted investigators to examine this interaction more closely. Under greenhouse conditions, 3-day-old seedlings were inoculated by dipping roots into various concentrations of zoospore suspensions of *S. graminicola* (Kumar et al., 1993). Above-ground parts had significantly less disease after challenge with the fungus 4 to 6 days later. There was no protective effect when challenged only 2 days after the root-dip treatment. Challenge inoculations of tillers or emerging inflorescences at later times demonstrated that these parts could also be protected by the prior root-dip treatment. Lower concentrations (3×10^3 to 1.2×10^4 zoospores per milliliter) were more effective than a higher concentration (1.5×10^4 zoospores per milliliter) at inducing resistance.

II. INDUCTION PROCESS

A. Plant-Pathogen Interactions Leading to Induced Resistance

In the previous section, the effect of prior pathogen infection on the induction of local and systemic resistance was described. In all plant families studied, the development of a pathogen-induced necrotic lesion appears to be necessary for the induction of resistance. In this section, some aspects of the intial disease phenotypes that lead to systemic resistance induction will be discussed.

1. Compatible Interactions

Systemic resistance induction was first reported in tobacco (Cruickshank and Mandryk, 1960) and cucumber (Kuc et al., 1975) as a result of a prior infection with a pathogen that caused a spreading, necrotic lesion. These are examples of how compatible or susceptible interactions can result in the expression of systemic resistance. The necrotic response, however, does appear to be specific to disease development as damage to tissue caused by localized dry-ice freezing (Hammerschmidt et al., 1982), or by wounds (Hammerschmidt, unpublished observations), or insect attack (Ajlan and Potter, 1991, 1992; Apriyanto and Potter, 1990) does not cause an increase in systemic disease resistance. Infection with pathogens that cause disease without necrosis (e.g., TMV on tobacco or cucumber) does not elicit systemic resistance. It is not known how the necrosis that develops during pathogenesis results in the generation of systemic signals that induce systemic resistance.

2. Incompatible Interactions

Incompatible reactions of plants to pathogens often result in a rapid and very localized death of one or a few cells at the infection site. This is known as the hypersensitive response. Ross (1961a,b) demonstrated that the hypersenstive response of TMV on N gene tobacco was effective in inducing both local and systemic resistance against a subsequent challenge inoculation by the same fungus. Further work by Kuc and colleagues showed that the TMV-induced resistance was also effective against disease caused by *P. tabacina*, the blue mold pathogen. The hypersenstive response has also been shown to induce systemic resistance in cucumber to anthracnose (Smith et al., 1991) and Arabidopsis (Cameron et al., 1994).

The utility of the hypersentive response in studying the early events leading to the expression of induced resistance was demonstrated by Smith et al. (1991). In this study, a nonpathogen of cucumber, *Pseudomonas syringae* pv. *syringae*, was shown to induce a rapid hypersenstive response in cucumber and to induce systemic resistance within 24 h. Generating mutants of the bacterium that were no longer able to elicit the hypersensitive response were unable to induce systemic resistance. Complementing these mutations to restore the ability to cause the hypersenstive response restored the ability to induce systemic resistance. Thus, the ability to cause the hypersensitive response was necessary to induce resistance. Time course studies, in which leaves inoculated with *P.s. syringae* were removed from the plants at 2-h intervals, demonstrated that the inoculated leaf only needed to remain on the plant for 6 h to induce a low level of systemic resistance. This was before visible signs of the hypersensitive response, demonstrating that the events preceding cell death appear to be important in the induction of systemic resistance.

B. Endogenous Signals for Induced Resistance

1. Induction Process and Systemic Signal Source

Because localized infection of one part of a plant can result in the systemic expression of resistance in other parts, it has been hypothesized that a signal or signals must be generated and mobilized from the site of initial infection. Several different chemicals have been suggested to play a role as signals conditioning disease resistance. These include jasmonic acid, abscisic acid, cytokinins, pectic fragments, and salicylic acid (for general reviews see Hammerschmidt, 1993; Stermer 1995; Pierpont, 1994; Farmer, 1994). One of these, salicylic acid (SA) does in fact appear to play an important role in the signaling mechanism of induced resistance. However, the response mechanism is far from being elucidated and is likely to vary between plant species and after different stimuli.

Dean and Kuc (1986a,b) were able to demonstrate that the infected leaf was the source of the signal. Removing the inoculated leaf prior to the onset of symptoms resulted in a lack of resistance expression in the rest of the plant. In addition, grafting a noninduced plant onto a rootstock with an inoculated leaf resulted in resistance expression in the scion. If, however, the inoculated leaf was removed from the rootstock prior to grafting, no resistance was induced in the scion. These experiments demonstrated that only the inoculated leaf was the source of the translocated signal and that other rootstock tissues were unable to generate more signal after an initial induction.

Grafting experiments with tobacco and cucurbits demonstrated that the signal moved internally. Grafting cucumber, watermelon, or muskmelon scions onto an inoculated cucumber rootstock (the signal source) resulted in the expression of resistance in all three types of scions (Jenns and Kuc, 1979). Similarly, the signal for systemic resistance in tobacco also moved through graft unions. Petiole girdling to kill the phloem further supported an internal route for signal transport, suggesting that the signal moves through the phloem (Tuzun and Kuc, 1985).

2. Role of Salicylic Acid

The first reported molecular candidate for the systemic signal was salicylic acid. Previous work had demonstrated that exogenous application of salicylic acid to many plants could induce resistance (see review by Raskin, 1992). This, and the wide distribution of salicylic acid throughout the plant kingdom provided support for the hypothesis that salicylic acid might be involved in some part of induced resistance.

Malamy et al. (1990) and Métraux et al. (1990) provided the first direct evidence that salicylic acid may play a role in induced resistance. In tobacco infected with TMV, Malamy et al. (1990) reported that endogenous salicylic acid levels rose prior to the onset of enhanced resistance to TMV and the expression of PR-1. In cucumber, Métraux et al. (1990) found that salicylic acid concentrations of phloem exudates increased prior to the expression of resistance to infection by *C. lagenarium*. Further support for salicylic acid as an endogenous inducer of resistance in cucumber was provided by the experimentation of Rasmussen et al. (1991), who found that induction of resistance was accompanied by a systemic increase in salicylic acid in phloem exudates and that exogenous application of salicylic acid would induce resistance to *C. lagenarium*. Infection of Arabidopsis with *P. syringae* pv. *syringae* also induced systemic accumulation of salicylic acid and was associated with resistance.

Following those initial reports, several papers provided additional evidence, although mostly correlative, that salicylic acid was involved in the expression of induced resistance. For example, feeding salicylic acid to tobacco leaves resulted in the induction of PR-1 (Ryals et al., 1994). Ward et al. (1991) expanded upon this and demonstrated that exogenously applied salicylic acid induced the expression of the same set of genes that are induced systemically by localized TMV infection.

Ryals and co-workers have provided some very convincing results for the role of salicylic acid in both systemic and localized resistance. Gaffney et al. (1993) transformed tobacco with the bacterial *nahG* gene that codes for the enzyme salicylic acid hydroxylase. This enzyme converts salicylic acid to catechol. Tobacco plants expressing *nahG* did not accumulate salicylic acid after infection with TMV nor did they exhibit expression of systemic-induced resistance. The TMV lesions that developed on the *nahG* plants were also larger than on the controls, suggesting that salicylic acid may also influence the expression of local defense responses. Delaney et al. (1994) reported that *nahG*-transformed tobacco and Arabidopsis plants were more susceptible to a number of pathogens. Friedrich et al. (1995) confirmed that *nahG* plants could not be systemically or locally induced. They also found that application of exogenous salicylic acid did not induce resistance or PR gene expression in *nahG* plants. Catechol, the product of salicylic acid hydroxylase action on salicylic acid, did not induce resistance nor did this compound have any effect on the induction of resistance in control plants treated with salicylic acid.

a. Salicylic Acid Is Not the Translocated Signal

Early evidence discussed previously suggested that salicylic acid might be the translocated signal for induced resistance. This hypothesis was challenged by Rasmussen et al. (1991) who suggested that salicylic acid accumulated in cucumber in response to a primary translocated signal. Inoculating one leaf of cucumber with *Pseudomonas syringae* pv. *syringae* resulted in high levels of salicylic acid in phloem exudates collected 24 h after inoculation. Since a previous report (Smith et al., 1991) found that the signal for systemic resistance was induced by 6 h after inoculation with *P.s. syringae*, Rasmussen et al. (1991) detached the inoculated leaf at 2-h intervals after inoculation and then analyzed the phloem exudate from the inoculated leaf petiole. Salicylic acid was first detected at 8 h after inoculation. Interestingly, when the phloem exudates from the second leaf of the plant were analyzed 24 h after inoculation, plants that had leaf 1 detached at 4 and 6 h had greatly elevated amounts of salicylic acid. Since these plants did not have increased amounts of salicylic acid in the phloem of the inoculated leaves, these results indicated that salicylic acid is not the systemic signal but that its synthesis is induced by another mobile signal that is released as early as 4 h after infection of the first leaf.

Systemic synthesis of salicylic acid in induced as compared to control cucumber plants has also been studied (Meuwly et al., 1995). In this work, the first true leaf was inoculated with tobacco necrosis virus. At 70 h after inoculation, leaf disks from induced and control plants were fed ^{14}C-benzoic acid. The induced tissues converted more of the benzoic acid into salicylic acid than did the controls, thus supporting the hypothesis that enhanced synthesis of salicylic acid may account for the systemic increases in this signal molecule.

Additional evidence that salicylic acid is not the translocated signal was reported in a recent study by Vernooij et al. (1994). Wild-type tobacco scions were grafted onto *nahG*-transformed tobacco rootstock. Inoculation of leaves of the rootstock, which could not accumulate salicylic acid, still resulted in the expression of systemic resistance and the accumulation of salicylic acid in the scion. The results of this work demonstrated that the translocated signal was chemically different from salicylic acid because salicylic acid produced in the rootstock of the *nahG* plants would be converted to the biologically inactive compound catechol.

Salicylic acid appears to function as a locally acting signal for resistance in Arabidopsis, tobacco, and cucumber. The role of salicylic acid in other plants, however, is less clear. Treatment of lower leaves of potato with arachidonic acid induced systemic resistance to *P. infestans* and *A. solani* without an increase in salicylic acid in the systemic tissue (Coquoz et al., 1995). Arachidonic acid did, however, induce localized accumulation of salicylic acid. Infection of rice with *Pseudomonas* sp. or *Magneporthe* sp. did not result in an increase in salicylic acid. However, there was a correlation between endogenous levels of salicylic acid in different rice cultivars and resistance to *Magneporthe* (Silverman et al., 1995).

b. Mode of Action of Salicylic Acid

All evidence presented thus far indicates that salicylic acid is an important signal in the induction of induced resistance. However, the mode of action of salicylic acid is controversial. In

1991, the presence of a soluble protein that bound salicylic acid was reported to be present in tobacco (Chen and Klessig, 1991). This protein was determined to be a catalase, and the enzymic activity of this catalase was inhibited by salicylic acid (Chen et al., 1993). This led to the hypothesis that salicylic acid might act to induce resistance by allowing hydrogen peroxide to accumulate. It was well known at that time that part of the initiation of plant defenses was associated with the production of activated oxygen species (the "oxidative burst") and that hydrogen peroxide could activate some defense processes (for example, see review by Dixon et al., 1994). Thus, it was hypothesized that resistance was induced by the increased levels of hydrogen peroxide arising from the inhibition of catalase by salicylic acid (Chen and Klessig, 1991). The hypothesis was supported by the observations that treatment of plant tissue with a synthetic catalase inhibitor or with hydrogen peroxide would lead to the induction of PR-1 expression. However, it was not determined if hydrogen peroxide would induce resistance or if there was a systemic increase in hydrogen peroxide that preceded the induction of PR-1 or resistance expression in inoculated plants.

Recent reports have, however, challenged the role of catalase and hydrogen peroxide in the expression of systemic-induced resistance (Neuenschwander et al., 1995; Bi et al., 1995). In addition, Ruffer et al. (1995) have provided evidence that the binding of salicylic acid to catalase is not specific. Thus, the role of catalase as the receptor of salicylic acid in the induced resistance response is not very clear.

c. *Synthesis of Salicylic Acid*

Localized infection of plant tissues is associated with the synthesis and accumulation of phenolic compounds (Nicholson and Hammerschmidt, 1992). This general observation suggested that salicylic acid might be biosynthesized at the site of infection and then translocated throughout the plant. The biosynthesis of salicylic acid in tobacco has been studied by Raskin and co-workers. They have shown that salicylic acid is derived from cinnamic acid via benzoic acid. Feeding of labeled benzoic acid to tobacco cell suspension cultures resulted in the production of labeled salicylic acid. A benzoic acid 2-hydroxylase was also induced in tobacco infected with TMV or treated with benzoic acid. Leon et al. (1993) characterized the benzoic acid 2-hydroxylase enzyme as a 160-kDa p450 enzyme that has the unique property of being soluble.

The synthesis of salicylic acid in cucumber appears to follow the same pathway (Meuwly et al., 1995). Feeding labeled phenylalanine to cucumber leaf disks from TNV- or *Pseudomonas*-infected plants resulted in higher amounts of labeled salicylic acid than feeding phenylalanine to control tissues. Treatment of leaf tissues with an inhibitor of phenylalanine ammonia lyase blocked the conversion of phenylalanine into salicylic acid. Labeled cinnamic and benzoic acids were also incorporated into salicylic acid in infected leaf tissues. Rice, which has a very high endogenous content of salicylic acid, also appears to synthesize salicylic acid by the same pathway (Silverman et al., 1995).

Knowing the enzymes involved in biosynthesis of salicylic acid will eventually lead to the cloning of the genes coding for these enzymes, Having these cloned genes will enable researchers to fully examine the role of salicylic acid in resistance induction.

III. INDUCED RESISTANCE DEFENSE RESPONSES

A. Systemically Induced Defense

Induction of resistance is accompanied by the simultaneous induction of several classes of genes and their protein products throughout the plant body. These include oxidases and five groups of proteins known as pathogenesis-related or PR proteins. Putative roles for many of these proteins have been determined (see review by Stermer, 1995) and several can be used as markers for the expression of the induced resistant state (Kessmann et al., 1994).

1. PR Protein Expression

The induction of systemic resistance in several plant species has been associated with the accumulation of proteins throughout the plant. These proteins, known as pathogenesis-related or PR proteins, have been most extensively studied in tobacco where five distinct classes have been described. The appearance of PR proteins is the result of new transcription and translation, and their synthesis is induced by salicylic acid.

Functions for several PR proteins have been elucidated during the last decade. For example, the PR-2 are β-1,3-glucanases and the PR-3 class are chitinases. Both of these enzymes may function in resistance by acting on cell walls of fungi. The PR1 and PR-5 classes have no known enzymic function, but both have antifungal activity. Transformation of plants with genes encoding PR-1, PR-5, or chitinase has been reported to increase resistance to some fungi (Alexander et al., 1993; Liu et al., 1994; Benhamou et al., 1993).

Although a function for each PR protein has not been fully described, the coordinate induction of PR proteins and the expression of systemic resistance has allowed researchers to use PR gene expression as a reliable marker of induced resistance (for example, see Ward et al., 1991). Extensive reviews on PR proteins and their functions have recently been published (for example, Bol et al., 1990; Linthorst, 1991; Stermer, 1995), and the reader is referred to these for further information.

2. Oxidative Enzymes

In addition to PR proteins, peroxidase, phenoloxidase, and lipoxygenase activities have been shown to increase systemically in induced plants (see review by Stermer, 1995). Of these, the peroxidases have received the most study.

The systemic induction of peroxidase has been extensively investigated in the cucumber system. Infection of one leaf with a resistance-inducing pathogen results in the systemic appearance of three acidic apoplastic peroxidases (Hammerschmidt et al., 1982). The cDNA for one of these peroxidases has been cloned and it was shown that the transcription of the peroxidase gene occurs prior to the appearance of the enzyme, thus suggesting that the enzymes are synthesized *de novo* (Rasmussen et al., 1995). A dual role for peroxidase in the defense response has been proposed. Its involvement in lignification, part of cucumber induced resistance, is likely, as it catalyzes the formation of free radicals of lignin monomer precursors, allowing these monomers to polymerize. The formation of free radicals such as hydrogen peroxide may inhibit pathogen growth directly. Experiments using antisense RNA-expressing cucumber may help elucidate the exact role and need for these enzymes in induced resistance.

Lipoxygenase has also been shown to be associated with systemic induced resistance in cucumber and rice (Avdiushko et al., 1993; cited in Kessmann et al., 1994). This enzyme may function through the generation of antimicrobial lipid oxidation products (Croft et al., 1993). Alternatively, it may act to generate another defense-related signal, jasmonic acid (Farmer, 1994).

3. Secondary Metabolites

Many antimicrobial secondary metabolites, either as preformed constituents or induced phytoalexins, have been implicated in the defense of plants to pathogens. The general role that these types of putative defenses may play in induced resistance is, however, unclear. Except in the few cases where more rapid induction of phytoalexins occurs in induced plant tissues after challenge (e.g., Elliston et al., 1977), and in the systemic accumulation of salicylic acid in tobacco (Malamy et al., 1990), cucumber (Métraux et al., 1990; Rasmussen et al., 1991), and green bean (Dan and Deverall, 1996), there is little reported research on changes in secondary metabolites in induced tissues. This is in stark contrast to studies on induced resistance responses to arthropod herbivores where systemic (e.g., nicotine in *Nicotiana sylvestris* [Baldwin et al., 1994]) and local induction (e.g., furanocoumarins in wild parsnip [Zangerl, 1990]) can result in relatively large increases in natural products that are involved in resistance to these pests. Thus, examining plant tissues with induced pathogen resistance for increases in bioactive natural products or precursors to putative defense compounds may be an area worthy of further study.

B. Cytology of Induced Resistance

Cytological studies of the interaction between plant tissues and pathogens can reveal much about potential mechanisms involved in resistance. Light and electron microscopy of the infection process and host responses in cucumber, green bean, and several solanaceous species have been described.

1. Prepenetration Process

In most cases, there is little effect of induced resistance on the initial stages of infection by pathogenic fungi into plant tissues. For example, in the interaction of pathogenic *Colletotrichum* species with cucumber and bean leaf tissue, there is little difference in the rate of spore germination and appressorium formation on induced as compared to noninduced tissues (Cloud and Deverall, 1987; Richmond et al., 1979; Jenns and Kuc, 1980; Hammerschmidt and Kuc, 1982). In the interaction of potato or tomato leaf tissue with *Phytophthora infestans*, there have been reports of preinfection effects associated with induced resistance. Induced potato leaves were reported to have a negative effect on cystospore germination as compared to controls (Chai and Doke, 1987; Doke et al., 1987). Contrary to that observation, Stromberg and Brishammer (1993) reported that more cystopores germinated on induced leaves as compared to controls. Part of this discrepancy may be due to the level of induced resistance. Kovats et al. (1991b) found that *P. infestans* cystospores germinated at a lower rate on tomato plants expressing a very high level of induced resistance as compared to controls.

2. Penetration Phenomena

Postgermination events have also been examined. In cucumber, penetration of *C. lagenarium* into induced host tissue is greatly reduced (Richmond et al., 1979; Jenns and Kuc, 1980). The inhibition of penetration is correlated with enhanced lignification (Hammerschmidt and Kuc, 1982; Siegrist et al., 1994) and autofluorescent papilla (Stumm and Gessler, 1986) directly under appressoria that did not penetrate the epidermal cell. Ultrastructural studies demonstrated that the epidermal cell directly underneath nonpenetrating appressoria were modified (Xuei et al., 1988). Ultrastructural histochemistry confirmed the presence of lignin-like materials associated with these modifications (Stein et al., 1993). Energy-dispersive X-ray analysis of these structures revealed the presence of silicon (Stein et al., 1993). Lignified papilla have also been observed more frequently under nonpenetrating appressoria of *Sphaerotheca fuliginea* on induced as compared to control cucumber leaves (Conti et al., 1990). Decreased penetration associated with nonlignified papilla formation was observed in the interactions between bean and *C. lindemuthianum* (Cloud and Deverall, 1987) and between potato or tomato and *P. infestans* (Chai and Doke, 1987; Kovats et al., 1991b; Stromberg and Brishammer, 1993).

3. Development of Pathogens in Induced Tissues

Cytological studies of induced resistance have shown that postpenetration events may also contribute to resistance. Successful penetrations of *C. lagenarium* into induced cucumber epidermal cells can result in lignification of the entire epidermal cell, which may then block further development of the fungus (Kovats et al., 1991a). Lignification of hyphae in induced cells may also contribute to the resistance. Stein et al. (1993) found that hyphae that had successfully penetrated into induced cells were encased in a lignin-like material. *Colletotrichum lindemuthianum* hyphae that successfully penetrated induced tissues were only found in epidermal cells and were shorter than hyphae found in control plants. The hyphae that were found in the induced plants became necrotic by 72 h after infection (Cloud and Deverall, 1987). In control plants, however, hyphae were observed in the mesophyll of control plants by 72 h after infection. Cystospores of *P. infestans* that penetrated induced potato (Stromberg and Brishammer, 1993) and tomato (Heller and Gessler, 1986; Kovats et al., 1991b) leaves were associated with rapid hypersenstive death of the plant cells.

There were no reported differences in the rates of penetration of *Peronsopora tabacina* into induced as compared to control tobacco plants (Stolle et al., 1988). This may be due to the fact that this pathogen penetrated through stomata rather than directly into the epidermal layer as for *Colletotrichum* sp. and *Phytophthora* sp. Within two days of inoculation, however, *Peronospora* sp. in induced plants had not developed to the same extent as in untreated tissue. Enhanced lignification was observed in the infected induced plants by two days after infection and host cell necrosis was seen at five days (Ye et al., 1992). The fungal haustoria observed in induced plants were also encased in a nonidentified material.

C. Postinfection Biochemical Changes

1. Phytoalexins

Phytoalexins are low molecular weight antimicrobial compounds that accumulate locally in plant tissues after infection or elicitation treatments (Paxton, 1981). Although phytoalexins have been hypothesized to play an important role in plant defense, the role of these compounds in induced resistance has not been extensively studied.

Infection of potato tuber tissue with an incompatible race of *Phytophthora infestans* results in the development of localized resistance of the tissue to infection by compatible races of *P. infestans* (Müller and Borger, 1940). This observation led to the development of the phytoalexin hypothesis of Müller. It is likely that phytoalexins do play a role in this localized induced resistance because the terpenoid phytoalexins, rishitin and lubimin, rapidly accumulate in this interaction (Kuc, 1982). However, other defenses, such as lignification, are also induced (Bostock et al., 1986). Thus, this localized defense is likely to be due to several factors.

Prior inoculation of green bean hypocotyls with incompatible *Colletotrichum lindemuthianum* races or nonpathogenic *Colletotrichum* species results in very localized resistance to subsequent infection by compatible races of *C. lindemuthianum*. Because the prior inoculation can elicit the accumulation of phytoalexins, it is likely that the phytoalexins may be a factor in the expression of the localized resistance. Similarly, localized resistance of carnation to *Fusarium* (van Peer et al., 1991) and soybean to *Phytophthora megasperma* (Paxton and Chamberlain, 1967) also appear to be related to the accumulation of phytoalexins at the infection site.

The role of phytoalexins in systemic induced resistance has been studied even less than for localized induced resistance. Elliston et al. (1977) reported that the induced resistance of bean hypocotyls to compatible *C. lindemuthianum* races at a short distance (1 cm) from the induction site was not the result of systemic accumulation of phytoalexins. The induced tissues did, however, accumulate phytoalexins rapidly after challenge inoculation with a compatible race. In tobacco, systemic resistance to *Peronsopora tabacina* was not correlated with the ability of the induced tissues to accumulate phytoalexins more rapidly after challenge infection.

2. Cell Wall Alterations

As described above, one of the mechanisms expressed in induced resistance is a decreased rate of fungal penetration into plant tissues. This inhibition of penetration has been attributed to the localized depostion of lignin (Hammerschmidt and Kuc, 1982; Stein et al., 1993; Conti et al., 1990; Siegrist et al., 1994), silicon (Stein et al., 1993), and callose (Kovats et al., 1991a) in the cell wall. As proposed by Ride (1978), these types of cell wall modifications may enhance the ability of the plant to arrest pathogen ingress by providing resistance to mechanical penetration, increase resistance to cell wall-degrading enzymes, or provide barriers that do not allow diffusion of toxins or enzymes from the pathogen to the host cell. Localized lignification has been shown to increase resistance to pectinases (Ride, 1978). In addition, an increase in wall-bound hydroxyproline-rich glycoproteins, which has been correlated with heat shock-induced resistance to fungal infection in cucumber, appears to increase resistance of isolated cell walls to pectinolytic enzymes.

The role of lignification in the induced resistance of cucumber to anthracnose is supported by the observation that OH-PAS, an inhibitor of the lignin biosynthetic enzyme cinnamyl alcohol dehydrogenase (Moersbacher et al., 1990), increased the rates of penetration by *C. lagenarium* into induced tissues. Associated with this increase in penetration was a decrease in the deposition of lignin at penetration sites and in infected cells (Yang-Cashman and Hammerschmidt, unpublished results).

D. Similarities between Induced Resistance and Other Types of Defense Responses

Heath (1995) has categorized resistance into four basic types: basic, parasite-specific, age- or organ-related, and induced. In each case where mechanism of defense have been studied, there appears to be a considerable overlap in the type of defense processes. For example, lignification has been associated with parasite-specific (Hammerschmidt et al., 1984) and basic (Hammerschmidt et al., 1985) resistance of cucumber as well as induced resistance (Hammerschmidt and Yang-Cashman, 1995). Similarly, phytoalexin accumulation is associated with both parasite-specific and induced resistance of bean hypocotyls (Elliston et al., 1977). These studies illustrate that induced resistance utilizes the same types of defense processes as other types of active defense.

IV. PRACTICAL APPLICATIONS OF INDUCED RESISTANCE

A. Biological Inducers of Resistance

1. Pathogens

As described earlier in the chapter, pathogens have been frequently used as inducers of systemic resistance to disease. These include both compatible interactions (for example, *Colletotrichum lagenarium* on cucumber and *C. lindemuthianum* on green bean) as well as incompatible, hypersensitive responses such as TMV on N-gene tobacco or *Psuedomonas syringae* pv. *syringae* on cucumber. Although these interactions have facilitated the development of systems to study the induction and expression of induced resistance under controlled conditions, the practical use of pathogens as inducers would appear to be limited because of application problems (that is, proper environmental conditions must be in place) as well as the potential spread of disease. Thus, alternative inducing agents are needed for the practical use of induced resistance.

2. Nonpathogens

The induction of disease resistance by nonpathogenic microorganisms which are natural inhabitants of leaf and root/soil environments has been reported. This raises the question of whether all plants grown under nonsterile conditions exist in an elevated state of resistance. This is a concept discussed in more detail by Heath (1995).

The potential of utilizing such microbes in crop protection has been recognized. There are few reports of the induction of resistance by phylloplane or leaf-inhabiting fungi and bacteria, and these are briefly reviewed elsewhere (Tuzun and Kloepper, 1995; Heath, 1995). It was suggested that rhizosphere agents may have a greater chance of effectively inducing resistance, as they are not as prone as phylloplane agents to the detrimental effects of UV irradiation, temperature fluctuations, etc. (Tuzun and Kloepper, 1995). This should not, however, deter investigations of phylloplane inducers.

Many of the soil-borne inducers may have beneficial associations with the plant, and this section will present recent reports of apparent resistance induction by these microorganisms, which include the plant growth-promoting rhizobacteria and fungi (PGPR and PGPF) and symbiotic vesicular arbuscular mycorrhizae (VAM). Most studies have concentrated on the induction of resistance by strains of PGPR, and have been reviewed recently (Tuzun and Kloepper, 1994, 1995).

Wei et al. (1991) describes results of the effect of pelleting (coating) cucumber seeds with PGP rhizobacteria on anthracnose disease after subsequent inoculation of second leaves with *C. lagenarium*. The greatest reduction in lesion number and size was observed after seed treatment with *Pseudomonas aureofaciens* strain 36-5, however an initial inoculation of first leaves with *C. lagenarium* 7 days prior to challenge (i.e., the standard induced resistance control) provided even greater protection. None of the PGPR strains could be reisolated from the petioles, and bacterized plants had fewer discolored root hairs compared with nonbacterized controls. Plant growth was stimulated by the PGPR pretreatments.

Pseudomonas fluorescens strain WCS417r pretreatments to roots of carnation *(Dianthus caryophyllus)* or radish *(Raphinus sativus)* resulted in decreased incidence of Fusarium wilt, caused by *Fusarium oxysporum* f. sp. *dianthi* (van Peer et al., 1991) or *F. o.* f. sp. *raphini* (Hoffland et al., 1995), respectively. In carnation, dianthramide phytoalexins in stems after *Fusarium* sp. inoculation accumulated to greater extents if plants had been exposed to *P. fluorescens*. Bacterization alone did not enhance phytoalexins and the apparent induced resistance response was greater in a naturally more resistant cultivar than in a susceptible one. In radish there was no accumulation of PR proteins in leaf or intercellular wash fluids (IWF) after *P. fluorescens* or salicylic acid root treatments (salicylic acid, but not INA, also induced resistance to the wilt disease). *P. syringae* pv. *tomato* inoculation of first leaves induced accumulation of PR-1, -2, and -5 proteins in noninoculated leaves. Effects of challenge-inoculating these *P. s. tomato*-inoculated plants with the wilt fungus were not reported. The authors concluded that the accumulation of PR proteins was not a prerequisite for the expression of SIR, a statement which is perhaps premature given the preliminary nature of the results in a system which has been little studied.

Experimentation in the well-known tobacco/TNV system demonstrated that exposure of roots to *P. fluorescens* strain CHA0 provided the same level of resistance to TNV on leaves as TNV-induced systemic resistance (Maurhofer et al., 1994). PR-1, endochitinases, β-1,3-glucanases and salicylic acid were induced to similar extents by both pretreatments, i.e., the same physiological changes occurred after *P. fluorescens* treatment to roots or TNV to leaves. This report has provided the most convincing evidence so far that protection observed after bacterization by PGPR strains is attributable to an induced resistance response and not some other mechanism.

Recent reports of the lesser-studied PGPF demonstrated apparent SIR in cucumber leaves to the anthracnose pathogen *Colletotrichum lagenarium*, after previous exposure of roots to several PGPF isolates (Meera et al., 1994, 1995a,b). PGPF-treated plants exhibited greater plant heights and biomass compared with nonbacterized plants (Meera et al., 1995a). Pretreatment with two isolates only resulted in similar extents of anthracnose suppression on second leaves as observed when first leaves had received a prior necrotizing inoculation with *C. lagenarium* (Meera et al., 1994). Interestingly, these isolates and others offering protection did not colonize the roots of plants, and the decrease in anthracnose in a susceptible cultivar was less than that observed in a resistant one (Meera et al., 1995b). These results suggest that further investigation is required to determine the mode of action responsible for disease suppression by PGP fungi.

The potential of arbuscular mycorrhizae (AM or VAM) associations to induce resistance upon colonization of host roots obviously exists, however there have been few studies undertaken in this area. Root rot disease in holly *(Ilex crenata)* caused by *Thielaviopsis basicola* was apparently suppressed if VAM colonization had occurred, increasing wound barrier formation (Dehne and Shonbeck, 1979). Similarly, root cells of tomato and cucumber exhibited increased lignification if colonized by mycorrhizae, which the authors speculated may have acted as a barrier to penetration by *Fusarium oxysporum* f. sp. *lycopersici*, thus accounting for the suppression of Fusarium wilt in tomato (Wick and Moore, 1984). Potato minitubers grown in a peat-based medium containing *Glomus intraradix*, an arbuscular mycorrhizal fungus, had significantly less dry rot disease after exposure to *Fusarium sambucinum* than minitubers grown in the absence of the AM fungus (Niemira et al., 1996).

Due to the universal nature of PGPR, PGPF, and mycorrhizal associations, it is likely that they contribute to the maintenance of the plants' natural state of resistance in a similar manner as that proposed for saprophytic microorganisms on the phylloplane (for example, see Heath, 1995), however this remains to be confirmed by more detailed investigations of the mechanisms responsible for the observed decrease in disease. Because of the apparent time lag required for resistance induction, it is likely that such microbes might be most beneficial in the maintenance of resistance after protection has been achieved early and relatively quickly by another means. The plant's growth-promoting characteristic as well as biocontrol properties also make these classes of potential inducers worthy of further investigation.

B. Chemical Inducers

A number of chemical compounds have been used to induce the defense responses in a number of plants. Some compounds have been synthetically manufactured, such as dinitroaniline herbicides (Cohen et al., 1992; Grinstein et al., 1984) or amino acid analogues (Cohen, 1994a,b), while others have been extracted from natural sources, for example oxalates from spinach (Doubrava et al., 1988), fatty acids (Cohen et al., 1991), and phenolic acids (Walters et al., 1993). Their list is extensive, and the reader is referred to reviews by Kessmann et al. (1994) and Deverall (1995) for more information.

The induction of disease resistance by 2,6-dichloroisonicotinic acid (INA) has been demonstrated under greenhouse and/or field conditions for a number of plants including cucumber (Métraux et al., 1991), tobacco (Ahl Goy et al., 1990; Ward et al., 1991; Métraux et al., 1991), rice, pepper, pear (Métraux et al., 1991), Arabidopsis sp. (Uknes et al., 1992), sugar beet (Nielsen et al., 1994), green bean (Dann and Deverall, 1995, 1996) and soybean (Hammerschmidt et al., unpublished). In the majority of studies, INA induced similar levels of protection as those afforded by microbial inoculation, and was associated with increased activities or enhanced mRNA transcript accumulation of a number of PR proteins (for example, see Ward et al., 1991; Uknes et al., 1992; Dann et al., 1996). For these reasons, INA has been a useful tool in the elucidation of aspects of the defense mechanism. Unlike microbial inoculation, INA treatment did not enhance the endogenous levels of salicylic acid (SA), a requirement for SIR (Gaffney et al., 1993; Vernooij et al., 1994) in tobacco (Vernooij et al., 1995), or in green bean (Dann et al., 1996), yet resistance was induced by INA in *nahG* tobacco (Vernooij et al., 1995) and Arabidopsis (Delaney et al., 1994) transformants lacking the ability to accumulate endogenous SA. Thus, it is thought that INA acts as a functional analogue of salicylic acid, i.e., enters the pathway at the same site of action as SA or downstream of it.

A new resistance-activating compound, benzo (1,2,3) thiadiazole-7-carbothioic acid S-methyl ester (BTH, CGA 245704) has recently been reported and released for commercial use (Kessmann et al., 1995; Ruess et al., 1995). BTH was found to be particularly effective in monocots. A single application to wheat resulted in significant protection lasting several weeks to powdery mildew caused by *Erysiphe graminis*, and also suppressed leaf spot and rust diseases caused by *Septoria* sp. and *Puccinia* sp., respectively. Similarly, a single application to rice provided long-lasting resistance to *Pyricularia oryzae*. The activator also induced resistance in dicotyledonous plants such as tobacco, cucumber, tomato, and green beans, however repeated applications were necessary for significant disease reductions. As for INA, the compound is readily translocated within the plant and has no direct antimicrobial effects. It appears to stimulate the same defense responses as observed after INA treatment, and is likely to be another functional analogue of SA.

Since the phenomenon of systemic-induced resistance was conclusively demonstrated in the early 1960s and a potential role for it as a tool in crop protection hypothesized, the release of BTH represents an exciting and significant advancement in disease control strategies. It is likely that disease resistance induced by chemical activators such as BTH or by other means will become a more important component of disease management programs, particularly in cases where current methods of disease control are less effective than they used to be.

C. Means to Evaluate Potential Resistance Inducers

As the phenomenon of systemic-induced resistance becomes more widely known and its potential in crop protection realized, there have been abundant reports claiming the induction of disease resistance following various diverse treatments. In order for such treatments to be classified as "resistance inducers" or "activators", however, it is imperative that their ability to activate the plants' defense mechanisms be demonstrated. Such mechanisms have already been discussed in detail, but include, for example, the induction of pathogenesis-related proteins, induction of the hypesensitive response, and an alteration of the infection and/or colonization processes which may be detrimental to the invading pathogen.

Conceivably, the enhancement of a particular crop's disease resistance by a chemical agent may be more readily incorporated into existing agronomic and disease management practices than other resistance-inducing methods. This is perhaps already reflected by the recent registration and commercial release of BTH, discussed in the previous section. The release of other chemical inducers is inevitable. Kessmann and colleagues (1994) have proposed three criteria which must be fulfilled before a chemical compound can be considered an inducer of the resistance response. Firstly, it must induce resistance to the same diseases as induced by a biological inoculation. Secondly, it must activate the expression of the same biochemical markers or resistance mechanisms as those activated in the biological model; and thirdly, it must have no antimicrobial activity nor be converted *in planta* to antimicrobial metabolites. These criteria provide a good set of rules that will help distinguish between control that results from induced resistance rather than some other mechanism.

V. CONCLUSIONS

Although the phenomenon of induced resistance has been known for many years, there have been few efforts to implement this as part of practical disease control. Some of this lack of interest in the application of induced resistance may have been due to the ready availability of pesticides and no easy means of inducing resistance in the field. However, the growing public concern over pesticide use and the loss of pesticide efficacy through resistance and/or legislation has led to interest in alternative disease control strategies. The recent successes in using nontoxic chemicals and nonpathogens as resistance-inducing agents in field and greenhouse trials demonstrate that induced resistance may be a significant inclusion in the integrated control of plant diseases.

As reviewed in this manuscript, induced resistance uses the same types of defenses that plants use in major gene or nonhost resistance. These results, and the use of biological agents or nontoxic chemicals, suggest that the use of induced resistance should be an environmentally safe means of control.

There are, however, a number of factors that have not been addressed in research on induced resistance. Our knowledge base is very limited, with most of the biological and molecular work being done with a few species within a small number of plant families. Furthermore, there is little information on the effect of the environment on the expression and efficacy of induced resistance. Finally, the effects of standard agronomic and horticultural practices on the expression of induced resistance have not been extensively studied. Although it would appear that induced resistance could be easily interfaced with other standard practices, this has not been tested.

In summary, induced resistance should provide a stable, environmentally safe type of disease control. The fact that induced resistance utilizes multiple defense responses suggests that it may also be more durable than major gene resistance. Thus, the future of induced resistance as part of practical integrated disease control appears promising.

REFERENCES

Ahl Goy, P., Métraux, J.-P., Speich, J., and Staub, T., Derivatives of isonicotinic acid induce resistance to viruses, bacterial and fungi in tobacco (Abstract), in *Proc. Fifth Int. Symp. the Molecular Genetic of Plant Microbe Interactions,* Kluwer Academic, Dordrecht, 1990.

Ajlan, A. M. and Potter, D. A., Does immunization of cucumber against anthracnose by *Colletotrichum lagenarium* affect host suitability for arthropods?, *Entomol. Exp. Appl.*, 58, 83, 1991.

Ajlan, A. M. and Potter, D. A., Lack of effect of tobacco mosaic virus-induced systemic acquired resistance on arthropod herbivores in tobacco, *Phytopathology,* 82, 647, 1992.

Alexander, D., Goodman, R. M., Gut-Rella, M., Glascock, C., Weyman, K., Friedrich, L., Maddox, D., Ahl-Goy, P., Luntz, T., Ward, E., and Ryals, J., Increased tolerance to two oomycete pathogens in transgenic tobacco expressing pathogenesis-related protein 1a, *Proc. Natl. Acad. Sci. U.S.A.*, 90, 7327, 1993.

Apriyanto, D. and Potter, D. A., Pathogen-activated induced resistance of cucumber: response of arthropod herbivores to systemically protected leaves, *Oecologia*, 85, 25, 1990.

Arora, Y. K. and Bajaj, K. L., Peroxidase and polyphenol oxidase associated with induced resistance of mung bean to *Rhizoctonia solani* Kuhn, *Phytopathol. Z.*, 114, 325, 1985.

Avdiushko, S. A., Ye, X. S., Hildebrand, D. F., and Kuc, J., Induction of lipoxygenase activity in immunized cucumber plants, *Physiol. Mol. Plant Pathol.*, 42, 83, 1993.

Baldwin, I. T., Schmelz, E. A., and Ohnmeiss, T. E., Wound-induced changes in root and shoot jasmonic acid pools correlate with induced nicotine synthesis in *Nicotiana sylvestris* Spegazzini and Comes, *J. Chem. Ecol.*, 20, 2139, 1994.

Benhamou, N., Broglie, K., Chet, I., and Broglie, R., Cytology of infection of 35S-bean chitinase transgenic canola plants by *Rhizoctonia solani*, cytochemical aspects of chitin breakdown *in vivo*, *Plant J.*, 4, 295, 1993.

Bi, Y. M., Kenton, P., Mur, L., Darby, R., and Draper, J., Hydrogen peroxide does not function downstream of salicylic acid in the induction of PR protein expression, *Plant J.*, 8, 235, 1995.

Biles, C. L. and Martyn, R., Local and systemic resistance induced in watermelons by formae specialis of *Fusarium oxysporum*, *Phytopathology,* 79, 856, 1989.

Bol, J. F., Linthorst, H, J. M., and Cornelissen, B. J. C., Plant pathogenesis-related proteins induced by virus infection, *Annu. Rev. Phytopathol.*, 28, 113, 1990.

Bostock, R. M., Schaeffer, D. A., and Hammerschmidt, R., Comparison of elicitor activities of aracidonic acid, fatty acids and glucans from *Phytophthora infestans* in hypersensitivity expression in potato tuber, *Physiol. Mol. Plant Pathol.*, 29, 349, 1986.

Brock, P. M., Inwood, J. R. B., and Deverall, B. J., Systemic induced resistance to *Alternaria macrospora* in cotton *(Gossypium hirsutum), Australas. Plant Pathol.*, 23, 81, 1994.

Cameron, R. K., Dixon, R. A., and Lamb, C. J., Biologically induced systemic acquired resistance in *Arabidopsis thaliana, Plant J.*, 5, 715, 1994.

Caruso, F. L. and Kuc, J., Protection of watermelon and muskmelon against *Colletotrichum lagenarium* by *Colletotrichum lagenarium*, *Phytopathology,* 67, 1285, 1977.

Chai, H. B. and Doke, N., Systemic activation of O_2-generating reaction, superoxide dismutase, and peroxidase in potato plants in relation to induction of systemic resistance to *Phytophthora infestans*, *Ann. Phytopathol. Soc. Jpn.*, 53, 645, 1987.

Chen, Z. and Klessig, D. F., Identification of a soluble salicylic acid-binding protein that may function in signal transduction in the plant disease resistance response, *Proc. Natl. Acad. Sci. U.S.A.*, 88, 8179, 1991.

Chen, Z., Silva, H., and Klessig, D. F., Active oxygen species in the induction of plant systemic acquired resistance by salicylic acid, *Science*, 262, 1883, 1993.

Cloud, A. M. E. and Deverall, B. J., Induction and expression of systemic resistance to the anthracnose disease in bean, *Plant Pathol.*, 36, 551, 1987.

Cohen, R., Cuppels, D. A., Brammall, R. A., and Lazarovits, G., Induction of resistance towards bacterial pathogens of tomato by exposure of the host to dinitroaniline herbicides, *Phytopathology,* 82, 110, 1992.

Cohen, Y., Gisi, U., and Mosinger, E., Systemic resistance of potato plants against *Phytophthora infestans* induced by unsaturated fatty acids, *Physiol. Mol. Plant Pathol.*, 88, 255, 1991.

Cohen, Y., Local and systemic control of *Phytophthora infestans* in tomato plants by DL-3-amino-n-butanoic acids, *Phytopthology*, 84, 55, 1994a.

Cohen, Y., 3-Aminobutyric acid induces resistance against *Peronospora tabacina, Physiol. Mol. Plant Pathol.*, 44, 273, 1994b.

Conti, G., Bassi, M., Carminucci, D., Gatti, L., and Bocci, A. M., Preinoculation with tobacco necrosis virus enahnces peroxidase activity and lignification in cucumber as a resistance response to *Sphaerotheca fuliginea, J. Phytopathol.*, 128, 191, 1990.

Cook, R. J. and Baker, K. F., *The Nature and Practice of Biological Control of Plant Pathogens*. APS Press, St. Paul, MN, 1983.

Coquoz, J. L., Buchala, A. J., Meuwly, P., and Métraux, J. P., Arachidonic acid induces local but not systemic synthesis of salicylic acid and confers systemic resistance in potato plants to *Phytophthora infestans* and *Alternaria solani, Phytopathology*, 85, 1219, 1995.

Croft, K. P. C., Juttner, F., and Slusarenko, A. J., Volatile products of the lipoxygenase pathway evolved from *Phaseolus vulgaris* L. leaves inoculated with *Pseudomonas syringae* pv. *phaseolicola, Plant Physiol.*, 101, 13, 1993.

Cruickshank, I. A. M. and Mandryk, M., The effect of stem infection of tobacco with *Peronospora tabacina* on foliage reaction to blue mold, *J. Aust. Inst. Agric. Res.*, 26, 369, 1960.

Dann, E. K. and Deverall, B. J., 2,6-Dichloroisonicotinic acid (INA) induces resistance in green beans to the rust pathogen (*Uromyces appendiculatus*) under field conditions, *Aust. J. Plant Pathol.*, 25, 199, 1996.

Dann, E. K. and Deveral, B. J., Effectiveness of systemic resistance in bean against foliar and soil-borne pathogens as induced by biological and chemical means, *Plant Pathol.*, 44, 458, 1995.

Dann, E. K., Meuwly, P., Métraux, J. P., and Deverall, B. J., Effect of pathogen inoculation or chemical treatment on activities of chitinase and β-1,3-glucanase and accumulation of salicylic acid in leaves of green bean, *Phaseolus vulgaris* L., *Physiol. Mol. Plant Pathol.*, 49, 307, 1996.

Dean, R.A. and Kuc, J., Induced systemic protection in cucumber: the source of the "signal", *Physiol. Mol. Plant Pathol.*, 28, 227, 1986a.

Dean, R. A. and Kuc, J., Induced systemic protection in cucumber: time of production and movement of the signal, *Phytopathology*, 76, 966, 1986b.

Dehne, H. W. and Schonbeck, F., The influence of endotropic mycorrhiza on plant diseases. II. Phenolmetabolism and lignification, *Phytopathol. Z.*, 95, 210, 1979.

Delaney, T. P., Uknes, S., Vernooij, B., Friedrich, L., Weymann, K., Negrotto, D., Gaffney, T., Gut-Rella, M., Kessmann, H., Ward, E., and Ryals, J., A central role of salicylic acid in plant disease resistance, *Science*, 266, 1247, 1994.

Deverall, B. J., Plant protection using natural defence systems of plants, *Adv. Plant Pathol.*, 11, 211, 1995.

Deverall, B. J. and Dann, E. K., Induced resistance in legumes, in *Induced Resistance to Disease in Plants*, Hammerschmidt, R. and Kuc, J., Eds., Kluwer Academic, Amsterdam, 1995, chap. 1.

Dixon, R. A., Harrison, M. J., and Lamb, C. J., Early events in the activation of plant defense responses, *Annu. Rev. Phytopathol.*, 32, 479, 1994.

Doke, N., Ramirez, A. V., and Tomiyama, K., Systemic induction of resistance in potato plants against *Phytophthora infestans* by local treatment with hyphal wall components of the fungus, *J. Phytopathol.*, 119, 232, 1987.

Doubrava, N. S., Dean, R. A., and Kuc, J., Induction of systemic resistance to anthracnose caused by *Colletotrichum lagenarium* in cucumber by oxalate and extracts from spinach and rhubarb leaves, *Physiol. Mol. Plant Pathol.*, 33, 69, 1988.

Elliston, J., Kuc, J., Williams, E. B., and Rahe, J. E., Relationship of phytoalexin accumulation to local and systemic protection of bean against anthracnose, *Phytopathol. Z.*, 88, 114, 1977.

Enkerli, J., Gisi, U., and Mosinger, E., Systemic acquired resistance to *Phytophthora infestans* in tomato and the role of pathogenesis-related proteins, *Physiol. Mol. Plant Pathol.*, 43, 161, 1993.

Farmer, E. E. and Ryan, C. A., Octadecanoid precursors of jasmonic acid activate the synthesis of wound-inducible proteinase inhibitors, *Plant Cell*, 4, 129, 1992.

Farmer, E. E., Fatty acid signalling in plants and their associated microorganisms, *Plant Mol. Biol.*, 26, 1423, 1994.

Friedrich, L., Vernooij, B., Gaffney, T., Morse, A., and Ryals, J., Characterization of tobacco plants expressing a bacterial salicylate hydroxylase gene, *Plant Mol. Biol.*, 29, 959, 1995.

Gaffney, T., Friedrich, L., Vernooij, B., Negrotto, D., Nye, G., Uknes, S., Ward., E., Kessmann, H., and Ryals, J., Requirement of salicylic acid for the induction of systemic acquired resistance, *Science*, 261, 754, 1993.

Grinstein, A., Lisker, N., Katan, J., and Eshel, Y., Herbicide-induced resistance to plant wilt diseases, *Physiol. Plant Pathol.*, 24, 347, 1984.

Hammerschmidt, R. and Kuc, J., Lignification as a mechanism for induced systemic resistance in cucumber, *Physiol. Plant Pathol.*, 20, 61, 1982.

Hammerschmidt, R., Nuckles, E. M., and Kuc, J., Association of enhanced peroxidase activity with induced systemic resistance of cucumber to *Colletotrichum lagenarium*, *Physiol. Plant Pathol.*, 20, 73, 1982.

Hammerschmidt, R., The nature and generation of systemic signals induced by pathogens, arthropod herbivores and wounds, *Adv. Plant Pathol.*, 10, 307, 1993.

Hammerschmidt, R., Lamport, D. T. A., and Muldoon, E. P., Cell wall hydroxyproline enhancement and lignin deposition as an early event in the resistance of cucumber to *Cladosporium cucumerinum*, *Physiol. Plant Pathol.*, 24, 43, 1984.

Hammerschmidt, R., Bonnen, A. M., Bergstrom, G. C., and Baker, K. K., Association of epidermal lignification with nonhost resistance of cucurbits to fungi, *Can. J. Bot.*, 63, 2393, 1985.

Hammerschmidt, R. and Kuc, J., *Induced Resistance to Disease in Plants*, Kluwer Academic, Amsterdam, 1995.

Hammerschmidt, R. and Yang-Cashman, P., Induced resistance in cucurbits, in *Induced Resistance to Disease in Plants*, Hammerschmidt, R. and Kuc, J., Eds., Kluwer Academic, Amsterdam, 1995, chap. 3.

Hammerschmidt, R. and Smith Becker, J., Acquired resistance to disease in plants, *Hortic. Rev.*, 10, 18, 247, 1997.

Heath, M. C., Thoughts on the role and evolution of induced resistance in natural ecosystems, and its relationship to other types of plant defenses against disease, in *Induced Resistance to Disease in Plants*, Hammerschmidt, R. and Kuc, J., Eds., Kluwer Academic, Amsterdam, 1995, chap. 6.

Heller, W. I. and Gessler, C., Induced systemic resistance in tomato plants against *Phytophthora infestans*, *Phytopathol. Z.*, 116, 323, 1986.

Hijwegen, T. and Verhaar, M. A., Induced resistance to *Peronospora parasitica* in red cabbage, *Neth. J. Plant Pathol.*, 99 (Suppl. 3), 103, 1993.

Hoffland, E., Pieterse, C. M. J., Bik, L., and van Pelt, J. A., Induced systemic resistance in radish is not associated with accumulation of pathogenesis-related proteins, *Physiol. Mol. Plant Pathol.*, 46, 309, 1995.

Jenns, A. E. and Kuc, J., Graft transmission of systemic resistance of cucumber to anthracnose induced by *Colletotrichum lagenarium* and tobacco necrosis virus, *Phytopathology*, 69, 753, 1979.

Jenns, A. E. and Kuc, J., Characteristics of anthracnose resistance induced by localized infection of cucumber with tobacco necrosis virus, *Physiol. Plant Pathol.*, 17, 81, 1980.

Jetiyanon, K., Backman, P. A., and Tuzun S., Immunization of cabbage for long-term resistance to black rot, APS Southern Div. Annu. Meet., Auburn, Alabama, March 20-23, 1994 (Abstr.).

Kessmann, H., Staub, T., Hofmann, C., Maetzke, T., Herzog, J., Ward, E., Uknes, S., and Ryals, J., Induction of systemic acquired disease resistance in plants by chemicals, *Annu. Rev. Phytopathol.*, 32, 439, 1994.

Kessmann, H., Ryals, J., Staub, T., Oostendorp, M., Ahl Goy, P., Hofmann, C., Friedrich, L., Delaney, T., Lawton, K., Weymann, K., Ligon, H., Vernooij, B., and Uknes, S., CGA 245704: mode of action of a new plant activator, Int. Plant Prot. Congr., The Hague, The Netherlands, 2-7 July, 1995.

Kogel, K.-H., Beckhove, U., Dreschers, J., Münch, S., and Rommé, Y., Acquired resistance in barley. The resistance mechanism induced by 2,6-dichloroisonicotinic acid is a phenocopy of a genetically based mechanism governing race-specific powdery mildew resistance, *Plant Physiol.*, 106, 1269, 1994.

Kogel, K.-H., Ortel, B., Jarosch, B., Atzorn, R., Schiffer, R., and Wasternack, C., Resistance in barley against the powdery mildew fungus *(Erysiphe graminis* f. sp. *hordei)* is not associated with enhanced levels of endogenous jasmonates, *Eur. J. Plant Pathol.*, 101, 319, 1995.

Kovats, K., Binder, A., and Hohl, H. R., Cytology of induced systemic resistance of cucumber to *Colletotrichum lagenarium*, *Planta*, 183, 484, 1991a.

Kovats, K., Binder, A., and Hohl, H. R., Cytology of induced systemic resistance of tomato to *Phytophthora infestans*, *Planta*, 183, 491, 1991b.

Kuc, J., Phytoalexins from the solanaceae, in *Phytoalexins*, Bailey, J. A. and Mansfield, J. W., Eds., Blackie, Glasgow, 1982, 81.

Kuc, J., Shockley, G., and Kearney, K., Protection of cucumber against *Colletotrichum lagenarium* by *Colletotrichum lagenarium, Physiol. Plant Pathol.*, 7, 195, 1975.

Kuc, J. and Richmond, S., Aspects of the protection of cucumber against *Colletotrichum lagenarium* by *Colletotrichum lagenarium, Phytopathology*, 67, 533, 1977.

Kumar, V. U., Meera, M. S., Hindumathy, C. K., and Shetty, H. S., Induced systemic resistance protects pearl millet against downy mildew disease due to *Sclerospora graminicola, Crop Prot.*, 12, 458, 1993.

Kutzner, B., Hellwald, K.-H., and Buchenauer, H., Systemic induction of resistance in *Phaseolus vulgaris* L. to tobacco necrosis virus by *Uromyces appendiculatis, J. Phytopathol.*, 138, 9, 1993.

Leon, J., Yalpani, N., Raskin, I., and Lawton, M. A., Induction of benzoic acid 2-hydroxylase in virus-inoculated tobacco, *Plant Physiol.*, 103, 323, 1993.

Linthorst, H. J. M., Pathogenesis-related proteins in plants, *Crit. Rev. Plant Sci.*, 10, 123, 1991.

Liu, D., Raghothama, K. G., Hasegawa, P. M., and Bressan, R. A., Osmotin overexpression in potato delays development of disease symptoms, *Proc. Natl. Acad. Sci. U.S.A.*, 91, 1888, 1994.

Malamy, J., Carr, J. P., Klessig, D. F., and Raskin, I., Salicylic acid — a likely endogenous signal in the resistance response of tobacco to tobacco mosaic virus, *Science*, 250, 1002, 1990.

Mandeel, Q. and Baker, R., Mechanisms involved in biological control of Fusarium wilt of cucumber with strains of nonpathogenic *Fusarium oxysporum, Phytopathology*, 81, 462, 1991.

Mauch-Mani, B. and Slusarenko, A., Systemic acquired resistance in *Arabidopsis thaliana* induced by a predisposing infection with a pathogenic isolate of *Fusarium oxysporum, Mol. Plant-Microbe Interact.*, 7, 378, 1994.

Maurhofer, M., Hase, C., Meuwly, P., Métraux, J. P., and Défago, G., Induction of systemic resistance of tobacco to tobacco necrosis virus by root-colonizing *Pseudomonas fluorescens* strain CHA0: influence of the *gacA* gene and of pyoverdine production, *Phytopathology*, 84, 139, 1994.

Meera, M. S., Shivanna, M. B., Kageyama, K., and Hyakumachi, M., Plant growth promoting fungi from zoysiagrass rhizosphere as potential inducers of systemic resistance in cucumbers, *Phytopathology*, 84, 1399, 1994.

Meera, M. S., Shivanna, M. B., Kageyama, K., and Hyakumachi, M., Persistence of induced systemic resistance in cucumber in relation to root colonization by plant-growth promoting fungal isolates, *Crop Prot.*, 14, 123, 1995a.

Meera, M. S., Shivanna, M. B., Kageyama, K., and Hyakumachi, M., Responses of cucumber cultivars to induction of systemic resistance against anthracnose by plant-growth promoting fungi, *Eur. J. Plant Pathol.*, 101, 421, 1995b.

Métraux, J. P., Signer, H., Ryals, J., Ward, E., Wyss-Benz, M., Gaudin, J., Raschdorf, K., Schmid, E., Blum, W., and Inverardi, B., Increase in salicylic acid at the onset of systemic acquired resistance in cucumber, *Science*, 250, 1004, 1990.

Métraux, J. P., Ahl Goy, P., Staub, T., Speich, J., Steinmann, A., Ryals, J., and Ward, E., Induced systemic resistance in cucumber in response to 2,6-dichloro-isonicotinic acid and pathogens, in *Advances in Molecular Genetics of Plant-Microbe Interactions*, Vol. 1, Hennecke, H. and Verma, D. P. S., Eds., Kluwer Academic, Dordrecht, 1991, 432.

Meuwly, P., Molders, W., Buchala, A., and Métraux, J. P., Local and systemic biosynthesis of salicylic acid in infected cucumber plants, *Plant Physiol.*, 109, 1107, 1995.

Moerschbacher, B. M., Noll, U., Gorrichon, L., and Reisener, H. J., Specific inhibition of lignification breaks hypersensitive resistance of wheat to stem rust, *Plant Physiol.*, 93, 465, 1990.

Müller, K. O. and Borger, H., Experimentelle Untersuchungen uber die Phytophtora-Resistanz der Kartoffel, *Arb. Biol. Anst. Reichsanst.*, 23, 189, 1940.

Neuenschwander, U., Vernooij, B., Friedrich, L., Uknes, S., Kessmann, H., and Ryals, J., Is hydrogen peroxide a second messenger of salicylic acid in systemic acquired resistance?, *Plant J.*, 8, 227, 1995.

Nicholson, R. L. and Hammerschmidt, R., Phenolic compounds and their role disease resistance, *Annu. Rev. Phytopathol.*, 30, 369, 1992.

Nielsen, K. K., Bijsen, K., Collinge, D. B., and Mikkelsen, J. D., Induced resistance in sugar beet against *Cercospora beticola*: induction by dichloroisonicotinic acid is independent of chitinase and β-1,3-glucanase transcript accumulation, *Physiol. Mol. Plant Pathol.*, 45, 89, 1994.

Niemira, B. A., Hammerschmidt, R., and Safir, G. R., Postharvest suppression of potato dry rot (*fusarium sambucinum*) in prenuclear minitubers by arbuscular mycorrhizal fungal inoculum, *Am. Potato J.*, 73, 509, 1996.

O'Neill, N. R., Elgin, J. H., Jr., and Baker, C. J., Characterization of induced resistance to anthracnose in alfalfa by races, isolates, and species of *Colletotrichum*, *Phytopathology*, 79, 750, 1989.

Owen, K., Investigations of Chemical and Biological Agents as Inducers of Systemic Resistance to Leaf Rust Infections in Wheat Seedlings, MScAgr Thesis, University of Sydney, 1995.

Ozeretskovskaya, O., Induced resistance in the solanaceae, in *Induced Resistance to Disease in Plants*, Hammerschmidt, R. and Kuc, J., Eds., Kluwer Academic, Amsterdam, 1995, chap. 2.

Paxton, J. D., Phytoalexins — a working redefinition, *J. Phytopathol.*, 101, 106, 1981.

Paxton, J. D. and Chamberlain, D. W., Acquired local resistance of soybean plants to *Phytophthora* spp., *Phytopathology*, 57, 351, 1967.

Pierpont, W. S., Salicylic acid and its derivatives in plants. Medicines, metabolism and messenger molecules, *Adv. Bot. Res.*, 20, 164, 1994.

Pospieszny, H., Chirkov, S., and Atabekov, J., Induction of antiviral resistance in plants by chitosan, *Plant Sci.*, 79, 63, 1991.

Rahe, J. E., Kuc, J., Chuang, Chien-mei, and Williams, E. B., Induced resistance in *Phaseolus vulgaris* to bean anthracnose, *Phytopathology*, 59, 1641, 1969.

Raskin, I., Role of salicylic acid in plants, *Annu. Rev. Plant Physiol. Plant Mol. Biol.*, 43,439, 1992.

Rasmussen, J. B., Hammerschmidt, R., and Zook, M. N., Systemic induction of salicylic acid accumulation in cucumber after inoculation with *Pseudomonas syringae* pv. *syringae*, *Plant Physiology*, 97, 1342, 1991.

Rasmussen, J. B., Smith, J. A., Williams, S., Burkhardt, W., Ward, E., Somerville, S., Ryals, J., and Hammerschmidt, R., cDNA cloning and systemic expression of acidic peroxidases associated with systemic acquired resistance to disease in cucumber, *Physiol. Mol. Plant Pathol.*, 46, 373, 1995.

Reuveni, R., Reuveni, M., and Agapov, V., Induction of growth increase and systemic resistance to *Exserohilum turcicum* in maize by foliar spray of phosphates, *J. Phytopathol.*, 141, 337, 1994.

Richmond, S., Elliston, J. E., and Kuc, J., Penetration of cucumber leaves by *Colletotrichum lagenarium* is reduced in plants systemically protected by previous infection with the pathogen, *Physiol. Plant Pathol.*, 14, 329, 1979.

Ride, J. P., The role of cell wall alterations in resistance to fungi, *Ann. Appl. Biol.*, 89, 302, 1978.

Ross, A. F., Localized acquired resistance to plant virus infection in hypersensitive hosts, *Virology*, 14, 329, 1961a.

Ross, A. F., Systemic acquired resistance induced by localized virus infections in plants, *Virology*, 14, 340, 1961b.

Ruess, W., Kunz, W., Staub, T., Müller, K., Poppinger, N., Speich, J., and Ahl Goy, P., Plant activator CGA 245704, a new technology for disease management, Int. Plant Prot. Congr., The Hague, The Netherlands, 2-7 July, 1995.

Ruffer, M., Steipe, B., and Zenk, M. H., Evidence against specific binding of salicylic acid to plant catalase, *FEBS Lett.*, 377, 175, 1995.

Ryals, J., Uknes, S., and Ward, E., Systemic acquired resistance, *Plant Physiol.*, 104, 1109, 1994.

Siegrist, J., Jeblick, W., and Kauss, H., Defense responses in infected and elicited cucumber (*Cucumis sativus* L.) hypocotyl segments exhibiting acquired resistance, *Plant Physiol.*, 105, 1365, 1994.

Silverman, P., Seskar, M., Kanter, D., Schweizer, P., Métraux, J. P., and Raskin, I., Salicylic acid in rice: Biosynthesis, conjugation, and possible role, *Plant Physiol.*, 108, 633, 1995.

Simms, E. L. and Vision, T. J., Pathogen-induced systemic resistance in *Ipomea purpurea*, *Oecologia*, 102, 494, 1995.

Skipp, R. A. and Deverall, B. J., Studies on cross-protection in the anthracnose disease of bean, *Physiol. Plant Pathol.*, 3, 299, 1973.

Smith, J. A., Fulbright, D. W., and Hammerschmidt, R., Rapid induction of systemic resistance in cucumber by *Pseudomonas syringae* pv. *syringae*, *Physiol. Mol. Plant Pathol.*, 38, 223, 1991.

Stein, B. D., Klomparens, K., and Hammerschmidt, R., Histochemistry and ultrastructure of the induced resistance response of cucumber plants to *Colletotrichum lagenarium*, *J. Phytopathol.*, 137, 177, 1993.

Steiner, U. and Schonbeck, F., Induced resistance to disease in monocots, in *Induced Resistance to Disease in Plants*, Hammerschmidt, R. and Kuc, J., Eds., Kluwer Academic, Amsterdam, 1995, chap. 4.

Stermer, B. A., Molecular regulation of induced systemic resistance, in *Induced Resistance to Disease in Plants*, Hammerschmidt, R. and Kuc, J., Eds., Kluwer Academic, Amsterdam, 1995, chap. 5.

Stolle, K., Zook, M. N., Shain, L., Heberd, F., and Kuc, J., Restricted colonization by *Peronospora tabacina* and phytoalexin accumulation in immunized tobacco leaves, *Phytopathology*, 78, 1193, 1988.

Stromberg, A. and Brishammar, S., Induction of systemic resistance in potato (*Solanum tuberosum* L.) to late blight caused by local treatment with *Phytophtora infestans* (Mont.) de Bary, *Phytophtora cryptogea* Pethyb and Laff., or potassium diphosphate, *Potato Res.*, 34, 219, 1991.

Stromberg, A. and Brishammar, S., A histological evalution of induced resistance to *Phytophtora infestans* (Mont.) de Bary in potato leaves, *J. Phytopathol.*, 137, 15, 1993.

Stumm, D. and Gessler, C., Role of papillae in the induced systemic resistance of cucumber against *Colletotrichum lagenarium*, *Physiol. Mol. Plant Pathol.*, 29, 405, 1986.

Summermatter, K., Sticher, L., and Métraux, J. P., Systemic responses in Arabidopsis thaliana infected and challenged with *Pseudomonas syringae* pv. *syringae*, *Plant Physiol.*, 108, 1379, 1995.

Sutton, D. C., Systemic cross protection in bean against *Colletotrichum lindemuthianum*, *Australas. Plant Pathol.*, 8, 4, 1979.

Sutton, D. C., Field protection of bean against *Colletotrichum lindemuthianum* by *Colletotrichum lindemuthianum*, *Australas. Plant Pathol.*, 11, 50, 1982.

Tuzun, S. and Kuc, J., Movement of a factor in tobacco infected with *Peronospora tabacina* which systemically protects against blue mold, *Physiol. Mol. Plant Pathol.*, 26, 321, 1985.

Tuzun, S. and Kuc, J., Induced systemic resistance to blue mold of tobacco, in *Blue Mold of Tobacco*, McKeen, W. E., Ed., American Phytopathological Society Press, St. Paul, MN, 1989, 177.

Tuzun, S. and Kloepper, J., Induced systemic resistance by plant growth-promoting rhizobacteria, in *Improving Plant Productivity with Rhizosphere Bacteria, Proc. 3rd Int. Workshop on Plant Growth-Promoting Bacteria*, Ryder, M. H., Stephens, P. M., and Bowen, G. D., Eds., CSIRO, Australia, 1994, 104.

Tuzun, S. and Kloepper, J., Practical application and implementation of induced resistance, in *Induced Resistance to Disease in Plants*, Hammerschmidt, R. and Kuc, J., Eds., Kluwer Academic, Amsterdam, 1995, chap. 7.

Uknes, S., Mauch-Mani, B., Moyer, M., Potter, S., Williams, S., Dincher, S., Chandler, D., Slusarenko, A., Wards, E., and Ryals, J., Acquired resistance in *Arabidopsis*, *Plant Cell*, 4, 645, 1992.

Uknes, S., Winter, A. M., Delany, T., Vernooij, B., Morse, A., Friedrich, L., Nye, G., Potter, S., Ward, E., and Ryals, J., Biological induction of systemic acquired resistance in *Arabidopsis*, *Mol. Plant-Microbe Interact.*, 6, 692, 1993.

van Peer, R., Niemann, G. J., and Schippers, B., Induced resistance and phytoalexin accumulation in biological control of Fusarium wilt of carnation by *Pseudomonas* sp. strain WCS417r, *Phytopathology*, 81, 728, 1991.

Vernooij, B., Friedrich, L., Morse, A., Reist, R., Kolditz-Jawhar, R., Ward, E., Uknes, S., Kessmann, H., and Ryals, J., Salicylic acid is not the translocated signal responsible for inducing systemic acquired resistance but is required in signal transduction, *Plant Cell*, 6, 959, 1994.

Vernooij, B., Friedrich, L., Ahl Goy, P., Staub, T., Kessmann, H., and Ryals, J., 2,6-Dichloroisonicotinic acid-induced resistance to pathogens without the accumulation of salicylic acid, *Mol. Plant-Microbe Interact.*, 8, 228, 1995.

Walters, D. R., Mitchell, A. F., Hampson, J., and McPherson, A., The induction of systemic resistance in barley to powdery mildew infection using salicylates and various phenolic acids, *Ann. Appl. Biol.*, 122, 451, 1993.

Ward, E., Uknes, S. J., Williams, S. C., Dincher, S. S., Wiederhold, D. L., Alexander, D. C., Ahl-Goy, P., Métraux, J. P., and Ryals, J. A., Coordinate gene activity in response to agents that induce systemic acquired resistance, *Plant Cell*, 3, 1085, 1991.

Wasternack, C., Atzorn, R., Jarosch, B., and Kogel, K. H., Induction of a thionin, the jasmonate-induced 6 kDa protein of barley by 2,6-dichloroisonicotinic acid, *J. Phytopathol.*, 140, 280, 1994.

Wick, R. L. and Moore, L. D., Histology of mycorrhizal and nonmycorrhizal *Ilex crenata* 'Helleri' challenged by *Thielaviopsis basicola*, *Can. J. Plant Pathol.*, 6, 146, 1984.

Wei, G., Kloepper, J. W., and Tuzun, S., Induction of systemic resistance of cucumber to *Colletotrichum obiculare* by select strains of plant growth-promoting rhizobacteria, *Phytopathology*, 81, 1508, 1991.

Xuei, X. L., Jarlfors, U., and Kuc, J., Ultrastructural changes associated with induced systemic resistance of cucumber to disease: host response and development of *Colletotrichum lagenarium* in systemically protected tissue, *Can. J. Bot.*, 66, 1028, 1988.

Ye, X. S., Jarlfors, J., Tuzun, S., Pan, S. Q., and Kuc, J., Biochemical changes in cell walls and cellular responses of tobacco leaves related to systemic resistance to blue mold *(Peronospora tabacina)* induced by tobacco mosaic virus, *Can. J. Bot.*, 70, 49, 1992.

Zangerl, A. R., Furanocoumarin induction in wild parsnip: evidence for an induced defense against herbivores, *Ecology*, 71, 1926, 1990.

CHAPTER 9

Plant Breeding

Martin L. Carson

I. INTRODUCTION

A. Considerations in the use of Genetic Resistance

Genetically resistant cultivars have been the mainstay for the economic control of plant diseases of many food, fiber, and forage crops around the world. When available, resistant cultivars provide an attractive alternative to the use of pesticides, less economically feasible crop rotations, or other management practices that are not easily adopted. Disease-resistant cultivars generally are not any more expensive than susceptible cultivars, and require no special management practices to use. In some instances, resistant cultivars may lack the yield potential or market quality of susceptible cultivars, so farmers must decide if the trade-off between reduced yields or quality and reduced losses to disease and reduced pesticide costs is economic. For many horticultural crops of high cash value in the developed world, quality considerations are often so great and pesticide costs are so low that the use of disease-resistant cultivars is not considered practical at this time. With increased concern about pesticides in the environment and residues in foodstuffs, and with increasing costs and restrictions on pesticide use, the development of disease-resistant cultivars in these high cash value commodities in the future is a distinct possibility (Dayton et al., 1984). It should also be noted that although high levels of resistance to specific diseases of many crops is available, few, if any, cultivars are resistant to all diseases that may afflict them. However, it is often possible to find cultivars resistant to the most destructive and common pathogens in a given agroecosystem.

B. Brief History

Probably the first recorded observation that varieties of a crop plant differed in their disease reaction was that of Theophrastus (371–286 B.C.) who noted not only differences in susceptibility to rust among crop species, but also noted that a kind of barley called "Achillean" was most susceptible of all barleys (Whetzel, 1918). How much emphasis was placed on these observed differences in disease susceptibility during the selection process by early plant breeders is not known, but it is likely that highly susceptible types were routinely discarded or lost favor over the years. It was not until the early 19th century, when there was widespread interest in plant breeding, that observations of differences in the disease reactions of crop cultivars became common. Differences in the reaction of wheat cultivars to rust, turnip varieties to clubroot, onion varieties to smudge, potato to late blight, and grape varieties to downy mildew, among others, were noted. Despite these observations, disease resistance was a not a major criteria or objective in most plant-breeding efforts at the time, although Cobb, Farrar, and others began to compile information on rust resistance of wheat collections used in breeding programs. Until the rediscovery of Mendel's laws of inheritance at the turn of the century, plant breeding remained more an art than science. Biffen (1905) was the first to report that disease resistance (yellow rust of wheat) was inherited as a simple Mendelian trait. These results were quickly greeted with skepticism, as it was pointed out that varieties resistant in one locale were often quite susceptible elsewhere. It was not until sometime later that the concept of physiological specialization or races was introduced, and the basis for these conflicting results was understood. At about the same time, Orton (1907, 1909) initiated studies on Fusarium wilts of cotton, watermelon, and cowpea, and was able to increase the level of resistance by selection for surviving plants grown in naturally infested soils. Similar success was achieved in breeding for resistance to Fusarium wilts of flax, tomato, and cabbage soon after Orton's initial success (Bolley et al., 1901; Essary, 1912; Whitaker, 1979). Furthermore, resistance to the Fusarium wilts appeared to be quite stable, and in many cases saved the crop. With these early successes, disease resistance increasingly became an objective of many programs of plant improvement. The problems posed by variation in pathogen populations for virulence on resistant varieties were also now recognized, and strategies to identify and use new sources of disease resistance in crops became standard practice. The influence of temperature and other environmental factors on expression of some disease resistance was also elucidated.

Today, disease resistance is the primary means of control for the major diseases of almost all field crops as well as many fruits and vegetables. For many of our major food crops, losses due to major epidemics are now an exceptional event, not the norm, due primarily to the successful development and use of resistant varieties. For example, the periodic, widespread, and destructive epidemics of wheat stem rust in North America are now considered largely history. Despite the overwhelming successes in controlling diseases via resistance, the need for vigilance remains. The southern corn leaf blight epidemic of 1970 served as a reminder of the potential of pathogens to continually adapt to the plants we breed, and the problems of genetic uniformity in our crops (National Academy of Sciences, 1972).

II. GENETICS AND EXPRESSION OF DISEASE RESISTANCE

A. Oligogenic Resistance

At the risk of oversimplification, it is often easiest to classify types of disease resistance by their mode of inheritance. Resistance can be classified as either oligogenic in inheritance, or as polygenically inherited. Oligogenic inheritance simply means that segregating plants can be readily classified into discrete phenotypes and the resulting ratios fit to Mendelian ratios. By far, most reports on the inheritance of disease resistance are reports of oligogenic types. Biffen first reported that resistance to yellow rust of wheat was inherited as a single recessive gene; that is F_2 plants from a cross of a resistant by a susceptible cultivar segregated three susceptible: one resistant. The most common type of oligogenic resistance are single, dominant genes, but recessive disease resistance genes are not uncommon. Frequently, resistant varieties may have several independent dominant genes, each of which may confer resistance by itself. This results in ratios of 15:1, 63:1, and 255:1 (resistant:susceptible) in F_2s of susceptible varieties crossed with resistant varieties having two, three, and four dominant genes, respectively. Other types of di- and trigenic epistatic F_2 ratios are possible as well. In several species, it is not uncommon for resistance genes to be tightly linked in gene clusters or act as multiple alleles at single locus (Hooker, 1967; Hooker and Saxena, 1971). The *Rp1* locus for resistance to common rust, *Puccinia sorghi,* located on the long arm of chromosome 10 in maize, is an example of such a locus (Saxena and Hooker, 1968).

Oligogenic types of resistance are usually considered qualitative in their expression, as the resistant phenotype is quite distinct from the susceptible. With foliar pathogens, it may consist of a chlorotic fleck type of reaction, a reduced, nonsporulating lesion, or a pustule with greatly reduced sporulation. With resistance to vascular wilts, oligogenic resistance may be expressed as simply the lack of any wilting with little or no vascular discoloration. Resistance to root knot nematodes may be expressed as the absence of root knots and the failure of the nematode to successfully reproduce on the resistant plant. There are, however, many examples of monogenically inherited resistance that are expressed in a quantitative manner. In maize, the *HtN* gene causes an increase in the incubation period (the period of time from infection to symptom development) of the foliar pathogen *Exserohilum turcicum* such that no lesions are apparent until after anthesis (Gevers, 1975; Raymundo et al., 1981). Other forms of monogenic resistance are only expressed at later growth stages on adult plants, thus the name adult plant resistance. Still other forms of oligogenic resistance are only expressed under certain temperatures.

Oligogenic forms of disease resistance have been found and exploited for a wide array of important diseases of cultivated plants. However, in many instances this resistance has been short-lived, as new races or strains of pathogens evolve that are capable of rendering them ineffective. Thus, plant breeders are often faced with the prospect of continually searching for new resistance genes in an attempt to stay at least one step ahead of the pathogen. There are, however, many examples of monogenic forms of resistance being widely used for many years and still remaining effective. Resistance to the soil-borne Fusarium wilt pathogens, for example, has remained effective in many cases in excess of 50 years.

B. Polygenic Resistance

Polygenically inherited resistance to plant diseases is also common in most economically important crops, although it has perhaps not been as well studied as oligogenic forms. Because polygenic resistance is commonly expressed in a quantitative manner as reduction in disease development, it is sometimes referred to as partial resistance or incomplete resistance. Also, because it is often considered to be effective against all strains or races of the pathogen, it is sometimes called general resistance or race nonspecific resistance. While it is often convenient to use all these terms synonymously, it should be kept in mind that exceptions to these generalizations do exist. There are forms of polygenically inherited resistance that produce a qualitatively different phenotype than a susceptible genotype. There are varieties of maize with such high levels of polygenic resistance to certain foliar diseases that they appear practically immune, so their resistance could hardly be called partial. Similarly, there are examples of polygenically inherited partial resistance to disease that is not effective against all races of the pathogen. The "slow-rusting" resistance of Suwan 85 wheat to leaf rust was not found to be effective against all races of *Puccinia recondita* (Lee and Shaner, 1984).

Disease resistance is usually considered "polygenic" if there is continuous variation for disease resistance in genetically segregating populations (typically the F_2 between a resistant and a susceptible variety), with no apparent means to classify individuals into discrete categories (Hooker and Saxena, 1971; Geiger and Heun, 1989). This continuous variation is assumed to be the result of the action of many genes acting together, with the effect of any one gene too small to be readily detected. While this type of resistance is amenable to classical quantitative genetic analysis, in almost all cases so-called polygenic disease resistance appears to be controlled by a relatively small number of loci acting in a mostly additive manner. Estimates of the numbers of loci controlling polygenic resistance are commonly less than four or five. This is in contrast with traits traditionally thought of as being polygenic, such as grain yield, seed weight, grain quality, etc., where the number of loci is thought to be much greater.

As stated earlier, polygenic resistance is often expressed as a quantitative reduction in disease development and as such pathologists and breeders often look at its expression in plant epidemiological terms. Polygenic resistance to polycyclic foliar diseases may be expressed and measured in a variety of ways. If disease development over a single growing season is compared, resistant and susceptible varieties will have dramatically different disease progress curves (Figure 1). These differences in disease progress curves may be expressed as differences in apparent infection rates or areas under the disease progress curve. While the effect of polygenic resistance is most dramatically seen as differences in disease development over an entire growing season, breeders are often interested in evaluating for resistance on young seedlings before anthesis, or on seedlings in the greenhouse in the off-season, therefore they are interested in specific epidemiological components of polygenic resistance. Oftentimes, polygenic resistance may be measured as an increase in the incubation or latent periods, a reduction in sporulation intensity in lesions or pustules, reduction in infection efficiency (number of lesions per pathogen propagule applied), or reduced lesion or pustule size (Parlevliet, 1979; Carson, 1995). If selection for a specific component of polygenic resistance is to be effective, it must be highly correlated with resistance observed in the field under epidemic conditions, it must be of a highly heritable character, and it should be readily measurable so that large numbers of individuals may be screened.

III. GENETICS AND EXPRESSION OF PATHOGENICITY, VIRULENCE, AND AGGRESSIVENESS IN PLANT PATHOGENS

A. Pathogenicity

Breeding for disease resistance differs from breeding for other traits in that expression of resistance is dependent not only on variability in the host plant but also in the pathogen. Within

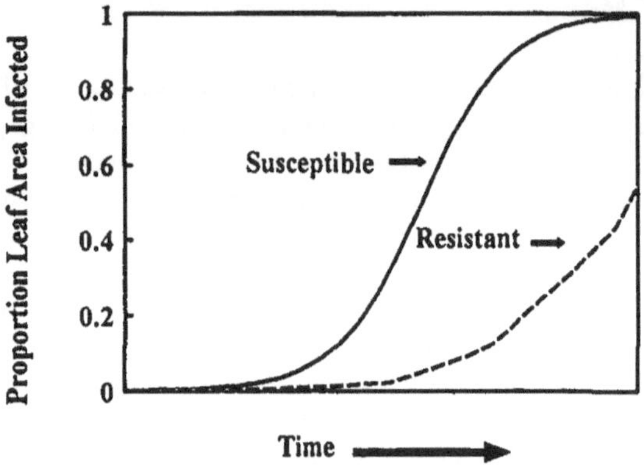

Figure 1 Progress curves of a typical polycyclic foliar disease on a susceptible and a polygenic resistant variety. The infection rate of the disease on the polygenic resistant variety is one half that of the susceptible variety.

a given species of plant pathogen, variation in pathogenicity, virulence, and aggressiveness are common. Variation in virulence (i.e., the presence of physiological races) is particularly troublesome to the plant breeder, as it means that pathogens may have the ability to cause disease on resistant plants.

Pathogenicity, as defined here, is the ability of an individual to cause disease on a given host species (Shaner et al., 1990). In other words, pathogenicity has to do with the host range of the pathogen. Some plant pathogens, such as *Sclerotinia sclerotiorum*, have very wide host ranges covering many plant genera and even families, without any clear evidence of host specificity of individual strains or isolates of the fungus. Still other species of plant pathogens are composed of many "form species" each of which may attack only a limited array or only a single host plant species. Examples of these types include the stem rust fungus, *Puccinia graminis,* which is composed of a number of form species such as *P. graminis* f. sp. *tritici,* pathogenic on wheats, and *P. graminis* f. sp. *avenea,* pathogenic on *Avena* spp., etc. In the few instances when two strains or form species with different host specificities have been successfully crossed, progenies are found that are either nonpathogenic to both original hosts, have the host range of one or the other parent, or are pathogenic on both hosts (Green, 1971). Progenies with the combined pathogenicities of the parent strains are usually less aggressive pathogens than the parents.

B. Virulence

From the practical viewpoint of a plant breeder, probably the most important variation in the pathogen he must deal with is variation in virulence, or physiological races of the pathogen that can overcome the resistance conferred by one or more host resistance genes. Physiological races are identified by means of differential cultivars (Figure 2). The number of races that can possibly be differentiated is determined by the number of genetically different differential cultivars. With n differential cultivars, it is possible to differentiate 2^n races of the pathogen. Thus with 5 differential cultivars, 32 physiological races can be determined.

In modern breeding programs aimed at controlling pathogens that are composed of many physiological races, it is very important to have well-defined systems of race identification and nomenclature to characterize the pathogen population one is seeking to control. Ideally, differential cultivars have known and well-described gene(s) for disease resistance that have been shown to be unique, both in the spectrum of resistance they confer and in terms of their relationship to other known resistance genes (Samborski and Dyck, 1968; Long and Kolmer, 1989). It is often desirable to backcross these genes into a common genetic background to develop a set of near-isogenic differentials. In order to be useful to the breeder, the differential set should be representative of the genes currently being used in cultivars and those used in breeding programs.

Differential Cultivar	Pathogen Isolate 0	1	2	3	4
A	−	+	−	+	+
B	+	−	−	+	−
C	+	+	+	+	+

Figure 2 The use of differential host cultivars to detect physiological races of a plant pathogen; +,− denote compatible and incompatible interactions between host and pathogen, respectively. Note that cultivar C is universally susceptible and that pathogen isolates 1 and 4 are identical physiological races based on this test.

Virulence and avirulence are usually expressed in a clear-cut, qualitative manner, but can vary greatly depending upon the particular pathosystem. For example, with the rust fungi virulence/avirulence may be expressed, respectively, as a full-sized, sporulating pustule vs. a chlorotic fleck, or with many of the necrotrophic leaf-spotting fungi, a necrotic, sporulating lesion vs. a chlorotic, nonsporulating one. In those systems where the inheritance of virulence to a specific resistance gene has been studied, it usually segregates as a single factor. In diploid or dikaryotic fungi, virulence is usually recessive to avirulence. Also, in pathogens with many virulence genes that have been well studied genetically, virulence loci are usually unlinked genetically (Flor, 1946, 1971).

By studying the frequency and distribution of physiologic races of a pathogen, the plant breeder can determine which gene or combination of genes will give him the resistance needed in a given geographic area. It is also often possible to detect new races of a pathogen while its frequency is still relatively low, and thus develop a breeding strategy to respond to the new race before it reaches damaging levels.

While the concept of physiological races has been most useful in understanding the nature of pathogen populations that breeders are hoping to develop resistance to, it should be kept in mind exactly what these races are genetically. Races are an abstraction. They represent a collection or group of genotypes that have in common only their ability to express the same pattern of virulence when inoculated upon a set of differential cultivars. When other characters or genetic markers are considered, or if a new set of differentials is used, it may be found that a "race" is actually a mixture of many unique individuals (Russell, 1978). Further, if the pathogen regularly undergoes sexual reproduction, progenies may differ quite dramatically in virulence from their parent races, and therefore races may not be true breeding.

Finally, it should be pointed out that, in most pathosystems, distinct physiologic races of the pathogen have not evolved. The presence of physiologic races is probably most characteristic of pathosystems where there has been a long history of coevolution of the crop and pathogen.

C. Aggressiveness or Parasitic Fitness

Pathogens also can express variation in their relative ability to cause disease on a compatible host. That is, some strains or individuals may be less damaging (cause less disease) than other individuals, even though both produce a compatible (susceptible) reaction on the host. They are said to differ in their levels of aggressiveness (sometimes called parasitic fitness) (Shaner et al., 1990). In the few instances this trait has been studied genetically, it appears to be a quantitative trait, with progeny exhibiting continuous variation (Caten et al., 1984; Kolmer and Leonard, 1986).

		Pathogen	
		AvrAvr	*avravr*
Host	*RR*	−	+
	rr	+	+

Figure 3 Host and pathogen genotypes in gene-for-gene interaction where the interaction is for resistance (*sensu* Flor); +,− denote compatible and incompatible interaction between host and pathogen, respectively.

One would intuitively think that aggressiveness would be an important fitness trait for a pathogen and would be highly selected against in nature, but it is not uncommon to find a significant amount of variation for aggressiveness among individuals in a pathogen population.

IV. GENETICS OF THE HOST-PATHOGEN INTERACTION

A. Gene-for-Gene Interaction: Oligogenic Resistance

Although the genetics of resistance in the host plant and virulence in the pathogen have been discussed already, it should be clear that the expression of resistance is dependent upon the genetics of not just the host, but also the pathogen. Conversely, the expression of pathogen virulence is also dependent upon the host genotype. Thus when we study disease resistance, we are studying the host-pathogen interaction, and not just a phenomenon of the host plant.

It was the meticulous pioneering work of H. H. Flor, working with flax rust in North Dakota, that led to his formulation of the gene-for-gene hypothesis to explain host/pathogen interactions (Flor, 1956, 1971). By simultaneously studying the inheritance of virulence in the flax rust fungus and the inheritance of resistance in flax, he was able to demonstrate that the only explanation for his data was that for specific genes for resistance in the host there were corresponding genes for virulence in the pathogen. In this system, the interaction between a specific host resistance gene and the corresponding pathogen avirulence gene results in an incompatible (resistant/avirulent) reaction; all other host/pathogen combinations result in compatible (susceptible/virulent) reactions (Figure 3). This type of gene-for-gene interaction is sometimes referred to as the "interaction for resistance" model. All the evidence to date, including the successful cloning of avirulence genes in several pathogens and host resistance genes, supports the concept that avirulence and resistance genes are the functional alleles in the pathogen and host, respectively, that lead to recognition of the pathogen by the host and to the subsequent activation of the host defense response (Newton and Andviron, 1995). When the gene-for-gene model is expanded to cover multiple corresponding loci in the host and pathogen, it only takes the combination of a single host resistance gene and the corresponding pathogen avirulence gene to give an incompatible phenotype, regardless if the host/pathogen genotypes at all other corresponding loci are all compatible. In some pathosystems, over 20 corresponding gene loci have been identified in the host and pathogen (Hooker, 1967; Hooker and Saxena, 1971; Flor, 1971). Flor's gene-for-gene interaction has been found to operate in a wide array of pathosystems, from bacterial diseases, to apple scab, to obligate parasites such as the cereal rusts, powdery mildews, and downy mildews.

In at least a few pathosystems, a different type of gene-for-gene system appears to operate. In several diseases of grasses caused by *Cochliobolus* species, the pathogen consists of strains that either have or lack the ability to produce a host-specific toxin, and the host either has genes for toxin sensitivity or lacks them (Scheffer and Walton, 1995). Thus, the specific interaction is

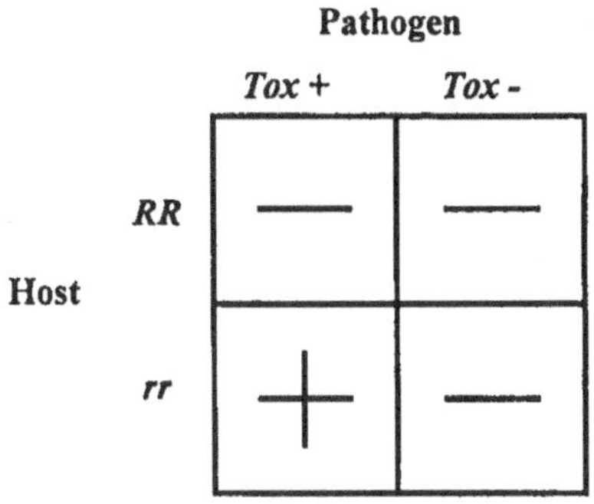

Figure 4 Host and pathogen genotypes in gene-for-gene interaction where the interaction is for susceptibility; +,– denote compatible or incompatible interaction between host and pathogen, respectively.

between the toxin-producing gene in the pathogen and the gene for toxin sensitivity in the host (Figure 4). This type of interaction is sometimes called the "interaction for susceptibility" model. The infamous race T of *Cochliobolus heterostrophus* (cause of the 1970 southern maize leaf blight epidemic in North America) has the ability to produce a polyketide toxin (T toxin) that can bind with the mitochondrial membrane of maize containing the Texas male sterile cytoplasm (cms T), causing mitochondria to lyse and cell death (Levings and Seidow, 1992). The gene responsible for sensitivity to T toxin (T-urf 13) is in the mitochondrial genome of cms-T maize. Other examples of this type of interaction include the *Cochliobolus victoriae*: *Avena sativum* and *Cochliobolus carbonum* race 1:*Zea mays* pathosystems (Scheffer and Walton, 1995; Briggs and Johal, 1994).

B. Genetics of Host-Pathogen Interaction: Polygenic Resistance

While gene-for-gene interaction appears to be widespread and common in pathosystems where monogenic race-specific qualitative resistance predominates, the evidence for gene-for-gene interaction in systems where the host and pathogen exhibit quantitative polygenic variation for resistance and aggressiveness, respectively, is ambiguous at best. One approach that has been used to detect specificity of pathogen isolates toward host cultivars has been the analysis of variance of experiments where a number of host cultivars are inoculated with a set of pathogen isolates. Vanderplank (1984) proposed the idea that "vertical" or race-specific resistance could be defined by the presence of a significant cultivar X isolate interaction in the analyses of such experiments, whereas "horizontal" resistance was simply the lack of such interactions. Such statistical evidence has been used to reach the conclusion that polygenic partial resistance also operates on a gene-for-gene basis and that even these forms of resistance may not be durable in the long run. This analysis of variance approach to detect the putative presence of gene-for-gene interaction in polygenic resistance, and the resulting interpretations, have been widely criticized on a number of points. Although it is common to detect small, albeit significant cultivar X isolate interactions when individual experiments are analyzed, these interactions are usually not repeatable when the experiments are duplicated (Kulkarni and Chopra, 1982). When the results of repeated experiments are analyzed in a combined analysis of variance, the cultivar X isolate X environment effects are significant and cultivar X isolate effects are not. It has also been pointed out that these small but significant cultivar X isolate effects may arise simply due to nonlinear scaling effects of the disease rating scale used in the experiments (Winer, 1984). Finally, several authors have demonstrated how different models of host-pathogen interaction could give rise to cultivar X isolate interactions, even though resistance would be predicted to be quite durable and the host-pathogen interaction may not follow Flor's interaction for a resistance model (Parlevliet and Zadoks, 1977; Fleming and Person, 1982; Carson, 1987). Clearly, the issue of gene-for-gene interaction in pathosystems where both host and pathogen exhibit quantitative variation for resistance and aggressiveness, respectively, is far from resolved. One exciting new approach to this problem may lie in the use of molecular markers which are being used to map quantitative trait loci (QTLs) for a wide array of quantitative traits, including partial resistance to plant diseases. By mapping QTLs for partial resistance to an array of different isolates of a pathogen, it may be possible to demonstrate that at least some of these QTLs are isolate specific in their expression.

V. INOCULATION AND SCREENING METHODS

A. Considerations in Selecting a Screening Method

For a resistance breeding program to be successful, at least two basic conditions must be met (Hooker, 1980). First, there must be a reliable method to consistently rate or score plants for disease severity or reaction; and secondly, there must be heritable variation for disease resistance in the materials being screened. Thus any resistance screening method must be able to successfully generate disease epidemics consistently and as uniformly as possible, and differences among host genotypes observed must reflect real differences in resistance observed during natural epidemics.

B. Natural Infection

In certain instances where natural epidemics occur consistently and uniformly, it may not be necessary to artificially induce epidemics by introducing inoculum into the research plots. The early success of Bolley (1901) in selecting for resistance to Fusarium wilt of flax simply by continual planting in a "flax sick" field and selecting genotypes that survived, is an example of how simply relying upon natural infection may be sufficient to initiate a program of breeding for disease resistance. Indeed, the Fusarium wilt nurseries established by Bolley in North Dakota are still in use today for screening flax cultivars for resistance. Similarly, several winter wheat breeding programs in the U.S. still rely on long-established soil-borne wheat mosaic nursery plots to conduct resistance screening. In order to insure high and uniform amounts of inoculum in these nurseries where screening for resistance to soil-borne pathogens occurs, breeders will often plant susceptible cultivars at regular intervals throughout the nursery. By observing disease development in these susceptible check plots, the breeder can also determine the uniformity of inoculum in the nursery. Although the use of natural infection in continuously cropped disease nurseries is commonly used in screening for resistance to soil-borne pathogens, it may also lend itself to certain foliar pathogens as well, particularly if the pathogen can consistently overwinter in crop residues.

C. Artificial Inoculation

In the vast majority of instances, however, most programs of breeding for disease resistance must rely to some degree on artificial inoculation methods to induce disease development in the nursery. In selecting an artificial inoculation, it is generally best to try to mimic as closely as possible natural inoculation processes and create conditions known to be favorable for disease development. For example, with most foliar diseases caused by fungi, inoculum consists of spores that require free moisture on the leaf surface to successfully infect the host. Therefore, plants in the nursery are typically sprayed with a suspension of pathogen spores, usually before favorable (wet) conditions are predicted or at least late in the evening before dew is anticipated. In some instances, supplemental sprinkler irrigation is applied immediately after inoculation. Still other pathogens require wounds to enter the host, so again the inoculation method must attempt to mimic nature. Once the inoculum is introduced, conditions favorable for continued development of the disease need to be provided as much as is feasible, whether it be through management of irrigation, soil pH and fertility, or maintenance of insect vector populations in the nursery.

Often it is possible and highly desirable to conduct disease screening in the greenhouse or growth chamber. This gives the breeder a chance to have more precise control over environmental conditions and may allow additional generations of selection during the off-season. A possible disadvantage of greenhouse screening is that many crop plants such as maize may not grow normally under the conditions of lower light and potbound root systems found in most greenhouses in the winter. Again, care should be taken to determine that disease reactions observed under greenhouse or growth chamber conditions reflect those seen in the field.

D. Disease Rating and Evaluation

Disease rating scales used to assess resistance should be repeatable, easy to use, compatible with the objectives of the experiment, and reflect the level of damage caused by the disease. In screening germplasm for monogenic, race-specific, qualitative forms of resistance it may suffice to simply classify plants as resistant or susceptible, based on knowledge of how the resistance gene(s) is expressed (chlorotic fleck vs. full-sized lesion, for example). In other types of diseases, such as vascular wilts and certain viral diseases, plants may be classified as simply either infected or symptomatic or asymptomatic. In these cases resistance may be expressed as a reduction in the percentage of plants infected or symptomatic. In many diseases, we see a continuous spectrum of reactions from highly resistant to high susceptibility. In these cases, it is common to develop some sort of descriptive quantitative scale to classify reactions (James, 1974). For many common important diseases of major crops, descriptive scales with visual keys have been published and are available to standardize ratings used by different researchers (James, 1971). For many foliar pathogens, it is common for researchers to estimate the percent of leaf area diseased. With practice, and with the help of disease assessment training software, it is possible for even an untrained worker to become reasonably accurate and consistent in assessing disease severities.

With partial, quantitative forms of resistance to foliar diseases, it is often possible to screen for specific components of resistance on seedling or young plants in the greenhouse that reflect the levels of resistance seen on adult plants in the field. Increased latent and incubation periods, reduced lesion size and numbers, reduced sporulation, etc., individually or collectively, may be major components of partial resistance (Parlevliet, 1979). Besides reflecting the levels of resistance seen in the field, these components must also be relatively easily measured, and of reasonably high heritability to be useful in an applied breeding program.

VI. SOURCES OF RESISTANT GERMPLASM

A. The Problem of Genetic Vulnerability

The problem of the lack of genetic diversity or genetic uniformity in most of our major crop species has been the subject of great concern and discussion among plant breeders, geneticists, and plant pathologists (National Academy of Sciences, 1972; Marshall, 1977). This genetic uniformity makes the entire crop vulnerable to any new diseases or new strain of an old pathogen. This uniformity greatly increases the likelihood that a new pathogen will be highly virulent on the entire crop due to the genetic relatedness of cultivars being grown in a given production area. This problem of genetic vulnerability is further compounded in many instances by increasingly intensive (short or nonexistent crop rotations) and extensive (lack of crop diversity and monoculture) cultivation of the crop in a given area. Thus the stage is often set for epidemics to develop rapidly and spread over a vast geographic region with often devastating consequences. The 1970 Southern maize leaf blight epidemic in North America was a perfect example of the potential problems posed by genetic uniformity (National Academy of Sciences, 1972; Ullstrup, 1972; Frankel, 1977). In this case, the problem was the uniformity of the mitochondrial genome of maize hybrids in North America, which all relied upon Texas male-sterile (cms-T) cytoplasm for pollination control in seed production. A new race, race T, of the "old" southern leaf blight pathogen *Cochliobolus heterostrophus* (anamorph = *Bipolaris maydis*) specifically virulent upon cms-T genotypes was first noticed late in the 1969 growing season. In 1970, when favorable conditions prevailed over much of North America, the pathogen was able to spread across the continent from Florida to Canada in a single growing season, and cause an estimated 15% reduction in the U.S.

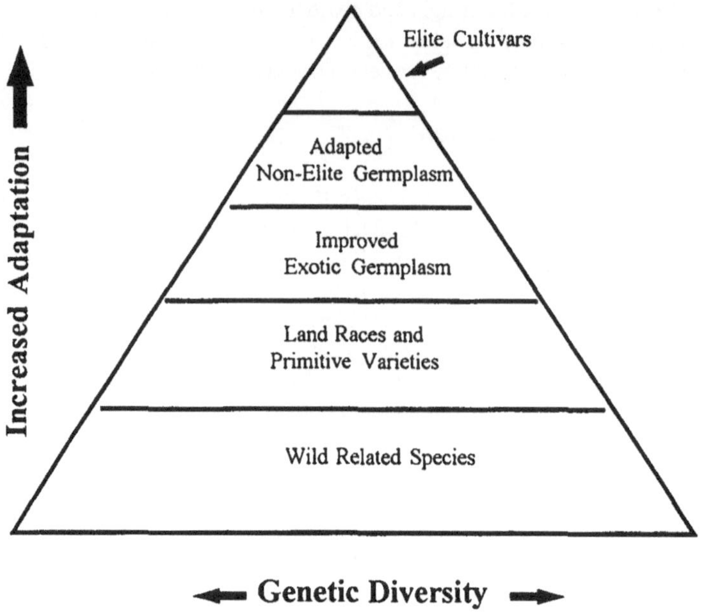

Figure 5 Diagrammatic representation of the distribution of genetic diversity or variation within a crop species.

maize crop, with some fields sustaining 100% yield losses to the disease. With world food reserves at near all time lows, it is not hard to envision the potentially devastating consequences a major disease epidemic might have in one of the world's major grain production regions.

The genetic diversity in a given crop species can conveniently be represented by a pyramid (Figure 5) where the width represents the relative amount of genetic diversity, and the height represents the relative agronomic worth or adaptation of the germplasm. In almost all crops, the amount of genetic diversity present among cultivars actually grown by farmers represents a tiny fraction of the genetic diversity present in the species as a whole (Marshall, 1977; Frankel, 1977). In some instances, a single cultivar may account for over 50% of the acreage of the crop in a major production region. In crops where quality and/or consumer preference are of paramount importance, there may be demand for a specific cultivar such that the use of any other cultivar is precluded. Even in the absence of quality concerns, farmers are often quick to recognize and demand those cultivars that are most high yielding in a given region, thus genetic uniformity is a natural result of this demand for the best cultivars.

B. Considerations in Searching for Resistance

When looking for sources of disease resistance, breeders often will begin their search at the top of the genetic diversity pyramid. The more adapted or elite the source of resistance, the less chance that undesirable agronomic or other traits will be transferred with the disease resistance. If resistance is present among elite cultivars already being grown, then a recommendation that farmers simply adopt that resistant cultivar may be a reasonable measure to control the disease in the short term. Furthermore, this resistant cultivar will probably be used extensively in future breeding, as it already has the desired combination of elite agronomic performance and disease resistance. In most instances, however, the breeder must look elsewhere for resistance, either because there is simply none among elite cultivars, the resistance present may not be high enough to prevent losses, or because new sources are needed since it is expected that the pathogen will develop virulence to the resistance present in current cultivars. As the search for resistance widens and the breeder goes further down the genetic diversity pyramid, there is greater likelihood that the sources of resistance will have many undesirable agronomic traits and be poorly adaptated. If resistance is simply inherited, these problems may not be so problematic, as resistance may be transferred via simple backcross breeding. However, if resistance is complexly inherited, then it is likely that a more long-term breeding effort will be needed to successfully develop a cultivar with acceptable agronomic performance and resistance. At the bottom of the genetic diversity pyramid are wild relatives of the crop species. Wild relatives have often been rich sources of new and unique disease resistance genes for a number of crops, including wheat, oats, sunflowers, and others (Stalker, 1980; Laferrier, 1986; Lenne and Wood, 1991). Fertility barriers and sterility in progenies from interspecific crosses may seriously complicate using them in short-term breeding programs, but methods to overcome these barriers, such as embryo rescue techniques, the use of bridging species crosses, and the use of artificial polyploids have been developed for many crop species.

In summary, there are no hard and fast rules regarding where to find sources of disease resistance. In other words, resistance is where you find it. It has been argued that breeders should look for resistance where the host and pathogen have coevolved for a long period, but there are many instances of high levels of resistance being found in germplasm from regions where the pathogen has never occurred. Past experience has, however, proven the value of collecting, preserving, and evaluating germplasm for resistance to disease and for other unique traits. Although there are extensive germplasm collections of most of our major crop species, collections of many of our so-called minor crops are not so extensive. Also, the maintenance and evaluation of these collections is sometimes spotty at best, so they are not being utilized as well as they perhaps could be. We should also avoid the trap of being satisfied that once we find and successfully utilize a source of disease resistance, our task is finished. History has taught us that plant pathogens are "shifty enemies" that have the capacity to continually adapt to our efforts to control them via plant breeding. It would be to our advantage to locate as many sources of resistance as possible and introgress them into adapted germplasm so we will have a complete arsenal at our disposal in this never-ending battle.

VII. DEVELOPMENT OF DISEASE-RESISTANT CULTIVARS

It is not the intent to present here a complete review of plant breeding methodology, but rather present a brief overview of how various common plant breeding methods might be used in a resistance breeding program. There are numerous excellent texts on plant breeding that the reader may refer to for more details on specific methods. Principle considerations to be made in the selection of an appropriate breeding method include: the natural mode of reproduction of the crop (self- vs. cross-pollinated, asexually propagated, apomictic, etc.), the ease with which controlled pollinations are made and the number of progeny expected per cross, resource limitations of the breeding program, the desired product of the program (pure line variety, synthetic variety, clone, F_1 hybrid, etc.), and the relative value of disease resistance vs. other traits such as yield, quality, etc. The answers to all these considerations will of course vary greatly, depending on the crop species. Readers are urged to consult specific literature on the breeding of crops of their particular interest to learn which methods and approaches are appropriate.

A. Backcross Breeding

In those situations where resistance is inherited in a monogenic dominant manner, backcross breeding is often the method of choice to transfer the gene into an elite cultivar, particularly if the elite cultivar is a homogeneous pure line variety or inbred line. Backcrossing consists of making an initial cross of the elite cultivar with the resistance source, crossing the resulting F_1 with the recurrent parent (the elite cultivar) and selecting resistant individuals in this initial backcross (BC1) generation to be backcrossed to the recurrent parent. The process of selecting resistant plants and backcrossing them to the recurrent parent continues for several generations, usually to at least the BC5 or BC6 generation, until the breeder is satisfied that the germplasm now has the desirable agronomic characteristics of the recurrent parent. At this point, resistant plants are self-pollinated, and the resulting progeny are tested to find those that are homozygous resistant. These homozygous resistant lines are then performance tested to insure that they have retained the desired performance of the recurrent parent, and the best line released as a new cultivar. Backcross breeding, although an easy way to transfer simply inherited forms of resistance, is inherently conservative. Oftentimes, by the time the backcross-derived cultivar is released, the recurrent parent may be considered obsolete in its agronomic performance. Although most commonly used to transfer single dominant genes, backcrossing can also be successfully used to transfer recessive genes, multiple dominant genes, and even some types of so-called polygenic partial resistance to disease with appropriate modifications (Carson, 1991).

B. Pure Line Breeding Methods

More commonly, breeding for disease resistance is incorporated into an overall program of cultivar development (Hooker, 1984). Disease resistance is often considered just one of numerous traits, such as yield, quality, lodging resistance, maturity, insect resistance, etc. that must be taken into account in developing breeding or source populations that are subjected to selection in the breeding process. When synthesizing a breeding population from which to initiate selection, the breeder usually tries to select parents with complementary traits, with the hopes of selecting individuals that have the proper combination of favorable traits from the parents. In crops where the desired end product consists of a pure line cultivar or inbred line for making F_1 hybrids, the source population may consist simply of a F_2 from a cross of two parent lines or be a much more complicated three, four, or multiway cross of selected parents. Usually, the less agronomically desirable or poorly adapted the source of disease resistance, the lower the percentage of parentage it represents in the source population. Selection for disease resistance is often initiated early in the inbreeding process to eliminate the most susceptible individuals. Selection may be simply on an individual plant basis if the resistance is oligogenically inherited or the heritability is high, or alternatively, selection may be on a family (F_3 or F_4, for example) basis if the heritabilities are lower. The breeder should attempt to maintain a reasonably high population size so that he still has sufficient numbers of families left for agronomic yield testing in later generations.

There are two basic methods used to develop pure line cultivars or inbred lines through the process of continued self-pollination: the bulk method and the pedigree method. In the latter method, the pedigree or identity of individual families is maintained throughout the inbreeding process, so that any F_5 family may be traced back to an individual plant in the origin source population. The pedigree method allows selection on an individual family basis at each stage of the inbreeding process, but it also entails substantial record keeping and may use more field space than the bulk method. In the bulk method, seeds from individual plants in each generation are bulked or combined such that it is impossible to trace the pedigree of any plant. These bulk populations are usually self-pollinated for several generations until they consist of a mixture of homozygous pure lines. Seeds from individual plants are then planted out in progeny rows for replicated testing and selection. The bulk method does allow selection on an individual plant basis during the inbreeding process, and some natural selection against more susceptible individuals should also occur if epidemics can be generated in the breeding nursery. The bulk method eliminates the need for extensive record keeping and does allow a breeder to work with a large number of breeding populations at once, as each population takes a minimal amount of space in the nursery.

C. Hybrid Crops

In crops where the final product is an F_1 hybrid between two homozygous inbred lines, the parental lines are commonly developed using the pedigree method as is used in self-pollinated crops such as the small grains and beans. In most crops where hybrids are used, there exist known heterotic patterns or breeding groups that exhibit high heterosis when crossed. For example, in maize hybrids grown in the U.S., almost all hybrids can trace the pedigree of their parental inbred lines to either the Iowa Stiff Stalk synthetic (typically used as female parents of the hybrid) or the open pollinated variety Lancaster Surecrop (typically used as the male parent). During the inbreeding process, selection for disease resistance and other traits of medium to high heritability is usually practiced, but yield can only be assessed in hybrid combination with a tester which may be an inbred line, a set of inbred lines, or a cross of two lines of the opposite heterotic group. Although the stage of inbreeding at which testcross performance is assessed varies greatly from crop to crop and among individual breeders, it should be emphasized that with hybrid crops the value or worth of an inbred line is judged by its performance in hybrid combination and not its per se performance. Thus in developing disease-resistant hybrids, it may suffice for only one parent of an F_1 hybrid to carry resistance if it transfers a sufficient level of resistance to the hybrid. Similarly, it may be possible to develop a hybrid with multiple disease resistance if the resistances of the two parental inbred lines are complementary.

D. Population Improvement Methods

In certain crops such as the forage legumes and grasses, the final product or variety consists not of a pure line or a F_1 hybrid, but rather a heterogeneous mixture of individuals or a synthetic variety. The simplest and oldest form of plant breeding is mass selection, which consists of saving seed from selected individuals and combining this seed for the next generation. As mentioned earlier, this method was very effective in improving resistance to Fusarium wilts in several crops at the turn of the century and is still used in breeding forages for disease resistance. Mass selection is a very effective method for selecting for disease resistance where resistance is highly heritable, but is relatively ineffective for traits of low heritability.

A more advanced set of population improvement techniques that can be highly effective in developing disease-resistant plant populations is generally lumped under the heading of recurrent selection schemes. Simply put, recurrent selection is evaluation of the performance of some sort of progeny from a population, followed by recombination of selected progenies to produce the next generation or cycle of the population. Progenies may be half-sib, full-sib, or selfed (S1 or S2) families. These may be evaluated per se or in testcrosses with a suitable tester genotype. Progenies or their testcrosses, or both, may be evaluated in replicated trials in multiple environments and selection is made on the basis of the progeny mean performance. Typically, at least 100 to 200 or more progenies are evaluated and the top 10 to 20% recombined to synthesize the next selection cycle. In order for it to be recurrent selection, at least two cycles need to have been completed. Because selection is made on the basis of replicated progeny tests and not individual plant phenotypes, recurrent selection may be effective for traits of lower heritability that would not respond to simple mass selection methods. Usually only a few cycles of recurrent selection are needed to achieve high levels of disease resistance, assuming that adequate genetic variation for resistance was present in the initial population. Recurrent selection in its various forms may be used to develop improved existing open-pollinated cultivars, improved synthetic varieties, or to improve breeding populations that will serve as sources of improved inbred lines developed by pedigree selection.

VIII. MONITORING AND COPING WITH VARIABLE PATHOGENS

A. Monitoring Pathogen Variability

From the very outset of modern efforts at breeding for disease resistance, we have had to cope with the problem of "shifty pathogens". With some diseases such as the Fusarium wilts, resistance has remained quite durable for many years, but with others such as the cereal rusts, resistance genes are often ineffective before they are even deployed in a cultivar. Even if an effective gene is found, it was not uncommon for it to remain effective for only five years or less. Breeders have tried to cope with these variable pathogens by constantly searching for new sources of resistance and selecting combinations of resistance genes that will give resistance to the spectrum of pathogen races present in a growing region. These efforts are of course dependent on breeders knowing the relative frequency, distribution, and virulence of races present where the crop is being grown. There are at least two basic approaches to monitoring pathogen populations: race surveys where cooperating scientists in a region collect isolates that are then identified as to race via the use of differential varieties, and the use of monitoring plots where plots of differential varieties are planted at various locations in a region and the occurrence and severity of disease on them monitored during the growing season. There are many examples of race surveys being used to monitor pathogen race structure in a given region. The USDA Cereal Rust Laboratory in St. Paul, MN regularly collects samples of leaf and stem rust from wheat throughout North America and race types them on differential varieties (Long et al., 1985; Roelfs et al., 1989). This information is then used by wheat breeders throughout the region to determine which combinations of resistance genes will be effective against the races prevalent at the time. The rust survey may also serve as an early warning system to alert breeders of the presence of new races virulent on existing cultivars, thus giving them a chance to respond before significant losses have occurred. The second approach, the use of monitoring plots, may be used alone or in conjunction with race surveys. By itself, information on disease development on differential varieties may give a rough estimate of the frequency of virulence to specific resistance genes, but may not provide information on specific virulence combinations present in races. However, samples of isolates from monitoring plots may be collected and later identified as to race and provide this information. Monitoring may also serve as effective early warning systems to alert breeders to the presence of new virulences in pathogen populations. The Dekalb Plant Genetics Maize Pathogen Monitoring Project in the U.S. is an excellent example of how a private plant-breeding company is using this method to gather useful information on variation in and the distribution of maize pathogens in North America (Smith, 1984).

B. Using More than One Effective Resistance Gene at a Time

Durable resistance is defined as resistance that remains effective during its prolonged and widespread use in an environment favorable to the disease. Several approaches are either being used or have been proposed to increase the durability of disease resistance. Probably the most common means employed to prolong the useful life of a resistant cultivar is to combine more than one effective resistance gene into a single cultivar. The rationale behind this approach is that if virulence toward a specific resistance gene is a very rare random mutational event, then the probability of multiple mutations conferring virulence to multiple resistance genes is so low that the resistance gene combination will have prolonged durability (Mundt, 1990, 1991). This probability explanation for the apparent durability of certain stem rust resistance gene combinations in wheat has been criticized. Kolmer et al. (1991) have pointed out that often it is only certain resistance gene combinations that exhibit this apparent durability, and their durability has more to do with the specific gene combinations. For example, certain combinations of *Sr* genes with *Sr2* (an adult plant resistance gene) and seedling *Lr* genes with *Lr13* or *Lr34* (adult plant resistance genes) appear to confer durable resistance to stem and leaf rust of wheat, respectively (Johnson, 1984). As a general rule, the use of multiple resistance gene combinations can be effective, but experience will dictate which combinations appear to give durability and which will not.

C. Gene Pyramiding

Another approach to durability that has been proposed by Nelson (1978) is the pyramiding of as many resistance genes as possible into a cultivar to give it durability. Nelson does not make the distinction between race-specific single genes and polygenes for partial resistance. Indeed, he and others consider these so-called polygenes for partial resistance to simply be defeated race-specific genes that now have a small incomplete effect on resistance. This alleged residual effect of defeated race-specific genes is sometimes referred to as "ghost gene" resistance. The evidence for ghost genes is contradictory at best, and does not appear to be of significance in many pathosystems (Johnson, 1984). The value of arbitrarily pyramiding resistance genes without regard to their specificity or record of durability is questionable. Perhaps a better and certainly more efficient approach would be to pyramid race-specific gene combinations that appear to be durable, as mentioned above, along with genes for partial resistance.

D. Multiline Varieties and Cultivar Mixtures

Multiline cultivars or mixtures have also been promoted as a means to achieving durable resistance to highly variable pathogens such as the cereal rusts (Jensen, 1965; Browning and Frey, 1969; Kush and Virmani, 1988). A multiline cultivar consists of a set of near-isogenic backcross-derived lines, each of which contains a different race-specific single gene for disease resistance. Because only a fraction of the pathogen population will be virulent on a single-component line of a multiline variety, the net effect is a reduction in disease development on the multiline as a whole. It is also assumed that stabilizing selection (selection against unnecessary genes for virulence, *sensu* Vanderplank) will prevent selection for complex races with virulence on multiple components of the multiline (Parlevliet, 1981). Multiline varieties have been developed and they do appear to have resistance in field tests, but their durability remains to be demonstrated. Theoretical simulations have shown that stabilizing selection must be relatively strong to prevent the buildup of complex races virulent on multilines (Groth, 1976; Barrat, 1980). Experimental evidence for strong stabilizing selection operating in agroecosystems is inconsistent (Parlevliet, 1981), so it is not clear if multilines will truly be as durable as promised. Also, the problems associated with developing multilines that are high yielding and uniform, and their maintenance, have limited their acceptance on a wide scale.

E. Polygenic Partial Resistance

Polygenically controlled partial disease resistance has been assumed to be more durable than monogenic forms of resistance (Hooker and Saxena, 1971; Geiger and Heun, 1989; Parlevliet, 1979; Russell, 1978; Fleming and Person, 1982; Nelson, 1978). Presumably this durability is the result of the polygenic nature of the resistance and also the incomplete nature of resistance, which allows some pathogen reproduction and thus would apply less selection pressure on the pathogen. There are, however, several examples of monogenic resistance that have proven to be quite durable. Although polygenic resistance appears to be controlled by several genes, each of which contributes a small portion to the total resistance, most experimental estimates of the numbers of loci controlling polygenic resistance are less than ten and commonly only two to four (Geiger and Heun, 1989). Thus, although called polygenic, the low numbers of genes involved differentiate polygenic resistance from traditional polygenic characters such as yield, where perhaps hundreds of loci may be involved. Polygenic resistance appears to be found in a wide array of crops to diverse pathogens, but its use depends on the crop and the ease with which it can be manipulated in a breeding program. Perhaps the best example of the successful use of polygenic forms of resistance is in maize, where hybrids rely almost exclusively on polygenic partial resistance for control of leaf blights (Hooker, 1978). This situation is a reflection of the ease with which maize can be manipulated by an array of breeding methods and the widespread availability of resistant germplasm. Although monogenic resistance is known to some of these diseases, it is little used. In other pathosystems such as the cereal rusts, polygenic resistance has not been widely exploited due to difficulties in effectively selecting for it in traditional pedigree breeding programs and the availability of easily scored and transferred monogenic forms of resistance. Once useable levels of polygenic resistance have been transferred into adapted, elite germplasm, there are fewer barriers to its use in many breeding programs. A final note regarding the presumption of the durability of polygenic resistance: although it has been effectively used for many years in crops such as maize without any evidence that resistance has significantly eroded, it would be wrong to assume that pathogens cannot eventually adapt to polygenic resistance. Indeed, there is some experimental evidence to suggest that pathogens do indeed have the capacity to at least partially overcome polygenic forms of resistance (Kolmer and Leonard, 1986). However, this adaptation process would probably be a slow, gradual process. If breeders were to continually locate and incorporate new sources of polygenic resistance (and presumably new polygenes) into new cultivars, then the adaptation by the pathogen to a given set of polygenes would be to little avail, as new genes are continually being deployed in the host.

F. Gene Deployment Strategies

Another strategy to prolong the durability of race-specific genes has been the regional or area deployment of those genes in a given epidemiological area. Specifically, Browning et al. (1969) proposed that different regions in the central North American "*Puccinia* pathway" utilize different genes for crown rust resistance in their oat varieties. Because crown rust, *Puccinia coronata*, and other cereal rusts overwinter in the southern U.S. and in northern Mexico and then move northward each summer with the developing crop, it is theorized that using different resistance genes in the southern states from those used in the midwest and Canada would prevent this yearly spread of the pathogen. Those races that would be selected for virulence to the resistance genes used in the southern states would be avirulent on the resistance genes used in the northern states. While this approach is appealing, it would require the cooperative agreement between breeders from several states and three countries. It is highly unlikely that breeders and growers in the southern states and Mexico would agree to not use certain resistance genes, particularly if faced with the prospect of races developing that were virulent on their varieties but not on those used in more northern states.

IX. RESISTANCE IN CONJUNCTION WITH OTHER CONTROL STRATEGIES

Although the emphasis in this chapter has been on the use of genetic resistance as a primary means of disease control, it should be pointed out that resistance is often but a single component in an integrated pest management system. In many cases, the level of resistance available is insufficient by itself for adequate control and must be enhanced by the use of proper agronomic, phytosanitary, and chemical controls. The level of partial resistance needed to control *Psuedomonas solanacearum*, a soil-borne bacterium causing a vascular wilt of a wide range of solanaceous crops, is dependent upon the length of rotation to nonhost crops such as maize or other grasses (Melton and Powell, 1991; Hartman and Elphinstone, 1994). In maize, the level of resistance needed to gray leaf spot is dependent upon tillage practices and crop rotations. The causal pathogen, *Cercospora zeae-maydis*, survives in maize residues, so continuous maize culture in conjunction with reduced tillage practices that leave residue on the soil surface is conducive to early and potentially destructive epidemics.

Resistance may also be used in to reduce the number and amount of fungicides needed for effective disease control. The number of fungicide sprays needed for effective control of late blight of potato *(Phytophthora infestans)* can be substantiallly reduced and the interval between sprays increased on a partially resistant compared to a susceptible variety (Fry, 1978). The use of fungicides in conjunction with race-specific resistance has also been advocated as a means of prolonging the useful life of race-specific single genes. In theory, fungicides would reduce the rate at which virulent races would increase in the population.

Despite the increased interest in "integrated" approaches to plant disease control in the last two decades, there is still relatively little hard experimental data on the successful integration of host resistance with other control measures in many crop species. With increasing costs and restrictions on pesticide use in much of the developed world, the demand for strategies to reduce or eliminate their use will undoubtedly fuel further work in this area and disease resistance will be a major component of any such strategy.

X. THE FUTURE OF DISEASE RESISTANCE

In this chapter, the discussion of the use of genetically engineered resistance has been avoided since it is covered in Chapter 14 in this book. Engineered resistances to several viral diseases of plants have already been commercially developed, and several groups are close to having commerciallly viable engineered resistance to certain plant pathogenic fungi and plant parasitic nematodes. These developments beg the question, will engineered resistances replace or make conventional plant breeding approaches obsolete? For several reasons, this scenario is highly unlikely in the foreseeable future. First, any resistance developed, regardless whether by conventional or by genetic engineering, must be thoroughly tested for its efficacy, specificity, and stability across a range of environments. Secondly, and perhaps more importantly, for any resistance mechanism to be commercially viable it must be successfully transferred to and expressed in a cultivar having all the other desirable agronomic traits such as yield, maturity, lodging resistance, quality, etc. This requires a conventional plant-breeding approach. Genetically engineered disease resistances are perhaps best viewed as just one more set of tools available to the plant breeder in the search for durable control of plant diseases. For commercial viability, it is imperative for programs developing engineered disease resistance be linked with and integrated into a program of overall plant improvement that exploits all the tools at its disposal including conventional plant breeding, plant molecular biology, classical genetic and cytogenetic, and other approaches. Finally, there is also the question of durability of engineered resistances. Plant pathogens have shown an uncanny ability to adapt to changes in crop plants brought by humans via plant breeding, so it would be naive to think that they will respond differently to the changes brought about via genetic engineering. Considerations about specificity and durability should apply equally to both types of resistance. By improving our understanding of the population dynamics and genetics of plant pathogens, and the mechanism(s) underlying the apparent durability of certain types of disease resistance, there is no reason to believe that we cannot specifically engineer disease resistance that will not only be effective in the short term, but also will be durable.

REFERENCES

Barrat, J. A., Pathogen evolution in multilines and variety mixtures, *Phytopathol. Z.*, 87, 383, 1980.

Biffen, R. H., Mendel's laws of inheritance and wheat breeding, *J. Agric. Sci.*, 5, 4, 1905.

Bolley, H. L., Flax wilt and flax sick soil, *N.D. Agric. Exp. Stn. Bull.*, 50, 1901.

Briggs, S. P. and Johal, G. S., Genetic patterns of plant host-parasite interactions, *Trends Genet.*, 10, 12, 1994.

Browning, J. A. and Frey, K. J., Multiline cultivars as a means of disease control, *Annu. Rev. Phytopathol.*, 7, 355, 1969.

Browning, J. A., Simons, M. D., Frey, K. J., and Murphy, H. C., Regional deployment for conservation of oat crown-rust resistance genes, in *Disease Consequences of Intensive and Extensive Culture of Field Crops*, Iowa Agric. Exp. Stn. Spec. Rep. 64, 1969, 49.

Carson, M. L., Inheritance of latent period length in maize infected with *Exserohilum turcicum*, *Plant Dis.*, 79, 581, 1995.

Carson, M. L., Assessment of six models of host-pathogen interaction in horizontal pathosystems, *Phytopathology*, 77, 241, 1987.

Carson, M. L., Relationship between leaf freckles and wilt severity and yield losses in closely related maize hybrids, *Phytopathology*, 81, 95, 1991.

Caten, C. E., Person, C., Groth, J. V., and Dhahi, S. J., The genetics of pathogenic aggressiveness in three dikaryons of *Ustilago hordei*, *Can. J. Bot.*, 62, 1209, 1984.

Dayton, D. F., Bell, R. L., and Williams, E. B., Disease resistance, in *Methods of Fruit Breeding*, Moore, J. N. and Janick, J., Eds., Purdue University Press, West Lafayette, IN, 1984, chap. 13.

Essary, S. H., Notes on tomato diseases with results of selection for resistance, *Tenn. Agric. Exp. Stn. Bull.*, 95, 1912.

Fleming, R. A. and Person, C. O., Consequences of polygenic determination of resistance and aggressiveness in nonspecific host:parasite relationships, *Can. J. Plant Pathol.*, 4, 89, 1982.
Flor, H. H., Genetics of pathogenicity in *Melampsora lini*, *J. Agric. Res.* 73, 337, 1946.
Flor, H. H., The current status of the gene for gene concept, *Annu. Rev. Phytopathol.*, 9, 275, 1971.
Flor, H. H., The complementary genetic systems in flax and flax rust, *Adv. Genet.*, 8, 29, 1956.
Frankel, O. H., Genetic resources, *Ann. N.Y. Acad. Sci.*, 287, 345, 1977.
Fry, W. E., Quantification of general resistance of potato cultivars and fungicide effects for integrated control of potato late blight, *Phytopathology*, 68, 1650, 1978.
Geiger, H. H. and Heun, M., Genetics of quantitative resistance to fungal diseases, *Annu. Rev. Phytopathol.*, 27, 317, 1989.
Gevers, H. O., A new major gene for resistance to *Helminthosporium turcicum* leaf blight of maize, *Plant Dis. Rep.*, 59, 296, 1975.
Green, G. J., Hybridization between *Puccinia graminis tritici* and *Puccinia graminis secalis* and its evolutionary implications, *Can. J. Bot.*, 49, 2089, 1971.
Groth, J. V., Multilines and "superraces": a simple model, *Phytopathology*, 66, 937, 1976.
Hartman, G. L. and Elphinstone, J. G., Advances in the control of *Pseudomonas solanacearum* race 1 in major food crops, in *Bacterial Wilt: the Disease and Its Causative Agent,* Pseudomonas solanacearum, Hayward, A. C., Ed., CAB International, Wallingford, U.K., 1994, chap. 12.
Hooker, A. L., Genetics of disease resistance in maize, in *Maize Breeding and Genetics*, Walden, D. B., Ed., John Wiley & Sons, New York, 1978, chap. 21.
Hooker, A. L., The genetics and expression of of resistance in plants to rusts of the genus *Puccinia*, *Annu. Rev. Phytopathol.*, 5, 163, 1967.
Hooker, A. L. and Saxena, K. M. S., Genetics of disease resistance in plants, *Annu. Rev. Genet.*, 5, 407, 1971.
Hooker, A. L., The pathological and entomological framework of plant breeding, *Proc. 16th Stadler Genet. Symp.*, Mo. Agric. Exp. Stn., Columbia, MO, 1984, 177.
Hooker, A. L., Breeding to control pests, in *Crop Breeding*, Wood, D. R., Rawal, K. M., and Wood, M. N., Eds., American Society Agronomy, Madison, WI, 1980, chap. 10.
James, W. C., A manual of disease assessment keys for plant diseases, Publ. 1458, Can. Department of Agriculture, Ottawa, 1971.
James, W. C., Assessment of plant diseases and losses, *Annu. Rev. Phytopathol.*, 12, 27, 1974.
Jensen, N. F., Multiline superiority in cereals, *Crop Sci.*, 5, 566, 1965.
Johnson, R., A critical analysis of durable resistance, *Annu. Rev. Phytopathol.*, 22, 309, 1984.
Kolmer, J. A. and Leonard, K. J., Genetic selection and adaptation of *Cochliobolus heterostrophus* to corn hosts with partial resistance, *Phytopathology*, 76, 774, 1986.
Kolmer, J. A., Dyck, P. L., and Roelfs, A. P., An appraisal of stem and leaf rust resistance in North American hard red spring wheats and the probability of multiple mutations to virulence in populations of cereal rust fungi, *Phytopathology*, 81, 237, 1991.
Kulkarni, R. N. and Chopra, V. L., Environment as the cause of differential interaction between host cultivars and pathogenic races, *Phytopathology*, 72, 1384, 1982.
Kush, G. S. and Virmani, S. S., Plant breeding strategies for biological control, in *Biocontrol of Plant Diseases*, Vol. II, Murkerji, K. G. and Garg, K. L., Eds., CRC Press, Boca Raton. FL, 1988, chap. 3.
Laferriere, J. E., Interspecific hybridization in sunflowers an illustration of the importance of wild genetic resources in plant breeding, *Outlook Agric.*, 15, 104, 1986.
Lee, T. S. and Shaner, G., Infection processes of *Puccinia recondita* in slow- and fast-rusting wheat cultivars, *Phytopathology*, 74, 1419, 1984.
Lenne, J. M. and Wood, D., Plant diseases and the use of wild germplasm, *Annu. Rev. Phytopathol.*, 29, 35, 1991.
Levings, C. S., III and Seidow, J. N., Molecular basis of disease susceptibility in the Texas cytoplasm of maize, *Plant Mol. Biol.*, 19, 135, 1992.
Long, D. L., Schafer, J. F., and Roelfs, A. P., Specific virulence of *Puccinia recondita* f. sp. *tritici* in the United States from 1978 through 1983, *Plant Dis.*, 69, 343, 1985.
Long, D. L. and Kolmer, J. A., A North American system of nomenclature for *Puccinia recondita* f. sp. *tritici*, *Phytopathology*, 79, 525, 1989.
Marshall, D. R., The advantages and hazards of genetic homogeneity, *Ann. N.Y. Acad. Sci.*, 287, 1, 1977.

Melton, T. A. and Powell, N. T., Effects of two-year crop rotations and cultivar resistance on bacterial wilt in flue-cured tobacco, *Plant Dis.*, 75, 695, 1991.

Mundt, C. C., Probability of mutation to multiple virulence and durability of resistance gene pyramids, *Phytopathology*, 80, 221, 1990.

Mundt, C. C., Probability of mutation to multiple virulence and durability of resistance gene pyramids: further comments, *Phytopathology*, 81, 240, 1991.

National Academy of Sciences, *Genetic Vulnerability of Major Crops*, National Academy of Science, Washington, D.C., 1972.

Nelson, R. R., Genetics of horizontal resistance to plant diseases,*Annu. Rev. Phytopathol.*, 16, 359, 1978.

Newton, A. C. and Andviron, D., Assumptions and implications of current gene-for-gene hypotheses, *Plant Pathol.*, 44, 607, 1995.

Orton, W. A., A study of disease resistance in watermelon, *Science*, 25, 288, 1907.

Orton, W. A., The development of farm crops resistant to disease, U.S. Department of Agriculture Yearbook, 1908, 453, 1909.

Parlevliet, J. E., Stabilizing selection in crop pathosystems: an empty concept or a reality?, *Euphytica*, 30, 259, 1981.

Parlevliet, J. E. and Zadoks, J. C., The integrated concept of disease resistance; a new view including horizontal and vertical resistance in plants, *Euphytica*, 26, 5, 1977.

Parlevliet, J. E., Components of resistance that reduce the rate of epidemic development, *Annu. Rev. Phytopathol.*, 17, 203, 1979.

Raymundo, A. D., Hooker, A. L., and Perkins, J. M., Effect of the gene *HtN* on the development of northern corn leaf blight, *Plant Dis.*, 65, 327, 1981.

Roelfs, A. P., Casper, D. H., Long, D. L., and Roberts, J. J., Races of *Puccinia graminis* in the United States and Mexico during 1987, *Plant Dis.*, 73, 385, 1989.

Russell, G. E., *Plant Breeding for Pest and Disease Resistance*, Butterworth Publishers, Boston, 1978, 21.

Samborski, D. J. and Dyck, P. L., Inheritance of virulence in wheat leaf rust on the standard differential wheat varieties, *Can. J. Genet. Cytol.*, 10, 24, 1968.

Saxena, K. M. S. and Hooker, A. L., On the structure of a gene for disease resistance in maize, *Proc. Natl. Acad. Sci.*, 61, 1300, 1968.

Scheffer, R. P. and Walton, J. D., Toxin-producing Cochlioboli as model pathogens of plants: ecological, evolutionary, and genetic considerations, in *Helminthosporia Metabolites, Biology, Plant Diseases*, Chelkowski, J., ed., Institute of Plant Genetics, Polish Academy of Sciences, Posnan, Poland, 1995, chap. 3.

Shaner, G., Stromberg, E. L., Lacy, G. H., Barker, K. R., and Pirone, T. P., Nomenclature and concepts of pathogenicity and virulence, *Annu. Rev. Phytopathol.*, 30, 47, 1990.

Smith, D. R., Monitoring corn pathogens nationally, in *Proc. 20th Annu. Illinois Corn Breeders School*, Alexander, D. E., Ed., University of Illinois, Champaign, IL, 1984, 101.

Stalker, H. T., Utilization of wild species for crop improvement, *Adv. Agron.*, 33, 111, 1980.

Ullstrup, A. J., The impacts of the southern corn leaf blight epidemics of 1970-71, *Annu. Rev. Phytopathol.*, 10, 37, 1972.

Vanderplank, J. E., *Disease Resistance in Plants*, 2nd ed., Academic Press, Orlando, FL, 1984, chap. 7.

Whetzel, H. H., *An Outline of the History of Phytopathology*, W. B. Saunders, Philadelphia, 1918, 16.

Whitaker, T. W., The breeding of vegetable crops: highlights of the past 75 years, *Hortscience*, 14, 359, 1979.

Winer, P., Additive and multiplicative models for resistance in plant pathology, *Euphytica*, 33, 963, 1984.

Part V
Cultural Practices

CHAPTER 10

Sanitation, Eradication, Exclusion, and Quarantine

Ann B. Gould and Bradley I. Hillman

1-8493-2627-3/97/$0.00+$.50

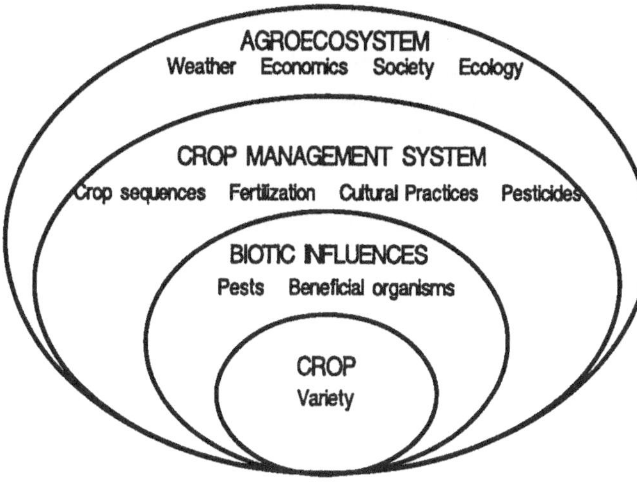

Figure 1 Relationships among the components of an integrated management system. (Adapted from Fry, W. E., *Principles of Plant Disease Management,* Academic Press, New York, 1982.)

I. INTRODUCTION

Successful disease management in most cropping systems involves an integrated program that includes cultural practices combined with the use of pesticides (Figure 1). Often these cultural measures are necessary to improve fungicide efficacy and, historically, were the only measures available to farmers prior to the advent of chemical controls. Indeed, cultural methods of disease management were often devised even before farmers knew the cause of the diseases themselves.

Although there are many types of cultural management strategies, most of them can be placed into one of four categories: exclusion, quarantine, eradication, and sanitation. In this chapter, we will define these strategies in broad terms and give brief examples.

II. DEFINITION OF TERMS

Historically, disease management practices were devised to manage existing epidemics and to prevent new ones. Simply stated, an epidemic is a rapid and widespread development of disease (Schumann, 1991). Establishing disease management strategies for monocyclic (for which pathogens have one generation per season) or polycyclic diseases (for which pathogens have more than one generation) are often different. Monocyclic diseases are most effectively controlled by reducing initial inoculum levels. Alternatively, polycyclic diseases are suppressed when initial inoculum levels are suppressed and the rate of disease development is limited (Fry, 1982). Cultural management strategies used to reduce the initial levels of inoculum include exclusionary measures and quarantines; strategies that suppress existing inoculum to retard disease development include eradication and sanitation (Fry, 1982; Van der Plank, 1963).

Exclusion and quarantine are methods that restrict the movement of a pathogen to regions in which the pathogen is not endemic. Whetzel (1929) defined the term exclusion as the ability to prevent "to a profitable degree the entrance and establishment of a pathogene [*sic*] in an uninfested area, as in a garden, field, region, state, or country." Exclusionary tactics are usually conducted on a small scale and include certification programs and planting pathogen-free propagation material. By their nature, these measures are often temporary. Alternatively, quarantines are conducted on a larger scale and require federal and/or state regulations to succeed. Quarantines may remain in effect until economic pressures or new information on the biology of the pathogens render them unnecessary, or when they fail and the pathogen becomes endemic in the quarantined region.

Once a pathogen becomes established in a given region, eradicative and sanitation measures are necessary to affect the survival of the pathogen, thereby preventing or delaying disease development. Eradication has been defined (Whetzel, 1929) as "the more or less complete elimination or destruction of a pathogene [*sic*] after it is established in a given area." Eradication of a disease usually requires the removal and destruction of all sources of inoculum, including infected host plants, plants in a predetermined area surrounding the infection foci, alternate hosts (as in rust diseases), or alternative hosts (for example, weeds and viral diseases).

The term sanitation is usually reserved for the physical measures that are used to eliminate sources of inoculum from plant residues, propagation tools, farm equipment, pots, benches, and other sources. Sanitary measures, for example, are extremely important for the control of certain greenhouse diseases such as Botrytis blight, which is easily spread in plant debris.

III. REDUCTION OF INITIAL INOCULUM: QUARANTINE AND EXCLUSION

A. Quarantine

When a foreign disease is introduced for the first time into an area, there is often minimal resistance to that pathogen in the local population of plants, so resistance may be of little importance as an exclusion mechanism. For these situations, quarantine as an exclusion mechanism is of great importance. Quarantine is the use of legal or regulatory means to prevent material, in this case plant material, from moving in an unrestricted fashion into or out of a specific area or jurisdiction. The term originally referred to the 40-day period of isolation for crew members and others aboard ships during which potential disease carriers could develop symptoms (if they were going to) prior to their release back into developed countries. As used today, the term quarantine does not necessarily imply a holding period, although such a holding period may be required. If movement of material is completely prevented, the quarantine is referred to as an embargo. More often, restrictions requiring certification of materials are imposed upon the importer or exporter of the materials. In order for quarantines to be effective, they must be put into law, and there must be mechanisms for their enforcement. There are many types and levels of certifications that may be required by a jurisdiction for movement of plant materials. Among the excellent resources available to those interested in a thorough coverage of plant protection and quarantine is the three-volume set written and edited by Kahn (Kahn, 1989 a,b,c).

There are different reasons for imposing quarantines. Probably the most clear-cut situation occurs when a plant pathogen has never been found in one region or jurisdiction, and there is a desire to import plant material of the same type into that area. In this case, it is relatively easy to justify a strong quarantine on the importation of plant materials. Examples of this kind of quarantine include restrictions against importation of plant material infected with plum pox virus or maize streak virus into the U.S. Because of the extreme risks they pose, these viruses may not even be brought in and propagated for research purposes. Even if a pathogen is already prevalent in an area, its movement through the U.S. is still generally regulated. For example, import or interstate movement of the chestnut blight fungus, *Cryphonectria parasitica*, is restricted in the U.S. even though the fungus is ubiquitous in the natural range of the American chestnut. On the other hand, tobacco mosaic virus (TMV) is among the pathogens that now can be moved across states' borders in an unregulated fashion.

1. Quarantine Procedures

a. Embargo

Embargo refers to the international prohibition on the movement of goods imposed on one country by another. Embargoes are often used on a seasonal basis as mechanisms to prevent the import of seasonal pests (Mathys and Baker, 1980). In addition to their uses as legitimate quarantine tools for preventing the movement of suspect plant material, embargoes have often been used as political tools, for example for punitive purposes, to enhance the value of local products, or to void contracts.

b. Inspection at Point of Origin

This type of quarantine inspection is useful for reducing unnecessary costs and situations associated with exportation of plant materials that subsequently are rejected by the importing country at the point of entry. Since the exporting country would be less inclined to inspect shipments for export as rigorously as the importer, preclearance by the importer is used to increase the reliability of such inspections. A major drawback to inspection at the point of origin is the possibility of the introduction of latently infected materials (Mathys and Baker, 1980). This risk is reduced by use of highly sensitive pathogen detection methods on site.

c. Inspection at Point of Entry

If inspection at the point of entry is chosen, the importing country retains the right to reject a contaminated shipment. These spot inspections alone suffer from many of the same drawbacks as the point of origin inspections, except that the importing country is likely to have more control of the inspection. Although the actual impact of such inspections is unclear, point of entry inspections remain an important mechanism for controlling the movement of potentially serious pathogens (Mathys and Baker, 1980).

d. Post-Entry Quarantines

The holding of materials in a true quarantine facility for a specified time following entry, but prior to release by the importing country, remains an effective means of preventing pathogen spread from relatively small consignments such as breeding stocks (Kahn, 1978). By such means, a time is provided for multiplication, signs, and symptom expression by latent pathogens. Facilities for post-entry quarantine in the U.S. are regulated at the federal level and are inspected by state inspectors (Rosenberg, 1989).

2. Brief History of Quarantines

The first record of a quarantine law passed specifically to exclude importation of plant materials into a country for pest control purposes was the 1873 quarantine act passed by Germany to prevent the import of Colorado potato beetles (Mathys and Baker, 1980). The first quarantine enacted in the U.S. was the 1891 initiation of a seaport inspection station at San Pedro, California (Mathys and Baker, 1980). The first national plant quarantine legislation enacted in the U.S. was the 1912 Plant Quarantine Act, which was a response to massive epidemics of chestnut blight, white pine blister rust, and citrus canker (Waterworth and White, 1982). This act authorized the government to institute quarantines to prevent the introduction and movement of pests, and laid the groundwork for other quarantine legislation. The next major legislation passed in the U.S. was the 1944 Organic Act, which authorized the Secretary of Agriculture to cooperate with smaller groups such as states, grower groups, or individuals to detect and control plant pests (Kahn, 1991). The Federal Plant Pest Act of 1957 extended coverage of the 1912 Quarantine Act to pests not mentioned in the earlier act, and authorized emergency action to prevent introduction or movement of such pests (Kahn, 1991). Within these three broad federal acts, many individual regulatory measures can and have been authorized to deal with specific situations. At the national level the Animal and Plant Inspection Service (APHIS) of the U.S. Department of Agriculture (USDA), launched in 1971 as APHS then renamed and redefined in 1972, monitors pest importation and movement.

Whereas the federal government regulates movement of plant materials across federal borders, individual states have their own departments of agriculture, which may regulate interstate and intrastate movement of plant materials. The 1944 Organic Act ensured a genuine interaction between federal and state governments and private groups in the process of regulating plant movement in a rational way. At its best, this decentralizing of the quarantine system helped to provide a much more efficient way to promote the importation and distribution of plant materials (Plucknett and Smith, 1989). Among state agencies, those that have been the most aggressive at enforcing quarantines include states in which agriculture is strong and major regions of biogeographical isolation exist. So, for example, California with its agriculturally rich Central, San Joaquin, and Imperial Valleys, protected by mountain ranges on one side and the ocean on the other, has had more aggressive border screening programs operated by the California Department of Food and Agriculture (CDFA) than most other states. At its peak in the 1970s, the CDFA had a staff of over 30 plant pathologists and nematologists (Rosenberg, 1989).

One of the strongest independent advisory groups that continues to help shape quarantine policy in the U.S. is the National Plant Board (NPB). The NPB is composed of regional plant boards, currently seven, which meet annually. In 1931, the NPB adopted a series of principles that has since been revised several times, and most recently was approved in 1995 as the National Plant Board Plant Quarantine and Nursery Inspection Guidelines (Siddiqui, 1995). These guidelines and the NPB itself are consulted before a quarantine is imposed by the USDA.

As international interest in biodiversity and the productive use of available germplasm rises, the need for effective quarantine procedures to regulate germplasm movement increases as well (Plucknett and Smith, 1989). The desire to incorporate genes from diverse sources into breeding programs often places breeders at odds with regulatory agents. In 1951, the International Plant Protection Convention was established to foster cooperation among countries in preventing the spread of plant pests and diseases (Mathys and Baker, 1980). This has since become a cornerstone in organizing international efforts at information exchange. The U.S. became a signatory to the Convention in 1972.

3. Successful Quarantines

Numerous quarantine measures have led to the restricted movement of important plant pests since their introduction in 1912. Some examples include quarantines against citrus canker, white pine blister rust, golden nematode, and ring rot of potato. Two of these will be discussed in more detail.

a. Citrus Canker

Among the most celebrated examples of successful quarantines is one of the early targets of the first quarantine, the citrus canker bacterium. Following its introduction into Florida in the early 1910s, the pathogen spread to other Gulf States. The initial quarantine was instituted, in conjunction with an eradication program to eliminate the disease from the southeast U.S. In 1947, citrus canker was declared eradicated following the destruction of 1.5 million planted citrus trees, 1.5 million alternate hosts, and 6.6 million nursery trees. When citrus canker was found again in Florida in 1984, an extraordinary emergency was declared by APHIS. For the first time, an indemnity was paid for trees that were destroyed under this program. Two bacterial diseases caused by different *Xanthomonas campestris* pathovars were subsequently found to be associated with Florida citrus: *X. c. citri* caused citrus canker, whereas *X. c. citrumelo* causes bacterial spot (Graham and Gottwald, 1991; Stall and Civerolo, 1991). Because there were two diseases associated with these bacteria, and neither was as serious as the one causing the original epidemic, the quarantine program was less effective than in the first half of the century. The eradication program was terminated in 1986, and the quarantine was discontinued in 1994.

b. Golden Nematode

Potato cyst nematodes, also known as golden nematodes (*Globodera* spp.) have been major problems in European potato growing areas and became epidemic on Long Island, New York during the middle of the twentieth century (Rich, 1983). A strict quarantine was imposed on the affected area of Long Island, preventing free movement of planting materials or soil and providing the appointed state administrator power to monitor all sanitation measures required under specified guidelines (McCubbin, 1954). These quarantine measures were challenged in the New York State Supreme Court, but upheld as reasonable and necessary. In 1948, the U.S. Congress passed The Golden Nematode Act, which authorized controls at the federal level. Due to this quarantine and associated control measures, the golden nematode has not become a national problem in the U.S.

4. Unsuccessful Quarantines

In many cases, quarantine measures are not sufficient to stop an epidemic. Often this is because measures are enacted too late and are associated with a pathogen that can spread very quickly or across long distances. Quarantine measures are particularly difficult to enact when administrators have less control of the affected plant population. Two examples of forest pathogens that became major problems despite quarantine efforts are presented here.

a. Chestnut Blight

Chestnut blight disease, caused by *Cryphonectria parasitica,* was first identified in the U.S. in 1904 at the Bronx Zoo. The first organization of resources by an affected state was by Pennsylvania in 1912, before the Federal Plant Quarantine Act. By 1915, the Federal Act was amended to include chestnut blight. This, however, was much too late to prevent the pathogen from spreading throughout the range of the chestnut — it had already reached as far south as North Carolina (McCubbin, 1954). In addition to the extraordinarily rapid spread of the pathogen, estimated at 23 miles per year by Shain (1982), subsequent research strongly suggests that the causal agent, *C. parasitica,* was most probably introduced into New Jersey in the 1880s, and was well established and spreading by the time it was noted by botanists in 1904 (Anagnostakis, 1993). Added to the fact that there were no effective measures to control chestnut blight, the quarantine measures imposed 11 years after the initial observation were much too late to be useful.

b. Dutch Elm Disease

Another pathogen of shade trees that has been the subject of quarantine measures is *Ophiostoma ulmi,* the causal agent of Dutch elm disease. Dutch elm disease came into the U.S. somewhat later than chestnut blight, around 1930. The pathogen most probably arrived at the East Coast in an elm shipment from Europe, which was then transported by rail to Ohio, where the disease was first noted. Dutch elm disease and chestnut blight share many features in common, but important differences include the intimate association of a bark beetle vector with the spread of Dutch elm disease. It was not until several years after the disease was established in Ohio that quarantine measures were undertaken (Sinclair and Campana, 1978). A Federal Quarantine was enacted in 1933 to prevent further importation of infested elm materials. In 1935, an attempt was made to prevent the spread of the pathogen from its point of establishment by Domestic Quarantine. The long-distance nature of spread of spores by the bark beetle made it impossible to stop Dutch elm disease using these late control measures. Large-scale control work was terminated in the early 1940s, and the quarantine was rescinded in 1947 (McCubbin, 1954). Local quarantine measures for the control of Dutch elm disease by eradication remain in effect in some areas, for example in California and Minnesota.

5. Conclusion

Quarantine is one of the important tools for preventing the large-scale spread of plant disease. Although it can be expensive and difficult to implement, recent streamlining of international, federal, and state regulations provides hope that we can avoid repeating some of the historical quarantine failures that resulted in major epidemics. The rapid action at all levels to contain the recent Karnal Bunt introduction in the southwest U.S. by quarantine provides an very useful example of effective implementation of quarantine regulations. Within months after identification of the pathogen in Arizona, a quarantine had been imposed. There has since been no evidence of spread outside of the Southwest. Similar rapid and targeted action against imported pathogens should be the goal of future quarantine efforts.

B. Exclusion

Exclusion practiced on a small scale includes the use of pathogen-free propagation stock. To be successful, growers must obtain and utilize this material at the beginning of each new production cycle. Pathogens can be disseminated long distances on infected propagative material, so sanitation practices and testing for pathogens must be especially rigorous. Examples of exclusion practices include seed certification, pathogen indexing systems for producing clean stock, and potato and fruit tree certification programs. Some of these systems are discussed below.

1. Seed Certification

Seed certification programs are necessary to ensure that crops grown from seed are vigorous, genetically pure, and are free from pathogens (Maloy, 1993). There are more than 30 types of seed certified in the U.S. Pathogens associated with seed can be found on seed surfaces mixed with debris, associated with seed coats, within the seed just under the seed coat, and in the embryo (Table 1). The effect of these pathogens on plant vigor ranges from 0 to 100% mortality. The incidence of seed-borne diseases can be reduced or eliminated through the use of pathogen-free seed (Maloy, 1993).

Detection of pathogens in infected seed may include direct macro- and microscopic examination, standard isolation techniques using nonselective or selective media, grow-out tests, seed soak assays, direct seed assays, and ergosterol tests. Pathogens can also be detected using serological tests such as enzyme-linked immunosorbent assay (ELISA) and indirect immunofluorescence (IF or IIF) staining (Maloy, 1993; McGee, 1981). A review of detection techniques for bacteria in seed and other plant materials is found in Saettler et al., 1989.

2. Indexing

Since most greenhouse crops and fruit trees are propagated vegetatively, undetected diseases in stock plants and budwood can severely affect plant quality and yield. Culture indexing for bacterial and fungal pathogens and indexing for viruses and viroids can greatly reduce losses due to these diseases in commercial operations.

a. Culture and Virus Indexing in Greenhouse Crops

In 1941, H. G. Johnson improved large-scale techniques for use in the floriculture industry to prevent losses due to Verticillium wilt (Raju and Olson, 1985), one of the most serious diseases of greenhouse and field-grown chrysanthemum (Dimock, 1940). Indexed plants were soon placed in trial blocks in greenhouses throughout the U.S. Later, indexing systems were developed to prevent losses due to other diseases such as vascular wilt of carnation (caused by *Fusarium oxysporum* f. sp. *dianthi*), carnation bacterial wilt (caused by *Pseudomonas caryophylli*), Verticillium wilt of geranium *(V. albo-atrum)*, and bacterial blight of geranium *(Xanthomonas campestris* pv. *pelargonii)* (Raju and Olson, 1985).

Bacterial blight of geranium attacks the leaves and stems of geranium and can move quickly throughout production greenhouses via splashing water and contaminated tools. Prior to the

Table 1 Diseases Associated With Seeds

Location of Pathogen	Disease
Mixed with debris on seed surface	Blind seed of ryegrass
	Fusarium solani chlamydospores with bean debris
	Microslerotia of *Verticillium*
	Nematode cysts
	Sclerotia of *Rhizoctonia, Sclerotium, Sclerotinia,* and ergot
	Smut spores
	Tobacco mosaic virus
On the seed surface	Angular leafspot of cotton
	Angular leafspot of cucumber
	Bacterial spot of tomato
	Bean halo blight
	Fusarium spores on asparagus and China aster
	Rust spores
	Smut spores
	Stem nematode
	Tomato bacterial canker
Shallow penetration of seed coat	Bean anthracnose
	Black leg of crucifers
	Downy mildew of lima bean
	Karnal smut of wheat
	Rice white tip nematode
	Tomato bacterial canker
Within the seed	*Ascochyta* in lentil and chickpea
	Bacterial blight of carrot
	Black chaff of wheat
	Black rot of cabbage
	Fusarium spp.
	Heterosporium on nasturtium
	Late blight of celery
In the embryo	Anther mold of red clover
	Barley stripe virus
	Epichloe typhina and other endophytes
	Lettuce mosaic virus
	Loose smut of wheat and barley
	Pea seed-borne mosaic virus
	Peanut stripe virus

Adapted from Maloy, O. C., *Plant Disease Control, Principles and Practice,* John Wiley & Sons, New York, 1993.

introduction of commercially available culture indexed cuttings, this disease frequently attained epidemic proportions. Even today, this disease causes extensive losses when sanitation is poor or indexed cuttings are not used for propagation. Geraniums and other greenhouse crops can be culture indexed by aseptically removing thin slices of tissue from the base of cuttings just prior to planting. The tissue is plated onto nutrient medium, and plates are examined for fungal or bacterial contamination after 14 days. Plants from which contaminated slices were obtained are discarded along with contaminated potting medium containers. New cuttings are later taken from plants that passed the first culture index, and the entire process is repeated two times before mother plants are culture-index (CI) certified (Horst and Nelson, 1985).

The development of virus indexing began in the late 1940s in response to the development of a stunt disease in carnation (Raju and Olson, 1985) that was later shown to be caused by a viroid (Diener and Lawson, 1973). Virus indexing, which is accomplished through grafting and mechanical transmission of sap to indicator plants (Hollings, 1965), is often part of a complete indexing program that includes culture indexing, heat therapy, and meristem tip culture (Lawson, 1981; Raju and Olson, 1985). Once plants are successfully indexed, they must be maintained in programs that employ good sanitation and pest control techniques to prevent reinfection.

Table 2 Disease Tolerances for Foundation and Certified Seed Potatoes in Maine[a] and Minnesota[b]

Disease	Maine Foundation %	Maine Certified %		Minnesota Foundation %	Minnesota Certified %
Spindle tuber	0	0.25		0	0.1
Severe mosaic				0.3	1.0
Leafroll				0.3	0.3
Total of severe mosiac and leafroll viruses	0.5	5	Total of all viruses	0.3	1.0

[a] Maine Seed Potato Certification, Department of Agriculture, Food and Rural Resources, Augusta, Maine, 1996.
[b] Minnesota Directory of Certified Seed Potatoes, Minnesota Department of Agriculture, St. Paul, Minnesota, 1996.

3. *Certification Programs*

a. *Seed Potato Certification*

Seed potato certification programs in the U.S. represent some of the most extensive certification efforts ever developed. The cut surfaces of tubers and their high water content make seed pieces vulnerable to infection. Certification programs represent voluntary agreements between state certifying agencies and growers, and were originally developed in the early 1900s to reduce the incidence of diseases caused by viruses (Maloy, 1993). Later efforts broadened the scope of these programs to include other tuber-borne diseases (Shepard and Claflin, 1975).

Seed potato certification programs set minimum standards or tolerance limits for diseases, which may vary from state to state. For example, in Maine the minimum tolerance at postharvest for spindle tuber viroid is 0.25% for certified seed and 0% for foundation seed (Table 2). In Minnesota, however, the tolerance level for this pathogen in certified seed is 0.10%. The current tolerance limit for *Phytophthora infestans*, the causal agent of late blight of potato, is 1%. Even at this level, *P. infestans* can spread rapidly in the field under the right environmental conditions, and officials are now concerned that this tolerance level may be too high (S. Johnston, personal communication).

Depending on the state, agency personnel inspect fields for diseased plants two or three times during the growing season. If tolerances are exceeded at the last inspection, the crop is not eligible for certification (Shepard and Claflin, 1975). Potatoes are also inspected in storage for bacterial ring rot and other disorders. In Maine and Minnesota, the tolerance for this disease is 0%.

IV. SUPPRESSION OF EXISTING INOCULUM: ERADICATION AND SANITATION

Should exclusion or quarantine measures fail, eradication and sanitation measures are necessary to remove or destroy as much primary inoculum as possible, thereby reducing disease development in subsequent disease cycles. Although these control strategies may also fail, they may serve to delay the onset of disease and provide time for other controls to be implemented (Maloy, 1993).

A. Eradication

Eradication strategies, as mentioned above, usually involve the destruction of infected host plants and alternate and alternative hosts. Eradication also includes destruction of the pathogen itself in the form of seed treatment and soil fumigation. A discussion of these last eradication strategies can be found in Chapter 11.

1. Removal of the Pathogen and Host

a. Roguing

In situations where exotic pathogens are recently introduced to an area, roguing, or the removal and destruction of infected plants, is justifiable and may serve to eliminate the pathogen. For example, the Asiatic strain of citrus canker was contained and completely eliminated in the state of Florida for over 50 years due to an aggressive eradication policy that included prompt removal of all trees in infected nurseries and all infected mature trees in groves.

Roguing may also serve to reduce the spread of diseases that are endemic or well established in a given area. Management strategies for Dutch elm disease include prompt removal of infested trees to reduce inoculum and the number of breeding sites available for the bark beetle vector (Campana and Stipes, 1981).

b. Eradicative Pruning

Eradicative pruning, or removal of infected branches, has been used for the management of Dutch elm disease with mixed results (Hart, 1970; Himelick and Ceplecha, 1976; Van Sickle and Sterner, 1976). Indeed, Hart (1970) concluded that the high cost of trimming large trees often renders this method of control neither effective nor practical. The success of eradicative pruning for Dutch elm disease control has been attributed to the length of lesions in infected branches (Banfield, 1937; Campana, 1978) and to early detection and removal of infected tissue. Campana (1978) concluded that eradicative pruning is most successful on large trees with lower levels of disease.

Pathological pruning (removal of lower green limbs on young trees), has also been evaluated for the management of white pine blister rust. In a nine-year study conducted in Wisconsin, Lehrer (1982) reported that pathological pruning was three times more effective in disease management than was removal of canker tissue only. Hunt (1982, 1991) reported similar successes in British Columbia. Due to environmental factors, most rust cankers are limited to lower branches (Van Arsdel et al., 1956). The pruning and disposal of branches most likely to become infected removes latent infections and prevents new cankers from occurring (Lehrer, 1982). Lehrer concluded that pathological pruning could be employed as a sole control measure in low-hazard regions where currant *(Ribes)* eradication is too expensive, and could supplement *Ribes* eradication programs in regions where the potential for disease is high.

A radical pruning technique called the recepa method was employed during an attempt to eradicate coffee rust in Nicaragua (Schieber and Zentmyer, 1984a). Infected plants were sprayed with a mixture that contained paraquat and a copper fungicide and then were pruned to 1 ft in height. This technique was also used to renew old coffee plantings in South and Central America (Schieber and Zentmyer, 1984a).

2. Indirect Elimination of the Pathogen

Elimination is defined as the indirect destruction of a pathogen through the removal of alternate or alternative hosts, crop rotation, cultivation techniques, and biological control.

a. Removal of Alternate Hosts

The removal of alternate hosts is a common strategy used in attempts to eradicate rust diseases such as white pine blister rust, stem rust of wheat, and cedar-apple rust. In theory, removal and elimination of an alternate host breaks the disease cycle and thus should significantly reduce disease incidence. In the Midwestern and Northern Plains states, however, eradication of the common barberry *(Berberis vulgaris)*, which is the principle alternate host of wheat stem rust *(Puccinia graminis)*, following disease epidemics was not entirely successful due to the transport of uredospores from southern regions. In this case, elimination efforts most likely reduced the severity of rust epidemics by delaying the onset of disease by 10 days and by reducing initial levels of inoculum (Roelfs, 1982).

Eradication efforts to remove the alternate host of white pine blister rust began in 1915, and by the mid-1930s, 20,000 workers were employed to remove currant and gooseberry bushes from surrounding woodlands (Maloy, 1993). Although tremendous resources were expended to eliminate the alternate host these efforts were largely unsuccessful, in part because the alternate host produces a large number of spores per plant that can, as in the case of wheat stem rust, be disseminated long distances.

Cedar-apple rust, caused by the fungus *Gymnosporangium juniperi-virginianae*, became a common problem on both of its alternate hosts in the 1800s. By 1940, nine states restricted the use of juniper species in the vicinity of apple orchards, which was unpopular with landowners with red cedar and juniper plantings. Once resistant varieties and effective fungicides were developed, however, eradication of the alternate host was discontinued (Maloy, 1993).

b. Removal of Alternative Hosts

Alternative hosts often serve as reservoirs of pathogens important to economic crops. For example, tomato spotted wilt virus (TSWV) and impatiens necrotic spot virus (INSV) thrive in weed hosts in and around greenhouses and are spread to bedding plants by thrips (Bitterlich and MacDonald, 1993; Brown, 1988). Management strategies designed to control TSWV and INSV includes the elimination of both weeds and thrips vectors (Johnston et al., 1990; Thomas, 1991). Alternative hosts also play an important role in the development of diseases caused by the xylem-limited bacterium *Xylella fastidiosa*. One such disease, Pierce's disease of grapevines, can limit grape production in some regions of California. To manage this disease, grapes should not be planted within 100 m of permanent water where riparian weeds (an alternative host) and leafhopper vectors persist (Hopkins, 1989; Raju et al., 1983).

c. Crop Rotation

Historically, crop rotation has been an important management tool for the suppression of soil-borne diseases caused by pathogens that persist in the soil for several growing seasons. For example, in ancient Incan culture, rotation of potatoes and other crops were regulated by law and could be grown only once every seven years. It was later hypothesized that this practice was likely initiated due to the golden nematode, which is native to South America (Schumann, 1991; Thurston, 1990). In the areas of New York state where the golden nematode has been introduced, crop rotation has played an integral role in the success of a management scheme that includes determination of threshold populations in the soil, the use of nematicides, and, when available, planting of resistant varieties (Brodie and Mai, 1989; Evans and Brodie, 1980).

d. Cultivation

As disease epidemics develop, growers frequently modify their cultivation practices to accommodate new control strategies or attempt to reduce environmental conditions conducive to disease development. The development of coffee rust in Brazil forced growers to increase spacing between planting rows to reduce humidity and accommodate spray equipment (Schieber 1975; Schieber and Zentmyer, 1984a).

Web blight of bean, caused by the fungus *Thanatephorus cucumeris*, is managed by traditional farmers in Central and South America by a practice known as *frijol tapado*, or "covered beans" (Thurston, 1990). Bean seeds are cast into selected patches of broadleaf weeds, which are then cut into a mulch that prevents weed growth and conserves soil moisture as the beans grow. The mulching effectively prevents movement of soil particles by splashing, which is the primary means of disseminating the pathogen. In the absence of disease pressure, *frijol tapado* generally produces lower yields than more modern, clean cultivation practices. The method requires minimal inputs, however, and is well suited to small farms with few resources.

e. Biological Control

In the broadest sense, biological control means to reduce disease development using biological mechanisms or organisms other than humans (Campbell, 1989). This includes the use of tillage systems that affect microbial activity, direct addition of antagonistic or growth-promoting organisms to production systems, the use of chemicals that change microflora, and plant breeding (Campbell, 1989). A more in-depth treatment of biological control methods can be found in Chapters 4, 5, and 6.

3. Destruction of Pathogens

Plant pathogens can be directly destroyed using seed treatments, pesticides, and soil fumigants. In an integrated program, these methods are used in combination with cultural practices to enhance disease management, and are discussed in Chapter 3.

B. Sanitation

Physical measures such as raking and removal of infested debris, burying infected plant material, field burning, and surface disinfection of propagation tools, equipment, pots, greenhouse benches, and storage areas are sanitation practices often used to eliminate sources of disease inoculum. Sanitation is often labor intensive, is difficult to automate, and requires a special "state of mind" by the grower to be employed effectively (Palti, 1981). Sanitation is of particular importance in the ornamental production industries where diseases are readily spread on plant debris, infested pots, and pruning tools. In the tropics, sanitation is important for the management of root diseases such as Armellaria root rot, which readily colonizes decaying timber and can be spread to economic hosts (Palti, 1981). Palti (1981) discusses several rules that can be followed to avoid disease spread:

1. Treatments should be applied to less severely infected fields first, followed by more severely infected fields;
2. No treatments should be performed when the foliage is wet;
3. Equipment should not be driven from infected fields to other areas when the ground is wet; and
4. All packing equipment should be cleaned of soil, debris, and disinfested.

1. Removal of Debris

The raking and removal of infested plant debris is a management strategy often recommended for the suppression of foliar diseases in landscape and fruit tree plantings. This sanitation measure is of primary value when inoculum is located in above-ground parts. For example, the causal agent of cherry leaf spot *(Blumeriella jaapii*; syn. *Coccomyces hiemalis)* overwinters on leaf debris and produces abundant inoculum in the spring (Sztejnberg, 1986). The beneficial effects of sanitation (removal of fallen leaves prior to budbreak) for control of this disease were demonstrated by Keitt in 1937. In separate studies, Stanosz (1990, 1992) assessed the impact of litter burning on disease development in cherry seedlings and also discussed the merits of sanitation in cherry nursery production. Sanitation is also important in the management of greenhouse diseases such as Botrytis leaf blight. *Botrytis* readily colonizes decaying and senescent tissue and produces abundant spores

Table 3 Diseases in Perennial Seed Crops Controlled by Field Burning in Oregon

Crop	Burning Started	Disease
Perennial ryegrass	1948–1949	Blind seed disease Ergot Silver top Multiple leaf diseases
Tall fescue	1949–1950	Blind seed disease Ergot Silver top Multiple leaf diseases
Bluegrasses	1950s	Ergot Silver top Multiple leaf diseases
Chewings and red fescue	1950s	Silver top Seed nematode Ergot Multiple leaf diseases
Colonial bentgrass	1954	Ergot Silver top Multiple leaf diseases
Orchardgrass	1958	Ergot Silver top Multiple leaf diseases

Adapted from Hardison, J. R., *Plant Dis.*, 64, 641, 1980.

that can spread throughout a greenhouse under environmental conditions conducive to disease development (Maude, 1980).

2. Burying

In theory, the burial of infected debris places inoculum below the plow layer where it can have minimal contact with host plants. Resting structures such as sclerotia, however, can survive in soil for extended periods, and may be infective if returned close to the soil surface by plowing (Palti, 1981).

3. Field Burning and Flaming

Removal of infected plant residue is sometimes accomplished by burning. This practice has been particularly useful in grass seed fields where other control measures are not feasible or commonly used (Hardison, 1980). In the 1940s and 1950s, burning as a disease control practice was routinely practiced by grass-seed producers in western sections of the U.S. and in Minnesota. For diseases such as blind seed disease and ergot, burning removes inoculum that survives on unharvested seed (Hardison, 1980; McGee, 1981). Unfortunately, open field burning in grass-seed production areas has been severely curtailed during the past decade due to safety and air pollution concerns (personal communication, W. Meyer). A list of some of the grass diseases controlled by burning can be found in Table 3.

Open fires have also been used to manage diseases such as brown spot needle blight of longleaf pine (caused by *Scirrhia acicola*). Burning also reduces the incidence of fusiform rust on southern pine by the killing of lower branches of saplings where most fusiform cankers occur. In addition, burning of logging slash destroys trees that remain after logging and which may be infected with dwarf mistletoe, a serious problem of western pines (Hardison, 1976).

Burning can be a cost-effective method of disease control when residues are dry enough to burn without the use of accelerants such as flame throwers, but burning rarely destroys all of the infective material present in the field (Palti, 1981). In addition, burning may be objectionable from an environmental standpoint by enhancing erosion on slopes, removing potential sources of organic matter (Palti, 1981), and increasing pollutants in the air.

C. Do Eradication Programs Work?

When quarantines and exclusion methods fail, eradication of newly introduced pathogens generally involves the extensive removal of all infected material as well as alternate or alternative hosts. In theory, eradication should eliminate the pathogen and prevent further economic loss. Often, however, these measures serve, at best, to delay the onset of an epidemic. Some examples of eradication programs conducted in the U.S., with varying success, are described below.

1. Wheat Stem Rust

About 200 years before Anton deBary associated stem rust of wheat with its alternate host, farmers in seventeenth-century France observed that cereals were more severely affected by rust when planted close to barberry bushes. The incidence of this disease decreased when, in 1660, the French government passed legislation that permitted the eradication of barberry plants. Similar local laws were passed from 1727 to 1766 in Connecticut, Massachusetts, and Rhode Island. Enforcement of these laws lapsed, however, as grain production moved to the western territories. Barberry plants were also introduced to these territories, and as a result, stem rust epidemics began to occur on a yearly basis (Schumann, 1991).

Eradication programs for stem rust of wheat were reemphasized during World War I as food supplies became limited. Following a particularly severe epidemic in 1916, a number of states passed local legislation that permitted the removal of barberry plants to reduce disease incidence (Roelfs, 1982). In 1918, Federal Domestic Quarantine 38 provided for eradication measures in 19 Midwestern and Plains states. This program was also designed to prevent the movement of barberry plants between states and established cultivar testing programs to screen germplasm for resistance to this disease.

During the early stages of the eradication program, barberry bushes in the immediate vicinity of small grains were felled at ground level and the cut crowns were sprayed or dusted with ammonium sulfamate (Ammate™) (Maloy, 1993). This had a great impact on reducing disease development. In later eradication efforts, bushes not in immediate contact with wheat or seedlings were removed. By 1955 almost half a billion barberry bushes had been destroyed (Maloy, 1993).

By 1977, the use of resistant wheat cultivars and a reduction in the number of barberry plants reduced disease incidence to the point where the federal program became unnecessary. Responsibility for the eradication program was then left up to individual states. Although the disease was never totally eliminated by the eradication program, eradication efforts have significantly affected stem rust epidemics by (1) delaying disease onset; (2) reducing the initial inoculum level; (3) decreasing the number of the polygenic races; and (4) stabilizing pathogenic races (Roelfs, 1982).

2. Citrus Canker

Citrus canker (known as Asiatic citrus canker or canker A) has been reported in the citrus-growing regions of Asia, South Africa, Australia, South America, and North America. In the U.S., the disease was first identified in a Texas grove in 1910 (Graham and Gottwald, 1991; Schoulties et al., 1987). By 1914, citrus canker had spread to citrus groves and nurseries throughout Florida. Although growers initially burned trees to eliminate the disease, the State of Florida and the federal government initiated a citrus canker eradication campaign in 1915. Similar programs were begun in other states affected by the disease (Schoulties et al., 1987).

Through the citrus canker eradication program, groves throughout the state of Florida were inspected, and infected trees were destroyed on site. If the disease was found in a nursery, all trees in the nursery were destroyed. Workers followed strict sanitation guidelines, and quarantines were enforced. By 1933, approximately $6 million had been spent on eradication measures, and

Florida was declared free of the disease. The U.S. was declared citrus canker-free in 1947 (Graham and Gottwald, 1991). Federal and state laws, however, continued to prohibit the import of citrus in any form (with the exception of seed) from canker-infested countries. Indeed, between 1971 and 1983, officials intercepted infected plant material 171 times at Florida ports (Schoulties et al., 1987).

Eradication efforts implemented in other parts of the world have also been successful to varying degrees. Australia and New Zealand eliminated the disease after 9 and 12 years, respectively. In Brazil, however, periodic outbreaks of citrus canker continued to occur, and Argentina abandoned its eradication program in 1978 after it became apparent that the disease had become endemic (Schoulties et al., 1987; Stall and Civerolo, 1991).

For 50 years, the citrus canker eradication program in Florida was a success. In 1984, however, citrus canker was identified at a 23-ha citrus nursery in southern Florida. Although it is now known that the nursery strain of this disease (now called citrus bacterial spot) differs from the Asiatic strain in symptom development and virulence, its diagnosis triggered the implementation of the Citrus Canker Disease Action Plan. This plan, developed in 1982 by USDA-APHIS in cooperation with citrus-producing states, was intended to ensure that regulatory personnel were prepared to eradicate canker should it ever reappear in the U.S. Under this plan (from 1984 to 1985), eradication measures for the nursery form of citrus canker included destruction of all trees within infected nurseries. In addition, all trees exposed to the nursery strain and sold prior to the eradication program were also destroyed. In all, nearly 20 million trees were destroyed at a cost of $94 million (Graham and Gottwald, 1991; Schoulties et al., 1987).

By late 1985, researchers observed that the nursery form of citrus canker did not infect trees in mature groves and that nursery trees themselves were not always seriously affected. As a result, a risk-assessment policy change was implemented to minimize the destruction of trees by evaluating factors such as the amount and location of infection, aggressiveness of the pathogen, worker activity in the infested area, and sanitation measures used by workers (Schoulties et al., 1987). Because of this, many trees that would have been destroyed under the earlier eradication policy were saved. When it became apparent that the nursery form of this disease was not a threat to commercial citrus production, eradication measures were discontinued.

In 1986, however, the Asiatic form (canker A) of citrus canker reappeared, affecting trees in residential plantings and groves in western Florida near the Gulf coast (Schoulties et al., 1987). Eradication policies formulated for the nursery form were immediately applied to this situation, resulting in the destruction of 1645 infected trees, the severe pruning of 1111 exposed trees within 15.2 m of the infected trees, and the voluntary destruction of an additional 3067 exposed grove trees. In addition, movement of fresh fruit was restricted, and lawn care services were under orders to practice proper sanitation techniques to prevent spread of the disease between landscapes (Schoulties et al., 1987).

Were the citrus canker eradication programs successful? The fact that the citrus industry survived several major outbreaks of this disease is proof that eradication programs can work if properly implemented. After the first eradication program was complete, the disease did not reappear in U.S. for 50 years. Although earlier eradication programs relied on the destruction of trees to eradicate the disease, improved information on disease development and spread have led to the enactment of more rational strategies based on quarantine and surveillance.

3. White Pine Blister Rust

In 1906, white pine blister rust was discovered in New York. By 1915, eradication crews removed currants *(Ribes)* and infected pines throughout pine forests in the northern U.S. and California. As many as 20,000 workers removed wild currant and gooseberry bushes prior to World War II. *Ribes* were removed at first by hand; later, plants were removed using bulldozers, hand-sprayed with sodium chlorate, or sprayed by helicopter with 2,4-D or 2,4,5-T (Maloy, 1993).

In the long run, the white pine blister rust eradication program was unsuccessful due to several factors. First, it was discovered that aeciospores were able to travel a mile or more to infect *Ribes* in distant locations. Second, a long latent period of infection made detection and removal of

Table 4 Detection Date and Possible Means of Introduction of Coffee Rust to Countries in the Western Hemisphere

Country	Detection Date	Means of Introduction
Brazil	1970	Infected plants from Africa
Paraguay	1972	Wind
Argentina	1972	Wind
Nicaragua	1976	Humans (migrant workers and others)
Peru	1979	Infected seedlings
Bolivia	1978	Infected seedlings
El Salvador	1979	Humans
Honduras	1980	Humans
Guatemala	1981	Humans and wind
Mexico	1981	Humans and wind

Adapted from Schieber, E. and Zentmyer, G. A., *Coffee Rust in The Americas,* Fulton, R. H., Ed., APS Press, St. Paul, MN, 1984.

infected pines difficult. Third, the pathogen was well established in pine-growing regions of the country long before the eradication program began. As a result, the federal eradication program for this disease was canceled in 1977 (Maloy, 1993).

4. Eradication Efforts for Coffee Rust in Latin America

Coffee rust, caused by *Hemileia vastatrix,* is a fungal disease that results in severe defoliation and yield reductions. Its only known host is coffee (*Coffea* spp.), a crop that accounts for more than $15 billion in international trade and, after oil, is the second most important commodity worldwide (Schieber and Zentmyer, 1984a).

Coffee rust most likely originated on wild coffee plants growing in the Lake Victoria region of East Africa and in Ethiopia (Schieber, 1972). It was first reported on cultivated coffee *(Coffea arabica)* in Ceylon (Sri Lanka) in 1869. Due to lack of effective disease control measures, by 1878 coffee production was reduced by more than 75% (Schieber, 1972). The disease had spread to coffee plantations in southeast Asia, India, and the Philippines by 1889, and by 1954 had moved through coffee-growing regions of East Africa. Coffee rust was reported in Nigeria and Angola in the mid-1960s.

In the Western Hemisphere, coffee rust was not reported until 1970, when plant pathologist Arnaldo Medeiros found the disease in the state of Bahia in Brazil (Schieber and Zentmyer, 1984b). By 1971, *H. vastatrix* had spread to Sao Paulo and Paraná, two major coffee growing areas in Brazil. *Hemileia vastatrix* is disseminated chiefly by movement of infected plant material, prevailing winds, and on a local level by rain water (Table 4). Original introduction of the fungus to Brazil may have been due to the shipment of infected plants from Africa. The disease moved quickly from Brazil into Paraguay and Argentina with prevailing winds. Due to geographical barriers, however, spread of the disease to Nicaragua and El Salvador was most likely due to human activity (migrant workers and others) (Schieber and Zentmyer, 1984a).

Plant pathologists had predicted the movement of *H. vastatrix* from Africa to the Western Hemisphere 20 years prior to its discovery in Brazil (Schieber, 1972). Although Riker proposed that certain eradication measures be enacted when the disease was first discovered, eradication was not practiced in Latin America until the disease was detected in the Carazo region of Nicaragua in 1976 (Schieber, 1972). Based on legislation passed in 1965, a law was enacted which required six preventive applications of copper fungicide annually in the coffee-producing Pacific region and five applications in the northern region (Llano, 1977). Infected plants and surrounding vegetation were sprayed with a mixture of paraquat, copper fungicide, and diesel fuel. Infected plants were also pruned to a height of 1 ft (a technique known as the recepa method). In addition, a 30-m buffer zone was sprayed with a mixture of a systemic and a protective fungicide (Llano, 1977). Although these measures contained the disease in the Caraco region of Nicaragua for the first year, within two years the rust was detected throughout the country and eradication programs were stopped (Schieber and Zentmyer, 1984a).

Although it was known that general eradication measures were usually not effective against wind-disseminated pathogens like rusts, they were enacted in Nicaragua to delay the relative spread and buildup of the pathogen. This allowed coffee growers in other central American countries time to utilize new planting and pruning techniques that were devised to permit more feasible control of this disease.

REFERENCES

Anagnostakis, S. L., Chestnuts and the introduction of chestnut blight, *North. Nut Grow. Assoc.*, 1993.

Banfield, W. M., Distribution of spores of wilt-inducing fungi throughout the vascular system of the elm by the sap stream, *Phytopathology*, 27, 121, 1937.

Bitterlich, I. and MacDonald, L. S., The prevalence of tomato spotted wilt virus in weeds and crops in southwestern British Columbia, *Can. Plant Dis. Surv.*, 73, 137, 1993.

Brodie, B. B. and Mai, W. F., Control of the golden nematode in the United States, *Annu. Rev. Phytopathol.*, 27, 443, 1989.

Brown, L. G., Tomato Spotted Wilt Virus in Ornamentals, Circ. 313, Department of Agriculture and Conservation Services, Tallhassee, FL, 1988.

Campana, R. J., Control tactics in research and practice. III. Eradicative pruning, in *Dutch Elm Disease Perspectives after 60 Years*, Sinclair, W. A. and Campana, R. J., Eds., Cornell Univ. Agric. Exp. Stn. Search (Agriculture), 8, 33, 1978.

Campbell, R., *Biological Control of Microbial Plant Pathogens*, Cambridge University Press, Cambridge, 1989, chap. 1.

Diener, T. O. and Lawson, R. H., Chrysanthemum stunt: a viroid disease, *Virology*, 51, 94, 1973.

Dimock, A. W., Importance of *Verticillium* as a pathogen of ornamental plants, *Phytopathology*, 30, 1054, 1940.

Evans, K. and Brodie, B. B., The origin and distribution of the golden nematode and its potential in the U.S.A., *Am. Potato J.*, 57, 79, 1980.

Fry, W. E., *Principles of Plant Disease Management*, Academic Press, New York, 1982.

Graham, J. H. and Gottwald, T. R., Research perspectives on eradication of citrus bacterial diseases in Florida, *Plant Dis.*, 75, 1193, 1991.

Hardison, J. R., Fire and flame for plant disease control, *Annu. Rev. Phytopathol.*, 14, 355, 1976.

Hardison, J. R., Role of fire for disease control in grass seed production, *Plant Dis.*, 64, 641, 1980.

Hart, J. H., Attempts to control dutch elm disease by pruning, *Plant Dis. Rep.*, 54, 985, 1970.

Himelick, E. B. and Ceplecha, D. W., Dutch elm disease eradication by pruning, *J. Arboric.*, 2, 81, 1976.

Hollings, M., Disease control through virus-free stock, *Annu. Rev. Phytopathol.*, 3, 367, 1965.

Hopkins, D. L., *Xylella fastidiosa*: xylem-limited bacterial pathogen of plants, *Annu. Rev. Phytopathol.*, 27, 271, 1989.

Horst, R. K. and Nelson, P. E., *Diseases of Geranium*, Cornell Coop. Extension Info. Bull. 201, Cornell University Press, Ithaca, NY, 1985.

Hunt, R. S., Operational control of white pine blister rust by removal of lower branches, *For. Chron.*, 67, 284, 1991.

Hunt, R. S., White pine blister rust in British Columbia. I. The possibilities of control by branch removal, *For. Chron.*, 58, 136, 1982.

Johnston, S. A., Ghidiu, G. M., and Vasvary, L. M., Prevent greenhouse spread of tomato spotted wilt virus from ornamentals to tomato and pepper transplants, *N.J. Grower*, 13, 1, 1990.

Kahn, R. P., Exclusion as a plant disease control strategy, *Annu. Rev. Phytopathol.*, *29*, 219, 1991.

Kahn, R. P., Ed., *Plant Protection and Quarantine: Biological Concepts*, Vol. 1, CRC Press, Boca Raton, FL, 1989a.

Kahn, R. P., Ed., *Plant Protection and Quarantine: Selected Pests and Pathogens of Quarantine Significance*, Vol. 2, CRC Press, Boca Raton, FL, 1989b.

Kahn, R. P., Ed., *Plant Protection and Quarantine: Special Topics*, Vol. 3, CRC Press, Boca Raton, FL, 1989c.

Kahn, R. P., Plant quarantine: principles, methodology, and suggested approaches, in *Plant Health and Quarantine in International Transfer of Genetic Resources*, Hewitt, W. B. and Chiarappa, L., Eds., CRC Press, Boca Raton, FL, 1977, 289.

Keitt, G. W., Blodgett, E. C., Wilson, E. E., and Magie, R. O., The epidemiology and control of cherry leaf spot, *Wis. Agric. Exp. Stn. Res. Bull.*, 132, 56, 1937.

Lawson, R. H., Controlling virus diseases in major international flower and bulb crops, *Plant Dis.*, 65, 780, 1981.

Lehrer, G. F., Pathological pruning: a useful tool in white pine blister rust control, *Plant Dis.*, 66, 1138, 1982.

Llano, A., The orange coffee rust in Nicaragua, *Plant Dis. Rep.*, 61, 999, 1977.

Maloy, O. C., *Plant Disease Control, Principles and Practice*, John Wiley & Sons, New York, 1993.

Mathys, G. and Baker, E. A., An appraisal of the effectiveness of quarantines. *Annu. Rev. Phytopathol.*, 18, 85, 1980.

Maude, R. B., Disease control, in *The Biology of Botrytis*, Coley-Smith, J. R., Verhoeff, K., and Jarvis, W. R., Eds., Academic Press, London, 1980, chap. 10.

McCubbin, W. A., *The Plant Quarantine Problem*, Vol. 11, Verdoorn, F., Ed., Ejnar Munksgaard, Copenhagen, 1954.

McGee, D. C., Seed pathology: its place in modern seed production, *Plant Dis.*, 65, 638, 1981.

Palti, J., *Cultural Practices and Infectious Crop Diseases*, Springer-Verlag, Berlin, 1981, chap. 2.

Plucknett, D. L. and Smith, N. J. H., Quarantine and the exchange of crop genetic resources, *Bioscience*, 39, 16, 1989.

Raju, B. C., Goheen, A. C., and Frazier, N. W., Occurrence of Pierce's disease bacteria in plants and vectors in California, *Phytopathology*, 73, 1309, 1983.

Raju, B. C. and Olson, C. J., Indexing systems for producing clean stock for disease control in commercial floriculture, *Plant Dis.*, 69, 189, 1985.

Rich, A. E., *Potato Diseases*, Academic Press, New York, 1983, chap. 5.

Roelfs, A. P., Effects of barberry eradication on stem rust in the United States, *Plant Dis.*, 66, 177, 1982.

Rosenberg, D. Y., The interaction of state and federal quarantines in the U.S., in *Plant Protection and Quarantine: Special Topics*, Vol. 3, Kahn, R. P., Ed., CRC Press, Boca Raton, FL, 1989, 59.

Saettler, A. W., Schaad, N. W., and Roth, D. A., Eds., *Detection of Bacteria in Seed and Other Planting Material*, APS Press, St. Paul, MN, 1989.

Schieber, E., Economic impact of coffee rust in Latin America, *Annu. Rev. Phytopathol.*, 10, 491, 1972.

Schieber, E., Present status of coffee rust in South America, *Annu. Rev. Phytopathol.*, 13, 375, 1975.

Schieber, E. and Zentmyer, G. A., Coffee rust in the Western Hemisphere, *Plant Dis.*, 68, 89, 1984a.

Schieber, E. and Zentmyer, G. A., Distribution and spread of coffee rust in Latin America, *Coffee Rust in the Americas*, Fulton, R. H., Ed., APS Press, St. Paul, MN, 1984b, 1.

Schoulties, C. L., Civerolo, E. L., Miller, J. W., Stall, R. E., Krass, C. J., Poe, S. R., and DuCharme, E. P., Citrus canker in Florida, *Plant Dis.*, 71, 388, 1987.

Schumann, G. L, *Plant Diseases: Their Biology and Social Impact*, APS Press, St. Paul, MN, 1991.

Shain, L., Strategies for enhancing dissemination of hypovirulence in *Endothia parasitica*, in *Proc. U.S. Forest Service American Chestnut Cooperators Meet.*, Smith, H. C. and MacDonald, W. L., Eds., West Virginia University Books, Morgantown, 1982, 175.

Shepard, J. F. and Claflin, L. E., Critical analyses of the principles of seed potato certification, *Annu. Rev. Phytopathol.*, 13, 271, 1975.

Siddiqui, I. A., *National Plant Board Plant Quarantine, Nursery Inspection, and Certification Guidelines*, National Plant Board, Glouster, OH, 1995.

Sinclair, W. A. and Campana, R. J., *Dutch Elm Disease: Perspectives After 60 Years*, Cornell Univ. Agric. Exp. Stn. Search (Agriculture), 8, 1978.

Stall, R. E. and Civerolo, E. L., Research relating to the recent outbreak of citrus canker in Florida, *Annu. Rev. Phytopathol.*, 29, 399, 1991.

Stanosz, G. R., Relationships among forest litter burning, seedling density, and black cherry leafspot severity in northern Pennsylvania, *Phytopathology*, 80, 991, 1990 (Abstr.).

Stanosz, G. R., Effect of cherry leaf spot on nursery black cherry seedlings and potential benefits from control, *Plant Dis.*, 76, 602, 1992.

Stipes, R. J. and Campana, R. J., Eds., *Compendium of Elm Diseases*, APS Press, St. Paul, MN, 1981, 18.

Sztejnberg, A., Etiology and control of cherry leaf spot disease in Israel caused by *Cercospora circumscissa*, *Plant Dis.*, 70, 349, 1986.

Thomas, S. H., TSWV management relies on control of western flower thrips, *Greenhouse Manager*, Sept. 1991, 116.

Thurston, H. D., Plant disease management practices of traditional farmers, *Plant Dis.*, 74, 96, 1990.

Van Arsdel, E. P., Riker, A. J., and Patton, R. F., The effects of temperature and moisture on the spread of white pine blister rust, *Phytopathology*, 46, 307, 1956

Vander der Plank, J. E., *Plant Diseases: Epidemics and Control*, Academic Press, New York, 1963, chap. 1.

Van Sickle, G. A. and Sterner, T. E., Sanitation: a practical protection against Dutch elm disease in Fredericton, New Brunswick, *Plant Dis. Rep.*, 60, 336, 1976.

Waterworth, H. E. and White, G. A., Plant introductions and quarantine: the need for both. *Plant Dis.*, 66, 87, 1982.

Whetzel, H. H., The terminology of phytopathology, in *Proc. Int. Congr. Plant Sciences*, Vol. 2, Duggar, B. M., Ed., George Banta Publishing, Menasha, WI, 1926, 1204.

CHAPTER 11

Agronomic Measures

Charles M. Rush, Giovanni Piccinni, and Robert M. Harveson

1-8493-2627-3/97/$0.00+$.50

I. INTRODUCTION

For the majority of people who have ever lived, plant diseases have been of little importance. It has been estimated that more than 90% of the population since the beginning of time survived by hunting, fishing, and gathering wild plants for food. Plant disease probably had minimal impact on hunters and gatherers because of the diversity of food types that comprised their diets; population densities were low, consisting of small groups of extended families; and land was not a limiting factor. Although these nomadic groups practiced various forms of environmental manipulation, their overall impact was small and their life style was essentially in equilibrium with their environment (Lawton and Wilke, 1979). Then approximately 10,000 years ago, things began to change drastically.

For 100,000 generations, human beings did not practice agriculture. However, approximately 10,000 years ago in both the old and new worlds, instead of merely gathering food humans began to produce it. Reasons for this change are uncertain and open to debate but the impact it had on the development of civilization was dramatic. As people began the cultivation of crops, even in its simplest form such as reseeding wild grasses, they recognized the efficiency and benefits of this type of food production. Previously, large areas of land were required to feed a single family, but agriculture allowed large numbers of people to be fed from a small area. The nomadic way of life declined and villages and towns sprang up. Excess food was easily stored by a sedentary people and populations increased (Lawton and Wilke, 1979). However, as more and more people became dependent on food which was produced in relatively small concentrated areas, their susceptibility to starvation due to disease and famine greatly increased. Therefore, those directly involved in farming spent considerable time and effort in learning and developing techniques that improved their chances of producing good crops. Eventually, farmers realized that the techniques they used could impact final yields. What they failed to realize until much later was that the techniques they used could also impact disease development. Even today, many farmers, including those in industrialized nations, do not realize that the agronomic practices they use can directly affect disease incidence and severity.

Our charge in writing this chapter was to review the effects of agronomic practices on plant disease. We chose to focus our coverage on the impact of agronomic practices on diseases caused by soil-borne fungal pathogens. We decided to further limit our coverage primarily to diseases of field crops, although diseases of some of the major vegetable crops are also included. It is our belief that in order to understand why various agronomic practices affect plant disease, it is necessary to understand why they were developed in the first place and how they affect soil characteristics and plant growth. Therefore, several of the following sections include an introductory overview of the history and importance of specific agronomic practices.

II. AGRONOMIC PRACTICES AND IMPACT ON DISEASE DEVELOPMENT

Unfortunately, as populations become more urbanized, the understanding and appreciation of agriculture declines. Many people, if they think about it at all, have a very simplistic view of agriculture: a farmer plants, the crop grows, it is harvested, and then one eats. Farming is considered a simple step-by-step process with a definite beginning and end. In reality, crop production is a complex business and correctly should be viewed as a continuum of interrelated processes. Agronomic practices such as type and length of rotation, tillage and residue management, application of fertilizer, planting date and depth, cultivar selection, frequency of irrigation, and use of pesticides all have the potential to impact crop quality and yield. Furthermore, when plant pathogens are present in the agroecosystem, these same factors have the potential to impact disease incidence and severity.

The impact of a particular agronomic practice on disease incidence and severity can result from a direct effect on the pathogen, such as providing an increased food base or physically placing the pathogen nearer an infection court. Alternatively, an agronomic practice may indirectly impact pathogen populations and, subsequently, disease severity by affecting the host or modifying the soil environment. However, more often than not, the impact of a single agronomic operation on plant disease cannot be simply classified. There is seldom a straightforward cause and effect relationship and, typically, disease development is a result of a set of complex interactions over time between the host, pathogen, and soil environment. The disease pyramid (Browning et al., 1977) is still a valid and valuable concept, and understanding the relationships among its components provides the basis for managing plant diseases through manipulation of the environment by various agronomic practices.

A. Crop Rotation and Cropping Systems

The first and perhaps most important decision any producer has to make is what crop or crops to grow on a particular piece of land and the sequence in which they will be grown. All subsequent farming operations are affected by the type and length of rotation a producer selects. Decisions concerning whether to rotate, and the crops to include in a rotation, are usually determined by ecological and economic factors (Crookston, 1984; Reeves, 1994; Unger, 1984).

The value of crop rotation has long been recognized (Bullock, 1992; Francis and Clegg, 1990; Metcalfe and Elkins, 1980; Reeves, 1994; Unger, 1984). Over 3000 years ago, the Chinese understood the value of including legumes in a rotation (Pieters, 1927), and around 300 B.C. Theophrastus the Greek reported that a bean crop enriched the soil (Metcalfe and Elkins, 1980). In the Book of Exodus 23:10, the nomadic Hebrews were instructed to sow their land for six years but to let it rest and lie fallow the seventh year. Lord Townsend of Norfolk County, England introduced a four-year rotation sometime in the 1730s that included wheat, turnip, barley, and red clover (Reeves, 1994). This rotation, known as the Norfolk rotation, increased average wheat yields in England from 540 to 1350 kg/ha by the early nineteenth century.

Although results can vary, crop rotations typically increase yields compared to continuous cropping (Crookston et al., 1991; Lund et al., 1993; Metcalfe and Elkins, 1980). Yield increases associated with crop rotation are usually attributed to factors such as increased soil nitrogen when a legume is included in the rotation, increased soil organic matter content and associated improvement of soil physical conditions, erosion control, better weed management, and a reduction in losses from diseases and pests (Metcalfe and Elkins, 1980; Reeves, 1994).

1. Continuous Monoculture Vs. Crop Rotation

For years, traditional wisdom dictated the necessity of crop rotation. Weed control and maintenance of soil fertility were two of the dominant reasons for rotating crops. However, following World War II and the development of inexpensive synthetic fertilizers and pesticides, the need for crop rotation was questioned (Arnon, 1972; Glynne, 1965; Reeves, 1994; Shipton, 1977). The use of synthetic fertilizers in monoculture systems has resulted in yields that can rival those obtained with long rotations. Furthermore, the opportunity to grow cash crops continuously or in shorter rotations offers the promise of greater economic returns per unit area of land over time. There is no question that monoculture systems can be highly profitable (Daberkow and Gill, 1989; Gill and Daberkow, 1991), and in the U.S. crops such as cotton and corn are often grown continuously without rotation. However, when crops are grown in the presence of soil-borne pathogens, the economic benefits of monoculture may be reduced or eliminated.

Yield reduction resulting from increased disease incidence and severity is often cited as the consequence of continuous monoculture (Altieri, 1987; Garrett, 1970; Shipton, 1977; Whiting and Crookston, 1993). Sewell and Melchers (1924) reported that foot rot became so severe in continuously cropped wheat (*Triticum aestivum* L.) that plots had to be abandoned because they were no longer suitable for agronomic studies. Wheat in adjacent plots that had been rotated showed no symptoms of foot rot. Take-all caused by *Gaeumannomyces graminis* (Sacc.) Arx and Olivier var. *tritici* Walker and common root rot caused by *Bipolaris sorokiniana* (Sacc.) Shoem. are diseases of wheat that are also typically worse with continuous cropping (Sutton and Vyn, 1990; Wilson and Hamblin, 1990). Sutton and Vyn (1990) conducted studies in which wheat was grown continuously or in rotation with a number of other crops. In both years of the study, incidence of take-all and yield loss was greatest in continuous wheat. Stem rot of peanut (*Arachis hypogaea* L.) (Brenneman et al., 1995), sudden death syndrome (Von Qualen et al., 1989) and brown stem rot (Whiting and Crookston, 1993) of soybean (*Glycine max* L.), black root rot of burley tobacco (*Nicotiana tabacum* L.) (Meyer et al., 1989), Verticillium wilt of cotton and potato (Wilhelm, 1975), and numerous other diseases (Shipton, 1977) are all more severe in monoculture. Although increased disease incidence associated with monoculture has been attributed to a number of biotic and abiotic factors, an increase in pathogen populations is perhaps the most frequently cited explanation.

It is not surprising that continuous cultivation of a particular crop species can increase populations of any organism capable of using that crop species as a food source. This is true for most soil-borne fungal plant pathogens (Brenneman et al., 1995; Garrett, 1970; Shipton, 1977; Whiting and Crookston, 1993). *Thielaviopsis basicola* (Berk. & Br.) Ferr. has a wide host range and causes a disease on tobacco called black root rot. Continuous cultivation of burley tobacco increased populations of *T. basicola,* and disease severity was found to be dependent on inoculum density (Meyer et al., 1989). When tobacco was rotated with corn, a nonhost to *T. basicola*, populations of the pathogen dropped and disease severity was reduced. Likewise, high populations of *Glomus macrocarpum*, which has been related to a stunting disease of burley tobacco, were associated with increased disease severity in tobacco monoculture (Hendrix et al., 1992). When tobacco was rotated with tall fescue, populations of *G. macrocarpum* and disease severity were reduced. The authors suggested that the decrease in disease incidence and severity was due to reduced inoculum density.

Without question, monoculture has the potential to increase pathogen populations, but a concomitant increase in disease incidence or severity is not guaranteed. Pathogen populations may not be strongly correlated to disease development because other factors such as rainfall, irrigation, soil temperature, or the presence of competitors or antagonists can be more directly related to disease development than pathogen populations (Ristaino, 1991; Schmitthenner, 1992; Shew, 1987; Specht and Leach, 1987; Sumner and Bell, 1994; Sumner et al., 1991). Alternatively, increased pathogen populations associated with monoculture may increase disease incidence and severity but have no impact on final yield. Whiting and Crookston (1993) studied the impact of a corn (*Zea mays* L.)-soybean (*Glycine max* L.) Merr. rotation on brown stem rot development. Disease severity increased each year in plots of continuous soybean, but disease severity did not explain differences in yield. They concluded, "Although both a yield decline and brown stem rot prevalence might coincide under soybean monoculture, it cannot be assumed that the pathogen is solely responsible for the yield decline." A similar conclusion was reached in a study evaluating the impact of common root rot on wheat yield (Wilson and Hamblin, 1990). Continuous wheat had more disease and a lower yield than a wheat-lupin rotation. However, yield differences were largely due to nitrogen levels and not strongly correlated to disease severity. Root rot of pea caused by *Fusarium solani* (Mart.) Sacc. f. sp. *pisi* (Jones) Snyd. & Hans. and charcoal rot of sorghum (*Sorghum bicolor* [L.] Moench.) caused by *Macrophomina phaseolina* (Tassi) G. Goid. are additional examples of diseases that are environmentally dominated and disease severity may be unrelated to final yield in the absence of abiotic stress factors.

Although crop rotation is often cited as a means of controlling pathogen populations and reducing disease severity, there are instances when it will do neither (Oyarzun et al., 1993; Specht and Leach, 1987; Sumner and Bell, 1986). Sumner and Bell (1986) compared a number of cropping systems in a three-year rotation for their effect on crown and brace root rot of corn caused by *Rhizoctonia solani* Kuhn, anastomosis group (AG) 2-2. Cucumber (*Cucumis sativus* L.) and turnip (*Brasseca campestris* subsp. *rapifera* [Metzg.] Sinsk.), which are nonhosts to *R. solani* AG2-2, were included in two of the rotations. Disease severity increased each year of the study and none of the rotation systems had any effect on disease incidence or pathogen populations. *R. solani* survived as well following cucumber and turnip in double crop with cowpea (*Vigna unguiculata* [L.] Walp.) and soybean as it did following peanut, a highly susceptible host. The authors concluded that rotation was ineffective in reducing disease severity because the pathogen was adapted to a wide variety of environments, i.e., hosts. They also suggested that crop rotation might not be effective in reducing inoculum potential after a pathogen is established in the soil. Results from a study by Huisman and Ashworth (1976) support this suggestion. They reported that crop rotation failed to reduce inoculum potential of *Verticillium albo-atrum* and that the rate of attrition of the pathogen was no different under a host or nonhost crop species. Because the rates of pathogen population increase are potentially greater than mortality rates, crop rotation should be most effective in "preventing the development of a large population rather than reducing large populations to small ones" (Fry, 1982). In general, diseases caused by pathogens that are saprophytically competitive and can survive in crop debris, and those that produce long-lived survival structures such as oospores or sclerotia, will be more difficult to manage by short crop rotations once pathogens are established in a field (Aycock, 1966; Brenneman et al., 1995; Oyarzun et al., 1993; Sumner and Bell, 1986).

In some instances, crop rotation can actually increase pathogen populations and/or disease incidence. Perhaps the most recognized example of this happening is with take-all decline of wheat. When wheat is rotated on a regular basis with crops such as pea and bean that are nonhosts to *G. graminis* var. *tritici*, take-all seldom becomes a problem. When wheat is grown continuously, incidence of take-all typically increases and wheat yields decrease. However, over time with continued monoculture, incidence of take-all may decrease or completely disappear even though the pathogen is still present in the field (Rothrock and Cunfer, 1986; Walker, 1975). This phenomenon has been termed take-all decline and is attributed to an increase in populations of competitive or inhibitory soil microorganisms that are selectively antagonistic to *Gaeumannomyces* (Cook and Rovira, 1976; Glynne, 1965; Shipton, 1975). Soils containing such populations are termed suppressive. Breaking the cycle of continuous wheat production with rotation crops such as alfalfa, oats, potatoes, and soybeans can reverse the "suppressive effect" and soils will become conducive to take-all (Cook, 1981; Cook and Baker, 1983; Rothrock and Cunfer, 1986; Shipton, 1977). Take-all decline has been documented in several countries and "declines" may occur with diseases other than take-all (Shipton, 1977). However, the economic consequences of producing a crop without rotation in hopes of eventually establishing a disease decline could be disastrous.

A final example of how crop rotation can increase disease severity is illustrated in a study conducted by Rovira (1986). The effect of rotation crops on incidence of bare patch caused by *R. solani* was evaluated over a three-year period in Australia. Wheat was grown continuously or rotated with peas (*Pisum sativum* L.), medic pasture (*Medicago scutellata* L.), or medic pasture + annual grasses. Rotation with peas or medic pasture reduced disease incidence, even though these crops were susceptible to *R. solani*. However, disease incidence was significantly higher and wheat yields significantly lower when wheat followed wheat or medic grass. Wheat following medic pasture + grass had the most disease and lowest yields of all rotations. Although not specifically stated, the implication was that disease incidence was related to total root biomass and crop residue. The relation between crop residue management and plant disease will be discussed at length in Section II.B.

2. Multiple Cropping Systems

Allocation of land for production of a single crop per year often makes for inefficient use of land, equipment, and financial resources (Triplett, 1986). Increasing land use efficiency and productivity can be achieved in different ways. One method is to maximize production of a major cash crop. Farmers producing cotton or corn in monoculture have taken this approach. In the absence of diseases and pests, continuous production of these crops has proven to be a very cost-effective system (Daberkow and Gill, 1989) although their long-term sustainability is questionable. A second method of increasing land use efficiency is to develop cropping systems in which the land is used throughout the year. Multiple cropping systems employ this approach to maximize productivity of the land.

Multiple cropping can be defined as any system that results in more than one crop being harvested from a single piece of land during one year. Over the last 15 to 20 years, the number of farmers implementing various multicropping systems has increased dramatically. This increase has been associated with high land costs, improved herbicide specificity, irrigation, development of short-season cultivars of several crops, poor prices for small grains, and the introduction of reduced tillage systems (Phillips and Thomas, 1984).

Although multicropping systems are introduced to increase productivity, disease pressure can eliminate the profitability of intensive cropping systems. In Georgia, grain yield in the spring crop of double-cropped corn declined 37% over a 5-year period (Sumner et al., 1990). Yield reductions were associated with stunted chlorotic plants and increased populations of *Pythium* spp. Higher populations of soil-borne fungal pathogens were attributed to increased levels of crop residue that did not have time to fully decay between crops. Changing from a double crop of corn to a monocrop eliminated symptoms of decline, demonstrating that it was the short interval between crops that impacted disease development and not the continuous production of corn per se. Soil fumigation eliminated disease symptoms and increased yields of double-cropped corn, but tillage practices, a winter crop of rye, and soil treatments with various fungicides had no effect. Other studies evaluating various disease management practices for use in multiple cropping systems have also demonstrated the difficulty of achieving adequate levels of disease control (Rothrock and Cunfer, 1986; Sumner et al., 1995a,b). Even when disease control is reported, the biotic complexity of multiple cropping systems and the probability of crop-pathogen-treatment interactions greatly complicates the interpretation of results (Minton et al., 1990; Sumner et al., 1981, 1983, 1985, 1986b, 1995a,b). Sumner et al. (1983) suggested that maximum soil pest control might not be economically feasible in some multiple-cropping systems because of the complexity of the systems.

3. Cover Crops

Cover crops are typically grown during the off-season with an annual cash crop. The main function of a cover crop is to cover the ground in order to protect soil from erosion and loss of plant nutrients through leaching and runoff (Reeves, 1994). Traditionally, cover crops have been turned under prior to planting the cash crop but, with the development of conservation tillage systems, the cash crop is frequently seeded directly into the cover with minimal or no tillage. The type of cover crop and the way it is managed can directly impact pathogen populations and disease development in the cash crop. However, the effect of a cover crop on disease can vary depending on which pathogens are present. Rothrock et al. (1995) conducted a study to evaluate the influence of cover crops on pathogen populations and incidence of seedling disease in cotton (*Gossypium hirsutum* L.). Populations of *Thielaviopsis basicola* (Berk. & Broome) Ferraris (*Chalara elegans* Nag Raj & Kendrick) and isolation frequency from diseased seedlings decreased following incorporation of a winter cover crop of hairy vetch (*Vicia villosa* Roth) compared to a winter fallow. Conversely, isolation of *R. solani* from cotton seedling and populations of *Rhizoctonia* spp. increased. Populations of *Pythium* spp. also increased, but frequency of isolation was not affected by cover crop. In this study, each of the three pathogens in the cotton seedling disease complex was affected differently by incorporation of the legume cover crop. Others have reported that legume cover crops increase populations of pathogens such as *Rhizoctonia* and *Pythium* spp. but not *Fusarium* spp. or *Macrophominia phaseolina* (Rothrock and Hargrove, 1988; Sumner et al., 1995a). In general, populations of pathogens with strong saprophytic capabilities are most likely to increase dramatically following incorporation of a nonhost cover crop.

Increased pathogen populations associated with incorporation of cover crops do not always result in increased disease incidence or severity (Specht and Leach, 1987; Sumner et al., 1995a,b). Increased populations of antagonistic fungi, bacteria, and actinomycetes following incorporation of cover crops may suppress parasitic activity of plant pathogens, resulting in less disease despite higher pathogen populations (Rickerl et al., 1992; Specht and Leach, 1987; Sumner et al., 1981, 1986a). Circumstantial evidence supporting this hypotheses has been provided, but it has not been proven experimentally.

Cover crops can impact disease in ways other than serving as a food base for plant pathogens. In the spring, soils under cover crops are typically cooler than fallowed ground at planting time, so potential for seedling diseases caused by pathogens, such as *Pythium* and *Rhizoctonia spp.*, is increased. Cover crops may also have an allelopathic effect directly on pathogens and actually decrease pathogen populations or have an inhibitory effect on seedling growth and development. A thorough review of allelopathic effects on plant growth and disease development are beyond the scope of this chapter. However, several informative articles and reviews have been published (Blum et al., 1992; Bradow, 1993; Bradow and Connick, 1990; Cook and Haglund, 1991; Fredrickson et al., 1987; Grossman, 1993; Halsall et al., 1995; Harper and Lynch, 1981; Hicks et al., 1989; Jarvis, 1989; Johnson et al., 1992; Kimber, 1973; Mayton et al., 1996; Perez and Ormeno-Nunez, 1991; Zucconi, 1993).

4. Cropping Sequence and Previous Crops

The sequence in which rotation crops are grown is often of primary interest to producers. However, from a disease standpoint, crop sequence may not be as important as the cropping interval. Long rotations are sometimes necessary to reduce disease incidence in the primary cash crop to an acceptable level. In Texas, commercial sugar beet (*Beta vulgaris* L.) production began in the early 1960s. Typically, beets were grown approximately every three years and were rotated with a variety of crops. However, in some fields, by the third sugar beet crop root rots caused by *R. solani* AG2-2, *F. oxysporum* Schlechtend. f. sp. *betae* Snyd. & Hans, and *Aphanomyces cochlioides* Drechs. were reducing yields to half of what they were originally. By lengthening the interval between beet crops from three years to six or seven, producers reduced losses to Rhizoctonia and Fusarium root rots significantly.

In other parts of the U.S. farmers have grown sugar beets every other year for decades without experiencing severe disease losses, even though pathogens were present. This type of situation is not uncommon, and one-year rotations have been reported to reduce losses caused by soil-borne fungal pathogens in a number of crops (Brenneman et al., 1995; Cook, 1981; Shipton, 1977; Sutton and Vyn, 1990; Von Qualen et al., 1989). It is unknown why short rotations are possible in some locations and not others, but it highlights the reality that any given host-pathogen combination may interact differently from one geographic location to the next, making generalized disease control recommendations risky. The large number of conflicting reports in the literature addressing the impact of specific agronomic practices on disease control supports this statement.

The crop immediately preceding a cash crop has the greatest potential to affect disease development in the cash crop. Furthermore, the potential is significantly increased if the two crops host common pathogens. Wheat, barley (*Hordeum vulgare* L.), rye (*Secale cereale* L.), and triticale (X *Triticosecale* Wittmack), but not oats (*Avena sativa* L.), all host *G. graminis* var. *tritici*. Wheat is most susceptible to infection and yield loss, followed by triticale, barley, and rye (Rothrock, 1988). Even though rye is susceptible to *G. graminis* var. *tritici*, yields are usually unaffected by take-all. However, if rye precedes wheat in rotation, significant losses to take-all can occur in the wheat. Even though rye is not severely affected by take-all, it can apparently support sufficient inoculum of *G. graminis* var. *tritici* to cause severe disease in the following wheat crop (Rothrock and Cunfer, 1991). A similar situation exists with grain sorghum, *Phymatotrichum omnivorum* (Shear) Duggar and Phymatotrichum root rot of cotton (Rush and Gerik, 1984, 1989).

Grain sorghum is generally not considered a host to *P. omnivorum* (Ezekiel and Fudge, 1938; Rogers, 1942; Taubenhaus and Ezekiel, 1932). However, Rush and Gerik (1984) found that sorghum could support sclerotial production. Strands of *P. omnivorum* which came in contact with sorghum roots formed small lesions, but the fungus did not form hyphal mantles around infected roots or develop new strands as it does after infecting roots of cotton. Sorghum was capable of supporting production of sclerotia, though not as efficiently as cotton. It was suggested that production of sclerotia by *P. omnivorum* in association with grain sorghum was the result of saprophytic survival of the fungus on root exudates (Rush and Gerik, 1984). Ectotrophic growth of fungi on roots of nonhost plants is common, and it has been well documented that pathogenic and nonpathogenic organisms are capable of surviving on root exudates (Deacon, 1974; Hale et al., 1978). Therefore, preceding a cash crop with a crop that is not host to a particular pathogen does not guarantee pathogen populations or disease incidence will decline.

In many parts of the U.S., wheat precedes sugar beets in rotation. Wheat roots deeply and does a good job extracting residual soil nitrogen. High levels of residual soil nitrogen reduce sucrose accumulation in the beet crop. This is why wheat is a favored crop to precede beets (Winter, 1984). However, wheat is susceptible to infection by *R. solani* AG4, which also causes damping-off of sugar beet seedlings. For this reason, it was suggested that wheat might not be the best crop to precede sugar beets in rotation (Rush et al., 1994). In another study, wheat increased the incidence of Rhizoctonia root rot caused by *R. solani* AG2-2 in a subsequent sugar beet crop even though it is not a host to AG2-2 (Rush and Winter, 1990). When beets were preceded by crops which produced low levels of crop residue, disease incidence was low, but high-residue crops (including wheat, sorghum, and alfalfa) resulted in high levels of disease. It was not determined whether disease incidence was more related to the previous crop per se or residue concentrations. Although crops in a rotation have been shown to impact disease incidence and, subsequently, crop quality and yield, the "rotation effect" is often impossible to separate from the impact of residue management or the "tillage effect."

B. Tillage

In the early stages of agriculture, plants that competed with food crops were flattened or pulled by hand. Later, crude implements of wood or stone were developed to facilitate planting and weed control. These tools left most organic materials on the soil surface and had minimal effect on soil structure (Unger, 1984). Over time, tillage implements became larger and had a greater impact on soil chemical, physical, and biological properties. The introduction of the moldboard plow in the 1830s ushered in a new era in which deep plowing with complete burial of surface organic matter became the "agronomically correct" way to farm. Advantages associated with this type of farming included weed control, burial of crop debris which facilitated planting, deep placement of fertilizers, and disruption of insect and disease cycles. Some believed plowing also improved rooting, soil aeration, and water storage (Phillips and Phillips, 1984; Sprague, 1986).

Certain benefits were definitely associated with moldboard plowing. However, it was expensive, energy and labor intensive, and promoted deterioration of soil structure. Indeed, the moldboard plow was cited as a major cause of the devastating soil erosion that occurred in the U.S. during the Dust Bowl days of the 1920s and 1930s (Phillips and Young, 1973). It became evident that tillage systems which left crop residues on the soil surface to serve as ground covers and reduce erosion were necessary. The synthesis of inorganic herbicides such as 2,4-D, population shifts to urban areas which depleted labor pools, and the introduction of specific no-till equipment capable of planting in uncultivated soils all sped the acceptance of various forms of conservation tillage (Phillips and Phillips, 1984). The evolution of tillage and crop residue management systems and their impact on soil physical, chemical, and biological properties has been thoroughly reviewed in numerous books and articles (Dao and Unger, 1995; Palti, 1981; Phillips and Phillips, 1984; Phillips and Young, 1973; Sprague, 1986; Stewart et al., 1993; Stewart and Moldenhauer, 1994; Unger, 1984, 1990, 1994; Unger and Kaspar, 1994; Unger et al., 1991).

1. Tillage and Residue Management

The tillage and residue management system a farmer implements will probably have a greater overall impact on the soil environment than any other facet of crop production. The manner in which a soil is tilled and crop residues managed affects a soil's physical, chemical, and biological properties which, in turn, directly affects crop growth. Tillage can also directly affect soil-borne pathogens and disease incidence and severity through changes in the soil environment (Muehlchen et al., 1990). Several in-depth reviews have been published on this subject (Boosalis et al., 1986; Rothrock, 1992b; Sumner et al., 1981, 1986b; Watkins and Boosalis, 1994).

The literature is replete with information reporting how various tillage methods affect disease severity and incidence. Often, the comparisons are between conventional tillage systems and some sort of conservation tillage system. Conventional tillage is the traditional tillage system for a given area and, typically, includes some form of deep tillage followed by one or more secondary tillage operations for seedbed preparation. Conventional tillage systems are often clean till, i.e., most or all of the previous crop's residue is buried. Conservation tillage is any tillage practice that reduces loss of soil or water compared to the conventional system by retaining a minimum of 30% surface cover with residues after planting. Minimum tillage and no-till systems both fall under the general category of conservation tillage. Minimum till is the minimum amount of tillage required to create the proper soil conditions for seed germination and plant growth, and no-till, as its name implies, is a method of planting that involves no plowing or seedbed preparation (Sumner et al., 1981). Each of these systems creates unique microenvironments which will vary from region to region due to differences in soil types, weather patterns, etc. Therefore, the impact of a particular tillage system on pathogen populations and disease development may differ from one location to the next. This variation in pathogen response to specific tillage systems helps explain conflicting reports in the literature and also greatly complicates disease control recommendations.

Individual tillage operations can have different effects on pathogen populations and disease development. For instance, incidence of take-all can be increased (Bailey et al., 1992; Rothrock and Cunfer, 1986) or decreased (Moore and Cook, 1984; Shipton, 1977) by conventional tillage. Rothrock (1987) attributed an increased incidence of take-all in conventionally tilled plots to distribution of inoculum over a wider area. There is no question that any operation that moves soil has potential to spread soil-borne pathogens and thereby increase disease incidence (Harveson et al., 1996; Subbarao et al., 1996). Alternatively, as tillage implements move through the soil, fungal hyphae and crop residues can be broken and desiccated, resulting in reduced disease incidence. Moore and Cook (1984) suggested that reduced incidence of take-all in conventionally tilled wheat plots was associated with fragmentation of infested crop residue. They believed the inoculum potential of *G. graminis* var. *tritici* was related to size of infested crop residue and showed that tillage was effective in breaking down residue into small pieces. Rovira (1986) suggested decreased incidence of bare patch of wheat, caused by *R. solani* AG8 following tillage, was also due to a breakdown of crop residues and a concomitant reduction in inoculum potential. In addition to *G. graminis* var. *tritici* and *R. solani*, *P. omnivorum*, *S. rolfsii*, and *M. phaseolina* have also been reported to be negatively impacted by tillage (Rogers, 1942; Rothrock and Hargrove, 1988; Sumner et al., 1981; Weller et al., 1986).

Incorporation of crop residues during tillage operations can increase, decrease, or not affect pathogen populations and disease. *Bipolaris sorokiniana*, which causes common root rot of wheat, is an example of a pathogen that is usually increased by tillage. Several studies have shown that inoculum density and disease incidence is significantly greater in conventional tillage plots than in no-till plots (Mathieson et al., 1990; Reis and Abrao, 1983; Tinline and Spurr, 1991). Mathieson et al. (1990) suggested that soil environmental conditions in conventional tillage plots were more conducive for common root rot development, while incorporation of residue with conventional tillage provided more substrate on which the pathogen could grow.

Populations of soil-borne fungal pathogens, especially those that are saprophytically competitive, are typically increased following incorporation of crop residues. However, this is not always the case. Deep incorporation can place residues out of reach of pathogens or expose pathogens within the tissue to antagonistic microorganisms during residue decomposition. These mechanisms were proposed as the reasons tillage reduced disease incidence of southern blight of peanut caused by *S. rolfsii* and crown rot of barley caused by *F. graminearum* Schwabe, respectively (Dodman and Wildermuth, 1989; Sumner et al., 1981). However, in a study conducted by Summerell *et al.* (1989), tillage did not reduce the incidence of crown rot of barley. They postulated that because environmental conditions were very dry, incorporated residue did not adequately decompose before planting the next crop and, therefore, incidence of crown rot was not reduced (Summerell et al., 1989). In another study, Rush et al. (1986) showed that incorporation of fresh wheat chaff immediately following wheat harvest could increase populations of *P. ultimum*. However, if chaff was left on the soil surface in the field for as little as three weeks before incorporation, colonization of the chaff and soil populations of *P. ultimum* were significantly reduced. Autoclaving the chaff before addition to soil restored its availability as a food source. It was suggested that under natural field conditions, *P. ultimum* was unable to use wheat chaff as a food source because of prior colonization by fungi such as *Penicillium*, *Mucor*, *Rhizopus*, and *Chaetomium*. It was also pointed out that soil environmental conditions during wheat harvest were hot and dry and unsuitable for the saprophytic activity of *P. ultimum*. Therefore, even if chaff was incorporated before colonization by saprophytic fungi, it probably would not have affected populations of *P. ultimum*. This study demonstrated that not only the amount of residue incorporated into a soil but also the quality of the residue and time of incorporation will impact whether the residue can be used as a food source by plant pathogenic fungi.

It should be pointed out that in the study by Rush et al. (1986), prior colonization by other fungi and hot, dry soil conditions hindered residue colonization by *P. ultimum* and disease potential was reduced. Similarly, in the study by Summerell et al. (1989), hot, dry soil conditions slowed colonization and decomposition of residue by soil microbes but disease potential was increased. These studies highlight the importance and necessity of understanding the ecology of soil-borne fungal pathogens before attempting to predict how a given tillage operation will impact disease incidence. They also show how soil environmental conditions can affect pathogen populations and disease development.

When crop residues are left on the soil surface in conservation tillage systems, soils are normally cooler and wetter than soils in conventional tillage systems that are not covered. This can influence plant growth, pathogen populations, and disease development. In numerous studies, the greater incidence and severity of disease in reduced or no-till plots has been attributed to cooler, wetter soils which were more conducive to pathogen activity and disease development (Bockus et al., 1994; Colyer and Vernon, 1993; Cook, 1992; Smiley and Wilkins, 1993; Wrather et al., 1995). However, as previously stated, crop performance and yields do not always correlate well with pathogen populations or disease incidence. Poor stand establishment and yields may result from excessive residue and poor seedbed preparation in no-till plots and be totally unrelated to pathogen populations (Colyer and Vernon, 1993). Conversely, greater water availability in no-till plots may more than compensate for increased disease severity and result in higher yields (Grant and Bailey, 1994; Smiley and Wilkins, 1993).

2. Tillage and Soil Structure

The potential for soil compaction increases as agriculture becomes more mechanized and equipment larger. Soil compaction can occur during any tillage operation (planting, cultivation, application of pesticides, or harvest) and, in some cases, it can occur naturally (Allamaras et al., 1988). Soil compaction can occur in no-till fields as well as conventionally tilled fields. Excessive soil compaction degrades soil aggregation, which results in reduced air-filled pore space, poor drainage, and restricted root growth. Regardless of how or why soil compaction occurs, it adversely affects plant growth and can also impact pathogen distribution, pathogen density, and disease incidence and severity (Burke et al., 1970; Delroy and Bowden, 1986; Gray and Pope, 1986; Kraft and Wilkins, 1989).

Kraft and Allmaras (1985) showed a relationship between tillage pans and propagule distributions of *F. solani* f. sp. *pisi* and *P. ultimum*. Tillage pans were detected at a depth of approximately 20 cm in all surveyed pea (*Pisum sativum* L.) fields in eastern Washington and northeastern Oregon. The tillage pan restricted water percolation, resulting in an excessively wet plow layer and a relatively dry subsoil. Propagules of *F. solani* f. sp. *pisi* were present throughout the upper 60 cm of the soil profile, while those of *P. ultimum* were restricted to the tillage layer. These distributions were correlated to the optimum soil moisture requirements for the two pathogen. Lesions on roots excavated to a depth of 60 cm were readily explained by the vertical distribution of the two pathogens.

Plants growing in soils which are compacted or have tillage pans become stressed more rapidly than plants growing in noncompacted soils. Tillage pans restrict rooting and plants become susceptible to moisture stress. Because compacted soils typically have inferior soil structure, inadequate soil aeration may also stress plants. Root systems growing in such conditions exhibit increased exudation of organic compounds which supports increased populations of soil-borne fungal pathogens (Allmaras et al., 1988; Bhatti and Kraft, 1992; Miller, 1986; Miller et al., 1980). The combination of these biotic and abiotic stresses frequently results in increased disease incidence and severity and reduced yields (Bhatti and Kraft, 1992; Glenn and Sivasithamparam, 1990;

Tu and Tan, 1988). Subsoiling can often significantly reverse these stress effects and improve crop yields, even though pathogens are still present and plants become infected (Burke and Miller, 1983; Kraft and Giles, 1979; Miller and Burke, 1986).

3. Cultivation

In conventional tillage systems, crops are cultivated during the growing season to break soil crusts, improve soil aeration, and control weeds. When done correctly, cultivation typically improves conditions for plant growth and helps reduce disease problems by improving plant vigor. However, under certain circumstances, cultivation can increase disease incidence. Schneider et al. (1982) demonstrated that deposition of soil infested with *R. solani* into crowns of sugar beets significantly increased the incidence and severity of Rhizoctonia root rot. This frequently happened when cultivators were driven at relatively high speeds of 6.4 to 12.8 km/h (4 to 8 mph). Cultivating at slower speeds threw less soil and resulted in less disease.

Any operation that injures growing plants during the season can also result in increased disease. Brenneman and Sumner (1990) reported that tractor traffic during the growing season increased damage to peanut stems. Losses to Rhizoctonia limb rot was correlated to tractor traffic and associated injuries to plants.

C. Crop Nutrition and Plant Disease

Properly balanced nutrition is a critical factor in allowing crops to realize their full yield potential. The application of fertilizer to accomplish this balance is a universal practice in commercial crop production. Macro- and microelements have long been recognized as being associated with size, quality, and yield of crops, and also with changes in levels of disease. Pathogens, as well as crops, have nutritional requirements of their own. Two major objectives of nutrient applications to crops for protection from pathogens can be summarized as follows:

1. Avoid plant stress, which may allow crops to better withstand pathogen attack,
2. Manipulate nutrients to the advantage of plants and disadvantage of the pathogen (Palti, 1981).

1. Types of Fertilizers

Three elements — namely nitrogen, phosphorus, and potassium — chiefly affect relative resistance of plants to diseases.

a. Nitrogen

Nitrogen has been the most intensely studied element because of its profound influence on plant growth, cell size, and wall thickness, and its limited availability in soil (Huber and Watson, 1974). High levels of nitrogen frequently increase disease severity (Huber and Watson, 1974), however, this does not apply to every host-pathogen interaction. It is generally believed that the form of nitrogen available to the host, rather than the rate of nitrogen applied, is most important to disease severity or host resistance (Brennan, 1993; Huber, 1989; Huber and Watson, 1974). Nitrogen can be applied in many different forms. The NH_4+, NO_2–, and NO_3– are the most important forms from a soil fertility standpoint. Nitrous oxide and nitric oxide are also important because they represent nitrogen forms that are lost for crop use through denitrification. The ammonium nitrite and nitrate forms arise either from normal aerobic decomposition of soil organic matter or from additions to soil from various commercial fertilizers (Tisdale and Nelson, 1975). The most commonly used forms in agriculture as fertilizers are ammoniacal (anhydrous, anhydrous-sulfur, NH_4NO_3, [NH_4NO_3.$(NH_4)_2SO_4$], NH_4PO_4, or NH_4Cl), and nitrate ($NaNO_3$, KNO_3, or $CaNO_3$) and various forms of urea (Tisdale and Nelson, 1975).

Nitrogen forms and soil pH are closely related, so many effects of nitrogen on plant growth and disease can ultimately be attributed to the effect of the N form on soil pH. NH_4-N will decrease soil pH because its positive charge binds to soil particles or organic matter during nitrification (conversion of ammonia to nitrite and nitrate) and H ions are released into the soil (Huber and Watson, 1974; Keinath and Loria, 1989). NO_3 forms of nitrogen increase soil pH and are freely mobile in soil solution. Therefore, they are more subject to leaching and denitrification losses, in addition to being more readily available for plant uptake (Huber, 1989; Tisdale and Nelson, 1975). Amine, amide, and protein contents are higher with NH_4-N than with NO_3-N, so it is assimilated by the plant more efficiently, thus forming amino acids quicker and preventing toxicity (Huber, 1989). NH_4-N also increases availability of micronutrients, but this has also been attributed to lowering soil pH.

Several studies involving seedling disease of sugar beet and lettuce have shown greater numbers of diseased plants when NH_4-N was applied compared to NO_3-N (Huber and Watson, 1974). This same trend has been observed for Fusarium wilt (Jones et al., 1989). They found that NO_3-N decreased virulence of *F. oxysporum* and reduced levels of disease in tomatoes and several other crops compared to NH_4. This is in contrast to take-all of wheat. NH_4-N was much more effective in controlling *G. graminis* var. *titici* than was NO_3-N (Brennan, 1993; Engel and Mathre, 1988; Huber, 1989). Disease caused by *Sclerotium rolfsii* appeared to be equally controlled by NH_4-N or NO_3-N. The higher the rate of N applied, the greater the level of disease suppression as long as nitrogen was not applied at phytotoxic levels (Punja, 1989). Root or cortical diseases caused by *Rhizoctonia, Aphanomyces*, and *Cercosporella* are reduced by NO_3-N or increased by NH_4-N, while those responding in the opposite manner include diseases induced by *Pythium, Diplodia,* and *Streptomyces* (Huber, 1989). In general, N increases vigor and crop growth rate, thereby greatly affecting susceptibility to many pathogens.

b. Phosphorus

Phosphorus, like nitrogen, is commonly deficient in wheat soils and its application is widely used in combination with N to reduce take-all. It is less effective in reducing take-all when an N deficiency is limiting crop growth. High levels of P may cause greater severity in N- or Mn-deficient soils (Huber, 1989). Its benefit in reducing take-all relates to stimulated root activity and increased host resistance. High levels of P have been consistently demonstrated to cause increased disease severity of Fusarium wilt of tomato and cotton (Jones et al., 1989).

c. Potassium and Chlorides

Few references specifically address K alone in relation to disease. However, K or Na applied as chloride salts generally act in a similar manner to reduce take-all losses if used in combination with NH_4-N, but not NO_3-N (Brennan, 1993; Engel and Mathre, 1988; Huber, 1989). This observation, however, was not consistent in all locations or experiments (Grant and Bailey, 1994). Another study similarly found that yield losses due to take-all were less severe when KCl was applied compared to K_2SO_4, and further concluded that this effect was more marked during water stress conditions (Trolldenier, 1985). Elmer (1990) also reported that NaCl was more effective than KCl at increasing root weights and lengths of asparagus transplants and reducing infection from fusarium crown and root rot. The benefit of chloride has been attributed to decreasing water potential and by inhibiting NO_3- uptake and nitrification in acid soils, thereby favoring take-all reduction.

d. Calcium

Ca^+ is often applied as a carbonate salt to raise the pH of acid soils, but it can also be used as a fertilizer to supply N or S in the forms of [$Ca(NO_3)_2$] and $CaSO_4$ (gypsum), respectively. It plays a critical role in plant metabolism by removing carbohydrates, neutralizing cellular acids, cell wall deposition, and middle lamella pectate formation (Huber, 1989). The addition of Ca^+ to soils increases severity of common scab of potato (*Streptomyces scabies* [Thaxter]) Waksman et Henrici, as it does for take-all (Huber, 1989; Keinath and Loria, 1989). Additions of Ca to raise soil pH have been very effective in combination with NO_3-N to reduce losses to the Fusarium wilts (Jones et al., 1989). The addition of Ca has also been shown to reduce disease severity caused by *Pythium* spp. (Kao and Ko, 1986) and *Plasmodiophora brassicae* (Campbell and Greathead, 1989; Webster and Dixon, 1991a). However, no effect of Ca was observed on disease severity of *Synchitrium endobioticum* (Schilb.) Perc., causal agent of wart disease of potato, when used with either urea or NH_4NO_3 as N sources (Hampson, 1985).

e. Sulfur

A deficiency of S, like other mineral deficiencies, increases severity of take-all. NH_4SO_4 has been a preferred fertilizer under heavy take-all disease pressure as much for the ammonium source as the sulfur (Huber, 1989). It has been known for nearly 100 years that elemental sulfur is also effective in controlling potato scab (Keinath and Loria, 1989). Sulfur has been thought to control both of these diseases through its acidifying effect on oxidation (Huber, 1989; Keinath and Loria, 1989).

f. Micronutrients

Mg tends to be deficient in acid soils due to its limited solubility. High K or Ca can also inhibit Mg uptake (Huber, 1989). Boron has also been demonstrated to reduce the maturation rate and eventual symptom expression of both primary and secondary stages of *P. brassicae* at 20 and 30 ppm (Webster and Dixon, 1991b). Other micronutrients are generally readily available to plants in acid soils that are suppressive to take-all and common scab, while being less available in take-all conducive soils after liming. Many other micronutrients, including Cu, Fe, Mo, Mn, and Zn have shown varying degrees of disease suppression or induction, depending upon host, pathogen, soil pH, and environmental conditions. Because no one form of nitrogen or combination of accompanying nutrients controls all pathogens or influences all crop plants, each disease must be considered on an individual basis (Huber, 1989). Therefore, growers and researchers must have an intimate knowledge of the relationship between crops and diseases in their areas in order to make nutritional management decisions.

2. Liming and Soil pH

Soil pH, which can be manipulated by the addition of various forms of nutrients, has profound effects upon many diseases caused by soil-borne pathogens. Two classic examples of diseases controlled by altering soil pH are common scab of potatoes, favored by alkaline conditions, and club root of crucifers, which is favored by acid soil conditions (Campbell and Greathead, 1989; Keinath and Loria, 1989; Webster and Dixon, 1991a,b). Several other examples of soil pH and its effect upon various diseases have also been recently reported. Additions of urea or NH_3NO_3 were much more effective than agricultural lime in controlling wart disease of potato *(Synchitricum endobioticum),* another pathogen favoring alkaline conditions (Hampson, 1985). Take-all responds in a similar manner — liming to raise the pH of soil results in increased levels of disease, whereas lower pH values reduce infection (Brennan, 1993; Engel and Mathre, 1988; Huber, 1989). Black root of tobacco *(Thielaviopsis basicola)* has similarly been suppressed by low Ca, pH below 5.0, and also by exchangeable aluminum (Meyer et al., 1989).

The addition of lime to raise soil pH is a widespread practice regardless of the existence of disease problems. Lime is a generic term often used interchangeably for several different forms of Ca or Mg. The different forms of lime affect the amount of Ca supplied per unit weight of lime, the rate of Ca release, and the neutralizing value (Campbell and Greathead, 1989). The most commonly occurring form of lime is $CaCO_3$. Other forms used for control of soil-borne pathogens include calcium oxide (quick lime), CaOH (slaked lime), $CaSO_4$ (gypsum), and $CaCO_3 \cdot MgCO_3$ (dolomitic limestone).

Raising soil pH by adding Ca or Mg salts has effectively decreased not only club root (Webster and Dixon, 1991a), but pythium damping-off of cucumbers (Kao and Ko, 1986) and various Fusarium wilt diseases when used in combination with NO_3-N forms (Jones et al., 1989). Allamaras et al. (1987) further showed that gypsum and lime added to acidic eastern Washington soils reduced propagule density of *Fusarium solani* f. sp. *pisi*, presumably by increasing exchangeable Ca.

3. Effects of Nutrition on Pathogens

Nutrition is obviously important in determining susceptibility or resistance of plants to pathogens. Mineral nutrients are essential metabolic regulators of plant growth, and most studies indicate that nutrition affects pathogens and diseases indirectly. The relationship between nutrition and pathogenesis is complex because the use of nutrients is associated with chemical and physical changes in the soil and soil microflora. Interactions among these factors can influence incidence and severity of disease positively or negatively, depending on different host-pathogen interactions.

Although most nutrient effects on disease are indirect, some ammoniacal forms of nitrogen can directly affect pathogens. NH_4-N and urea break down in soil to form ammonia, which has been shown to be toxic to several fungal pathogens including *S. rolfsii*, *P. omnivorum*, and possibly *P. brassicae* (Palti, 1981; Rush and Lyda, 1982). Cu has also been suspected of being directly toxic to the take-all and potato scab pathogens: it is interesting to recall that the first widely used fungicide was Bordeaux mixture ($CuSO_4$); perhaps the Cu component is partially responsible for its activity. Recent evidence also points toward Ca as having some fungistatic activity. It may decrease germination and germ-tube growth of *Pythium splendens* (Kao and Ko, 1986). Under high inoculum levels, Ca has inhibited sporangial maturation of *P. brassicae* in crucifer root hairs, while prohibiting sporangial development and dehiscence at lower inoculum levels (Webster and Dixon, 1991a). However, the mechanism of control when Ca compounds are used to increase soil pH is not known for certain.

As previously stated, evidence indicates more indirect effects of nutrients on pathogenic fungi and plant disease. This is expressed in plants as improvement of overall health and vigor, and the increase of various soil microorganisms that may play an antagonistic role against pathogens. There are several ways that nutrient applications may influence diseases indirectly.

First, Ca is important to the structural stability of plant cell walls. Allamaras et al. (1987) suggested that additions of Ca may have improved resistance of pea root membranes or impaired saprophytic growth and survival of *F. solani*. In other studies, increased levels of Ca inhibited *F. oxysporum* and *P. splendens* (Jones et al., 1989; Kao and Ko, 1986). It was suggested that high levels of Ca could inhibit polygalacturonase production and other cellular-degrading enzymes often produced by invading pathogens (Kao and Ko, 1986). However, other studies indicate that this mechanism is not the only factor in inhibition (Jones et al., 1989; Webster and Dixon, 1991b).

Secondly, high soil pH has been shown to make many micronutrients (Cu, Fe, Mo, Mn, and Zn) unavailable. *Fusarium oxysporum* has a high requirement for these micronutrients, and its inability to grow and sporulate when micronutrients are deficient may help explain observations of reduced disease at high soil pH (Jones et al., 1989).

High pH increases take-all severity in wheat by increasing populations of Mn-oxidizing organisms, which make Mn unavailable to plants (Huber, 1989). Populations of fluorescent pseudomonads, reported to decrease take-all, increase in numbers after NH_4-N is added and reduce formation of Mn oxides, thereby increasing Mn availability for plant uptake. Increased resistance to take-all is most pronounced with a balanced nutrient supply of NH_4-N and Mn. Mn is readily available in acid soils and, together with N, is involved with carbohydrate synthesis and synthesis of phenols and other compounds associated with plant defense (Huber, 1989).

Lastly, balanced nutrition helps plants stay more vigorous, increase the growth rate, make better use of soil water, and improve anatomical or histological characteristics (Palti, 1981). These factors enable plants to produce greater numbers of roots, allowing more surface area for root absorption and nutrient uptake and, possibly, shortening susceptible stages of plant growth. This allows plants to function more efficiently even when some roots are infected. Histological changes strengthen plants and possibly create barriers more difficult for pathogens to breach. A healthier plant may also increase quantity or quality of exudates which can stimulate increases in populations of antagonistic microorganisms. These, in turn, may compete with pathogens for nutrients or possibly produce toxins that directly affect pathogen development and survival.

D. Planting

There are a number of cultural practices that producers can employ at planting that influence levels of disease. At planting, most agronomic practices are intended to modify the environment to the benefit of the crop and to the detriment of the pathogen. Since conditions vary extensively between crops, pathogens, and locations, an intimate knowledge and careful consideration of local problems is needed before choosing appropriate measures for control in individual situations.

1. Cultivar Selection

Cultivar selection is one of the first decisions that needs to be made at planting time. Even without considering resistance to diseases, selection of varieties which are locally adapted is important to successful crop production. In a recent study using several rhizomania-tolerant cultivars, Harveson and Rush (1994) showed that the performance of locally adapted entries, although not resistant to any specific disease, was better than most other varieties used in the test presumably because the rhizomania-tolerant varieties were adapted to and bred for other locations. Most of the rhizomania-tolerant cultivars were highly susceptible to some of the other pathogens in a disease complex, while the locally adapted cultivars had a high level of general field resistance. This study demonstrated the danger associated with bringing in an unadapted, disease-tolerant cultivar to a location where the disease had been newly identified. The presence of insects or other plant pathogens can cause the "resistant cultivar" to perform very poorly. Furthermore, in the absence of disease, disease-tolerant cultivars frequently underperform locally adapted, high-yielding cultivars. Still, the use of cultivars which are resistant or tolerant to specific diseases is one of the best disease management options available to producers.

Resistance to specific diseases is probably the most cost-effective disease control measure available to producers. In some instances, resistant cultivars may be the only means to accomplish disease control (Parry, 1990). Resistance is becoming even more important today because more emphasis is being placed on reducing costs of farming. The public is also closely scrutinizing the use of chemicals for disease control because of fears for public safety and environmental contamination. The development of a chemical-insensitive or -resistant pathogen populations is also associated with overuse of pesticides (Parry, 1990).

Genetic resistance can be based on a single gene or it can be polygenic. In general, resistance due to polygenic inheritance is more complex and difficult to achieve because minor or modifying genes may interfere with identifying and isolating major genes for resistance (Hecker and Ruppel, 1975). This is often the case when multiple pathogen resistance is desired. In breeding programs, development of cultivars resistant to single diseases can be accomplished relatively quickly. Therefore, most cultivars are bred for resistance against one or very few pathogens (Nene, 1988; Palti, 1981). However, real farm situations often require cultivars with combined resistance to multiple diseases and other biotic and abiotic stress factors (Harveson and Rush, 1993, 1994; Nene, 1988).

One way to compensate for the lack of multiple disease and stress resistance is the use of gene-pyramiding, cultivar mixtures, or multilines (Maloy, 1993; Parry, 1990; Pedersen and Leath, 1988; Trutmann et al., 1992; Wolfe, 1985). Gene pyramiding or gene stacking is the accumulation of several genes into single lines or varieties, which helps to provide more durable disease resistance (Pedersen and Leath, 1988).

The use of cultivar mixtures can also be a beneficial disease management strategy. It helps avoid genetic monoculture and allows diversification within a single field. This procedure involves planting mixtures of cultivars that have similar agronomic properties but with different sources of disease resistance. Cultivar mixtures can be beneficial even in the absence of disease by mixing varieties with slightly different growth characteristics. Studies have shown that mixtures frequently yield as well or better than individual cultivars (Harveson and Rush, 1995; Parry, 1990). Vilich (1993) observed the effects of mixtures of wheat and barley with pure stands as precrops. Disease incidence due to take-all on barley decreased from 44% on the pure stand to 9% when grown after the mixture. However, no significant difference in take-all incidence on wheat was seen with different precrops.

Multilines, in practice, are very similar to mixtures. They contain almost identical breeding lines (isolines) which have the same agronomic characteristics but differ in disease resistance. Multilines or cultivar mixtures are especially useful in underdeveloped countries (Trutmann et al., 1992) or in situations where multiple disease resistance is needed but not available (Wolfe, 1985).

2. Seed Quality

Good quality seed is an essential component to successful crop production. The approximate germination percentage and purity of seed should be known before planting. In general, seed quality usually can be directly related to the quality and vigor of seedlings. Seed that is contaminated with seed-borne pathogens or mixed with weed seed can similarly be problematic.

Seed certification is a procedure that regulates the quality of seed placed on the market. This is an effort to counteract problems that may arise due to use of poor quality seeds and insures a consistent, quality product for farmers to use. Breeder seed is the initial source of a genotype and is controlled by the sponsoring breeder or institution. Foundation seed is produced from breeder seed and is also controlled by the certifying agency. Registered seed is produced from foundation seed and also must be approved by the certifying agency. Certified seed can be produced from any of the three previously mentioned seeds.

Some form of seed testing prior to planting may help to reduce losses due to diseases. This is illustrated by Zorrilla et al. (1994). They found that field emergence of soybeans declined as seed infection by Phomopsis seed decay increased. They utilized seed quality tests to predict field performance of soybean seed lots with different incidences of infection. Rush (1987) also reported that severe hollow heart, a physiological disease of peas, increased seed electrolyte loss and reduced growth and vigor of seedlings, possibly predisposing peas to infection by root-rotting pathogens such as *Pythium ultimum* and *F. solani*. A significant correlation was observed between plant growth characteristics and electroconductivity (EC) readings, with seed affected by hollow heart producing significantly higher EC readings than healthy seed. He concluded that seed lots could be divided into different EC categories to estimate potential problems due to hollow heart prior to planting.

A foundation seed health program was developed in Idaho to eradicate black chaff *(Xanthomonas campestius* pv. *translucens)* from wheat seed by treating the seed with various chemicals. To produce pathogen-free seed, the chemically treated seed was assayed and grown for increase, and repeated two additional years. No disease was detected in any year, so the authors concluded that this seed health program could be a viable control for black chaff of wheat in Idaho (Forster and Schaad, 1988).

In general, it is thought that the higher the quality of seed, the better the economic return (Martin et al., 1976). Poor quality seeds (broken, immature, small size) generally produce smaller, less vigorous seedlings. This may make them more susceptible to either pathogen attack or harm due to adverse environmental conditions (Martin et al., 1976). Seed size has been reported to be important to establishing adequate stands of vegetable crops. Heather and Sieczka (1991) found that stand, dry weight, and final yield of several broccoli cultivars significantly increased as seed size increased. The opposite effect was observed with seed potatoes planted into soils infested with *Heminthosporium solani* Dur. & Mont. and *R. solani,* causal agents of silver scurf and black scurf, respectively. In all tests, incidence of both diseases increased as seed size increased (Firman and Allen, 1995). This is presumably due to the larger surface area of the tubers as infection sites.

Numerous studies have evaluated seed size of small grains and their effects upon stand establishment and yield with varying results. Douglas et al. (1994) found no significant differences in soft white winter wheat regarding seed size and density and final grain yield. In a separate study, seed size had no effect upon germination efficiency in the laboratory, but large seed emerged more rapidly in the field and plants grown from large seed were taller and produced more tillers (Chastain et al., 1995).

Studies using hard red winter wheat further confirm these observations. No differences were observed from fall-planted wheat for stand establishment or yield from four significantly different wheat seed quality categories in two of three years (Rush and Vaughn, 1992). A similar result was obtained with hard red wheat planted in the spring. Variation in seed size within a plot had no impact on total grain yield per plot (Gan and Stobbe, 1995). However, conflicting results with disease incidence and severity in relation to seed size have also been reported. Duczek and Piening (1982) observed no differences between seed size or seeding date and symptoms of common root rot of barley. Shriver et al. (1995) found no effect of seed size on disease severity of common root rot of wheat in early-planted wheat, but in later sown crops, plants grown from larger seed had significantly lower disease indices.

3. Seed Treatments

Seed treatments can be used on almost any crop for a myriad of purposes. A whole new agricultural industry is developing because seed treatment is now seen as a way to substantially increase the value of seed and, ultimately, improve plant vigor, growth, and profitability (Taylor and Harman, 1990). The most common goal of seed treatment technology is to exploit the seed as a delivery system or deployment agent for desired materials. For the purposes of this review, seed treatments will be divided into three basic categories: (1) chemical seed treatments, (2) seed priming, and (3) biological seed treatments.

a. Chemical Seed Treatments

Perhaps the most widespread use of seed treatments is coating seeds with various pesticides to inhibit seed decay or delay seedling disease by various pathogens. Imazilil seed treatments have been demonstrated to lower disease severity in barley due to common root rot and increase grain yield 6% (Herrman et al., 1990). Hymexazol-treated seed has increased sugar beet emergence and reduced seedling disease caused by *Aphanomyces* and *Pythium* (Harveson et al., 1994; Payne and Williams, 1990). Brassica seeds treated with a mixture of metalaxyl and fenpropimorph successfully increased seedling stands. This suggested that metalaxyl controlled postemergence damping-off by *Pythium sylvaticium* and seed-borne pathogens were eliminated by fenpropimorph (White and Ho, 1988). Chemical seed treatments may also provide more flexibility in terms of planting dates. In Texas, the use of fungicides protected wheat seedlings from damping-off and delayed infection when seed was planted early in the fall for forage production (Mathieson and Rush, 1991). It has also been useful in third-world countries in traditional varietal mixtures without eroding the genetic diversity of crops, thereby increasing the effective life of the resistance genes (Trutmann et al., 1992). It was further determined that treating sunflower seeds with a number of fungicides virtually eliminated seed-borne *Sclerotinia sclerotioium*. The authors concluded that if this practice became routine, the introduction of the pathogen into uninfested land could be prohibited (Herd and Phillips, 1988). The relatively low cost of chemical seed treatment, low volume of chemical usage, and capability of being incorporated into new seed coatings also make this technique attractive, both from an environmental and economic standpoint. For these reasons it will undoubtedly continue to occur.

b. Priming

Seed priming, or osmoconditioning, are terms used to describe a physiological preplant treatment to improve seedling emergence and establishment. The materials used for this process include polyethylene glycol (PEG) or various inorganic salts. They serve as osmiticums to start the germination process but stop it before radicle emergence. This makes germination and emergence quicker and more uniform in the field under possibly less than favorable conditions for normal emergence. Priming has been demonstrated to reduce early infection from numerous soil-borne pathogens by shortening juvenile susceptibility periods of growth. Hall and Sumner (1994) reported that planting primed watermelon seed enhanced stand establishment in soils infested with *Pythium irregulare* Buis when soil temperatures were suboptimal for watermelon seed germination and growth. Osmopriming sugar beet seed reduced preemergence damping-off of *P. ultimum* by 50% with NaCl and 65% in polyethylene glycol (PEG), compared to nontreated controls (Osburn and Schroth, 1989). Rush and Vaughn (1993) observed that solid matrix priming of sugar beet seed reduced Aphanomyces seedling disease by allowing seedlings to become established in soils too low in moisture for movement of zoospores of *A. cochlioides* (<100 J/kg). Priming is also beneficial because it can be combined with new seed technology (coating and pelleting) and antagonistic biological agents (Callan et al., 1990; Harmon and Taylor, 1988; Taylor and Harmon, 1990).

c. Biological Seed Treatments

Biological organisms can be coated onto seeds in a number of ways to provide an alternative to chemical pesticide treatments (Windels et al., 1985). Biologicals have been used primarily as competitors for seedling pathogens such as *Rhizoctonia* and *Pythium* spp. Biopriming sweet corn seeds with *Pseudomonas aureofaciens* AB254 and *P. fluorescens* AB254 have both been effective in increasing seedling emergence and reducing damping-off by *P. ultimum.* Both treatments were as effective as control provided by seed treatment with metalaxyl (Callan et al., 1990). Park et al. (1991) and Bowers and Park (1993) have further demonstrated that *Pseudomonas* spp. used as seed treatments can also reduce seedling infection of peas by *Aphanomyces euteiches.* In choosing the biological organism for protection, a great deal of preliminary information about those organisms is needed. Factors such as the ability to grow at temperatures that may be encountered and the inherent ability to control a particular pathogen under varying environmental conditions are critical. The treatment selected should also consider the environment most conducive for beneficial growth of the organism of interest, but also not favoring competitive microorganisms (Taylor and Harmon, 1990).

4. Transplants

The use of transplants instead of direct seeding is another approach to control soil-borne diseases. This practice also modifies the environment by shortening that period of the plant's growth stage when it is most susceptible to infection. Planting onion sets as transplants reduces disease by the smut pathogen *Urocystis cepulae* Frost. The pathogen is able to cause infection only for a brief time during development of the seedling. This period extends from about the second day after germination until the seedling is in the first leaf stage (Sherf and MacNab, 1986). Use of transplants has also been utilized with lettuce to reduce corky root severity in California (Van Bruggen and Rubatzky, 1992). However, reduction in losses only occurred when corky root inoculum levels were high.

Transplantation has also been used commercially in Japan and experimentally in California to protect sugar beets from rhizomania (Gerik and Temple, 1991). It has not become routine in U.S. sugar beet production, most likely due to its high cost and labor-intensive nature. Sugar beet transplants, even if infected with BNYVV at a later point in the season, are still often capable of producing acceptable yields, although quality and sugar content may be reduced (Rush and Heidel, 1995). Use of transplants may be a viable alternative for other crops if the cash value is high enough to justify the additional costs or if fumigation or use of resistant varieties is not possible.

5. Planting Rate

The rate at which crops are sown can significantly affect disease development. Some of the principle adverse effects of dense stands are etiolation, delay in lignification of young tissues, and earlier senescence of individual organs or whole plants (Palti, 1981). These effects can reduce yields by favoring conditions for damping-off organisms or other pathogens that attack crops before stems or crowns harden (Palti, 1981). An excess of water and nutrients on crowded stands can cause many of the same physiological effects. If water is limited, crowded stands may also increase plant susceptibility to diseases such as Diplodia stalk rot of maize or Sclerotium rot of cotton (Palti, 1981).

In general, however, losses due to soil-borne diseases are reduced by higher density plantings, while foliar diseases are increased with high plant populations. Greater numbers of plants create more of a foliar canopy, allowing less wind circulation, which in turn provides a more humid microclimate within the canopy. This favors foliar diseases by providing more plant area within shorter distances for infection to occur. In addition, many foliar diseases are polycyclic, which increases exponentially as a function of number and distance between plants available for infection and reinfection (Burdon and Chilvers, 1982; Maloy, 1993). Soil-borne diseases are mainly monocyclic, therefore, disease severity is related to initial infection sites available. Increasing the number of plants may reduce drastic losses by providing a surplus (Maloy, 1993). An exception to this generalization was observed in a study by Firman and Allen (1995). They reported that incidence of black scurf of potatoes caused by *R. solani* was significantly increased by high-density plantings.

Soil-borne diseases caused by zoosporic pathogens are often not monocyclic; therefore, plants may not benefit from dense plantings. This idea was demonstrated with *P. irregulare* and stands of cress seedlings. In this case, it was not mass numbers of plants that inhibited losses but the distance between plants (Burdon and Chilvers, 1982). Spread of the pathogen, and, therefore disease increase, was halted when plant spacing exceeded 6 cm.

Crowded stands may also enable many systemic diseases to be less severe. Infection of *Verticillium albo-atrum* and *V. dahlia* on cotton decreased when plants were planted thickly. This was attributed to reduction of effective inoculum per plant in proportion to the increase in plants per acre in high-density croppings (Wilhelm et al., 1976). It was suggested that other wilt diseases should exhibit this same relationship (Palti, 1981).

Although overplanting may decrease root disease, it has been shown to create conditions favoring foliar disease increases. It can also create other costly problems. Many soil-borne pathogens occur in clusters in fields. Those plants surviving to harvest may also be concentrated in groups. This can cause yield reductions due to competition between surviving infected plants and healthy uninfected plants. One example of this involved planting potato tubers infected with gangrene (*Phoma foreata* Foister) to observe the effect of neighboring plants adjacent to infected ones. Yield increase from healthy plants located between two missing plants was twice that of a healthy one next to one neighboring plant (Hide et al., 1995). However, if diseased plants survived, they were less able than healthy plants to benefit from small decreases in competition (dead neighboring plants) and were actually more sensitive to increasing competition. The authors further concluded that yields were decreased up to 19% when 15% of the plants died and the gaps in the rows occurred at random. There was minimum yield loss when the gaps occurred systematically (spacing plants at twice the normal distance) (Hide et al., 1995). A higher density of plants in the absence of expected disease levels may also require further expense in thinning stands to suit optimum production. Again, this illustrates the importance for producers to be intimately aware of their disease problems and cropping systems in order to take full advantage of management options available for their particular situations.

6. *Time of Planting*

Annually, growers are faced with the difficult decision of when to plant. Factors, such as the amount of time required for a crop to reach maturity and the time when soil moisture and temperature are optimal for plant growth must all be considered. However, in relation to disease management, the choice of a planting date has one principal goal, i.e., to reduce to a minimum the time a pathogen has access to susceptible plant tissue. Local experience and knowledge of individual situations are needed to select optimal planting dates which will minimize disease and maximize crop growth potential.

Planting dates of winter wheat have been particularly instrumental in influencing the severity of soil-borne fungal diseases. When wheat is planted in late August or early September in Texas to maximize forage production for cattle grazing, soil-borne diseases caused by *Rhizoctonia* and

Bipolaris are frequently observed to increase on seedlings (Mathieson and Rush, 1991; Rush and Mathieson, 1990). Early infection can be effectively avoided by planting in late October or November. However, this practice reduces forage production. In Texas, *R. solani* AG4 is the most prevalent anastomosis group found infecting wheat. In the Pacific Northwest, AG8 is the primary AG involved in wheat losses (Smiley and Uddin, 1993). It was determined that high soil temperatures were more likely to increase infection by *R. solani* in both locations. However, if planting of winter wheat is too late, problems can be encountered with pathogens such as *Pythium* which often cause more damage in cool, wet soils. Similar results have been observed with late-planted canola crops (Parry, 1990).

Growth chamber studies have further supported these field observations. A series of *R. solani* isolates collected from wheat and sugar beet were characterized in terms of growth rates and pathogenicity to different crops at different temperatures (Rush et al., 1994). Growth of most isolates was optimal at temperatures between 20 to 30°C. AG4 isolates were most pathogenic to wheat and sugar beet seedlings, but not mature sugar beet roots. All isolates were severely restricted at temperatures below 20°C. Since sugar beets often follow wheat in rotation in Texas, it would be logical to assume that because AG4 attacks both wheat and sugar beet seedlings, and because it was the predominant AG found in wheat, sugar beet damping-off would be a prevalent problem. However, this is not normally the case because sugar beets are typically planted in late March and early April when soil temperatures are usually less than 20°C. These data explain the lack of sugar beet seedling disease caused by *R. solani* in Texas and illustrate how time of planting can impact development of seedling disease.

Planting dates can also impact disease incidence and severity later in the growing season. Several soil-borne pathogens routinely cause yield reductions to Texas sugar beet growers (Harveson and Rush, 1994). Since most of these pathogens require warm temperatures to effectively parasitize plants, planting early allows sugar beet seedlings to easily germinate and grow. Plants escape early infection and become established while temperatures are below optimal ranges for pathogen activity.

This same relationship has been studied in England with sugar beets and incidence and severity due to rhizomania. Although the disease is caused by a virus (beet necrotic yellow vein virus, BNYVV), it is completely dependent upon the obligate fungal parasite *Polymyxa betae* for transmission and dissemination (Rush and Heidel, 1995). Since the optimum temperature for growth of *P. betae* is 25°C (Blunt et al., 1991), a series of planting dates was evaluated for their effects on infection of sugar beet seedlings by *P. betae*. Infection occurred sooner and rate of growth of the fungus was more rapid in later-planted crops than in earlier ones. The early sown plants allowed more sugar beet growth at temperatures too low for fungal infection and development. By delaying *P. betae* infection, subsequent disease severity caused by BNYVV was also reduced (Blunt et al., 1992).

However, the principle of early planting to avoid disease is not consistent with all pathogen-host systems. Survival of *Thielaviopsis basicola* propagules was greater at 16°C than at 24 or 28°C. Disease severity from this pathogen on cotton seedlings was also lower at 28°C than at 24 or 16°C (Rothrock, 1992a). This implies that black root rot of cotton is more problematic at cooler soil temperatures. In this instance, later planting would be more beneficial to avoid infection by the black root rot pathogen. In general, seedling disease can be minimized by planting when temperatures are optimum for plant growth. Therefore, seedling diseases in cotton are less severe when soil temperatures are high and seedling diseases of winter wheat are reduced when the crop is planted in late fall when soil temperatures are cool.

7. Planting Depth

Seeding depth can influence severity and incidence of certain diseases. In general, shallow seeding is recommended to reduce disease in most cropping systems. Often, pathogens infect seedlings only in the early stages of development. Deep planting increases the time of emergence, thereby increasing the period of susceptibility (Maloy, 1993).

Two examples involve common root rot, *B. sorokiniana*, on wheat and barley. This fungus is a stress pathogen and is primarily restricted to infection of subcrown internodes of wheat plants (Wiese, 1987). Grain yields were reduced and disease severity by the pathogen increased with deeper seeding depth of barley (Duczek and Piening, 1982) and variable seeding depths of wheat (Gan and Stobbe, 1995). Planting deeply causes greater elongation of the subcrown internode, giving the pathogen not only more time to infect, but also more surface area of the host plant for colonization. Both studies concluded that to obtain maximum grain yields, seed should be planted at uniform, shallow depths (Duczek and Piening, 1982; Gan and Stobbe, 1995).

E. Irrigation

Water is the most important compound in an active plant. It constitutes more than 80% of the growing tissue and is essential for plant growth, metabolism, physiology, and morphology. All of these are affected by the amount of water applied during irrigation and the time and method of water application.

The amount of water necessary to maintain foliar temperature or relative water content within an unstressed level may exceed 10 kg m^{-2} on a sunny day. The inability of a plant root system to supply such demands is one of the principal constraints of plant productivity (Baker et al., 1992).

Crop plants rarely attain their full genetic potential for yield because of the limitations imposed by the environment, and a common environmental limitation is lack of water. About one-third of all potentially arable land suffers from inadequate water supply, and water deficits affect every aspect of plant growth. Yield losses due to water deficit exceed those from all other causes combined (Kramer, 1980).

Irrigation is the application of water to crop plants by artificial means, designed to permit farming in arid regions and to offset drought in semiarid or semihumid regions. Even in areas where the total rainfall is sufficient, it is often unevenly distributed during the growing seasons, therefore, only irrigation can provide a stable system of production. Irrigation prolongs growing seasons in dry areas, thus permitting multiple cropping. With the security of cropping under irrigation, additional inputs needed to increase production become economically feasible. However, when fields are infested with soil-borne pathogens, the benefit of irrigation can be eliminated by increased disease incidence associated with elevated soil moisture.

1. *Impact of Irrigation on the Soil Environment*

A well-managed irrigation system is one that optimizes the spatial and temporal distribution of water to promote crop growth and yield and enhances the economic efficiency of crop production. The aim is not necessarily to obtain the highest yield per unit area of land or per unit volume of water, but to maximize the net return in the long run. Unbalanced irrigation tends to waste water, nutrients, and energy, and to cause soil degradation by water logging and salinization (Pierce and Rice, 1988). An increase in disease caused by soil-borne pathogens can also result.

In an economically profitable environment, irrigation scheduling is one of the more important factors affecting yield. Good irrigation practices increase profits by saving water and energy and reduce yield losses due to excessive wetness or drought stress. Plant water stress is due to excessive evapotranspiration, and because it is difficult to regulate evapotranspiration the simplest way to optimize plant water status is by means of irrigation.

In recent years, irrigation science and technology have evolved to the extent that a new comprehensive understanding of the interactive relationships among soil, plant, and climate in regards to the disposition and utilization of water, has evolved. These scientific developments have been paralleled by a series of technological innovations in the methodology of water control which have made it possible to establish new soil-plant-atmosphere relationships. Among these innovations are techniques for low-volume, high-frequency water applications to the root zone in precise and timely response to changing crop needs. Unfortunately, while irrigation reduces water stress problems, it often creates a complete new set of problems. Waterlogging of the soil creates asphyctic conditions for roots and can provide an environment conducive to certain soil-borne pathogens. Overirrigation can also lead to salt accumulation, depletion of aquifers, and overcompetition with urban and industrial users. The occurrence and severity of many plant diseases are influenced by soil moisture-related factors. However, the conditions necessary to minimize disease potential may or may not be those required to achieve good yields. Properly used, irrigation methods such as sprinkler or drip-irrigation can raise yields while avoiding waste, reducing drainage requirements, and minimizing losses due to soil-borne pathogens.

Irrigation, by definition, creates a more humid environment, thus promoting the development of some diseases (Camprag, 1976). Many diseases develop more rapidly in wet soils, especially when irrigation results in patches of excessive wetness. Soil saturation is favorable for the release, spread, and germination of spores of many pathogenic fungi and offers a favorable environment for some kinds of nematodes. In a series of microplot studies, Cappaert et al. (1992) compared the effect of different irrigation amounts on the severity of potato early dying. The authors found greater disease severity in plants grown under excessive irrigation than in those grown under moderate or deficit irrigation regimes. Moreover, as inoculum density increased, reductions in tuber yields were greater under excessive irrigation than under moderate or deficit irrigation regimes. Similar results were obtained by Campbell and Copeland (1954). They determined that the root system of citrus trees growing in soils frequently saturated by irrigation were prone to Phytophthora root rot.

Wet soil may also indirectly affect disease development by allowing the maintenance of higher host tissue water potentials. Lund and Kelman (1977) found that an increase in tuber water potential from –0.8 to –0.67 MPa dramatically increased the susceptibility of tubers to blackleg infections caused by *Erwinia carotovora* var. *atroseptica* (L. R. Jones) Holland.

Diseases such as Fusarium wilt of tomato and Verticillum wilt of cotton are caused by pathogens that block the plant's vascular system. Such diseases result in reduced transpiration, wilting, and often death. Farmers are frequently deceived by such diseases and tend to irrigate more, thinking that the crop is under water stress. Irrigation then tends to make the problem more serious.

Although soil-borne diseases are generally favored by high levels of soil moisture, roots that have experienced a period of drought are often more affected by disease. Duczek (1989) determined that common root rot symptoms were greater on wheat plants with fewer number of tillers, indicative of stressed plants, than on unstressed plants with a higher number of tillers. Diseases such as charcoal rot of grain sorghum caused by *M. phaseolina* and Fusarium root rot of beans and peas are also more severe when host plants are stressed.

2. *Water Quality and Disease Relationships*

Surface-water and groundwater contamination results when water containing salts, trace elements, or other contaminants runs off irrigated fields. This contaminated runoff either flows directly to surface waters or infiltrates groundwater aquifers. Salts and trace elements often exist in agricultural lands before irrigation, and irrigation concentrates these contaminants as evaporation and crop transpiration occurs.

Crop production in saline soils or irrigation with brackish waters may increase plant stress leading to increased susceptibility to certain soil-borne pathogens and, consequently, increased disease incidence and severity.

Duczek (1993) addressed the effect of soil salinity on common root rot of spring wheat and barley. Among plant samples collected from commercial fields with varying salinity levels, a significant positive correlation existed between salinity and common root rot. However, no significant correlation was found between inoculum level and disease. Moreover, laboratory testing showed that the growth of *Cochliobolus sativus* (Ito & Kurib.) Drechsl. ex Dastur, one of the soil-borne fungi responsible for common root rot in wheat and barley, was unaffected on solid and liquid media up to conductivity levels of 30 to 50 dS/m. Thus, disease severity in wheat and barley were attributed mostly to high levels of salinity, leading to an increase in plant susceptibility to the pathogen and not to an increase in inoculum density. Rasmussen and Stanghellini (1988) also reported that the primary effect of increased salinity was apparently on the plant. They indicated that mycelial growth of *Pythium aphanidermatum* was only slightly stimulated by increased salinity levels up to 7.1 dS/m at temperatures ranging between 15 and 30°C, and zoospore production was completely inhibited at electrical conductivity levels of 14 dS/m. The relationship between salinity and disease is consistent with the theory that stress increases plant susceptibility to soil-borne diseases (Duczek, 1986). This concept is also supported by the results obtained by Piening (1973) and Grey and Mathre (1984), in which healthy crops were produced in pathogen-infested fields when plants were grown under unstressed conditions.

MacDonald (1984) reported that hydroponically grown chrysanthemum plants were predisposed to Phytophthora root rot when exposed to salinity stress prior to inoculation, similar to other plant species that are susceptible to infection by Phytophthora when exposed to drought or prolonged soil flooding (Blaker and MacDonald, 1981; Duniway, 1977; Kuan and Erwin, 1980). One effect of salinity stress was to greatly increase the number of zoospores that attached to the roots. The increase of inoculum at the root surface increased the number of successful infections of the stressed root system and resulted in more severe disease.

In another experiment, the author tested if salinity stress increased plant susceptibility to disease. Unstressed plants were inoculated and salinity stress was imposed. Even though the same number of zoospores attached to the root after inoculation, severe root rot developed only on roots subsequently exposed to salinity stress. MacDonald concluded that shortly after inoculation defensive mechanisms are activated in unstressed roots, but subsequent salinity stress suppresses host response to inoculation, allowing successful establishment of the pathogen. Similar results were obtained in experiments with several citrus and tomato cultivars infected by *Phytophthora parasitica* (Blaker and MacDonald, 1986; Snapp and Shennan, 1994).

It has been suggested that the primary mechanism by which salinity stress increases a root's susceptibility to pathogen invasion is through the inhibition of biosynthesis of compounds that are toxic to the invading pathogen (Bell, 1981). Biosyntheses of such compounds are strongly inhibited in a saline environment (Campbell and Pitman, 1971; Poljakoff-Mayber, 1975).

The effect of salinity on disease development can be contradicting due to the fact that different pathogens react differently to salinity in different environmental conditions or with different cultivation practices. The application of chloride fertilizer as a common practice is one of the most common factors enhancing root zone salinity. Although increased root zone salinity can intensify plant stress, several studies have been conducted on the use of salt-type fertilizers for the control of soil-borne pathogen. Shefelbine et al. (1986) studied the effect of two chloride fertilizers on the control of common root rot disease on barley caused by *B. sorokiniana.* Use of KCl as a fertilizer significantly reduced disease ratings, but use of K_2SO_4 had no effect, indicating that the chloride anion was responsible for disease control. Similar effects have been observed by Christensen et al. (1981) for take-all disease in wheat.

Elmer (1990) reported beneficial effects of salt application in controlling Fusarium crown and root rot of asparagus caused by *F. oxysporum* and *F. moniliforme.* The author found that applications of NaCl decreased the disease incidence in infested soil better than KCl applications. He suggested that growers, by traditionally applying rock salt to asparagus plantings, may have unknowingly improved plant tolerance to this pathogen in certain areas.

3. *Relationship between Irrigation Methods and Disease Development*

Irrigation methods can be divided into four major categories: surface, sprinkler, micro-, and subirrigation. With surface irrigation, the water is applied at one edge of the field and flows across the soil surface by gravity, infiltrating the soil as the stream advances across the field. Surface irrigation is widely used for delivering water at relatively low cost. The graded furrow method of furrow irrigation is commonly used for annual row crops as well as for some tree and vine crops. Prior to the first irrigation, furrows, which convey the water across the field, are formed between the crop rows. Water is introduced at the upper end of the field to each furrow by syphon tubes, cutouts in the ditch, or from other pipe outlets.

Sprinkler irrigation is the application of water to the soil using a device that directs a stream of water through the air onto the soil. With sprinkler irrigation, the water is applied in a pressurized network system and released from sprinkler heads mounted on either fixed or moving supports. Water sprayed from a pressurized pipeline into the air breaks into drops which fall to the soil like rain. The size of the drops, their uniformity, and rate with which they fall are affected by the system design.

Microirrigation includes methods commonly known as low-pressure systems. Water is distributed in plastic pipes and released by drippers, tricklers, or sprayers. Such a method allows the water to infiltrate near the point it is applied. This is considered one of the best irrigation methods for enhancing water use efficiency.

Subirrigation systems provide water to crops by controlling the water-table level. Roots can use the capillary water available above the water table and extract what is needed without incurring in asphyctic situations.

All of these irrigation systems have advantages and disadvantages and each can impact disease development. Davis and Everson (1986) investigated the relation of Verticillium wilt of potato and irrigation methods. In their study, the authors found that the incidence and severity of Verticillium wilt was increased with furrow irrigation compared to sprinkler irrigation. They also detected more wilt problems where fields were irrigated with subirrigation systems than in fields that were sprinkler irrigated. These results agree with those reported by Guthrie (1960) and McMaster (1959). Because nitrogen availability is the variable most commonly associated with severity of Verticillium wilt of potato, Davis and Everson (1986) found that with furrow irrigation nitrogen often accumulated in the upper 7.5 to 15 cm of the soil profile. In contrast, with sprinkler irrigation, nitrogen was more uniformly distributed throughout the soil profile. When nitrogen was less available to the plants root system because of leaching and poor distribution, disease incidence and severity increased.

Biles et al. (1992) studied the effects of alternate and every-row furrow irrigation on disease incidence of Phytophthora root rot in chili pepper (*Capsicum annuum* L.). The disease, caused by *P. capsici* Leonian, is more severe when water stands in fields for a long time (more than 6 h), when late-season rains occur after irrigation in heavy soils, or in low areas in the field where water puddles (Ristaino, 1991; Shannon, 1989). They found that chili pepper plants in plots receiving alternate-row irrigation had lower disease incidence than plants in plots receiving every-row irrigation. This suggests that alternate-row irrigation may be an effective technique for management of Phytophthora root rot of chili pepper by providing adequate water for plant growth while avoiding soil saturation which favors disease.

Daniell and Falk (1994) conducted an economic comparison of Phytophthora root rot control methods following the Biles et al. (1992) study. They compared the net return of disease reduction strategies including every-row irrigation, alternate-row irrigation, and metam-sodium, copper sulfate, or metalaxyl application with alternate-row irrigation. Results indicated that red chili pepper producers should irrigate every row because it did not pay to hire labor to create borders to contain water within alternate rows. However, green chili producers were advised to water alternate rows because in every year of the study the model indicated that it paid to build borders. When chemicals were applied in conjunction with the alternate-row irrigation method, use of metalaxyl and copper sulfate were recommended to reduce disease risk in red and green chili pepper, respectively.

"Ebb and flow" or "flood and drain" are common names for a type of subirrigation system widely adopted for greenhouse production in Europe and is being more widely adopted in the U.S. Potted plants are placed on specially designed benches or concrete floors which are periodically flooded with water or fertilizer solution. The water enters the potting soil by capillary action, moistening the roots. The water or solution is then drained and kept in a tank for reuse when the next irrigation is needed.

Sanogo and Moorman (1990) reported that *Pythium* species can be transported by recycled nutrient solution to crops grown in the ebb-and-flow system, particularly if the irrigation water or nutrient solution is reused for many crops over a long period of time and becomes heavily infested.

They also determined that pathogen movement does not occur through the growing media, therefore, pathogen spread from pot to pot is not a threat to production where water is not recirculated.

Sprinkler-type irrigation systems are more flexible than others in terms of being able to deliver a combination of water and fertilizer or other chemicals to the plant. This is referred to as chemigation, and is usually used with sprinkler or microirrigation. Chemigation that incorporates fungicides into the system and is used with overhead sprinkler systems provides excellent results in controlling leaf diseases; however, good results have also been obtained in controlling *Rhizoctonia solani* Kühn, the cause of fruit rot in cucumber and tomato, crown rot in sugar beet, and limb blight and pod and peg rot in peanut (Csinos, 1987; McCarter and Barksdale, 1977; Potter and Schneider, 1981; Sumner and Smittle, 1976).

Sumner and Littrell (1989) studied the effect of chemigation using different compounds on soil fungi affecting peanut. They found that chlorothalonil applied by chemigation suppresses diseases on lower branches and protects pods and pegs against invasion by *R. solani*. In addition, chemigation benefits included reduced soil compaction and vine injury from tractor traffic for ground application of chemicals.

Ristaino and Averre (1992) analyzed the effects of irrigation and chemical treatments on *Streptomyces ipomoeae* (Person and W. J. Martin) Waksman and Henrici on sweet potato (*Ipomoea batatas* [L.] Lam.). This fungus develops on fibrous roots, leading to root rot. Scablike lesions that develop on storage roots reduce the marketability of the crop. Disease management includes the use of combined cultural practices such as soil fumigation, crop rotation, and irrigation management. In their study, they found that drip irrigation reduced disease severity. Disease reduction with drip irrigation was equivalent to the one obtained by treating the soil with sulfur, but it was not as effective as soil fumigation.

4. Relationship between Irrigation Amount, Frequency, and Disease Development

Water management is one of the more effective cultural practices for avoiding disease development in areas where pathogen populations are low, and for limiting damage and increasing net returns in heavily infested fields.

In the previous section, the effectiveness of different irrigation methods in controlling disease was discussed. It is important to choose the irrigation method that gives the most flexibility in controlling the amount of water delivered to a field and still remain cost effective. Surface irrigation methods, then, are probably less effective because they typically create saturated environments conducive to the increase and spread of many pathogens.

Several studies have been conducted to determine the effects of different irrigation regimes on the development of soil-borne pathogens and their disease incidence. Cappaert et al. (1994) reported on the impact of pre- and posttuberization irrigation on symptom expression of potato early dying (*Verticillium dahliae* Kleb.). They found a positive relationship between disease incidence and irrigation. Plants had earlier and greater disease symptoms with excessive irrigation prior to tuber initiation. Disease symptoms were not severe until 6 to 8 weeks after the irrigation treatment. This was probably due to the time necessary for the pathogen to develop and colonize the tuber. The authors concluded that altering irrigation practices during the period between emergence and tuber initiation may be a viable option to manage the disease.

Although high soil water content usually promotes high disease incidence, there are some exceptions to this rule. For example, high soil moisture in the upper 0.25 m of the soil profile from 5 days before to 6 weeks after tuber initiation greatly reduced the incidence of common scab (Davis et al., 1974). This was due to the beneficial effect of wet soil on the development of bacteria antagonistic to *Streptomyces scabies*, a fungus that prefers dry soils. Similar results were obtained by Ristaino (1993) with infection of *Streptomyces ipomoeae* on sweet potato. Disease severity was lower in plants grown in pots kept at field capacity or slightly below than in plants kept at low soil moisture content. These examples are significant in emphasizing the importance of taking soil assays before planting and/or during the growing season in order to determine which pathogens are affecting the crop and to plan management practices accordingly.

Variation in irrigation frequency and duration can also influence the rate at which soil-borne pathogens develop and cause yield losses. Ristaino et al. (1988) reported on the influence of furrow irrigation frequency and duration on the development of Phytophthora root rot on tomatoes. They found a significant positive correlation between irrigation frequency and duration and disease progress. Prolonged irrigation of inoculated plots caused an increase in disease severity compared to other treatments. Symptoms of disease appeared 8 days after irrigation treatment, whereas symptoms developed later in plants in the lower-duration-longer-interval treatment. Fruit yields followed the same pattern, with greater yield reduction found in plants in the more frequent and longer-duration irrigation treatment. The authors concluded that in Phytophthora-infested fields, avoidance of prolonged irrigation early in the season and application of well-timed, less frequent irrigations of shorter duration would increase yield.Similar results were obtained by Ristaino (1991) with bell peppers (*Capsicum annuum* L.) grown in fields infested with *Phytophthora capsici* Leonian, the pathogen responsible for Phytophthora root and crown rot. Disease development was significantly higher in fields that were frequently drip irrigated. A positive significant correlation was also found between the amount of rainfall and disease incidence. The amount of rain in a given rainfall can affect both the extent and duration of soil saturation and the amount of secondary inoculum spread; less secondary spread was found in fields where less rainfall occurred. Thus, the total amount of rain in a growing season may be less important than the amount of rain in a single event.

To better define the impact of irrigation on pathogen development, Hord and Ristaino (1992) determined in a pot experiment that a saturation period of at least 2 h was required for infection of pepper seedlings by oospores of *P. capsici*. Plants did not develop symptoms and did not become infected when held for 15 days at a soil moisture content lower than saturation. This suggested that, without a saturation period, either oospore germination did not occur or germinated oospores did not cause infection.

Similar trends were observed by Rush and Vaughn (1993) on sugar beet seeds germinated in miniplots infested with *A. cochlioides*. Irrigation treatments were significantly correlated with disease development. Plots irrigated for emergence after planting had a greater disease incidence than ones that received only preplanting irrigation. The authors attributed the lower disease incidence in the preplanting irrigation treatment to low soil matric potential. Zoospores of *A. cochlioides* typically infect the seedling hypocotyl after emergence. In preirrigation treatments, the soil was wet enough for seed germination but too dry for zoospore movement.

F. Pesticide Applications

Although the yield potential of all major crop species can be dramatically increased by genetic improvement and careful cultivar selection, the development of specific inorganic pesticides to control weeds, insects, and diseases has had a greater impact on crop yields over the last 50 years. The use of pesticides is now commonplace in most modern agricultural systems. The stability and broad-spectrum activity of many of the early pesticides resulted in soil and water pollution and gave pesticides in general a bad name, even though only a few were responsible for the majority of the problems. Most modern pesticides are formulated to be more target specific and short-lived in the environment. Therefore, with proper application, pesticides can now be safely used for the purpose for which they were designed with minimal environmental impact. However, any time a pesticide is applied to a crop, there is potential for nontarget effects. The following discussion will focus on the effects of soil fumigation and herbicides on disease incidence. Fungicides are covered in Chapter 3.

1. Fumigation

Soil fumigation with compounds such as 1,3-dichloropropene, metam-sodium, methyl bromide, and chloropicrin has been an effective method of reducing losses to nematodes and soil-borne fungal pathogens for many years. Depending on which chemical is used and the rate of application, fumigation can be very expensive. However, with high-value cash crops, the level of disease control makes the treatment economically feasible. This is especially true in multiple cropping systems where a single fumigation treatment controls disease in successive crops (Sumner and Phatak, 1988).

Fumigation is frequently used in conjunction with other disease management practices such as use of disease-tolerant cultivars or controlled irrigation. Ristaino and Averre (1992) used fumigation, sulfur, and irrigation management to control Streptomyces soil rot of sweet potato (*Ipomoea batatas* L.). Soil fumigation significantly reduced disease severity and increased root yields compared with nonfumigated plots. They found that fumigation reduced disease to a greater extent than either of the other control practices and that combination of treatments did not improve disease control. Harveson and Rush (1994) evaluated the effectiveness of rhizomania-tolerant cultivars and fumigation of sugar beets for control of a disease complex of soil-borne pathogens which included *R. solani*, *F. oxysporum* Schlechtend.:Fr. f. sp. *betae* (Stewart) Snyd. & Hans., *A. cochlioides* Drechs., and *P. betae* Keskin. Fumigation reduced disease severity and increased root yield and gross sucrose with most but not all cultivars. The use of locally adapted cultivars without fumigation gave better disease control than nonadapted cultivars with fumigation. The highest yield, however, was achieved with fumigation and a locally adapted cultivar with a high level of nonspecific disease tolerance.

Soil fumigation is typically very effective in reducing disease severity and increasing yields, but the effect of soil fumigants on nontarget organisms can have adverse effects on disease control and crop yields. Anderson (1993) reported that fumigation reduced inoculum density of *Chalara elegans* Nag Raj and Kendrick, the cause of black root rot of tobacco, but also reduced populations of the beneficial mycophagous amoeba *Vampyrella lateritia* Leidy. The population of *C. elegans* returned to prefumigation levels within a year, but populations of *V. lateritia* did not recover and remained low. Jawson et al. (1993) reported that fumigation increased yields in monoculture corn in only one of six site years. Furthermore, in rotated corn, fumigation reduced grain yield at 7 of 12 sites. Plants in fumigated plots were phosphorus deficient early in the growing season, and reduced yields were attributed to reduced mycorrhizal infection. They concluded that fumigation could have a considerable impact on corn growth and yield by reducing both deleterious and beneficial microorganisms.

2. Herbicides

Since the introduction of 2,4-D in the early 1940s, use of herbicides has steadily increased and are now a standard component of most agricultural production systems. Use of herbicides has especially increased as reduced tillage systems have replaced conventional tillage. As herbicide usage has increased so has the potential to impact nontarget organisms including plant pathogens. Several articles addressing interactions between herbicides and plant pathogens have been published (Altman, 1985; Altman and Campbell, 1977; Altman and Rovira, 1989; Moustafa-Mahmoud and Sumner, 1993).

Altman and Campbell (1977) suggested four primary herbicide effects that increased plant disease: (1) reduction of host structural defenses, (2) inhibition of antagonist microflora competing with pathogens, (3) direct stimulation of the pathogen, and (4) increased exudation from host plants of amino acids, carbohydrates, and other compounds. Supporting the exudation theory, Rovira and McDonald (1986) reported that chlorsulfuron (Glean) increased severity of Rhizoctonia bare patch and decreased yields in wheat, but had no effect on incidence of take-all. It was suggested that the herbicide increased the susceptibility of the host, possibly through increasing root exudations, rather than affecting virulence of the pathogen. Smiley and Wilkins (1992) also reported that chlorsulfuron increased the incidence and severity of Rhizoctonia root rot in eastern Oregon, but grain yields were not significantly affected. No explanation was offered for why chlorsulfuron increased disease severity.

Another way in which herbicides impact disease development is by elimination of alternate hosts. Control of weeds, especially wild grasses and volunteer wheat or barley, can reduce losses to Rhizoctonia root rot and take-all of wheat (Smiley et al., 1992; Wong et al., 1993). In the U.S. Pacific Northwest spring barley is frequently seeded directly into reduced or no-till wheat fields. Typically, nonselective herbicides such as glyphosate and 2,4-D are sprayed 1 to 3 days before barley is sown. Smiley et al. (1992) demonstrated that increasing the interval between herbicide application and planting reduced losses to Rhizoctonia root rot. They suggested that the inoculum potential of *R. solani* AG8 was strongly influenced by elimination of volunteer cereal and weeds and that careful timing of herbicide applications could minimize crop damage and maximize yields.

Rush and Gerik (1989) reached a similar conclusion with regard to Phymatotrichum root rot of cotton and grain sorghum. In central Texas, cotton is frequently rotated with grain sorghum. After the sorghum harvest, fields are lightly disked but many plants are not killed and regrowth occurs. When sorghum was sprayed with glyphosate soon after harvest instead of allowing regrowth to occur until frost, incidence of Phymatotrichum root rot in the following cotton crop was significantly reduced. Grain sorghum is not severely affected by *P. omnivorum* but can support mycelial strand growth and sclerotial production (Rush and Gerik, 1984). Rush and Gerik (1989) concluded that the increase in disease incidence in plots not sprayed with glyphosate was due to an increase in inoculum density resulting from prolonged availability of sorghum roots as substrate for saprophytic growth of *P. omnivorum.*

G. Harvest

Harvest marks the completion of a given production period. However, as previously stated, crop production should be viewed as a continuum of interrelated processes and not as a single operation with a specific beginning and end. Therefore, harvest can be considered as a point in a cycle and, just as agronomic practices that proceed that point can impact disease, agronomic practices at that point can also impact disease.

1. Timing

As with other agronomic practices, timing can have a significant impact on disease at harvest. Merida et al. (1994) evaluated the effects of cultivar and time of harvest on severity of silver scurf of potato (*Solanum tuberosum* L.) caused by *Helminthosporium solani* Kurieu & Mont. As the harvest date was delayed, disease severity increased in all cultivars. They concluded that disease was directly correlated to the time that tubers were exposed to inoculum in the field. Conversely, in some situations, delaying harvest can result in increased yields. Rhizomania of sugar beets can cause extreme reductions in both root yield and sucrose content. The disease is more severe under conditions of high temperature and soil moisture. Delaying harvest in the fall and allowing the beet crop to grow in cooler soils that are inopportune for disease development can result in increased sucrose content and total sucrose yields (Rush and Heidel, 1995).

Delaying harvest until crops are fully ripe, especially vegetables, may predispose them to a number of postharvest diseases. As vegetables ripen, sugars and other soluble solids increase, and this may partially account for the greater susceptibility of ripe fruit to postharvest fungal pathogens (Fallik et al., 1993). Furthermore, as harvests are delayed and crops ripen, some become more susceptible to mechanical damage which also predisposes them to postharvest diseases (Punja and Gaye, 1993). Conversely, harvesting grain crops early while the moisture content is still relatively high increases the likelihood of postharvest disease problems occurring in storage.

2. Crop Residue Disposal

Large amounts of soil and crop debris are often picked up by digging equipment during harvesting of root and tuber crops. This is especially problematic when fields are wet and soil adheres to harvested roots. The mixture of soil, leaf debris, and small or broken roots separated from sugar beets at harvest is called tare soil. Tare soil is frequently infested with pathogens and, in some instances, is collected and returned to the farmer for disposal. In the past, it was a common practice to dump tare soil back on the field that was being harvested or on an adjacent field. If the field being harvested was not obviously diseased, farmers believed this was a safe and convenient way to dispose of the tare soil. What they failed to consider was that the tare soil they received probably was mixed with soils from other farms which could have had severe disease problems. Because of the potential for spreading pathogens and infesting fields, this practice is now strongly discouraged. Any practice that results in movement of soil from one field to another has the potential to increase disease incidence. Disease management through sanitation is covered in Chapter XX.

III. CONCLUDING THOUGHTS

When human beings first began practicing agriculture, their impact on the land was small and was essentially in equilibrium with the environment. It is ironic and somewhat humbling to realize that, after thousands of years of crop production, present-day farmers are striving to achieve some of the same conditions that the earliest farmers started out with. In recent history, there have been those who envisioned a world without hunger or pestilence. Larger tractors would plow more land, herbicides would kill all unwanted weeds, and pesticides would eliminate insect and disease problems; humans would conquer the environment. Instead, excessive plowing resulted in devastating soil erosion, pesticide residues polluted land and water, and insects and plant pathogens developed resistance and became more problematic than ever. Definitely not a sustainable situation! Now scientists are being asked whether it is possible to develop agriculture systems that meet the needs of an ever-expanding population, but at the same time preserve our natural resources in a sustainable manner (Fulkerson, 1987; Teng and Yang, 1993). The best way to achieve this goal is to take full advantage of modern technologies and strive not to dominate the environment but develop agronomic practices that complement and strengthen natural processes.

One difficulty for plant pathologists involved in production agriculture is that disease management is only one aspect of the total crop production system farmers must deal with. Often, agronomic practices that are known to reduce disease incidence and severity, even if they are environmentally sound, are not economically feasible. In the Texas Panhandle, take-all of wheat can be managed by planting in mid to late October when soils temperatures are too cool for infection by *G. graminis* var. *tritici.* However, wheat must be planted in late August or early September to maximize forage production for cattle grazing. Because farmers frequently make more profit with a combination of winter grazing and grain production compared to grain production alone, they are unwilling to plant in October even though they know their losses to take-all will be less. The biggest challenge for plant pathologists may be developing a disease management strategy that fits into an existing farming system as opposed to suggesting the farmer change his entire system just to control one disease on one crop.

A second major problem in making disease management recommendations that involve agronomic practices is that there are few absolutes in disease control. Some pathogens are controlled by deep tillage, while others are not. Some diseases can be managed by irrigation, while irrigation makes other diseases worse. Extreme variability in pathogen or crop response to any given agronomic practice can occur from region to region, field to field, or acre to acre. However, with the continuing development of precision farming techniques, farmers are gaining the ability to apply specific treatments to smaller and smaller units of land (Robert et al., 1995). In the future, with the use of expert systems and precision farming practices, the problem of variability in crop response to recommended agronomic practices may be reduced. Expert systems for disease control will have the best chance for success if plant pathologists are involved with their development (Travis and Latin, 1991). At present, much of the basic ecological and epidemiological data required to develop expert systems are not available for many diseases and agronomic systems. However, expanded collaboration between plant pathologists, production agronomists, and those who develop the models for expert systems should speed the development of such systems.

REFERENCES

Allamaras, R. R., Kraft, J. M., and Miller, D. E., Effects of soil compaction and incorporated crop residue on root health, *Annu. Rev. Phytopathol.*, 26, 219, 1988.

Allamaras, R. R., Kraft, J. M., and Pikul, J. L., Lime and gypsum effects on pea-root-pathogen inoculum and related factors in a wheat-pea rotation, *Agron. J.*, 79, 439, 1987.

Altieri, M., Crop rotation and minimum tillage, in *Agroecology: The Scientific Basis of Alternative Agriculture*, Altieri, M., Ed., Westview Press, Boulder, CO, 1987.

Altman, J., Impact of herbicides on plant diseases, in *Ecology and Management of Soilborne Plant Pathogens*, Parker, C. A., Rovira, A. D., Moore, K. J., Wang, P. T. W., and Kollmorgen, J. F., Eds., American Phytopathological Society, St. Paul, MN, 1985.

Altman, J. and Campbell, C. L., Effect of herbicides on plant diseases, *Annu. Rev. Phytopathol.*, 15, 361, 1977.

Altman, J. and Rovira, A. D., Herbicide-pathogen interactions in soil-borne root diseases, *Can. J. Plant Pathol.*, 11, 166, 1989.

Anderson, T. R., Soil populations of the root rot fungal pathogen *Chalara elegans* and the mycophagous amoeba *Vampyrella lateritia* following soil fumigation, *Soil Biol. Biochem.*, 25, 223, 1993.

Arnon, I., *Crop Production in Dry Regions*, Vols. 1-2, Leonard Hill, London, 1972, 650.

Aycock, R., *Stem Rot and Other Diseases Caused by* Sclerotium rolfsii, North Carolina Agric. Exp. Stn. Tech. Bull., Raleigh, NC, 174, 1966.

Bailey, K. L., Mortensen, K., and Lafond, G. P., Effects of tillage systems and crop rotations on root and foliar diseases of wheat, flax, and peas in Saskatchewan, *Can. J. Plant Sci.*, 72, 583, 1992.

Baker, J. M., Wraith, J. M., and Dalton, F. N., Root function in water transport, in *Advances in Soil Science*, Vol. 19, Springer-Verlag, New York, 1992, 53.

Bell, A. A., Biochemical mechanisms of disease resistance, *Annu. Rev. Plant Physiol.*, 32, 21, 1981.

Bhatti, M. A. and Kraft, J. M., Influence of soil bulk density on root rot and wilt of chickpea, *Plant Dis.*, 76, 960, 1992.

Biles, C. L., Lindsey, D. L., and Liddell, C. M., Control of *Phytophthora* root rot of chile peppers by irrigation practices and fungicides, *Crop Prot.*, 11, 225, 1992.

Blaker, N. S. and MacDonald, J. D., Predisposing effects of soil moisture extremes on the susceptibility of rhododendron to Phytophthora root and crown rot, *Phytopathology*, 71, 831, 1981.

Blaker, N. S. and MacDonald, J. D., The role of salinity in the development of phytophthora root rot of citrus, *Phytopathology*, 76, 970, 1986.

Blum, U., Gerig, T. M., Worsham, A. D., Holappa, L. D., and King, L. D., Allelopathic activity in wheat-conventional and wheat-no-till soils. Development of soil extract bioassays, *J. Chem. Ecol.*, 18, 2191, 1992.

Blunt, S. J., Asher, M. J. C., and Gilligan, C. A., Infection of sugar beet by *Polymyxa betae* in relation to soil temperature, *Plant Pathol.*, 40, 257, 1991.

Blunt, S. J., Asher, M. J. C., and Gilligan, C. A., The effect of sowing date on infection of sugar beet by *Polymyxa betae*, *Plant Pathol.*, 41, 148, 1992.

Bockus, W. W., Davis, M. A., and Norman, B. L., Effect of soil shading by surface residues during summer fallow on take-all of winter wheat, *Plant Dis.*, 78, 50, 1994.

Boosalis, M. G., Doupnik, B. L., and Watkins, J. E., Effect of surface tillage on plant diseases, in *No-Tillage and Surface-Tillage Agriculture, The Tillage Revolution*, Sprague, M. A. and Triplett, G. B., Eds., John Wiley & Sons, New York, 1986, 389.

Bowers, J. H. and Parke, J. L., Epidemiology of Pythium damping-off and aphanomyces root rot of peas after seed treatment with bacterial agents for biological control, *Phytopathology*, 83, 1466, 1993.

Bradow, J. M., Inhibitions of cotton seedling growth by volatile ketones emitted by cover crop residues, *J. Chem. Ecol.*, 19, 1085, 1993.

Bradow, J. M. and Connick, W. J., Jr., Volatile seed germination inhibitors from plant residues, *J. Chem. Ecol.*, 16, 645, 1990.

Brennan, R. F., Effect of ammonium chloride, ammonium sulfate, and sodium nitrate on take-all and grain yield of wheat grown on soils in south-western Australia, *J. Plant Nutr.*, 16, 349, 1993.

Brenneman, T. B. and Sumner, D. R., Effects of tractor traffic and chlorothalonil applied via ground sprays or center pivot irrigation systems on peanut diseases and pod yields, *Plant Dis.*, 74, 277, 1990.

Brenneman, T. B., Sumner, D. R., Baird, R. E., Burton, G. W., and Minton, N. A., Suppression of foliar and soilborne peanut diseases in bahiagrass rotations, *Phytopathology,* 85, 948, 1995.

Browning, J. A., Simons, M. D., and Torres, E. A., The components of disease — the disease pyramid, in *Plant Dis., An Advanced Treatise,* Vol. 1, Horsfall, J. G. and Cowling, E. B., Eds., Academic Press, New York, 1977, 196.

Bullock, D. G., Crop rotation, *Crit. Rev. Plant Sci.,* 11, 309, 1992.

Burdon, J. J. and Chilvers, G. A., Host density as a factor in plant disease etiology, *Annu. Rev. Phytopathol.,* 20, 143, 1982.

Burke, D. W., Hagedorn, D. J., and Mitchell, J. E., Soil conditions and distributions of pathogens in relation to pea root rot in Wisconsin soils, *Phytopathology,* 60, 403, 1970.

Burke, D. W. and Miller, D. E., Control of fusarium root rot with resistant beans and cultural management, *Plant Dis.,* 67, 1312, 1983.

Callan, N. W., Mathre, D. E., and Miller, J. B., Bio-priming seed treatment for biological control of *Pythium ultimum* preemergence damping-off in Sh2 sweet corn, *Plant Dis.,* 74, 368, 1990.

Campbell, L. C. and Pitman, M. G., Salinity and plant cells, in *Salinity and Water Use,* Talsma, T. and Phillips, J. R., Eds., Wiley Interscience, New York, 1971, 296.

Campbell, R. N. and Greathead, A. S., Control of clubroot of crucifers by liming, in *Soilborne Pathogens: Management of Diseases with Macro- and Microelements,* Engelhand, A. W., Ed., APS Press, St. Paul, MN, 1989.

Campbell, W. A. and Copeland, O. L., Little Leaf Disease of Shortleaf and Loblolly Pines, Circular 940, U.S. Department of Agriculture, Washington, D.C., 1954.

Camprag, D., The effect of irrigation on the occurrence of some important sugar beet pests in Yugoslavia and other country of southeastern Europe, Proc. 39th Winter Congr. Int. Inst. Sugar Beet Research, 1976, 449.

Cappaert, M. R., Powelson, M. L., Christensen, N. W., and Crowe, F. J., Influence of irrigation on severity of potato early dying and tuber yield, *Phytopathology,* 82, 1448, 1992.

Cappaert, M. R., Powelson, M. L., Christensen, N. W., Stevenson, W. R., and Rouse, D. I., Assessment of irrigation as a method of managing potato early dying, *Phytopathology,* 84, 792, 1994.

Chastain, T. G., Ward, K. J., and Wysocki, D. J., Stand establishment responses of soft white winter wheat to seedbed residue and seed size, *Crop Sci.,* 35, 213, 1995.

Christensen, N. W., Taylor, R. G., Jackson, T. L., and Mitchell, B. L., Chloride effects on water potentials and yield of winter wheat infected with take-all root rot, *Agron. J.,* 73, 1053, 1981.

Colyer, P. D. and Vernon, P. R., Effect of tillage on cotton plant populations and seedling diseases, *J. Prod. Agric.,* 6, 108, 1993.

Cook, R. J., The influence of rotation crops on take-all decline phenomenon, *Phytopathology,* 71, 189, 1981.

Cook, R. J., Wheat root health management and environmental concern, *Can. J. Plant Pathol.,* 14, 76, 1992.

Cook, R. J. and Baker, K. F., Agricultural practices and biological control, in *The Nature and Practice of Biological Control of Plant Pathogens,* American Phytopathological Society, St. Paul, MN, 1983, 390.

Cook, R. J. and Haglund, W. A., Wheat yield depression associated with conservation tillage caused by root pathogens in the soil not phytotoxins from the straw, *Soil Biol. Biochem.,* 23, 1125, 1991.

Cook, R. J. and Rovira, A. D., The role of bacteria in the biological control of *Gaeumannomyces graminis* by suppressive soils, *Soil Biol. Biochem.,* 8, 269, 1976.

Crookston, R. K., The rotation effect, *Crops Soil Mag.,* 36, 12, 1984.

Crookston, R. K., Kurle, J. E., Copeland, P. J., Ford, J. H., and Lueschen, W. E., Rotational cropping sequence affects yield of corn and soybean, *Agron. J.,* 83, 108, 1991.

Csinos, A. S., Control of southern stem rot and rhizoctonia limb rot of peanut with flutolanil, *Peanut Sci.,* 14, 55, 1987.

Daberkow, S. and Gill, M., Common crop rotations among major field crops, in Agricultural Resources: Inputs Situation and Outlook Report, AR-15, Economic Research Service, USDA, Washington, D.C., 1989.

Daniell, I. R. and Falk, C. L., Economic comparison of phytophthora root rot control methods, *Crop Prot.,* 13, 331, 1994.

Dao, T. H. and Unger, P. W., Agronomic practices in relation to soil amendments and pesticides, in *Soil Amendments and Environmental Quality,* Rechcigl, J. E., Ed., Lewis Publishers, Boca Raton, FL, 1995, chap. 11.

Davis, J. R. and Everson, D. O., Relation of *Verticillium dahliae* in soil and potato tissue, irrigation method, and N-fertility to verticillium wilt of potato, *Phytopathology,* 76, 730, 1986.

Davis, J. R., McMaster, G. M., Callihan, R. H., Garner, J. G., and McDole, R. E., The relationship of irrigation timing and soil treatments to control potato scab, *Phytopathology,* 64, 1404, 1974.

Deacon, J. W., Further studies on *Phialophora radicicola* and *Gaeumannomyces graminis* on roots and stem bases of grasses and cereals, *Trans. Br. Mycol. Soc.,* 63, 307, 1974.

Delroy, N. D. and Bowden, J. W., Effect of deep ripping, the previous crop, and applied nitrogen on the growth and yield of a wheat crop, *Aust. J. Exp. Agric.,* 26, 469, 1986.

Dodman, R. L. and Wildermuth, G. B., The effect of stubble retention and tillage practices in wheat and barley on crown rot caused by *Fusarium graminearum* Group 1, *Plant Prot. Q.,* 4, 98, 1989.

Douglas, C. L., Wilkins, D. E., and Churchill, D. B., Tillage, seed size, and seed density effects on performance of soft white winter wheat, *Agron. J.,* 86, 707, 1994.

Duczek, L. J., The effect of flooding at early growth stage and mechanical damage of foliage on common root rot of spring wheat and barley, *Can. J. Plant Pathol.,* 8, 418, 1986.

Duczek, L. J., Relationship between common root rot [*Cochliobolus sativus*] and tillering in spring wheat, *Can. J. Plant Pathol.,* 11, 39, 1989.

Duczek, L. J., The effect of soil salinity on common root rot of spring wheat and barley, *Can. J. Plant Sci.,* 73, 323, 1993.

Duczek, L. J. and Piening, L. J., Effect of seeding depth, seeding date, and seed size on common root rot of spring barley, *Can. J. Plant Sci.,* 62, 885, 1982.

Duniway, J. M., Predisposing effect of water stress on the severity of Phytophthora root rot of safflower, *Phytopathology,* 67, 884, 1977.

Elmer, W. H., Suppression of fusarium crown and root rot of asparagus with chloride and different forms of nitrogen fertilizers, *Acta Hortic.,* 271, 323, 1990.

Engel, R. E. and Mathre, D. E., Effect of fertilizer nitrogen source and chloride on take-all of irrigated hard red spring wheat, *Plant Dis.,* 72, 393, 1988.

Ezekiel, W. N. and Fudge, J. F., Studies on the cause of immunity of monocotyledonous plants to phymatotrichum root rot, *J. Agric. Res.,* 56, 773, 1938.

Fallik, E., Klein, J., Grinberg, S., Lomaniec, E., Lurie, S., and Lalazar, A., Effect of postharvest heat treatment of tomatoes on fruit ripening and decay caused by *Botrytis cinerea, Plant Dis.,* 77, 985, 1993.

Firman, D. M. and Allen, E. J., Effects of seed size, planting density and planting pattern on the severity of silver scurf *(Helininthosporium solani)* and black scurf *(Rhizoctonia solani)* diseases on potatoes, *Ann. Appl. Biol.,* 127, 73, 1995.

Forster, R. L. and Schaad, N. W., Control of black chaff of wheat with seed treatment and a foundation seed health program, *Plant Dis.,* 72, 935, 1988.

Francis, C. A. and Clegg, M. D., Crop rotations in sustainable production systems, in *Sustainable Agricultural Systems,* Edwards, C. A., Ed., Soil and Water Conservation Society, Ankeny, IA, 1990.

Fredrickson, J. K., Elliott, L. F., and Engibous, J. C., Crop residues as substrate for host-specific inhibitory pseudomonads, *Soil Biol. Biochem.,* 19, 127, 1987.

Fry, W. E., *Principles of Plant Disease Management,* Academic Press, New York, 1982.

Fulkerson, J. F., Biological impact and assessment: biological potential over time, *BioScience,* 37, 187, 1987.

Gan, Y. and Stobbe, E. H., Effect of variation in seed size and planting depth on emergence, infertile plants, and grain yield of spring wheat, *Can. J. Plant Sci.,* 75, 565, 1995.

Garrett, S. D., *Pathogenic Root Infecting Fungi,* Cambridge University Press, Cambridge, 1970, 294.

Gerik, J. S. and Temple, S. R., Comparison of direct seeding and seedling transplanting on yield loss in sugar beet due to rhizomania, *J. Sugar Beet Res.,* 28, 71, 1991 (Abstr.).

Gill, M. and Daberkow, S., Crop sequences among 1990 major field crops and associated farm program participation, in Agricultural Resources: Inputs Situation and Outlook Report, AR-24, Economic Research Service, USDA, Washington, D.C., 1991.

Glenn, O. F. and Sivasithamparam, K., The effect of soil compaction on the saprophytic growth of *Rhizoctonia solani, Plant Soil,* 121, 282, 1990.

Glynne, M. D., Crop sequence in relation to soil-borne pathogens, in *Ecology of Soil-borne Plant Pathogens: Prelude to Biological Control,* Baker, K. F. and Snyder, W. C., Eds., University of California Press, Berkeley, 1965.

Grant, C. A. and Bailey, L. D., The effect of KC1, KNO_3, and $CaCl_2$ fertilization under conventional and zero-till systems on common root rot, dry matter yield and grain yield of Heartland barley, *Can. J. Plant Sci.*, 74, 1, 1994.

Gray, L. E. and Pope, R. A., Influence of soil compaction on soybean stand, yield, and phytophthora root rot incidence, *Agron. J.*, 78, 189, 1986.

Grey, W. E. and Mathre, D. E., Reaction of spring barleys to common root rot and its effect on yield components, *Can. J. Plant Sci.*, 64, 245, 1984.

Grossman, J., Brassica alternatives to herbicides and soil fumigants, *IPM Practitioner*, 15, 1, 1993.

Guthrie, J. W., *Early Dying* (Verticillium wilt) *of Potatoes in Idaho*, Univ. Idaho Res. Bull. 45, University of Idaho Press, Moscow, ID, 1960.

Hale, M. G., Moore, L. D., and Griffin, G. J., Root exudates and exudation, in *Interactions between Non-pathogenic Soil Microorganisms and Plants*, Dommengues, Y. R. and Krupa, S. V., Eds., Elsevier, New York, 1978.

Hall, M. R. and Sumner, D. R., Influence of cultivar and primed or germinated seed on stand establishment of watermelon in soil infested with *Pythium irregulare* or *Rhizoctonia solani* AG-4, *Crop Prot.*, 13, 443, 1994.

Halsall, D. M., Leigh, J. H., Gollasch, S. E., and Holgate, M., The role of allelopathy in legume decline in pastures. II. Comparative effects of pasture, crop and weed residues on germination, nodulation and root growth, *Aust. J. Agric. Res.*, 46, 189, 1995.

Hampson, M. C., Pathogenesis of *Synchitrium endobloticium*. V. Wart disease suppression in potato soils amended with urea and/or ammonium nitrate in relation to soil pH, *Plant Soil*, 87, 241, 1985.

Harmon, G. E. and Taylor, A. G., Improved seedling performance by integration of biological control agents at favorable pH levels with solid matrix priming, *Phytopathology*, 78, 520, 1988.

Harper, S. H. T. and Lynch, J. M., The kinetics of straw decomposition in relation to its potential to produce the phytotoxin acetic acid, *J. Soil Sci.*, 32, 627, 1981.

Harveson R. M. and Rush, C. M., An environmentally controlled experiment to monitor the effect of aphanomyces root rot and rhizomania on sugar beet, *Phytopathology*, 83, 1220, 1993.

Harveson, R. M. and Rush, C. M., Evaluation of fumigation and rhizomania-tolerant cultivars for control of a root disease complex of sugar beets, *Plant Dis.*, 78, 1197, 1994.

Harveson, R. M. and Rush, C. M., Evaluation of cultivar blends and irrigation frequnecy for control of multiple root rot pathogens of sugar beet, *Biol. Cult. Tests Control Plant Dis.*, 10, 19, 1995.

Harveson, R. M., Rush, C. M., and Vaughn, K. M., Seed treatment with Tachigaren for controlling black root of sugar beet, *Fungicide Nematicide Tests*, 49, 277, 1994.

Harveson, R. M., Rush, C. M., and Wheeler, T. A., The spread of beet necrotic yellow vein virus from point source inoculations as influenced by irrigation and tillage, *Phytopathology*, 86, 1242, 1996.

Heather, D. W. and Sieczka, J. B., Effect of seed size and cultivar on emergence and stand establishment of broccoli in crusted soil, *J. Am. Soc. Hortic. Sci.*, 116, 946–949, 1991.

Hecker, R. J. and Ruppel, E. G., Inheritance of resistance to rhizoctonia root rot in sugarbeet, *Crop Sci.*, 15, 487, 1975.

Hendrix, J. W., Jones, K. J., and Nesmith, W. C., Control of pathogenic mycorrhizal fungi in maintenance of soil productivity by crop rotation, *J. Prod. Agric.*, 5, 383, 1992.

Herd, G. W. and Phillips, A. J. L., Control of seed-borne *Sclerotinia sclerotioum* by fungicidal treatment of sunflower seed, *Plant Pathol.*, 37, 202, 1988.

Herrman, T. J., Forster, R. L., and Martin, J. M., Imazalil seed treatment reduces common root rot and increases yield of barley under commercial conditions, *Plant Dis.*, 74, 246, 1990.

Hicks, S. K., Wendt, C. W., Gannaway, J. R., and Baker, R. B., Allelopathic effects of wheat straw on cotton germination, emergence, and yield, *Crop Sci.*, 29, 1057, 1989.

Hide, G. A., Welham, S. J., Read, P. J., and Ainsley, A. E., Influence of planting seed tubers with gangrene *(Phoma foveata)* and of neighboring healthy, diseased and missing plants on the yield and size of potatoes, *J. Agric. Sci.*, 125, 51, 1995.

Hord, M. J. and Ristaino, J. B., Effect of the matric component of soil water potential on infection of pepper seedlings in soil infested with oospores of *Phytophthora capsici*, *Phytopathology*, 82, 792, 1992.

Huber, D. M., The role of nutrition in the take-all disease of wheat and other small grains, in *Soilborne Pathogens: Management of Diseases with Macro- and Microelements*, Engelhand, A. W., Ed., APS Press, St. Paul, MN, 1989, 46.

Huber, D. M. and Watson, R. D., Nitrogen form and plant disease, *Annu. Rev. Phytopathol.*, 12, 139, 1974.

Huisman, O. C. and Ashworth, L. J., Influence of crop rotation on survival of *Verticillium albo-atrum* in soils, *Phytopathology*, 66, 978, 1976.

Jarvis, W. R., Allelopathic control of *Fusarium oxysporum* F. sp. *radicis-lycopersici*, NATO ASI Series, Vol. H28, *Vascular Wilt Diseases of Plants*, Tjamos, E. C. and Beckman, C., Eds., Springer-Verlag, Berlin, 1989, 479.

Jawson, M. D., Franzluebbers, A. J., Galusha, D. K., and Aiken R. M., Soil fumigation within monoculture and rotations: Response of corn and mycorrhizae, *Agron. J.*, 85, 1174, 1993.

Johnson, A. W., Golden, A. M., Auld, D. L., and Sumner, D. R., Effects of rapeseed and vetch as green manure crops and fallow on nematodes and soil-borne pathogens, *J. Nematol.*, 24, 117, 1992.

Jones, J. P., Engelhand, A. W., and Woltz, S. S., Management of fusarium wilt of vegetables and ornamentals by macro- and microelement nutrition, in *Soilborne Pathogens: Management of Diseases with Macro- and Microelements*, Engelhand, A. W., Ed., APS Press, St. Paul, MN, 1989, 18.

Kao, C. W. and Ko, W. H., The role of calcium and microorganisms in suppression of cucumber damping-off caused by *Pythium splendens* in a Hawaiian soil, *Phytopathology*, 76, 221, 1986.

Keinath, A. P. and Loria, R., Management of common scab of potato with plant nutrients, in *Soilborne Pathogens: Management of Diseases with Macro- and Micro-elements*, Engelhand, A. W., Ed., APS Press, St. Paul, MN, 1989, 152.

Kimber, R. W. L., Phytotoxicity from plant residues. III. The relative effect of toxins and nitrogen immobilization on the germination and growth of wheat, *Plant Soil*, 38, 543, 1973.

Kraft, J. M. and Allmaras, R. R., Pea root pathogen populations in relation to soil structure, compaction, and water content, in *Ecology and Management of Soilborne Plant Pathogens*, Parker, C. A., Rovira, A. D., Moore, K. J., Wong, P. T. W., and Kollmorgen, J. F., Eds., American Phytopathological Society, St. Paul, MN, 1985, 203.

Kraft, J. M. and Giles, R. A., Increasing green pea yields with root-rot resistance and subsoiling, in *Soil-Borne Plant Pathogens*, Schippers, B. and Gams, W., Eds., Academic Press, New York, 1979.

Kraft, J. M. and Wilkins, D. E., The effects of pathogen numbers and tillage on root disease severity, root length, and seed yields in green peas, *Plant Dis.*, 73, 884, 1989.

Kramer, P. J., Drought stress, and the origin of adaptations, in *Adaptations of Plants to Water and High Temperature Stress*, Turner, N. C. and Kramer, P. J., Eds., John Wiley & Sons, New York, 1980, 7.

Kuan, T. L. and Erwin, D. C., Predisposition effect of water saturation of soil on Phytophthora root rot of alfalfa, *Phytopathology*, 70, 981, 1980.

Lawton, H. W. and Wilke, P. J., Ancient agricultural systems in dry regions, in *Agriculture in Semi-Arid Environments*, Hall, A. E., Cannell, G. H., and Lawton, H. W., Eds., Springer-Verlag, New York, 1979, 1.

Lund, B. M. and Kelman, A., Determination of the potential for development of bacterial soft rot *(Erwinia carotovora)* of potatoes, *Am. Potato J.*, 54, 211, 1977.

Lund, M. G., Carter, P. R., and Oplinger, E. S., Tillage and crop rotation affect corn, soybean, and winter wheat yields, *J. Prod. Agric.*, 6, 207, 1993.

MacDonald, J. D., Salinity effects on the susceptibility of chrysanthemum roots to *Phytophthora cryptogea*, *Phytopathology*, 74, 621, 1984.

Maloy, O. C., *Plant Dis., Control*, John Wiley & Sons, Inc., New York, 1993.

Martin, J. H., Leonard, W. J., and Stamp, D. L., *Principles of Field Crop Production*, 3rd ed., Macmillan, New York, 1976.

Mathieson, J. T. and Rush, C. M., Influence of temperature and five fungicides on rhizoctonia root rot of hard red winter wheat, *Plant Dis.*, 75, 983, 1991.

Mathieson, J. T., Rush, C. M., Bordovsky, D., Clark, L. E., and Jones, O. R., Effects of tillage on common root rot of wheat in Texas, *Plant Dis.*, 74, 1006, 1990.

Mayton, H. S., Olivier, C., Vaughn, S. F., and Loria, R., Correlation of fungicidal activity of *Brassica* species with allyl isothiocyanate production in macerated leaf tissue, *Phytopathology*, 86, 267, 1996.

McCarter, S. M. and Barksdale, T. H., Effectiveness of selected chemicals in controlling rhizoctonia fruit rot of tomato in greenhouse and field tests, *Plant Dis. Rep.*, 61, 811, 1977.

McMaster, G., *A New Concept in Sprinkler Irrigation*, Univ. Idaho Prog. Rep. 17, University of Idaho, Moscow, ID, 1959.

Merida, C. L., Loria, R., and Halseth, D. E., Effects of potato cultivar and time of harvest on the severity of silver scurf, *Plant Dis.*, 78, 146, 1994.

Metcalfe, D. S. and Elkins, D. M., Rotations and cropping systems, in *Crop Production Principles and Practices*, 4th ed., Macmillan, New York, 1980, chap. 14.

Meyer, J., Shew, H. D., and Shoemaker, P. B., Populations of *Thielaviopsis basicola* and the occurrence of black root rot on burley tobacco in Western North Carolina, *Plant Dis.*, 73, 239, 1989.

Miller, D. E., Root systems in relation to stress tolerance, in *Root Systems of Vegetable Crops*, Kahn, B. A., Ed., *Horticultural Sci.*, 21, 963, 1986.

Miller, D. E. and Burke, D. W., Reduction of fusarium root rot and sclerotinia wilt in beans with irrigation, tillage, and bean genotype, *Plant Dis.*, 70, 163, 1986.

Miller, D. E., Burke, D. W., and Kraft, J. M., Predisposition of bean roots to attack by the pea pathogen, *Fusarium solani* f. sp. *pisi*, due to temporary oxygen stress, *Phytopathology*, 70, 1221, 1980.

Minton, N. A., Csinos, A. S., and Morgan, L. W., Relationship between tillage and nematicide, fungicide, and insecticide treatments on pests and yield of peanuts double-cropped with wheat, *Plant Dis.*, 74, 1025, 1990.

Moore, K. J. and Cook, R. J., Increased take-all of wheat with direct drilling in the Pacific Northwest, *Phytopathology*, 74, 1044, 1984.

Moustafa-Mahmoud, S. M. and Sumner, D. R., Interaction of fungicides, herbicides, and planting date with seedling disease of cotton caused by *Rhizoctonia solani* AG-4, *Plant Dis.*, 77, 79, 1993.

Muehlchen, A. M., Rand, R. E., and Parke, J. L., Evaluation of crucifer green manures for controlling aphanomyces root rot of peas, *Plant Dis.*, 74, 651, 1990.

Nene, Y. L., Multiple-disease resistance in grain legumes, *Annu. Rev. Phytopathol.*, 26, 203, 1988.

Osburn, R. M. and Schroth, M. N., Effect of osmopriming sugar beet seed on germination rate and incidence of *Pythium ultimum* damping-off, *Plant Dis.*, 73, 21, 1989.

Oyarzun, P., Gerlagh, M., and Hoogland, A. E., Relation between cropping frequency of peas and other legumes and foot and root rot in peas, *Neth. J. Plant Pathol.*, 99, 35, 1993.

Palti, J., *Cultural Practices and Infectious Crop Diseases*, Springer-Verlag, New York, 1981.

Parke, J. L., Rand, R. E., Joy, A. E., and King, E. B., Biological control of pythium damping-off and aphanomyces root rot of peas by application of *Pseudomonas cepacia*, or *P. fluorescens* to seed, *Plant Dis.*, 75, 987, 1991.

Parry, D. W., *Plant Pathology in Agriculture*, Cambridge University Press, 1990, Cambridge, 1990.

Payne, P. A. and Williams, G. E., Hymexazol treatment of sugar-beet seed to control seedling disease caused by *Pythium* spp. and *Aphanomyces cochlioides*, *Crop Prot.*, 9, 371, 1990.

Pedersen, W. L. and Leath, S., Pyramiding major genes for resistance to maintain residual effects, *Annu. Rev. Phytopathol.*, 26, 369, 1988.

Perez, F. J. and Ormeno-Nunez, J., Difference in hydroxamic acid content in roots and root exudates of wheat (*Triticum aestivum* L.) and rye (*Secale cereale* L.). Possible role in allelopathy, *J. Chem. Ecol.*, 17, 1037, 1991.

Phillips, R. E. and Phillips, S. H., *No-Tillage Agriculture: Principles and Practices*, Van Nostrand Reinhold, New York, 1984.

Phillips, S. H. and Thomas, G. W., Multicropping, in *No-Tillage Agriculture: Principles and Practices*, Phillips, R. E. and Phillips, S. H., Eds., Van Nostrand Reinhold, New York, 1984, 231.

Phillips, S. H. and Young, H. M., Jr., *No-Tillage Farming*, Reiman Associates, Milwaukee, WI, 1973.

Piening, L., Differential yield responses of ten barley cultivars to common root rot, *Can. J. Plant Sci.*, 53, 763, 1973.

Pierce, F. J. and Rice, C. W., Crop rotation and its impact on efficiency of water and nitrogen use, in *Cropping Strategies for Efficient Use of Water and Nitrogen*, Publ. 51, Hargrove, W. L., Ed., American Society of Agronomy, Madison, WI, 1988.

Pieters, A. J., *Green Manuring Principles and Practices*, John Wiley & Sons, New York, 1927.

Poljakoff-Mayber, A., Morphological and anatomical changes in plants as a response to salinity stress, in *Plants in Saline Environments*, Poljakoff-Mayber, A. and Gale, J., Eds., Springer-Verlag, New York, 1975.

Potter, H. S. and Schneider, C. L., Control of rhizoctonia crown rot diseases of sugarbeet with fungicides applied by sprinkler irrigation, *J. Am. Soc. Sugar Beet Technol.*, 21, 50, 1981.

Punja, Z., Influence of nitrogen and calcium compounds on development of disease due to *Sclerotium rolfsii*, in *Soilborne Pathogens: Management of Diseases with Macro- and Microelements*, Engelhand, A. W., Ed., APS Press, St. Paul, MN, 1989.

Punja, Z. K. and Gaye, M.-M., Influence of postharvest handling practices and dip treatments on development of black root rot on fresh market carrots, *Plant Dis.*, 77, 989, 1993.

Rasmussen, S. L. and Stanghellini, M. E., Effect of salinity stress on development of Pythium blight in *Agrostis palustris*, *Phytopathology*, 78, 1495, 1988.

Reeves, D. W., Cover crops and rotations, in *Crops Residue Management*, Lewis Publishers, Boca Raton, FL, 1994, 125.

Reis, E. M. and Abrao, J. J. R., Effect of tillage and wheat residue management on the vertical distribution and inoculum density of *Cochliobolus sativus* in soil, *Plant Dis.*, 67, 1088, 1983.

Rickerl, D. H., Curl, E. A., Toughton, J. T., and Gordon, W. B., Crop mulch effects on *Rhizoctonia* soil infestation and disease severity in conservation-tilled cotton, *Soil Biol. Biochem.*, 24, 553, 1992.

Ristaino, J. B., Influence of rainfall, drip irrigation, and inoculum density on the development of *Phytophthora* root and crown rot epidemics and yield in bell pepper, *Phytopathology*, 81, 922, 1991.

Ristaino, J. B., Infection of sweetpotato fibrous roots by *Streptomyces ipomoeae*. Influence of soil water potential, *Soil Biol. Biochem.*, 25, 185, 1993.

Ristaino, J. B. and Averre, C. W., Effects of irrigation, sulfur, and fumigation on streptomyces soil rot and yield components in sweetpotato, *Phytopathology*, 82, 670, 1992.

Ristaino, J. B., Duniway, J. M., and Marois, J. J., Influence of frequency and duration of furrow irrigation on the development of phytophthora root rot and yield in processing tomatoes, *Phytopathology*, 787, 1701, 1988.

Robert, P. C., Rust, R. H., and Larson, W. E., Site-Specific Management for Agricultural Systems, in *Proc. Second Int. Conf.*, American Society Agronomy, Crop Science Society of America, and Soil Science Society of America, Madison, WI, 1995.

Rogers, C. H., *Cotton Root Rot Studies with Special Reference to Sclerotia, Cover Crops, Rotations, Tillage, Seeding Rates, Soil Fungicides, and Effects on Seed Quality*, Texas Agric. Exp. Stn. Bull., College Station, TX, 614, 1942.

Rothrock, C. S., Take-all of wheat as affected by tillage and wheat-soybean doublecropping, *Soil Biol. Biochem.*, 19, 307, 1987.

Rothrock, C. S., Relative susceptibility of small grains to take-all, *Plant Dis.*, 72, 883, 1988.

Rothrock, C. S., Influence of soil temperature, water, and texture on *Thielaviopsis basicola* and black root rot of cotton, *Phytopathology*, 82, 1202, 1992a.

Rothrock, C. S., Tillage systems and plant disease, *Soil Sci.*, 154, 308, 1992b.

Rothrock, C. S. and Cunfer, B. M., Absence of take-all decline in double-cropped fields, *Soil Biol. Biochem.*, 18, 113, 1986.

Rothrock, C. S. and Cunfer, B. M., Influence of small grain rotations on take-all in a subsequent wheat crop, *Plant Dis.*, 75, 1050, 1991.

Rothrock, C. S. and Hargrove, W. L., Influence of legume cover crops and conservation tillage on soil populations of selected fungal genera, *Can. J. Microbiol.*, 34, 201, 1988.

Rothrock, C. S., Kirkpatrick, T. L., Frans, R. E., and Scott, H. D, The influence of winter legume cover crops on soilborne plant pathogens and cotton seedling diseases, *Plant Dis.*, 79,167, 1995.

Rovira, A. D., Influence of crop rotation and tillage on rhizoctonia bare patch of wheat, *Phytopathology*, 76, 669, 1986.

Rovira, A. D. and McDonald, H. J., Effects of the herbicide chlorsulfuron on rhizoctonia bare patch and take-all of barley and wheat, *Plant Dis.*, 70, 879, 1986.

Rush, C. M., The relationship between hollow heart of pea and seed electrolyte loss, disease susceptibility, and plant growth, *Phytopathology*, 77, 1533, 1987.

Rush, C. M., Carling, D. E., Harveson, R. M., and Mathieson, J. T., Prevalence and pathogenicity of anastomosis groups of *Rhizoctonia solani* from wheat and sugar beet in Texas, *Plant Dis.*, 78, 349, 1994.

Rush, C. M. and Gerik, T. J., Interactions between *Phymatotrichum omnivorum* and *Sorghum bicolor*, *Plant Dis.*, 68, 500, 1984.

Rush, C. M. and Gerik, T. J., Relationship between postharvest management of grain sorghum and phymatotrichum root rot in the subsequent cotton crop, *Plant Dis.*, 73, 304, 1989.

Rush, C. M. and Heidel, G. B., Furovirus diseases of sugar beet in the United States, *Plant Dis.*, 79, 868, 1995.
Rush, C. M. and Lyda, S. D., Levels of anhydrous ammonia toxic to mycelium and sclerotia of *Phymatotrichum omnivorum, Phytopathology*, 72, 1085, 1982.
Rush, C. M. and Mathieson, J. T., Effects of common root rot on winter wheat forage production, *Plant Dis.*, 74, 982, 1990.
Rush, C. M., Ramig, R. E., and Kraft, J. M., Effects of wheat chaff and tillage on inoculum density of *Pythium ultimum* in the Pacific Northwest, *Phytopathology*, 76, 1330, 1986.
Rush, C. M. and Vaughn, K. M., Seed quality effects on emergence and yield of hard red winter wheat, *Phytopathology*, 82, 1072, 1992 (Abstr.).
Rush, C. M. and Vaughn, K. M., Effect of irrigation, soil matric potential, and seed priming on sugar beet seed germination and damping-off caused by *Aphanomyces cochlioides, Phytopathology*, 83, 202, 1993.
Rush, C. M. and Winter, S. R., Influence of previous crops on rhizoctonia root and crown rot of sugar beet, *Plant Dis.*, 74, 421, 1990.
Sanogo, S. and Moorman, G. W., Behavior of *Pythium* sp. in a flood and drain subirrigation system, *Phytopathology*, 80, 123, 1990 (Abstr.).
Schmitthenner, A. F., Relationship between the amount of *Phytophthora* in soil and yield loss from Phytophthora root rot of soybean in Ohio, *Phytopathology*, 82, 1105, 1992 (Abstr.).
Schneider, C. L., Ruppel, E. G., Hecker, R. J., and Hogaboam, G. J., Effect of soil deposition in crowns on development of rhizoctonia root rot in sugar beet, *Plant Dis.*, 66, 408, 1982.
Sewell, M. C. and Melchers, L. E., The effect of rotation and tillage on foot rot of wheat in Kansas, 1920–1924, *Agron. J.*, 16, 768, 1924.
Shannon, E., *Chile Disease Control*, Cooperative Extension Service, Guide H-219, New Mexico State University, Las Cruces, NM, 1989.
Shefelbine, P. A., Mathre, D. E., and Carlson, G., Effects of chloride fertilizer and systemic fungicide seed treatments on common root rot of barley, *Plant Dis.*, 70, 639, 1986.
Sherf, A. F. and MacNab, A. A., *Vegetable Diseases and Their Control*, 2nd ed., John Wiley & Sons, New York, 1986.
Shew, H. D., Effect of host resistance on spread of *Phytophthora parasitica* var. *nicotianae* and subsequent development of tobacco black shank under field conditions, *Phytopathology*, 77, 1090, 1987.
Shipton, P. J., Take-all decline during cereal monoculture, in *Biology and Control of Soil-Borne Plant Pathogens*, American Phytopathological Society, St. Paul, MN, 1975, 137.
Shipton, P. J., Monoculture and soilborne plant pathogens, *Annu. Rev. Phytopathol.*, 15, 387, 1977.
Shriver, J. M., Rush, C. M., and Vaughn, K. M., Effects of planting date and seed size on common root rot severity in hard red winter wheat in the Texas Panhandle, *Phytopathology*, 85, 1155, 1995 (Abstr.).
Smiley, R. W., Ogg, A. G., Jr., and Cook, R. J., Influence of glyphosate on rhizoctonia root rot, growth, and yield of barley, *Plant Dis.*, 76, 937, 1992.
Smiley, R. W. and Uddin, W., Influence of soil temperature on rhizoctonia root rot *(R. solani* AG-8 and *R. oryzae)* of winter wheat, *Phytopathology*, 83, 777, 1993.
Smiley, R. W. and Wilkins, D. E., Impact of sulfonylurea herbicides on rhizoctonia root rot, growth, and yield of winter wheat, *Plant Dis.*, 76, 399, 1992.
Smiley, R. W. and Wilkins, D. E., Annual spring barley growth, yield, and root rot in high- and low-residue tillage systems, *J. Prod. Agric.*, 6, 270, 1993.
Snapp, S. S. and Shennan, C., Salinity effects on root growth and senescence in tomato and the consequences for severity of phytophthora root rot infection, *J. Am. Soc. Hortic. Sci.*, 119, 458, 1994.
Specht, L. P. and Leach, S. S., Effects of crop rotation on *Rhizoctonia* disease of white potato, *Plant Dis.*, 71, 433, 1987.
Sprague, M. A., Overview, in *No-Tillage and Surface-Tillage Agriculture. The Tillage Revolution*, Sprague, M. A. and Triplett, G. B., Eds., John Wiley & Sons, New York, 1986, 1.
Stewart, B. A., Jones, O. R., and Unger, P. W., Moisture management in semiarid temperate regions, in *Agriculture and Environmental Challenges*, Proc. 13th Agric. Sector Symp., Srivastava, J. P. and Alderman, H., Eds., World Bank, Washington, D.C., 1993.

Stewart, B. A. and Moldenhauer, W. C., Crop Residue Management to Reduce Erosion and Improve Soil Quality, Southern Great Plains, USDA-ARS Conserv. Res. Rep. 37, U.S. Department of Agriculture, Washington, D.C., 1994.

Subbarao, K. V., Koike, S. T., and Hubbard, J. C., Effects of deep plowing on the distribution and density of *Sclerotinia minor* sclerotia and lettuce drop incidence, *Plant Dis.*, 80, 28, 1996.

Summerell, B. A., Burgess, L. W., and Klein, T. A., The impact of stubble management on the incidence of crown rot of wheat, *Aust. J. Exp. Agric.*, 29, 91, 1989.

Sumner, D. R. and Bell, D. K., Influence of crop rotation on severity of crown and brace root rot caused in corn by *Rhizoctonia solani*, *Phytopathology*, 76, 248, 1986.

Sumner, D. R. and Bell, D. K., Survival of *Rhizoctonia* spp. and root diseases in a rotation of corn, snap bean, and peanut in microplots, *Phytopathology*, 84, 113, 1994.

Sumner, D. R., Brenneman, T. B., and Harrison, G. W., Populations of fungi in soil after chemigation with chlorothalonil and tebuconazole via center-pivot irrigation, *Plant Dis.*, 75, 999, 1991.

Sumner, D. R., Doupnik, B., Jr., and Boosalis, M. G., Effects of reduced tillage and multiple cropping on plant diseases, *Annu. Rev. Phytopathol.*, 19, 167, 1981.

Sumner, D. R., Dowler, C. C., Johnson, A. W., and Baker, S. H., Conservation tillage and seedling diseases in cotton and soybean double-cropped with triticale, *Plant Dis.*, 79, 372, 1995a.

Sumner, D. R., Dowler, C. C., Johnson, A. W., Chalfant, R. B., Glaze, N. C., and Phatak, S. C., Effect of root diseases and nematodes on yield of corn in an irrigated multiple-cropping system with pest management, *Plant Dis.*, 69, 382, 1985.

Sumner, D. R., Dowler, C. C., Johnson, A. W., Glaze, N. C., Phatak, S. C., and Chalfant, R. B., Root diseases of cucumber in irrigated multiple-cropping system with pest management, *Plant Dis.*, 67, 1071, 1983.

Sumner, D. R., Gascho, G. J., Johnson, A. W., Hook, J. E., and Threadgill, E. D., Root diseases, populations of soil fungi, and yield decline in continuous double-crop corn, *Plant Dis.*, 74, 704, 1990.

Sumner, D. R. and Littrell, R. H., Effects of chemigation with chlorothalonil and diniconazole on soil fungi and pod, peg, and stem diseases of peanut, *Plant Dis.*, 73, 642, 1989.

Sumner, D. R. and Phatak, S. C., Efficacy of metam-sodium applied through overhead sprinkler irrigation for control of soilborne fungi and root diseases of vegetables, *Plant Dis.*, 72, 160, 1988.

Sumner, D. R., Phatak, S. C., Gay, J. D., Chalfant, R. B., Brunson, K. E., and Bugg, R. L., Soilborne pathogens in a vegetable double-crop with conservation tillage following winter cover crops, *Crop Prot.*, 14, 495, 1995b.

Sumner, D. R. and Smittle, D., Etiology and control of fruit rot of cucumber in single harvesting for pickles, *Plant Dis. Rep.*, 60, 304, 1976.

Sumner, D. R., Smittle, D. A., Threadgill, E. D., Johnson, A. W., and Chalfant, R. B., Interactions of tillage and soil fertility with root diseases in snap bean and lima bean in irrigated multiple-cropping systems, *Plant Dis.*, 70, 730, 1986a.

Sumner, D. R., Threadgill, E. D., Smittle, D. A., Phatak, S. C., and Johnson, A. W., Conservation tillage and vegetable diseases, *Plant Dis.*, 70, 906, 1986b.

Sutton, J. C. and Vyn, T. J., Crop sequences and tillage practices in relation to diseases of winter wheat in Ontario, *Can. J. Plant Pathol.*, 12, 358, 1990.

Taubenhaus, J. J. and Ezekiel, W. N., Resistance of monocotyledons to phymatotrichum root rot, *Phytopathology*, 22, 443, 1932.

Taylor, A. G. and Harmon, G. E., Concepts and technologies of selected seed treatments, *Annu. Rev. Phytopathol.*, 28, 321, 1990.

Teng, P. S. and Yang, X. B., Biological impact and risk assessment in plant pathology, *Annu. Rev. Phytopathol.*, 31, 495, 1993.

Tinline, R. D. and Spurr, D. T., Agronomic practices and common root rot in spring wheat: Effect of tillage on disease and inoculum density of *Cochliobolus sativus* in soil, *Can. J. Plant Pathol.*, 13, 258, 1991.

Tisdale, S. L. and Nelson, W. L., *Soil Fertility and Fertilizers*, Macmillan, New York, 1975.

Travis, J. W. and Latin, R. X., Development, implementation, and adoption of expert systems in plant pathology, *Annu. Rev. Phytopathol.*, 29, 343, 1991.

Triplett, G. B., Crop management practices for surface-tillage systems, in *No-Tillage and Surface-Tillage Agriculture. The Tillage Revolution*, Sprague, M. A. and Triplett, G. B., Eds., John Wiley & Sons, New York, 1986, 149.

Trolldenier, G., Effect of potassium chloride vs. potassium sulfate fertilization at different soil moisture on take-all of wheat, *Phytopathology Z.*, 112, 56, 1985.

Trutmann, P., Paul, K. B., and Cishabayo, D., Seed treatments increase yield of farmer varietal field bean mixtures in the central African highlands through multiple disease and beanfly control, *Crop Prot.*, 11, 458, 1992.

Tu, J. C. and Tan, S. C., Soil compaction effect on photosynthesis, root rot severity, and growth of white beans, *Can. J. Soil Sci.*, 68, 455, 1988.

Unger, P. W., Tillage Systems for Soil and Water Conservation, Food and Agriculture Organization of the United Nations, Rome, FAO Soils Bull. 54, World Health Organization, Geneva, 1984.

Unger, P. W., Conservation tillage systems, *Adv. Soil Sci.*, 13, 27, 1990.

Unger, P. W., *Managing Agricultural Residues*, Lewis Publishers, Boca Raton, FL, 1994.

Unger, P. W. and Kaspar, T. C., Soil compaction and root growth: a review, *Agron. J.*, 86, 759, 1994.

Unger, P. W., Stewart, B. A., Parr, J. F., and Singh, R. P., Crop residue management and tillage methods for conserving soil and water in semi-arid regions, *Soil Tillage Res.*, 20, 219, 1991.

Van Bruggen, A. H. and Rubatzky, V. E., Use of transplants instead of direct seeding to reduce corky root severity and losses due to corky root in iceberg lettuce, *Plant Dis.*, 76, 703, 1992.

Vilich, V., Crop rotation with pure stands and mixtures of barley and wheat to control stem and root rot diseases, *Crop Prot.*, 12, 373, 1993.

Von Qualen, R. H., Abney, T. S., Huber, D. M., and Schreiber, M. M., Effects of rotation, tillage, and fumigation on premature dying of soybeans, *Plant Dis.*, 73, 740, 1989.

Walker, J., Take-all diseases of Gramineae: A review of recent work, *Rev. Plant Pathol.*, 54, 113, 1975.

Watkins, J. E. and Boosalis, M. G., Plant disease incidence as influenced by conservation tillage systems, CRC Press, Boca Raton, FL, 1994, chap. 13.

Webster, M. A. and Dixon, G. R., Boron, pH, and inoculum concentration influencing colonization by *Plasmodiophora brassicae*, *Mycol. Res.*, 95, 74, 1991a.

Webster, M. A. and Dixon, G. R., Calcium, pH, and inoculum concentration influencing colonization by *Plasmodiophora brassicae*, *Mycol. Res.*, 95, 64, 1991b.

Weller, D. M., Cook, R. J., MacNish, G., Bassett, E. N., Powelson, R. L., and Peterson, R. R., Rhizoctonia root rot of small grains favored by reduced tillage in the Pacific Northwest, *Plant Dis.*, 70, 70, 1986.

White, J. G. and Ho, R. K., The effects of *Pythium sylvaticum*, seed-borne pathogens, and seed treatments on the emergence of brassica seedlots of high vigor, *Ann. Appl. Biol.*, 112, 661, 1988.

Whiting, K. R. and Crookston, R. K., Crop ecology, production and management. Host-specific pathogens do not account for the corn-soybean rotation effect, *Crop Sci.*, 33, 539, 1993.

Wiese, M. V., *Compendium of Wheat Diseases*, APS Press, St. Pual, MN, 1987.

Wilhelm, S., Sources and nature of *Verticillium* wilt resistance in some major crops, in *Biology and Control of Soil-Borne Plant Pathogens*, American Phytopathological Society, St. Paul, MN, 1975.

Wilhelm, S., Sagan, J. E., and Tietz, H., Optimizing verticillium wilt resistance in upland cotton with short-branching, arching, plonotypes, and in-row crowding, *Second Int. Verticillium Symp.*, University of California, Berkeley, 1976, 41.

Wilson, J. M. and Hamblin, J., The effects of fumigation, rotation with lupins, and nitrogen application on plant growth and common root rot of wheat at Geraldton, Western Australia, *Aust. J. Agric. Res.*, 41, 619, 1990.

Windels, C. E., Kommendahl, T., Sarbini, G., and Wiley, H. B., The role of seeds in the delivery of antagonists into the rhizosphere, in *Ecology and Management of Soilborne Plant Pathogens*, Parker, C. A., Rovira, A. D., Moore, K. J., Wong, P. T. W., and Kollmorgen, J. F., Eds., APS Press, St. Paul, MN, 1985.

Winter, S. R., Cropping systems to remove excess soil nitrate in advance of sugarbeet production, *J. Am. Soc. Sugar Beet Technol.*, 22, 285, 1984.

Wolfe, M. S., The current status and prospects of multiline cultivars and variety mixtures for disease resistance, *Annu. Rev. Phytopathol.*, 23, 251, 1985.

Wong, P. T. W., Dowling, P. M., Tesoriero, L. A., and Nicol, H. I., Influence of preseason weed management and in-crop treatments on two successive wheat crops. 2. Take-all severity and incidence of rhizoctonia root rot, *Aust. J. Exp. Agric.*, 33, 173, 1993.

Wrather, J. A., Kendig, S. R., and Anand, S. C., Effects of tillage, cultivar, and planting date on percentage of soybean leaves with symptoms of sudden death syndrome, *Plant Dis.*, 79, 560, 1995.

Zorrilla, G., Knapp, A. D., and McGee, D. C., Severity of phomopsis seed decay, seed quality evaluation, and field performance of soybean, *Crop Sci.,* 34, 172, 1994.

Zucconi, F., Allelopathies and biological degradation in agricultural soils: An introduction to the problem of soil sickness and other soil-borne diseases, *Acta Hortic.,* 324, 11, 1993.

CHAPTER 12

Physical Methods of Soil Sterilization for Disease Management Including Soil Solarization

Robert J. McGovern and Robert McSorley

I. INTRODUCTION

Survival of plant pathogens such as fungi, bacteria, nematodes, and, less commonly, viruses in association with soil particles or undecomposed plant matter in soil can provide an important source of inoculum for disease outbreaks. Soil "sterilization" to eliminate plant pathogens has, therefore, been an essential component of disease management for many years (Johnson, 1919).

1-8493-2627-3/97/$0.00+$.50

Sterilization, when strictly defined, refers to the nonselective inactivation of all organisms in a given substrate. In actual practice, the term is more broadly interpreted to include partial sterilization or disinfestation techniques which attempt to selectively kill or reduce specific pathogens. Soil pasteurization is an example of partial sterilization which uses lower than maximum temperatures (60 to 70°C) to destroy pathogens while preserving many beneficial microorganisms. The practice of flooding soil for disease management may also be viewed as an attempt at partial sterilization which relies on the inability of certain pathogens to tolerate anaerobic conditions.

Most physical sterilization procedures, whether they attempt total or partial disinfestation, rely on the input of some type of energy, most commonly thermal energy. Direct elevation of soil temperature beyond the heat tolerance threshold of pathogens provides the basis, in whole or part, for soil sterilization techniques which utilize dry sources of heat, steam, hot water, solarization, and composting. Thermal energy produces molecular states in cells that cause irreversible denaturation and/or degradation of essential proteins and nucleic acids, leading, in turn, to fatal disruption of critical cellular processes (Brock, 1978; Pflug and Schmidt, 1968; Sundarum, 1986). Membrane lipid characteristics also appear to be involved in heat sensitivity (Sundarum, 1986). Except for certain heat-resistant viruses such as tobacco mosaic, the majority of soil-borne plant pathogens are inactivated by heating at 70°C or less for 30 min (Table 1).

However, experimental results have indicated that the temperature and duration of treatment necessary for thermal inactivation may vary widely among different isolates of the same pathogen, and may be affected by the survival substrate utilized (Table 1). For example, it appears that higher temperatures may be required to inactivate the nematodes *Aphelenchoides fragariae* and *Radopholus similis* when they are associated with plant material than in water. Thermal inactivation may also depend on the temperature range to which a microorganism has adapted (Pflug and Schmidt, 1968). Isolates of *Phytophthora cinnamomi* from different areas of the world have been found to differ in their minimum, optimum, and maximum temperatures for growth (Zentmeyer, 1980). In addition, the level of dormancy and moisture content of microorganisms affect the lethal time-temperature relationship; inactivity and decreased hydration generally lower thermal sensitivity (Baker, 1962; Bollen, 1974; Johnson, 1946). Juveniles of *Heterodera glycines* and *Meloidogyne* sp. were significantly more heat sensitive than eggs (Endo, 1962; Hoshino and Godfrey, 1933). Therefore, it appears that actively growing pathogens are commonly more heat sensitive than their resting stages because of the inherently higher moisture contents of the former. The minimum exposure and lethal temperature for the fungus *Verticillium albo-atrum* was 5 min at 47°C for hyphae and conidia (active stages) and 10 min at 50°C for microsclerotia (resting stage) (Nelson and Wilhelm, 1958). A higher range of temperatures (55 to 65°C) was required for inactivation of resting sporangia than for zoospores (55°C) of *Olpidium brassicae* (Campbell and Grogan, 1964). The wide range of thermal inactivation temperatures and durations reported for *Verticillium albo-atrum* and *V. dahliae* (Nelson and Wilhelm, 1958; Bollen, 1969, 1985a; Pullman et al., 1981a) may have been based on the presence or absence of sclerotia, inherent differences in thermal tolerance among isolates, or experimental variation. Variation in thermal sensitivity among isolates of *Phytophthora megasperma* appears to be related to oogonial size; those with large oogonia have low optimal temperatures while those possessing small oogonia are more thermophilic (Wilcox and Mircetich, 1987).

Soil sterilization procedures should not be evaluated in isolation from other aspects of plant production systems, since they cannot by themselves guarantee disease-free plants. It is especially important that precautions be taken to avoid reintroduction of pathogens into disinfested soil. Therefore, to ensure effective plant disease suppression, soil disinfestation must be integrated with other practices including the use of disease resistance, pathogen-free propagative material, management of other pests such as weeds and insects which often serve as reservoirs of pathogens, maintenance of optimal cultural practices (irrigation, fertilization, and sanitation), and the judicious use of chemical controls. Combining different methods of soil sterilization can also be advantageous.

Table 1 Thermal Inactivation of Selected Plant Pathogens

Pathogen	Survival Substrate	Lethal Temp. (°C)	Exposure Time (min)	Ref.
Nematodes				
Ditylenchus dipsaci	Plant tissue	53.0	15	Baker and Roistacher, 1957b
Aphelenchoides fragariae	Water	44.4	15	Qiu et al., 1993
	Bulbs	45.0	180	Southey, 1993
Heterodera glycines	Water (eggs)	49.0	25	Endo, 1962
	(Juveniles in cysts)	49.0	30	Endo, 1962
	(Free juveniles)	49.0	2	Endo, 1962
Helicotylenchus multicinctus	Water	47.5	10	McSorley and Parrado, 1983
Meloidogyne incognita	Plant tissue	48.0	10	Baker and Roistacher, 1957b
Meloidogyne sp.	Water (juveniles)	47.0	3	Hoshino and Godfrey, 1933
	(eggs)	47.0	11	Hoshino and Godfrey, 1933
Pratylenchus penetrans	Plant tissue	49.0	10	Baker and Roistacher, 1957b
Radopholus similis	Water	50.0	1.5	Blake, 1961
	Banana corms	55.0	20	Blake, 1961
Fungi				
Botrytis cinerea	—	55.0	15	Baker and Roistacher, 1957b
Colletotrichum coccodes (2 isolates)	Soil	45.0–50.0	30	Bollen, 1985a,
Cylindrocarpon destructans	Soil and plant tissue	50.0	30	Bollen, 1969
Cylindrocarpon destructans (2 isolates)	Soil	45.0–47.5	30	Bollen, 1985a
Didymella lycopersici	Soil and plant tissue	50.0	30	Bollen, 1969
Fusarium avenaceum	Soil	57.5–60.0	30	Bollen, 1985a
F. oxysporum f.sp. *dianthi*	Soil	60.0	30	Bollen, 1969
F. oxysporum f.sp. *dianthi* (4 isolates)	Roots in soil	55.0–65.0	30	Bollen, 1985a
F. oxysporum f.sp. *gladioli*	Corms	57.0	30	Baker and Roistacher, 1957b
F. oxysporum f.sp. *gladioli*	Roots in soil	57.5–60.0	30	Bollen, 1985a
F. oxysporum f.sp. *lycopersici* (4 isolates)	Roots in soil	57.5–60.0	30	Bollen, 1985a
F. oxysporum f.sp. *melongenae* (2 isolates)	Roots in soil	57.5–60.0	30	Bollen, 1985a
Olpidium brassicae	Roots in soil	55.0–57.5	30	Bollen, 1985a
O. brassicae (zoospores)	Water	55.0	10	Campbell and Grogan, 1964
O. brassicae (resting sporangia)	Water	55.0–65.0	10	Campbell and Grogan, 1964
Phialophora cinerescens	Soil	55.0	30	Bollen, 1969
Phytophthora cactorum	Soil and plant tissue	45.0	30	Juarez-Palacios et al., 1991
P. capsici	Soil	40.0–50.0	30	Bollen, 1985a
P. cinnamomi (chlamydospores)	Soil	38.0	30	Barbercheck and von Broembsen, 1986
P. cinnamomi (chlamydospores)	Soil and plant tissue	40.0	20	Juarez-Palacios et al., 1991
P. cryptogea	Soil and plant tissue	50.0	30	Bollen, 1969

Table 1 (continued) Thermal Inactivation of Selected Plant Pathogens

Pathogen	Survival Substrate	Lethal Temp. (°C)	Exposure Time (min)	Ref.
P. cryptogea	Soil	40.0–45.0	30	Bollen, 1985a
P. megasperma (oospores, heat sensitive)	Soil and plant tissue	45.0	20	Juarez-Palacios et al., 1991
P. megasperma (oospores, heat tolerant)	Soil and plant tissue	>45.0	30	Juarez-Palacios et al., 1991
Pythium aphanidermatum	Soil	50.0–52.5	30	Bollen, 1985a
P. aphanidermatum (oospores)	Soil	55.0	30	Kulkarni et al., 1991
P. irregulare	Soil	52.5	30	Bollen, 1969
P. sylvaticum	Soil	42.5–45.0	30	Bollen, 1985a
P. ultimum	Plant tissue	46.0	20–40	Baker and Roistacher, 1957b
P. ultimum	Soil	50.0	30	Bollen, 1969
P. ultimum	Potato dextrose agar	50.0	35	Pullman et al., 1981a
Pythium sp.	Soil and plant tissue	52.5	30	Bollen, 1969
Plasmodiophora brassicae	Roots in soil	50.0–60.0	30	Bollen, 1985a
Rhizoctonia solani	Soil and plant tissue	52.5	30	Bollen, 1969
R. solani	Potato dextrose agar	50.0	12	Pullman et al., 1981a
Rhizoctonia sp.	Plant tissue	52.0	30	Baker and Roistacher, 1957b
Rhizoctonia sp.	Soil and plant tissue	52.5	30	Bollen, 1969
Sclerotinia sclerotiorum	—	50.0	5	Baker and Roistacher, 1957b
Sclerotium rolfsii	Tubers and rhizomes	50.0	30	Baker and Roistacher, 1957b
Synchytrium endobioticum	Tubers in soil	50.0–60.0	30	Bollen, 1985a
Thielaviopsis basicola	Soil	47.5	30	Bollen, 1969
T. basicola	Potato dextrose agar	50.0	35	Pullman et al., 1981a
Verticillium albo-atrum	Soil	52.5	30	Bollen, 1969
V. albo-atrum (hyphae and conidia)	Plant stems	47.0	5	Nelson and Wilhelm, 1958
V. albo-atrum (microsclerotia)	Plant stems	50.0	10	Nelson and Wilhelm, 1958
V. dahliae	Soil	57.5	30	Bollen, 1969
V. dahliae (2 isolates)	Soil	45.0–47.5	30	Bollen, 1985a
V. dahliae	Soil	40.0–45.0	30	Bollen, 1985a
V. dahliae	Soil	50.0	11	Pullman et al., 1981a
Bacteria				
Agrobacterium tumefasciens	—	51.0	30	Elliott, 1951
Clavibacter michiganense	—	53.0	30	Elliott, 1951
Erwinia carotovora	—	48.0–51.0	30	Elliott, 1951
Erwinia sp.	Plant tissue	52.0	30	Baker and Chandler, 1957
Ralstonia *(Pseudomonas solanacearum)*	—	49.0–50.0	30	Elliott, 1951
Viruses				
Cucumber Green Mottle Virus	Peat	100	10	Runia, 1986
Tomato Bushy Stunt Virus	Plant sap	80–90	10	Hollings and Stone, 1970
Tobacco Mosaic Virus	Soil	92.0	10	Johnson, 1946
Tobacco Necrosis Virus	Plant tissue	75.0	10	Lopez-Real and Foster, 1985

II. SOIL DISINFESTATION PROCEDURES BASED ON THE USE OF HEAT

A. Dry Sources of Heat

1. *Fire*

a. *Open Burning*

The earliest attempts to eliminate soil-borne plant pests through heat were probably the indirect and unplanned result of primitive types of slash-and-burn farming in which litter is burned prior to cultivation. Enhanced plant growth following burning has been attributed to increased availability of nitrogen, phosphorus, and other nutrients (Chambers and Attiwill, 1994; Hardison, 1976). Burning of crop residue has also been reported to reduce the subsequent incidence of diseases of mainly graminaceous hosts (Das et al., 1989; Hardison, 1976; Heydayetullah and Saha, 1941), but these results are most likely attributable to removal of pathogen propagules in plant tissue.

Burning plant residue was not effective for general sterilization of the subsoil or for nematode management (Bridge, 1987; Chinanzvavana et al., 1986). Similarly, burning Kentucky bluegrass debris by the open-field method or by means of an experimental burning machine generated peak soil-surface temperatures of 120 to 240°C and 280 to 320°C, respectively, but failed to reduce germination of *Claviceps purpurea* sclerotia except at the soil surface (Johnston et al., 1996). Root-knot nematodes were killed to a depth of only 9 cm as a result of crop burning (Maas, 1987). Following controlled burning in a pine forest, only 1 of 9 genera of plant-parasitic nematodes was significantly reduced in the upper 20 cm of soil (McSorley, 1993). In addition, emissions from open burning may pollute the atmosphere, and removal of crop residue can adversely affect soil erosion and fertility (Palti, 1981). Heating the soil to 75 to 82°C to a depth of 15 cm by means of a flame gun was suggested as a method for controlling Rhizoctonia root rot in *Piper betle* (Chodhury, 1944), however, the rising cost of fossil fuels as well as air quality issues would appear to make large-scale use of such devices impractical.

b. *Flame-Heated Roasters and Ovens*

According to Baker and Roistacher (1957a), dry heat, including heat derived from inflammation, is most appropriately used in devices (ovens and roasters) which treat small soil volumes or those through which soil is moved. Roasting soil in flame-heated pans to control Thielaviopsis root rot and weeds was widely practiced in tobacco seed beds in the southern U.S. starting in the late nineteenth and continuing through the early twentieth centuries (Newhall, 1955). Various soil roasters have been developed to pasteurize soil by moving it through tumbling containers heated with burners (Newhall and Schroeder, 1951; Romano and D'Errico, 1979; Shaw and Mitchell; 1978).

Using dry heat to sterilize soil has several drawbacks when compared with steam disinfestation: it is necessary to apply a high intensity of dry heat to raise a given amount of soil to the desired temperature, its distribution through the soil is usually poor, and its effectiveness varies greatly with soil moisture content (soil which is wet requires more heat to raise the temperature to the required sterilization point) (Baker and Roistacher, 1957a). Shaw and Mitchell (1978) found in preliminary experiments that it was necessary to moisten treated soil to prevent it from charring due to excessive heating in a burner-powered pasteurizer.

2. *Electromagnetic Heating*

a. *Generation of Heat by Electric Current*

Electromagnetic energy has been adapted for direct sterilization of soil by passing an electrical current between two or more electrodes; thermal inactivation is achieved by heat buildup resulting from the soil's inherent resistance to electricity. The current actually passes through a film of water containing dissolved salts (electrolytes) which surround soil particles. The resistance, and hence heat, created in the soil is directly proportional to the distance between electrodes and inversely proportional to their area (Blauser, 1935; Tavernetti, 1935). An indirect convection and conductance of heat to soil from insulated electrical heating elements has also been utilized to sterilize soil (Newhall, 1955).

Numerous electrical systems for "dry heat" soil sterilization have been developed including cabinets (Carney, 1932; Horsfall, 1935; Newhall and Nixon, 1935), devices through which soil is moved (Newhall, 1940; Tavernetti, 1942; Diprose et al., 1988; Ulyanov and Zmeev, 1978), or arrays of electrodes which are either stationary or mobile (Baranov et al., 1983; Nikitin and Nikitin, 1990). However, the cost of electrical consumption and difficulty in penetrating large soil volumes appear be limiting factors for the widespread application of both direct and indirect electrical soil sterilizers, and potential shock hazards are of concern with the former method.

b. Microwaves and Other Forms of Electromagnetic Radiation

Electromagnetic radiation in the form of microwaves (2450 MHz) has been used primarily as a research tool to sterilize small volumes of soil but has not proven efficacious for field use due to its shallow depth of penetration and variability with water content and heterogeneous soil types (Baker and Fuller, 1969; Van Wambeke et al., 1981, 1984). Water molecules selectively absorb and are heated by microwaves (Maas, 1987). Failure of microwave radiation to reduce bacteria and fungi in soil was attributed to a suboptimal soil moisture level (–30 kPa) for adequate soil heating (Wolf et al., 1989). Therefore, the sterilizing effects of microwave radiation appear to be primarily thermic, although nonthermal mechanisms have also been suggested by certain studies (Baker and Fuller, 1969; Mayers and Habeshaw, 1973; Vela and Wu, 1979). Either the treatment time or energy level used in microwave disinfestation must increase with increasing soil water content to achieve adequate sterilization (Ferriss, 1984; Moosman and Koch, 1988; Van Wambeke et al., 1981).

Treatment of 1 kg or 4 kg of soil at 7 to 37% water content for 150 or 425 s, respectively, in a commercial microwave oven (2450 MHz/625 W) eliminated populations of *Pythium*, *Fusarium*, but allowed survival of *Heterodera glycines* cysts, *Rhizoctonia*, and mycorrhizal fungi (Ferriss, 1984). One 3-min exposure to 2450 MHz radiation at 675 W completely eliminated fungi in 100 and 400 g of soil, while it required three consecutive 9- and 12-min irradiations to achieve comparable reductions of bacteria (Ou et al., 1985). *Meloidogyne incognita* was eliminated from small volumes of soil (90 to 400 cm^3) exposed to 2450 MHz radiation at 600 W power for 30 s in a microwave oven (O'Bannon and Good, 1971). Field application of microwave technology has been less successful. *Rotylenchulus reniformis* was reduced in only the top 10 cm of soil in field plots by microwave radiation (Heald et al., 1974; Heald and Wayland, 1975). Microwave radiation reduced weed seeds in the field, but had no effect on resident populations of bacteria and fungi (Vela et al., 1976), and it failed to eliminate tobacco mosaic virus and cucumber green mottle virus from rock wool (Runia, 1986).

Two nonheat-producing forms of electromagnetic radiation, ultraviolet light and ionizing radiation (gamma rays), have also been used experimentally to sterilize soil and other substrates (Evans, 1970; Green and Plumb, 1967; Lemanova et al., 1983; Parker and Vincent, 1981; Rattink, 1982; Runia, 1986; Stroetmann et al., 1994; Wolf et al., 1989). Ultraviolet radiation at 260 nm directly inactivates DNA by thymine dimerization while ionizing radiation kills indirectly by producing free radicals (Brock, 1979). Although UV light is effective in the disinfestation of water in closed irrigation systems (Ewart and Chrimes, 1980; Runia, 1994), and gamma irradiation has been used on a limited basis to sterilize a commercial cactus medium in the Netherlands (Rattink, 1982), neither method appears to be practical for large-scale commercial soil sterilization due to economic and/or public health considerations (Myers and Dropkin, 1959; Rattink, 1982; Runia, 1986).

B. Hot Water

Hot water (about 100°C) is often used effectively to rid plant propagative material such as seeds, bulbs, and nursery stock of pathogens (Agrios, 1988). However, attempts to disinfest soil with hot water have been less successful than steam sterilization (Bewley, 1923; Newhall, 1930). Upon cooling, water at 100°C releases only one sixth the thermal energy of steam, and copious quantities of the former may be needed to achieve adequate soil disinfestation, leading, in turn, to waterlogging. Bewley (1923) indicated that when applied as a surface drench, 7.6 million liters of hot water per hectare were necessary to achieve nematode control. Another study calculated that a prototype mobile device used to deliver hot water to a plastic mulch-covered bed required 114,000 to 265,000 l per treated hectare to affect nematodes in the surface layers of a fine, sandy soil (Noling, 1995a). Solar-heated water (75 to 90°C) failed to reduce populations of *Meloidogyne javanica* but did enhance the effectiveness of soil solarization in reducing root galling (Abu-Gharbieh et al., 1991a). Root-knot nematodes were eliminated from soil in small pots by immersion in near boiling (98°C) water for >5 min, but the effects of pouring hot water into potted soil varied, depending on the pot size and the amount of water used (Byars and Gilbert, 1920). At present, delivery of hot water for nematode control on a field scale is problematic due to economics, logistical constraints of application, and lack of consistent efficacy (Noling, 1994, 1995b). The use of hot water for soil sterilization would also obviously be hindered where there are water shortages.

C. Steam

Steam has excellent distribution characteristics in soil and releases large quantities of heat after condensation (Baker and Roistacher, 1957a). The first commercial application of steam for disinfestation of soil appears to have been made in an Ohio glasshouse about 100 years ago (Rudd, 1893). Steam was widely used to disinfest soil for plant production in glasshouses but because of the rising cost of fuel, has largely been replaced by fumigation (Bollen, 1985a; Garibaldi and Tamietti, 1984; A. J. Overman, personal communication). Over the years, many systems have been devised to steam-disinfest planting beds including grids of buried perforated pipes which are either permanent or transportable, arrays of hollow spikes for steam injection, surface application under inverted pans or reinforced sheets ("sheet steaming"), or mobile rakes (Baker and Roistacher, 1957c; Grainger and Kennedy, 1964; Heijina, 1966; Nederpel, 1972, 1979; Newhall, 1955). Disinfestation of potting soil by steam may be accomplished using portable bins, stationary structures, or devices through which the soil is moved (Baker and Roistacher, 1957c; Newhall, 1955). The downward movement of surface-applied steam was enhanced by the creation of negative pressure within buried pipes (Heijina, 1966; Runia, 1983). A steam auger used *in situ* significantly reduced apple replant disease and increased tree growth (Moyls, 1993; Moyls and Hocking, 1993), but because of inherent logistics and temporal factors, on-site use of steam to sterilize field soil has been limited.

Following the development of autoclaves for elimination of human pathogens from various substrates, plant pathologists initially employed steam for planting-bed disinfestation at or near 100°C (Johnson, 1946; Baker, 1962). Although very effective in eradicating plant pathogens, such temperatures also completely eliminate beneficial microorganisms, including those which may suppress the resurgence of introduced or residual pests (Baker, 1962; Bollen, 1967, 1974; Kreutzer, 1965; Olsen and Baker, 1968). In addition, excessively high soil temperatures may also lead to phytotoxicity through the accumulation of ammonia, nitrites, and elevated release of micronutrients, particularly manganese (Fujimoto and Sherman, 1948; Heald, 1987; Maas, 1987; Sonneveld and Voogt, 1975). To minimize these effects, aerated steam may be employed to achieve the lower temperatures used in pasteurization (Aldrich and Nelson, 1969; Baker, 1970; Baker and Olsen, 1960; Brazelton, 1968; Dawson et al., 1965). Steaming at lower temperatures also preserves populations of thermotolerant pathogen antagonists such as *Bacillus subtilis* and actinomycetes (Katznelson and Richardson, 1943; Kreutzer, 1965; Olsen and Baker, 1968). It has therefore become common practice to apply aerated steam so that a uniform temperature of 60 to 70°C is reached and maintained throughout the soil mass to be treated for 30 min (Maas, 1987). As can be observed in Table 1, such temperatures are sufficient to eliminate all but the most heat-resistant soil-borne viruses. For steam to be most effective, the soil should be of good tilth and free of clods, large pieces of plant debris, and excessive moisture. Reevaluation of the environmental impact of fumigation use may lead to renewed interest in steam disinfestation of soil, at least on a glasshouse scale.

D. Solarization

1. Background

Soil solarization, also referred to as solar pasteurization or solar heating, primarily employs clear plastic sheeting or mulch which is transparent to short-wavelength radiation from the sun but opaque to longer-wavelength terrestrial radiation. These physical characteristics result in a greenhouse-like effect which elevates the temperature of mulched soil. The goal of solarization is, therefore, to sufficiently raise soil temperature so that soil-borne pathogens as well as other pests (weeds and arthropods) will be eliminated or suppressed. Solarization results in maximum temperatures of 35 to 60°C during the summer months in the upper soil horizon, depending on soil characteristics, mulch type, and climate (Katan, 1981; Stapleton, 1991). As can be seen in a representative thermal profile of solarized soil, the greatest increase in temperature under clear mulch occurs near the soil surface (Figure 1). Therefore, the most effective reductions in pest densities due to solarization have occurred in the upper 10 to 30 cm of soil (Katan, 1987; Stapleton and DeVay, 1986).

Over the past ten years, increasing attention has been focused on soil solarization because of its relatively benign environmental impact in comparison with other disinfestation techniques such as fumigation (Heald, 1987; Katan and DeVay, 1991; Maas, 1987; McSorley and Duncan, 1995; Noling and Becker, 1994). Antecedents of the currently understood practice include the use of solar heat to reduce root-knot nematodes in pineapple, and the fungus *Thielaviopsis basicola* in tobacco and sesame (Hagan, 1933; Grooshevoy, 1939; Adams, 1971). However, the principles of solarization were first clearly delineated and the technique optimized by Katan and colleagues in Israel (Katan et al., 1976).

2. Mechanisms of Pathogen Reduction and Yield Increase

Decreased pathogen densities following soil solarization have been attributed to both direct and indirect mechanisms. Pathogens may be directly inactivated by high temperatures, or they may be weakened (decreased growth rate or infectability, decreased sporulation, delayed germination or egg hatch, increased susceptibility to biological and chemical control) by sublethal temperatures (DeVay and Katan, 1991; Heald and Robinson, 1987; Lifshitz et al., 1983; Phillips, 1990; Porter and Merriman, 1983; Pullman et al., 1981b; Stapleton and DeVay, 1986; Vannaci et al., 1988). Indirect reduction of pathogens may take place because of the buildup of antagonistic microorganisms, including those which are thermophilic/thermotolerant or can rapidly recolonize solarized soil including actinomycetes, fluorescent pseudomonads, *Bacillus* spp., *Pasteuria penetrans*, *Talaromyces flavus*, and *Trichoderma harzianum* (Antoniou et al., 1995; Gamliel and Katan, 1991; Stapleton and DeVay, 1982, 1984; Walker and Wachtel, 1988; Tjamos et al., 1991; Elad et al., 1980). Changes in the physical structure and chemical composition of soil due to solar heating and increased soil moisture levels resulting from mulching may also be possible mechanisms of disease control (Katan, 1981). Decreased nematode densities following mulching with partially shaded transparent film implicated the activity of accumulated volatiles and/or changes in the soil gas composition (Stapleton and DeVay, 1983). Removal of nutrient sources, e.g., plants, may also help to lower pathogen densities.

The mechanisms underlying increased yields accompanying solarization may be equally complex and involve elimination of major or minor root-damaging pests, increased nutrient availability, and stimulation of beneficial microorganisms (Katan, 1981; Kaewruang et al., 1989; Stapleton and

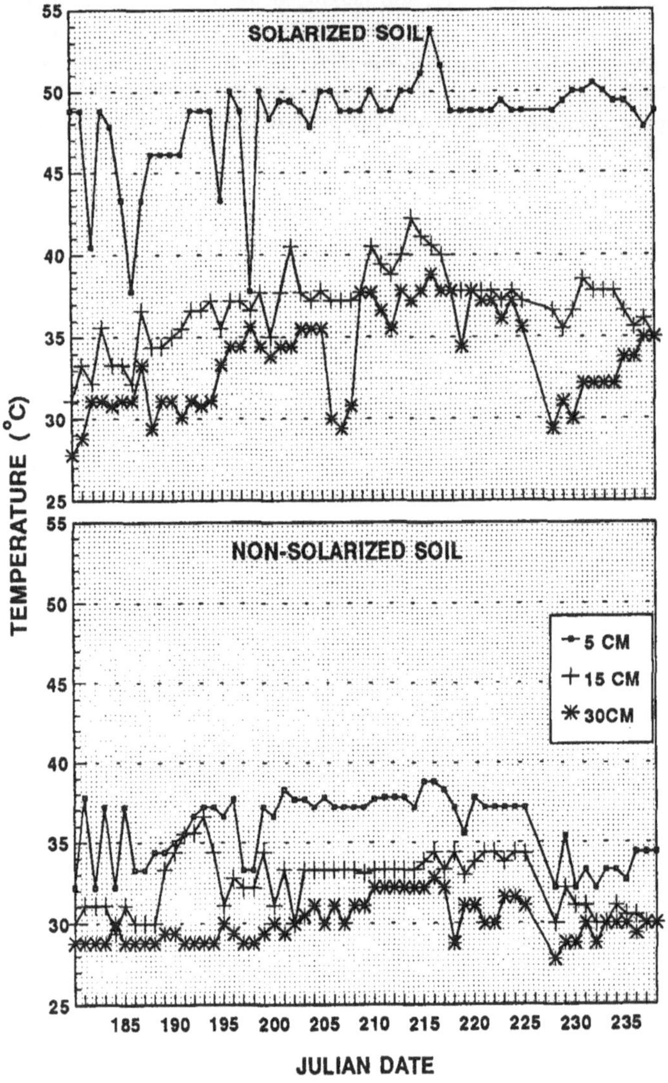

Figure 1 Maximum daily soil temperatures recorded at 2:30 to 3:00 p.m. in solarized soil (mulched with 25 μm transparent polyethylene) and nonsolarized soil (nonmulched) in Devonshire, Bermuda, 29 June through 26 August, 1995.

DeVay, 1986; Stapleton, 1991; Gamliel and Katan, 1991; Chen and Katan, 1980; Chen et al., 1991). Enhanced growth, termed the increased growth response, has been observed even in the absence of major pathogens following both solarization and fumigation (Chen et al., 1991; Gruenzweig et al., 1993; Altman, 1970). Multiple season suppression of pathogens such as *Fusarium oxysporum, Phytophthora cinnamomi, Pyrenochaeta lycopersici, Verticillium dahliae*, and *Rosellinia necatrix* has also been attributed to solarization (Katan et al., 1983; Kassaby, 1985; Abdel-Rahim et al., 1988; Morgan et al., 1991; Tjamos and Paplomatas, 1988; Tjamos et al., 1991; Sztejnberg et al., 1987).

3. Technology of Soil Solarization

Best results may be obtained by conducting solarization prior to planting for at least four weeks during the summer when high temperatures and intense solar radiation prevail. Longer solarization periods increase overall effectiveness by increasing the cumulative exposure of pests to heating and by enabling a greater depth of soil to be treated. The soil should be moist to increase the thermal sensitivity of pathogens and to aid in heat conduction, and should be covered with the thinnest clear mulch possible (a thickness of 25 to 30 μm has generally been recommended) (Katan, 1981). It is also important that the soil be free of debris, and as level as possible to ensure maximum mulch-soil contact by removing cool air pockets which may impede solarization.

To maximize the "greenhouse effect", a mulch should be highly transparent to solar radiation (visible light and short-infrared radiation) but maximally block heat loss from the soil (medium-long infrared radiation). Transparent mulch materials evaluated for soil solarization have largely included low-density polyethylene (PE), polyvinyl chloride (PVC), and copolymer of ethyl vinyl acetate (EVA). Although the transmission and insulation properties of PVC and EVA are superior, higher tensile strength, lower minimum thickness, and thus lower cost have made PE films the most commonly used mulches for soil solarization in most locations except Japan, where PVC predominates (Brown et al., 1991a; Grinstein and Hetzroni, 1991; Lamberti and Basile, 1991). The insulation properties of PE to infrared soil radiation can be enhanced by various additives and through manufacturing by coextrusion (De Luca et al., 1994; Lamberti and Basile, 1991). The condensation of water as droplets on the inner surface of the mulch during solarization can impede transmission of sunlight (Jaffrin and Makhlouf, 1990). Treatment of mulches with hydrophilic products which allow trapped moisture to condense in a thin continuous film maintains the thermic (insulatory) effect of the water without reducing light transmission (Jaffrin and Makhlouf, 1990; Lamberti and Basile, 1991). An additional consideration when selecting mulch for solarization is the presence or absence of stabilizers which inhibit UV degradation and help to determine mulch longevity (Bell and Laemmlen, 1991; Brown et al., 1991a).

Research comparing mulch thicknesses (25 or 50 μm vs. 100 μm, and 75 vs. 150 μm) determined that thinner single sheets were more effective in heating the soil due to increased solar transmission (Arun and Mathew, 1993; Meti and Hosman, 1994; Pullman et al., 1981b). However, single and double layers of transparent PE mulch (40 to 50 μm) have been reported to produce comparable reductions of Sclerotinia drop *(Sclerotinia sclerotiorum)* and plant weight increases in lettuce (Scannavini et al., 1993), reduced incidence of *Rhizoctonia solani* and increased biomass in bean (Garibaldi and Tamietti, 1989), suppression of Fusarium wilt *(Fusarium oxysporum* f. sp. *lycopersici)* in tomato (El-Shami et al., 1990), and reduction of soft rot (*Erwinia* sp.) and yield increases in Chinese cabbage (AVRDC, 1990a). Although light transmission is reduced, double layers of PE generally achieve higher soil temperatures than single sheets due to increased insulation afforded by heating the intervening air space (Ben-Yephet et al., 1987; Cenis, 1987; Duff and Barnaart, 1992; Duff and Connelly, 1993; Garibaldi and Tamietti, 1989; Lamberti and Basile, 1991; Raymundo and Alcazar, 1986). Double-layered PE (5 mm) manufactured with an interspace of trapped air bubbles, or EVA (150 μm) were more effective than a single layer of PE (150 μm) in heating the soil, and solarization using this "bubble mulch" significantly lowered densities of fungi and nematodes (Berninger et al., 1985). The separation of two clear PE sheets to create a tunnel raised 80 cm over a mulched bed resulted in significant reductions of *Fusarium* spp., *Pythium* spp., and *Rhizoctonia solani,* and seedling mortality (Annesi and Motta, 1994). The economic advantage accruing from the use of double as opposed to single layers of PE should obviously outweigh the added cost of the mulch as well as its disposal. Areas marginally suited for solarization may be the primary beneficiaries of double mulching including the use of transparent mulches within glass and plastic houses (Garibaldi and Tamietti, 1989).

Relatively few studies have compared the effectiveness of transparent and opaque mulches, primarily black mulch, in terms of pathogen suppression. Transparent PE mulch was more effective than black in reducing Fusarium wilt in tomato (El-Shami et al., 1990; Raj and Kapoor, 1993), southern blight *(Sclerotium rolfsii)* in peanut (Khandar and Bhomik, 1990), and galling of roots (*Meloidogyne* sp.) in eggplant (Braun et al., 1985). Although the two mulches produced equivalent reductions of Sclerotinia blight *(S. sclerotiorum)* in eggplant, transparent mulch was more effective in reducing populations of *Fusarium oxysporum, F. solani,* and *Meloidogyne javanica* (Abu-Gharbieh et al., 1991b). Similar results were observed in a comparison of the effectiveness of clear and black mulch in reducing soil-borne pests of potato, cauliflower, and cucumber (Abu-Blan and Abu-Gharbieh, 1994). Preemergence damping-off disease of *Pinus radiata* and *Eucalyptus obliqua* seedlings was significantly decreased, and survival of *P. radiata* increased, by solarizing with clear but not by black 50 μm PE (Kassaby, 1985). Solarization with double clear PE was slightly more effective in reducing *Meloidogyne incognita* densities than double black PE (Javed et al., 1994). In contrast, black PE mulch was as effective as transparent mulch in reducing Verticillium wilt in apricot and almond trees, and superior to transparent mulch in promoting tree growth in a postplant application (Stapleton et al., 1993). Superior performance by black mulch in postplant settings has been attributed to avoidance of the excessively high growth-inhibiting temperatures that may accompany the use of clear mulch in areas with intense solar radiation during the summer (Stapleton, 1991). Preplant application of either black or white PE for two weeks significantly reduced populations of *Hirschmanniella mucronata* and *Helicotylenchus* sp., with the highest temperature being recorded under the white mulch (Sivakumar and Marimuthu, 1987).

Plastic mulch applied in strips has been reported to be less effective than mulching the entire planting site, including intervening drive-rows where narrow bed culture is practiced because of the cooling "border effect" of the surrounding nonsolarized soil (Antoniou et al., 1995; Asworth and Gaona, 1982; Grinstein and Hetzroni, 1991; Grinstein et al., 1995; Hetzroni et al., 1983; Jacobsohn et al., 1980; Mahrer and Katan, 1981). This edge effect may also allow greater opportunity for reinfestation of the treated soil from surrounding, nontreated areas (Katan et al., 1976; Katan et al., 1980). Nevertheless, soil solarization of narrow beds ("row solarization") decreased pink root rot in onion *(Pyrenochaeta terrestris)* (Katan et al., 1980), suppressed infection by *Rhizoctonia solani* and *Fusarium* spp. in kale (McGovern and Harper., 1996), decreased inoculum densities of *R. solani* and reduced the incidence and severity of Rhizoctonia belly rot in pickling cucumber (Keinath, 1995), and reduced inoculum densities of *Fusarium oxysporum*, *F. solani*, and *Meloidogyne javanica* and increased yield in tomato (Abu-Gharbieh et al., 1991a). This method of solarization also decreased buried populations of *Phytophthora cactorum* and *Verticillium dahliae* and increased yield in strawberry (Hartz et al., 1993), decreased indigenous populations of *Hirschmanniella mucronata*, *Helicotylenchus* sp., and *Rotylenchulus reniformis* (Sivakumar and Marimuthu, 1987; Heald and Robinson, 1987; McSorley and Parrado, 1986), and increased yields in pepper, muskmelon, tomato, and watermelon (Hartz et al., 1985; Hartz and Bogle, 1989). Row solarization would seem to have particular merit for short-term crops, and for raised beds in climates where whole-field solarization would be unsuitable because of water pooling.

4. Disinfestation of Container Media

Disinfestation of container media using solarization has shown potential. Covering 20-cm-high 2.0 × 1.5 m piles of nursery potting soil with transparent PE (50 μm) film for two months was as or more effective than steam in reducing preemergence damping-off of *Eucalyptus obliqua* (Kassaby, 1985). *Pythium myriotylum*, *Phytophthora nicotianae* var. *nicotianae*, and *Sclerotium rolfsii* buried in a mound of potting medium (fine peat, sand, composted pine bark, 1:1:1) were killed, respectively, within 7, 3, and 7 days at 10 cm deep and within 7, 7, and 10 days at 25 cm deep, by mulching with a double layer of 50 μm clear PE film (Duff and Banaart, 1992). Solarizing a potting mix (sand:*Eucalyptus marginata* fines, 1:1) in 60 × 65 cm clear PE (200 μm) bags, turned weekly for 3 or 4 weeks, eliminated *Phytophthora crytogea*, *Fusarium oxysporum*, and *Rhizoctonia solani* (Kaewruang et al., 1989). No active juveniles of *Meloidogyne incognita* could be recovered, nor evidence of egg hatching detected from infested soil solarized in PE bags for 2 days (Kamra and Gaur, 1995). Densities of *Fusarium* spp., *Pythium* spp., and *Rhizoctonia solani*, and damping-off of *Pinus nigra* were significantly decreased by solarizing nursery soil for 7 weeks using two layers of transparent PE, one (50 μm) in contact with the soil and the other (150 μm) raised 80 cm above the first layer to create a tunnel (Annesi and Motta, 1994). A similar methodology was extensively illustrated and described for use in producing pathogen-free potato tubers (Aquilar et al., 1990). Solarization of a container medium for 40 to 55 days in pots arranged in

9.0 × 0.9 m plots under clear PE improved the growth and yield of *Gypsophila paniculata* (Gamliel et al., 1993). Solarization may be useful for the disinfestation of difficult-to-fumigate potting soils containing high levels of organic matter (40% or higher) and to enable routine recycling of media (Gamliel et al., 1993; Raviv et al., 1991).

5. Postplant Applications

Although solarization has been used predominantly for preplant soil disinfestation, a number of postplant applications have been identified. The technique successfully suppressed Verticillium wilt in established pistachio tree groves (Asworth and Gaona, 1982) and olive orchards (Tjamos et al., 1991), and reduced root rot *(Rosellinia necatrix)* in an existing apple orchard (Stzejnberg et al., 1987). When used alone or in combination with metam sodium, season-long mulching with clear polyethylene reduced buried inocula of *Phytophthora cactorum* and *Verticillium dahliae*, and produced yield increases in strawberry (Hartz et al., 1993). Full-season solarization with clear mulch also reduced infection of tomato plants by *V. dahliae* for two years and increased yields in tomato and eggplant the year after mulch application (Morgan et al., 1991). Postplant treatment of established almond and peach groves with clear mulch for one month reduced nematode densities but failed to increase tree growth after one year (Stapleton and DeVay, 1983). Verticillium wilt symptoms were reduced 86 to 100% by mulching apricot and almond trees at planting with either transparent or black PE. But mulching with black film gave better overall results than clear film because of the poor survival of trees mulched with clear PE (Stapleton at al., 1993). Mexican lime trees mulched with black PE out-yielded both nonmulched trees and those covered with transparent film (Stapleton and Garza-Lopez, 1988).

6. Climatic Effects on Solarization

Solarization has been most successful in suppressing soil-borne pests in hot, arid to semiarid temperate or tropical regions with limited rainfall and cloud cover during the summer characterized by high solar radiation and daily ambient temperatures which reach or exceed 32°C (DeVay, 1991; Katan, 1980; Katan et al., 1987), such as Australia, Egypt, Israel, India, Morocco, South Africa, southern Spain, and the southwestern U.S. (California and Texas) (Kassaby, 1985; Porter and Merriman, 1983; Abdel-Rahim et al., 1988; Satour et al., 1991; Katan et al., 1976; Sharma and Nene, 1990; Besri, 1991; Phillips, 1990; Jimenez-Diaz et al., 1991; Stapleton and DeVay, 1983; Heald and Robinson, 1987; Martyn and Hartz, 1986). The method has also been evaluated with varied results in marginally suitable climates such as cooler temperate areas, and subtropical or tropical regions having abundant summer rains. Mulching soil with clear plastic in central New York significantly reduced natural populations of *Globodera rostochiensis* to a depth of 10 cm during one summer but failed to show similar effects during the subsequent, unseasonably cool summer (LaMondia and Brodie, 1984). Solarization likewise reduced densities of *Verticillium dahliae* and *Pratylenchus penetrans* in the upper 10 cm of soil in central Ontario (Lazarovits et al., 1991), consistently decreased Rhizoctonia belly rot in pickling cucumber in South Carolina (Keinath, 1995), and produced variable reduction of southern blight *(Sclerotium rolfsii)* in tomato in North Carolina (Ristaino et al., 1991). Although solarization increased stand establishment and yield in sugar beet in central Michigan, it was viewed as impractical because of the necessary removal of land from production for a growing season (Halloin, 1993).

Numerous studies in cool climates, largely conducted in Europe and Japan, have determined that the effectiveness of the technique can be enhanced by conducting solarization in glass or plastic houses during the summer (Garibaldi and Lodovica-Gullino, 1991; Horiuchi, 1991). For example, solarization in an open field failed to reduce corky root *(Pyrenochaeta lycopersici)* in tomato in northern Italy when conducted for 5 to 6 weeks in July through August, but controlled the disease when employed within a plastic house or glasshouse during the same period. It was additionally determined that PVC film was more effective than PE in this case, and that soil temperatures generated by solarization were higher and soil disinfestation more consistent in a glasshouse than a plastic house (Garibaldi and Tamietti, 1984). The use of soil solarization during the summer in glass- or plastic-covered structures decreased lettuce drop *(Sclerotinia minor)*, reduced populations of *Meloidogyne* spp. and Phytophthora blight *(Phytophthora capsici)* in pepper, and increased yields in these crops in Italy (Fiume, 1994; Cosentino et al., 1993; Noto, 1994; Polizzi et al., 1994). This methodology also reduced clubroot in crucifers *(Plasmodiophora brassicae)* in the Czech Republic (Rod, 1994), corky root in tomato in France and Portugal (Goisque et al., 1984; Moura and Palminha, 1994), orange spot and lettuce big vein viruses in lettuce in France (Thicoipe and Despujols, 1992), the brown root rot complex *(Colletotrichum coccodes, Fusarium oxysporum* f. sp. *radicis-lycopersici*, and other fungi) and *Clavibacter michiganensis* subsp. *michiganensis* in tomato in Greece (Bourbos and Skoudridakis, 1991; Antoniu et al., 1995), and Fusarium wilt *(F. oxysporum* f. sp. *fragariae)* in strawberry in Japan (Kodama and Fukui, 1982).

Research conducted in subtropical and tropical locations with frequent rainfall has shown that solarization could be effective against some pathogens but was ineffective or produced inconsistent results against others. Solarization of soil during the rainy summer months in Florida reduced buried inocula of *Phytophthora nicotianae*, *Pseudomonas solanacearum*, *Fusarium oxysporum* f. sp. *lycopersici*, *F. oxysporum* f. sp. *radicis-lycopersici*, and naturally occurring densities of the reniform nematode *(Rotylenchulus reniformis)* to depths of 25, 15, 5, 5, and 20 cm, respectively (Chellemi et al., 1993,1994). The technique consistently decreased Verticillium wilt *(Verticillium dahliae)* and increased yield in tomato, but had a variable effect on the incidence of root-knot nematode *(Meloidogyne incognita)* (Overman, 1985; Overman and Jones, 1986; McSorley and Parrado, 1986), effectively reduced population levels of the sting nematode *(Belonolaimus longicaudatus)* and increased yields in strawberry (Overman et al., 1987), and failed to reduce Fusarium crown and root rot *(F. oxysporum* f. sp. *radicis-lycopersici)* in field-grown tomatoes (McGovern et al., 1995). Surprisingly, Phytophthora blight *(Phytophthora nicotianae)* of Madagascar periwinkle was also reduced in Florida by application of transparent mulch during the cooler but drier period (October) following the rainy season (McGovern and Begeman, 1995). Solarization for two months during the summer significantly reduced infection of broccoli and kale by *Rhizoctonia solani* and *Fusarium* spp. in Bermuda despite the close proximity of two hurricanes (McGovern and Harper, 1996). Summer solarization also significantly reduced the incidence of southern blight in pepper in Guadeloupe (J. A. Toribio, personal communication). Soft rot was reduced and yields increased in Chinese cabbage by solarizing soil in Taiwan for 4 to 14 days (AVRDC, 1990a,b). Additional research in Taiwan demonstrated that *Sclerotium rolfsii* could be controlled in tomato by solarization for 15 days (Tu et al., 1992).

Solarizing the soil prior to the onset of the rainy season in Mexico increased yields in sesame despite failure to control the primary target pathogen, *Macrophomina phaseolina* (Stapelton and Garza-Lopez, 1988). Other research in Mexico indicated that the practice reduced Fusarium wilt *(Fusarium oxysporum* f. sp. *melonis)* in melon (Vidales et al., 1987) and root-knot nematode *(Meloidogyne incognita)* in bean (Salgado et al., 1988), and increased yields in strawberry (Davalos and Castro, 1987).

7. Factors Affecting Commercial Adoption

Solarization is currently being used in a limited number of commercial settings. The majority of commercial onion growers in Israel have adopted the practice for management of pink root rot, and solarization has gained acceptance among producers of organic carrots (Grinstein and Ausher, 1991). A significant proportion of the strawberry and cucumber growers in Japan are routinely using solarization as a preplant disinfestation practice for plastic house culture. However, the practice has not been as widely adapted for tomato production because of apparent grower dissatisfaction with the level of control achieved against Fusarium wilt (Horiuchi, 1991). Since 1988, solarization has been applied commercially in 10 ha of plastic houses in northern Italy with success against Verticillium wilt in tomato, *Rhizoctonia solani* and *Sclerotinia sclerotiorum* in lettuce, and *Fusarium oxysporum* f. sp. *basilicum* in basil, but has been ineffective in controlling

Fusarium crown rot in tomato (Garibaldi and Lodovica-Gullino, 1991). Commercial application of solarization against bacterial canker *(Clavibacter michiganensis* pv. *michiganensis),* corky root, and Verticillium wilt in plastic house-grown tomatoes is expanding in Greece (Antoniou et al., 1995; Tjamos, 1991).

Such technical difficulties as improper and poorly timed mulch application and premature mulch degradation or removal have been cited as hindrances to commercial implementation of solarization (Bell and Laemmlen, 1991). Inability to fallow land during periods conducive to solarization will also obviously prevent its use (Grinstein and Ausher, 1991; Halloin, 1993; Morgan et al., 1991). Another issue which may impede the routine adoption of solarization is the disposal of mulches in an environmentally sound manner (Brown et al., 1991b). Remedies for this concern may be presented by the development of recycling programs for mulch or by the development and use of effective, biodegradable films (Brown et al., 1991b; De Luca et al., 1994; Lamberti and Basile, 1991).

Although numerous examples of disease reduction and yield increase following solarization have been experimentally demonstrated, only a small proportion of these studies compared the effectiveness of solarization with fumigation, currently the most commonly used commercial soil disinfestation practice. Solarization was as effective as methyl bromide in reducing Fusarium wilt of tomato and Verticillium wilt in potato, and equal to metam sodium in decreasing Fusarium wilt in watermelon (El-Shami et al., 1990; Elad et al., 1980; Gonzalez-Torres et al., 1993), but was ineffective in suppressing Panama disease *(Fusarium oxysporum* f. sp. *cubense)* of banana (Herbert and Marx, 1990). Neither methyl bromide nor solarization resulted in significant reductions in Fusarium crown rot or bacterial wilt (Ralstonia [*Pseudomonas*] *solanacearum*) in tomato, but the former did significantly increase total yield in fields infested with *F. oxysporum* f. sp. *radicis-lycopersici* (Sivan and Chet, 1993). Reduction of clubroot in crucifers by solarization surpassed or equaled that achieved by dazomet (Rod, 1994), but this technique has, in general, been inferior to dazomet, methyl bromide, or metam sodium in reducing corky root and increasing yields in onion and tomato (Porter et al., 1989, 1991; Hartz et al., 1989; Tjamos, 1984). Solarization equaled or surpassed methyl bromide in reducing symptoms caused by *Rhizoctonia solani* and *Verticillium dahliae* in potato, but failed to equal the fumigant in yield (Elad et al., 1980).

Solarization may be as effective as some fumigant nematicides such as dazomet and 1,3-dichloropropene (1,3-D) for nematode management, and may occasionally equal the nematode suppression achieved by methyl bromide, especially in hot, arid areas (Cartia et al., 1989; Raymundo and Alcazar, 1986; Stapleton et al., 1987; Stapleton and DeVay, 1983; Stephan et al., 1991). Often, results depend on the nematode species present. Relatively large decreases in numbers of certain nematodes such as *Rotylenchulus reniformis* in response to solarization are reported by several authors (Chellemi et al., 1993; Heald and Robinson, 1987; McSorley and Parrado, 1986; Sharma and Nene, 1990). On the other hand, *Meloidogyne* spp. are not always adequately controlled by solarization (Katan, 1980; Saleh et al., 1989; Stapleton and Heald, 1991; Stapleton et al., 1987). Although reduction of root-knot nematodes has been reported even under subtropical conditions, solarization has generally not been as effective in these climates as fumigants containing methyl bromide in decreasing densities of *Meloidogyne* spp. (Chellemi et al., 1993; McSorley and Parrado, 1986; Noto, 1994; Overman, 1985; Overman and Jones, 1986).

8. Integration of Solarization with Other Pest Management Practices

Combination of solarization with other pest management practices may be necessary to ensure acceptable reduction of difficult-to-control pests, especially in suboptimal climates. Katan (1980) forwarded this concept early in the development of the technique when he stated: "Soil solarization should not be regarded as a universal method of disease control but, rather, as an additional method that, used correctly, has many advantages." Solarization has been experimentally integrated with biological control, organic amendments, fumigants, hot water, and host resistance.

Combining the biological control agent *Trichoderma harzianum* with soil solarization increased its efficacy against *Rhizoctonia solani* and *Verticillium dahliae* in potatoes (Elad et al.,

1980). This combination also resulted in more substantial decreases in Fusarium crown rot and greater yields in tomato than either treatment alone, and exceeded the performance of methyl bromide (Sivan and Chet, 1993). However, *T. harzianum* did not increase the effectiveness of solarization against *Rosselina necatrix* in apple (Sztejnberg et al., 1987) or *Rhizoctonia solani* in bean (Garibaldi and Tamietti, 1989). Tzortzakakis and Gowen (1994) suggested that integrating solarization with the nematode parasite *Pasteuria penetrans* could result in improved nematode control.

Evolution of ammonia and sulfur-containing compounds from organic and inorganic soil amendments has been implicated in plant pathogen suppression (Rodriguez-Kabana, 1986; Zakaria et al., 1980; Lewis and Papavizas, 1975). Incorporation of dry cabbage residue in the soil prior to solarization produced a synergistic reduction in the severity of cabbage yellows *(Fusarium oxysporum* f. sp. *conglutinans);* tarped but shaded cabbage residues were ineffective (Ramirez-Villapudua and Munnecke, 1987). Solarization of soil amended with dry cabbage residue slowed the development of gummy stem blight *(Didymella bryoniae)* and increased yield in watermelon (Keinath, 1996). The use of cabbage residues in a simulated solarization system also increased the effectiveness of sublethal heating in reducing the viability of *Pythium ultimum* and *Sclerotium rolfsii* (Gamliel and Stapleton, 1993a). Amending soil prior to solarization with urea and manure prolonged the effectiveness of solar heating against *Fusarium oxysporum* f. sp. *cumini* and *Macrophomina phaseolina* (Lodha, 1995). Solarization of soil amended with chicken compost was more effective against *Meloidogyne incognita* and resulted in soil temperatures 2 to 3°C higher than solarization alone (Gamliel and Stapleton, 1993b), but amendment with farmyard manure did not improve the nematicidal efficacy of solarization (Gaur and Dhingra, 1991).

Synergistic reductions in soil-borne pests have resulted from combining solarization and fumigation, often using reduced fumigant rates. Solar heating of soil plus metam sodium resulted in enhanced reduction of delimited shell spot in peanut, *Fusarium* spp. and *Rhizoctonia solani* in kale, *Fusarium oxysporum* f. sp. *radicis-lycopersici* in tomato, and inoculum densities of *Verticillium dahliae* and *Phytophthora cactorum* (Frank et al., 1986; McGovern and Harper, 1996; McGovern et al., 1995; Hartz et al., 1993). In two of these studies, integration of solarization and metam sodium resulted in control of targeted pests equivalent to formulations of methyl bromide (Hartz et al., 1993; McGovern et al., 1995). Additive effects were also observed in control of clubroot of cauliflower by combining solarization with reduced rates of either dazomet or a methyl bromide-chloropicrin formulation (Porter et al., 1991). Solarization plus a reduced rate of methyl bromide was effective in controlling Verticillium wilt *(Verticillium dahliae)* in artichoke and strawberry (Tjamos et al., 1987; Cebolla et al., 1995). When fumigation with 1,3-D was used in combination with soil solarization, nematode numbers were usually lower than in plots treated with the fumigant alone (Stapleton and DeVay, 1983). However, in other experiments, combinations of solarization with various nematicides (including 1,3-D) were no more efficient in reducing nematode levels than solarization alone (Stapleton et al., 1987). Alternation of solarization in summer with methyl bromide application in winter resulted in effective control of *Meloidogyne* sp. in tomato (Calabretta et al., 1991), while fumigation with 1,3-D followed by solarization improved the control of *Heterodera carotae* in carrot and *Ditylenchus dipsaci* in onion (Greco et al., 1992).

Application of solar-heated hot water (75 to 90°C) either at the beginning or end of the solarization period produced more consistent reductions in root galling of tomato at harvest than either treatment alone, while the addition of hot water after solarization appeared to reduce populations of *Fusarium* spp. (Abu-Gharbieh et al., 1991a). A combination of heated irrigation water and solarization resulted in higher soil temperatures, but was not as effective as methyl bromide fumigation in reducing numbers of, and damage by, *M. incognita* in tomato (Noling, 1994). Integration of solarization with resistance in *Catharanthus roseus* produced consistent synergistic increases in plant biomass and enhanced suppression of dieback and collar rot *(Pythium aphanidermatum)* (Kulkarni et al., 1992). Soil solarization followed by the use of resistant plants reduced Fusarium wilt in cotton and Verticillium wilt in potato in subsequent plantings of susceptible cultivars or lines (Katan et al., 1983; Davis and Sorenson, 1986). While integration of solarization with other pest management strategies has been promising, additional work is needed.

E. Composting

Growers may choose to compost used container media along with plant debris in an effort to eliminate the environmental impact of their disposal and for resource recovery. The same rationale is causing a dramatic increase in the availability of composted municipal solid waste products for potential agricultural use (Smith, 1994). Commercial composting involves the carefully controlled microbial degradation of organic wastes into stable, usable forms. The process is most commonly conducted aerobically, and its efficiency, including heat evolution, is affected by such factors as C/N ratio, aeration, and moisture content (Hoitink and Fahy, 1986). It can be divided into three phases: (1) an initial phase of 1 to 2 days in which readily decomposable materials break down and temperature rapidly rises; (2) a thermophilic phase lasting a number of months in which cellulose is degraded at a relatively constant temperature; and (3) a stabilization phase when the process slows, temperature declines, and mesophilic microorganisms recolonize the compost (Hoitink and Fahy, 1986). Besides heat, aerobic composting produces ammonia, carbon dioxide, and water, while anaerobic decomposition produces methane, carbon dioxide, and numerous intermediate organic compounds (Golueke, 1972). Unless controlled through deliberate heat removal by ventilation, the temperature of composting masses typically peaks at 80°C. However, the optimization of microbial activity and substrate decomposition rate for large-scale operations requires that temperatures of compost piles be maintained below 60°C (Finstein and Miller, 1985).

Much is known about the fate of mammalian pathogens during the composting process (Burge et al., 1981; Finstein et al., 1978; Golueke, 1976; Wiley and Westerberg, 1969), but there is a dearth of information on the survival of plant pathogens in this environment. While not inherently a soil disinfestation process, composting may result in inactivation of plant pathogens through the development of high temperature, accumulation of toxic degradation products such as ammonia, buildup of antagonistic microorganisms including antinomycetes and *Bacillus* spp., or through combinations of these mechanisms (Bollen, 1985b; Hoitink and Fahy, 1986; Yuen and Raabe, 1984). Composting for 3 weeks, during which the pile was turned every 2 to 3 days, inactivated fungi such as *Armillaria mellea, Rhizoctonia solani, Sclerotium rolfsii,* and *Verticillium dahliae.* Temperatures at the center of the 0.7-m^3 bin reached 70°C, and persisted above 45°C for 10 to 14 days (Yuen and Raabe, 1984). Composting infected plant material for 10 to 12 weeks at 40 to 60°C eliminated the fungi *Botrytis cinerea, Rhizoctonia solani, Phytophthora cinnamomi,* and *Pythium irregulare,* and the bacterium *Erwinia chrysanthemi* (Hoitink et al., 1976). *Plasmodiophora brassica* was not detected after compost treatment for 24 h at an average temperature of 54°C (Lopez-Real and Foster, 1985).

Phytoparasitic nematodes and bacteria (excluding spore-formers) are generally more sensitive to heat than fungal pathogens and would not be expected to survive the temperatures typically achieved during composting (Hesling, 1972; Bollen, 1985b; Hoitink and Fahy, 1986). Potato cyst nematodes *(Globodera rostochiensis)* were killed by composting sludge from potato processing at 32 to 34°C (Sprau, 1967). The bacterium *Pseudomonas phaseolica* was eliminated from compost at 35°C after 4 days (Lopez-Real and Foster, 1985). The few research reports on virus sensitivity to composting present somewhat conflicting data. Tobacco plant residue infected with TMV was still infectious after composting with bark for 6 weeks at 50 to 70°C (Hoitink et al., 1982). However, TMV was apparently eliminated by composting a tobacco seed bed at 49 to 63°C (Grushevoi and Levykh, 1940). Two other tobamoviruses, tomato mosaic virus and cucumber green mottle mosaic virus, were undetectable in soil containing infected plant residues after composting at 47°C or less for 70 days, and for 3 days at 72°C, respectively (Agvelis and Manios, 1989, 1992). Tobacco necrosis virus was eliminated from compost after 4 days at an average temperature of 54°C (Lopez-Real and Foster, 1985).

Plant pathogens may escape thermal inactivation by colonizing the edges of compost piles where temperatures are cooler (Esser, 1977; Sewell et al., 1962; Yuen and Raabe, 1984). It is therefore very important that material from outer layers of piles be exposed to consistent heating by thoroughly mixing with interior layers early in the composting process (Bollen, 1985b). The use of diseased plants in compost is particularly risky if uniformly high temperatures cannot be ensured (Esser, 1977).

III. SOIL DISINFESTATION BY FLOODING

Flooding, or flood fallowing, for control of soil-borne plant pathogens is an ancient practice whose early manifestations included natural flooding of agricultural areas such as those along the Nile river and routine rotation of flooded rice with other crops (Cook and Baker, 1983a; Palti, 1981). Proposed mechanisms for pathogen elimination by submersion in water include prolonged anaerobiosis (Stover, 1954, 1955), elevation of carbon dioxide levels (Newcombe, 1960), and the activity of antagonistic microorganisms (Cook and Baker, 1983b; Moore, 1949). Effectiveness of flooding for pathogen elimination appears to be highest at warm temperatures, presumably due to more rapid depletion of available oxygen by submerged microorganisms. The survival times of several *Fusarium* species were substantially shortened by elevating the temperature to 24 to 30°C during flooding (Stover, 1954; Tu, 1991). Likewise, survival of *Chalara elegans* (*Thielaviopsis basicola*) in flooded soil at 4°C was comparable to that in nonflooded soil, but greatly reduced when the temperature of flooded soil was increased to 15 to 25°C (Chittaranjan and Punja, 1994).

Flooding soil for up to 6 months was used as a management strategy for Panama disease or banana wilt *(F. oxysporum* f. sp. *cubense)* (Stover et al., 1952) until superseded by the development of wilt-resistant cultivars (Palti, 1981). A one year rotation with lowland rice reduced populations of *F. oxysporum* f. sp. *pisi* and *F. solani* f. sp. *pisi*, but the field was recolonized after a pea crop (Wang and Lin, 1985). Verticillium wilt *(Verticillium dahliae)* in cotton was controlled by long-term (≥6 weeks) summer soil flooding, and lowland rice culture reduced the time necessary to eliminate the fungus compared to flooding alone (Pullman and DeVay, 1981). Flooding for a shorter period (10 to 40 days) reduced, but failed to eliminate sclerotial production by *V. dahliae*, possibly due to an inability to reach total anaerobiosis (Ioannou et al., 1977). On the other hand, sclerotia of *Sclerotinia sclerotiorum* decayed completely after 23 to 45 days of flooding in a variety of Florida soil types with a compact subsoil, whether flooding was continuous or imposed at 3-day intervals (Moore, 1949). It is interesting to note that *Phytophthora parasitica* var. *nicotianae* (syn. *Phytophthora nicotianae*), which causes diseases normally favored by waterlogged soils, was eliminated from a tobacco production site after 3 to 4 months of rotation with lowland rice (Van Schreven, 1948). A few reports have indicated that flooding could suppress bacterial pathogens including Ralstonia *solanacearum* (bacterial wilt in tobacco) (Van Schreven, 1948), *Xanthomonas campestris* pv. *oryzae* (bacterial blight of rice) (Thri Murty et al., 1982), and *Xanthomonas malvacearum* (bacterial blight of cotton) (Verma et al., 1977).

Flooding is often mentioned as a means for reducing nematode population densities (Maas, 1987; Trivedi and Barker, 1986). Caution must be used in interpreting reports about the effects of flooding on nematodes, since some critical studies (e.g., Hollis and Rodriguez-Kabana, 1966) were not carried out in the field but in containers in which byproducts from decomposition accumulated. In practice, celery following rice which had been flooded for two months had only slight galling from root-knot nematodes (*Meloidogyne* spp.), in comparison with celery following rice which had not been flooded (Thames and Stoner, 1953). Nematodes colonized bananas much more slowly in a site flooded for three months prior to planting than in a site which was not flooded (Maas, 1969). While nematode densities in soil may be reduced after 2 to 3 months of flooding, longer periods of time may be required to eliminate nematode eggs (Maas, 1987). In an early trial in California, eggs of root-knot nematodes remained viable for 12 to 22.5 months after flooding was initiated (Brown, 1933). Moreover, several species of plant-parasitic nematodes were recovered from cranberry bogs which had been flooded for two years (Bird and Jenkins, 1965). Flooding for nematode control was generally more successful at higher temperatures, and interrupted flooding alternated with dry periods was more effective than continuous flooding (Rhoades, 1964, 1982). Results obtained by flooding vary with aeration, season, soil type, level of the water table, presence of toxic byproducts, and other environmental conditions (Maas, 1987; Trivedi and Barker, 1986).

Flooding has become the soil disinfestation method of choice in a number of commercial situations. Flood fallowing is a common, though not universal, practice in the Everglades agricultural area of Florida which encompasses about 26,000 ha. This practice is used by growers for general soil disinfestation, to improve soil tilth, and to reduce soil loss from biological oxidation

Table 2 Plant Pathogens Suppressed by Flooding Soil

Pathogen	Disease/Common Name	Crop	Ref.
Nematodes			
Ditylenchus dipsaci	Stem Nematode	Ornamental bulbs	Muller and Van Aartrijk, 1989
Meloidogyne incognita	Root Knot Nematode	Celery, kenaf	Thames and Stoner, 1953; Fishler and Winchester, 1964; Johnson and Berger, 1972
Tylenchorynchus sp.	Stubby Root Nematode	Celery	Johnson and Berger, 1972
Fungi			
Fusarium oxysporum f. sp. *cubense*	Panama Disease	Banana	Stover, 1954, 1955
F. oxysporum f. sp. pisi	Root Rot and Wilt	Pea	Wang and Lin, 1985
F. solani f. sp. *pisi*	Root Rot	Pea	Wang and Lin, 1985
Leptosphaeria maculans	Blackleg	Rape	Petrie, 1995
Phytophthora parasitica var. *parasitica* (syn. *P. nicotianae*)	Black Shank	Tobacco	Van Schreven, 1948
Sclerotinia sclerotiorum	White Mold (sclerotia eliminated)	Bean, pepper, lettuce, etc.	Moore, 1949
Sclerotium cepivorum	White Rot	Onion	Banks and Edgington, 1989
Thielaviopsis basicola	Black Root Rot	Cotton	Holtz and Weinhold, 1994
Verticillium daliae	Verticillium Wilt	Cotton	Butterfield et al., 1978; Pullman and DeVay, 1981
Bacteria			
Ralstonia *(Pseudomonas) solanacearum*	Bacterial Wilt	Tobacco	Van Schreven, 1948
Xanthomonas campestris pv. *oryzae*	Bacterial Blight	Rice	Thri Murty et al., 1982
Xanthomonas malvacearum	Bacterial Blight	Cotton	Verma et al., 1977

(Snyder, 1987). Flooding of cotton fields during the summer has become a fairly common strategy against black rot *(Thielaviopsis basicola)* on about 16,200 ha in the Tulare Lake Basin area of the San Joaquin Valley (Holtz and Weinhold, 1994). The effectiveness of flooding has often been increased by combination with crop rotation. Lowland rice is commonly used every third or fourth year in rotation with cotton to control Fusarium wilt *(F. oxysporum* f. sp. *vasinfectum)* in cotton in the People's Republic of China (Cook, 1981). Tobacco and lowland rice culture are frequently used in a two-year rotation in Java (Van Schreven, 1948). In many cases, however, flooding for pathogen management may not be practical, especially as emphasis on water conservation increases (Noling and Becker, 1994). Also, it is important to remember that water can be an important agent for passive dispersal of plant pathogens (Newhall, 1955; Norton, 1978). A list of pathogens suppressed by flooding is presented in Table 2.

ACKNOWLEDGMENTS

The authors wish to thank A. J. Overman and P. D. Roberts for their critical review of this manuscript. Florida Agricultural Experiment Station Journal Series No. N-01379.

REFERENCES

Abdel-Rahim, M. F., Satour, M. M., Mickail, K. Y., El-Erakis, S. A., Grinstein, A., and Katan, J., Effectiveness of soil solarization in furrow-irrigated Egyptian soils, *Plant Dis.*, 72, 143, 1988.

Abu-Blan, H. A. and Abu-Gharbieh, W. I., Effect of soil solarization on winter planting of potato, cauliflower and cucumber in the central Jordan Valley, *Dirasat., Ser. B, Pure Appl. Sci.*, 21, 203, 1994.

Abu-Gharbieh, W. I., Saleh, W. I. H., and Al-Banna, L., Application of solar-heated water for soil solarization, in *Soil Solarization*, DeVay, J. E., Stapleton, J. J., and Elmore, C. L., Eds., FAO Plant Production and Protection Paper 109, World Health Organization, Geneva, 1991a, chap. 3.

Abu-Gharbieh, H. Saleh and Abu-Blan, H., Use of black plastic for soil solarization and post-plant mulching, in *Soil Solarization*, DeVay, J. E., Stapleton, J. J., and Elmore, C. L., Eds., FAO Plant Production and Protection Paper 109, World Health Organization, 1991b, chap. 7.

Adams, P. B., Effect of soil temperature and soil amendments on Thielaviopsis root rot of sesame, *Phytopathology*, 61, 93, 1971.

Agrios, G. N., *Plant Pathol.*, 3rd ed., Academic Press, New York, 1988, chap. 9.

Agvelis, A. D. and Manios, V. I., Elimination of tomato mosaic virus by composting tomato residues, *Neth. J. Plant Pathol.*, 95, 167, 1989.

Agvelis, A. D. and Manios, V. I., Elimination of cucumber green mottle mosaic tobamovirus by composting infected cucumber residues, *Acta Hortic.*, 302, 311, 1992.

Aldrich, R. A. and Nelson, P. E., Equipment for aerated steam treatment of small quantities of soil and soil mixes, *Plant Dis. Rep.*, 53, 784, 1969.

Altman, J., Increased and decreased plant growth responses resulting from soil fumigation, in *Root Diseases and Soil-borne Pathogens*, Tousson, T. A., Bega, R.V., and Nelson, P. E., Eds., University of California Press, Berkeley, 1970, 216.

Annesi, T. and Motta, E., Soil solarization in an Italian forest nursery, *Eur. J. For. Pathol.*, 24, 203, 1994.

Antoniu, P. P., Tjamos, E. C., Andreou, M. T., Panagopoulos, C. G., and Vanachter, A., Effectiveness, modes of action and commercial application of soil solarization for control of *Clavibacter michiganensis* subsp. *michiganensis* of tomatoes, *Acta Hortic.*, 382, 119, 1995.

Aquilar, J., Vittorelli, C., Molina, J., and Santisteban, O., Disinfest the plant substrate using the solarization method to produce basic category potato seed tubers, INIAA-COTESU-CIP Agreement, International Potato Center, Lima, Peru, 1990, 1.

Arun, A. and Mathew, D. S., Studies on rhizosphere microflora of pigeonpea: qualitative and quantitative incidence of microorganisms after solarization, *Indian Phytopathol.*, 46, 151, 1993.

Asworth, L., Jr. and Gaona, S., Evaluation of clear polyethylene mulch for controlling *Verticillium* wilt in established pistachio nut groves, *Phytopathology*, 72, 243, 1982.

AVRDC, Soft rot control in cabbage using solarization techniques, *1987 Progr. Rep.*, Asian Veg. Res. Dev. Ctr., Shanhua, Tainan, 1990a, 300.

AVRDC, Solarization against soft rot in Chinese cabbage, *1988 Progr. Rep.*, Asian Veg. Res. Dev. Ctr., Shanhua, Tainan, 1990b, 266.

Baker, K. F., The principles of heat treatment of soil and planting material, *J. Austr. Inst. Sci.*, 28, 118, 1962.

Baker, K. F., Selective killing of soil microorganisms by aerated steam, in *Root Diseases and Soil-Borne Pathogens*, Toussoun, T. A., Bega, R. V., and Nelson P. E., Eds., University of California Press, Berkeley, 1970, 234.

Baker, K. F. and Chandler, P. A., Development and maintenance of healthy planting stock, in *The U. C. System for Producing Healthy Container-Grown Plants*, Baker, K. F., Ed., Calif. Agric. Exp. Stn. Manual No. 23, University of California, Berkeley, 1957, 13.

Baker, K. F. and Fuller, W. H., Soil treatment by microwave energy to destroy plant pathogens, *Phytopathology*, 59, 193, 1969.

Baker, K. F. and Olsen, C. M., Aerated steam for soil treatment, *Phytopathology*, 50, 82, 1960.

Baker, K. F. and Roistacher, C. N., Heat treatment of soil, in *The U. C. System for Producing Healthy Container- Grown Plants*, Baker, K. F., Ed., Calif. Agric. Exp. Stn. Manual No. 23, University of California, Berkeley, 1957a, 8.

Baker, K. F. and Roistacher, C. N., Principles of heat treatment of soil, in *The U. C. System for Producing Healthy Container-Grown Plants*, Baker, K. F., Ed., Calif. Agric. Exp. Stn. Manual No. 23, University of California, Berkeley, 1957b, 9.

Baker, K. F. and Roistacher, C. N., Equipment for heat treatment of soil, in *The U. C. System for Producing Healthy Container-Grown Plants*, Baker, K. F., Ed., Calif. Agric. Exp. Stn. Manual No. 23, University of California, Berkeley, 1957c, 10.

Banks, E. and Edgington, L. V., Effect of integrated control practices on the onion white rot pathogen in organic soil, *Can. J. Plant Pathol.*, 11, 268, 1989.

Baranov, L. A., Kalamkaliev, M. K., and Ilyukhin, G. P., Electrothermic disinfection of soil in greenhouses, *Mekhan. Electr. Sel. Khoz.*, 5, 42, 1983 (in Russian).

Barbercheck, M. E. and von Broembsen, S. L., Effects of soil solarization on plant-parasitic nematodes and *Phytophthora cinnamomi* in South Africa, *Plant Dis.*, 70, 945, 1986.

Bell, C. E. and Laemmlen, F. F., Soil solarization in the Imperial Valley of California, in *Soil Solarization*, Katan, J. and DeVay, J. E., Eds., CRC Press, Boca Raton, FL, 1991, chap. 18.

Ben-Yephet, Y., Stapleton, J. J., Wakeman, R. J., and DeVay, J. E., Comparative effects of soil solarization with single and double layers of polyethylene film on survival of *Fusarium oxysporum* f. sp. *vasinfectum*, *Phyoparasitica*, 15, 181, 1987.

Berninger, E., Pionnat, J.-C., and Scotto La Massese, C., Essai de desinfection solaire des sols dans le Sud-Est de la France, *Agronomy*, 5, 505, 1985 (in French, with English summary).

Besri, M., Solarization of soil and agricultural materials in Morocco for control of Verticillium wilt and Didymella stem canker in tomato, in *Soil Solarization*, DeVay, J. E., Stapleton, J. J., and Elmore, C. L., Eds., FAO Plant Production Paper 109, FAO, Rome, 1991, 237.

Bewley, W. F., *Diseases of Glasshouse Plants*, Benn, London, 1923, chap. 8.

Bird, G. W. and Jenkins, W. R., Effect of cranberry bog flooding and low dissolved oxygen concentrations on nematode populations, *Plant Dis. Rep.*, 49, 517, 1965.

Blake, C. D., Root rot of bananas caused by *Radopholus simils* (Cobb) and its control in New South Wales, *Nematologica*, 6, 295, 1961.

Blauser, I. P., Sterilizing soil with electricity, *Agric. Eng.*, 16, 436, 1935.

Bollen, G. J., Kolonisatie van kasgrond door pathogen na gedeeltelijke sterilisatie met warme lucht, *Neth. J. Plant Pathol.*, 73, 66, 1967 (in Dutch, with English summary).

Bollen, G. J., The selective effect of heat treatment on the microflora of greenhouse soil, *Neth. J. Plant Pathol.*, 75,157, 1969.

Bollen, G. J., Fungal recolonization of heat-treated glasshouse soils, *Agro-Ecosys.*, 1, 139, 1974.

Bollen, G. J., Lethal temperatures of soil fungi, in *Ecology and Management of Soilborne Plant Pathogens*, (Part V) Parker, C. A., Rovira, A. D., Moore, K. G., and Wong, P. T. W., Eds., APS Press, St. Paul, MN, 1985a, 191.

Bollen, G. J., The fate of plant pathogens during composting of crop residues, in *Composting of Agricultural and Other Wastes*, Proc. Semin. Commission Eur. Communities, Oxford., U.K., 19–20 March, 1984, Gasser, J. K. R., Ed., Elsevier, London, 1985b, 282.

Bourbos, V. A. and Skoudridakis, M. T., Lutte contre la pourriture des racines de tomate en serre par la solarisation du sol, *Bull. SROP*, 14, 172, 1991.

Braun, M., Burgstaller, H., and Eldin, A. M. G. F., Approaches for the control of the parasitic weed *Orobanche ramosa* L. with special references to the use of glyphosate spraying and solar heating techniques on eggplant under the conditions of Sudan, *Acta Hortic.*, 158, 335, 1985.

Brazelton, R. W., Sterilizing soil mixes with aerated steam, *Agric. Eng.*, July, 400, 1968.

Bridge, J., Control strategies in subsistence agriculture, in *Principles and Practice of Nematode Control in Crops*, Brown, R. H. and Kerry, B. R., Eds., Academic Press, Orlando, FL, 1987, chap. 12.

Brock, T. D., *Thermophylic Microorganisms and Life at Higher Temperatures*, Springer-Verlag, New York, 1978.

Brock, T. D., *Biology of Microorganisms*, 3rd ed., Prentice-Hall, Englewood Cliffs, NJ, 1979, chap. 8.

Brown, L. N., Flooding to control root-knot nematodes, *J. Agric. Res.*, 47, 883, 1933.

Brown, J. E., Stevens, C., Khan, V. A., Hochmuth, G. J., Splittstoesser, W. E., Granberry, D. M., and Early, B. C., Improvement of plastic technology for soil solarization, in *Soil Solarization*, DeVay, J. E., Stapleton, J. J., and Elmore, C. L., Eds., FAO Plant Production Paper 109, FAO, Rome, 1991a, chap. 8.

Brown, J. E., Stevens, C., Khan, V. A., Hochmuth, G. J., Splittstoesser, W. E., Granberry, D. M., and Early, B. C., Current limitations to commercial uses of plastic mulches for soil solarization, in *Soil Solarization*, DeVay, J. E., Stapleton, J. J., and Elmore, C. L., Eds., FAO Plant Production Paper 109, FAO, Rome, 1991b, chap. 10.

Burge, W. D., Colacicco, D., and Cramer, W. N., Criteria for achieving pathogen destruction during composting, *J. Water Pollut. Control Fed.*, 53, 1683, 1981.

Butterfield, E. J., DeVay, J. E., and Garber, R. H., The influence of several crop sequences on the incidence of Verticillium wilt of cotton and on the population of *Verticillium dahliae* in field soil, *Phytopathology*, 68, 1217, 1978.

Byars, L. P. and Gilbert, W. W., Soil disinfection with hot water to control the root-knot nematode and parasitic soil fungi, Bull. No. 818, U.S. Department of Agriculture, Washington, D.C., 1920.

Calabretta, C., Colombo, A., Nucifora, S., and Privatera, S., Soil solarization and methyl bromide alternated treatments as new method in the disinfestation of soil in protected crops, *Bull. SROP*, 14, 153, 1991.

Campbell, R. N. and Grogan, R. G., Acquisition and transmission of lettuce big-vein virus by *Olpidium brassicae. Phytopathology, 54, 681, 1964.*

Carney, L. B., Electric soil sterilization, *Agric. Eng.*, 13, 95, 1932.

Cartia, G., Cipriano, T., and Greco, N., Effect of solarization and fumigants on soil borne pathogens of pepper in greenhouse, *Acta Hortic.*, 255, 111, 1989.

Cebolla, V., Martinez, P. F., Del Busto, A., Gomez De Barreda, D., Tuset, J. J., Del Busto, A., and Vanachter, A., Dosage reduction of methyl bromide in Spanish Mediterranean coast, *Acta Hortic.*, 382, 156, 1995.

Cenis, J. L., Double plastic sheet for improving soil solarization efficiency, in Proc. 7th Congr. Meditt. Phytopathol. Union, Cairo, Oct. 1, 1987, 73 (abstract).

Chambers, D. P. and Attiwill, P. M., The ash-bed effect in *Eucalyptus regnans* Forest. Chemical, physical and microbiological changes in soil after heating or partial sterilization, *Aust. J. Bot.*, 42, 739, 1994.

Chellemi, D. O., Olson, S. M., Scott, J. W., Mitchell, D. J., and McSorley, R., Reduction of phytoparasitic nematodes on tomato by soil solarization and genotype, *J. Nematol.*, Suppl. 25, 800, 1993.

Chellemi, D. O., Olson, S. M., and Mitchell, D. J., Effects of solarization and fumigation on survival of soilborne pathogens of tomato in northern Florida, *Plant Dis.*, 78, 1167, 1994.

Chen, Y. and Katan, J., Effect of solar heating of soils by transparent polyethylene mulching on their chemical properties, *Soil Sci.*, 130, 271, 1980.

Chen, Y., Gamliel, A., Stapleton, J. J., and Aviad, T., Chemical, physical, and microbial changes related to plant growth in disinfested soil, in *Soil Solarization*, Katan, J. and DeVay, J. E., Eds., CRC Press, Boca Raton, FL, 1991, chap. 8.

Chinanzavavana, S., Grosshandler, W. L., and Davis, D. C., Analysis of heat transfer in soil during open field burning, *Trans. ASAE*, 29, 1797, 1986.

Chittaranjan, S. and Punja, Z. K., Factors influencing the survival of phialospores of *Chalara elegans* in organic soil, *Plant Dis.*, 78, 411, 1994.

Chodhury, S., Diseases of Pan (*Piper betle*) in Assam. IV. Rhizoctonia root rot, *Proc. Indian Acad. Sci. B*, 20, 229, 1944.

Cook, R. J., Fusarium diseases in the Peoples Republic of China, in *Fusarium: Diseases, Biology, and Taxonomy*, Nelson, P. E., Toussoun, T. A., and Cook, R. J., Eds., Pennsylvania State University Press, University Park, PA, 1981, chap. 5.

Cook, R. J. and Baker, K. F., *Nature and Practice of Biological control of Plant Pathogens*, APS Press, St. Paul, MN, 1983a, chap. 2.

Cook, R. J. and Baker, K. F., *Nature and Practice of Biological control of Plant Pathogens*, APS Press, St. Paul, MN, 1983b, chap. 10.

Cosentino, S., Sortino, O., Columbo, A., and Nucifora, A., Pomodoro in serra fredda: la solararizzazione del terreno, *Colt. Prot.*, 22, 87, 1993 (in Italian, with English summary).

Das, K. M., Gangopadhyay, S., and Mishra, A. K., Cultural control of stem rot disease of rice incited by *Magnaporthe salvinii* (Catt.) Krause and Webster under field conditions in India, *Trop. Pest Manage.*, 35, 39, 1989.

Davalos, G., Castro, P. A., and Castro, F. J., La solarizacion como un medio de control para la "secadera" de la fresa en Irapuato, Gto., Mem. XIV Congr. Nac. Fitopatol., Sociedad Mexicana de Fitopatología, Morelia, Mich., Mexico, 1987, 68. (In Spanish)

Davis, J. R. and Sorenson, L. H., Influence of soil solarization at moderate temperatures on potato genotypes with differing resistance to *Verticillium dahliae*, *Phytopathology*, 76, 1021, 1986.

Dawson, J. R., Johnson, R. A. H., Adams, P., and Last, F. T., Influence of steam/air mixtures, when used for heating soil, on biological and chemical properties that affect seedling growth,*Ann. Appl. Biol.*, 56, 243, 1965.

DeLuca, V., Manera, C., Mazza, S., Immirzi, B., Malinconico, M., and Martuscelli, E., Impiego in agricoltura di plastiche innovative biodegradabili per la solarizzazione del terreno, 13th Int. Congr. on Plastics in Agric., (Proc. conf. Verona, Italy, 8-11 March, 1994), Vol. 1, 1994 (in Italian, English summary).

DeVay, J. E., Historical review and principles of soil solarization, in *Soil Solarization*, DeVay, J. E., Stapleton, J. J., and Elmore, C. L., Eds., FAO Plant Production Paper 109, FAO, Rome, 1991, chap. 2.

DeVay, J. E. and Katan, J., Mechanisms of pathogen control in solarized soils, in *Soil Solarization*, Katan, J. and DeVay, J. E., Eds., CRC Press, Boca Raton, FL, 1991, chap. 7.

Diprose, M. F., Evans, G. H., and Cox, S. W. R., Soil partial sterilization by dielectric heating, in *Engineering Advances for Agriculture and Food*, Proc. 1938-1988 Jubilee Conf. Inst. Agric. Engineers., Robinson College, Cambridge, 1988, 363.

Duff, J. D. and Barnaart, A., Solarization controls soilborne fungal pathogens in nursery potting mixes, *Aust. Plant Pathol.*, 21, 20, 1992.

Duff, J. D. and Connelly, M. I., Effect of solarization using single and double layers of clear plastic mulch on *Pythium*, *Phytophthora* and *Sclerotium* species in a nursery potting mix, *Aust. Plant Pathol.*, 22, 28, 1993.

Elad, Y., Katan, J., and Chet, I., Physical, biological, and chemical control integrated for soilborne diseases in potatoes, *Phytopathology*, 70, 418, 1980.

Elliott, C., *Manual of Bacterial Plant Pathogens*, 2nd ed., Chronica Botanica, Waltham, Mass., 1951.

El-Shami, M. A., Salem, D. E., Fadl, F. A., Ashour, W. E., and El-Zayat, M. M., Soil solarization and plant disease management. II. Effect of solarization in comparison with soil fumigation on the management of Fusarium wilt in tomato,*Agric. Res. Rev.*, 83, 601, 1990.

Endo, B. Y., Lethal time-temperature relations for*Heterodera glycines*, *Phytopathology*, 52, 992, 1962.

Esser, R. P., How soil borne nematodes enter and disperse in Florida nurseries, Nematology Circular No. 33, Florida Department of Agriculture and Consumer Services, Gainesville, FL, 1977.

Evans, K., The effects of gamma radiation on *Heterodera rostochiensis*, *Nematologica*, 16, 284, 1970.

Ewart, J. M. and Chrimes, J. R., Effects of chlorine and ultra-violet light in disease control in NFT, *Acta Hortic.*, 98, 317, 1980.

Ferriss, R. S., Effects of microwave oven treatments on microorganisms in soil, *Phytopathology*, 74, 121, 1984.

Finstein, M. S. and Miller, F. C., Principles of composting leading to maximization of decomposition rate, odor control, and cost effectiveness, in *Composting of Agricultural and Other Wastes*, Proc. Semin. Commission Eur. Communities, Oxford, U.K., 19-20 March, 1984, Gasser, J. K. R., Ed., Elsevier, London, 1985, 13.

Finstein, M. S., Wei-Ru Lin, K., and Fischler, G. E., Sludge composting and utilization: a review of the literature on temperature inactivation of pathogens, Rep. New Jersey Agric. Exp. Stn., Project 03543, 1978.

Fishler, D. W. and Winchester, J. A., The effects of flooding on root-knot nematodes in organic soils, *Soil Crop Sci. Soc. Fla. Proc.*, 24, 150, 1964.

Fiume, F., L'impiego dell'apprestamento protettivo per la solarizzazione del terreno nelle colture protette in Italia meridionale, *Informat. Fitopatol.*, 44, 52, 1994 (in Italian, with English summary).

Frank, Z. R., Ben-Yephet, Y., and Katan, J., Synergistic effect of metham and solarization in controlling delimited shell spots of peanut pods, *Crop Prot.*, 5, 199, 1986.

Fujimoto, C. K. and Sherman, G. D., Manganese availability as influenced by steam sterilization of soils, *J. Am. Soc. Agron.*, 40, 527, 1948.

Gamliel, A. and Katan, J., Involvement of fluorescent pseudomonads and other microorganisms in increased growth response of plants in solarized soils, *Phytopathology*, 81, 894, 1991.

Gamliel, A. and Stapleton, J. J., Characterization of antifungal volatile compounds evolved from solarized soil amended with cabbage residues, *Phytopathology*, 83, 899, 1993a.

Gamliel, A. and Stapleton, J. J., Effect of chicken compost or ammonium phosphate and solarization on pathogen control, rhizosphere microorganisms, and lettuce growth, *Plant Dis.*, 77, 886, 1993b.

Gamliel, A., Hadar, E., and Katan, J., Improvement of growth and yield of *Gypsophila paniculata* by solarization or fumigation of soil or container medium in continuous cropping systems, *Plant Dis.*, 77, 933, 1993.

Garibaldi, A. and Lodovica-Gullino, M., Use of solarization in marginally suitable climates, in *Soil Solarization*, DeVay, J. E., Stapleton, J. J., and Elmore, C. L., Eds., FAO Plant Production Paper 109, FAO, Rome, 1991, chap. 7.

Garibaldi, A. and Tamietti, G., Attempts to use soil solarization in closed glasshouses in northern Italy for controlling corky root of tomato, *Acta Hortic.*, 152, 237, 1984.

Garibaldi, A. and Tamietti, G., Solar heating: recent results obtained in Northern Italy, *Acta Hortic.*, 255, 125, 1989.

Gaur, H. S. and Dhingra, A., Management of *Meloidogyne incognita* and *Rotylenchulus reniformis* in nursery- beds by soil solarization and organic soil amendment, *Rev. Nematol.*, 14, 189, 1991.

Goisque, M. J., Louvet, H., Marten, C., Lagier, J., Davet, P., Couteaudier, Y., and Louvet, J., The solar sterilization of greenhouses, *Plasticulture*, 64, 32, 1984.

Golueke, C. G., *Composting. A Study of the Process and its Principles*, Rodale, Emmaus, PA, 1972.

Golueke, C. G., Composting: A review of rationale, principles and public health, *Comp. Sci.*, 17, 11, 1976.

Gonzalez-Torres, R., Melero-Vara, J. M., Gomez-Vazquez, J., and Jimenez Diaz, R. M., The effects of soil solarization and soil fumigation on fusarium wilt of watermelon grown in plastic houses in southeastern Spain, *Plant Pathol.*, 42, 858, 1993.

Grainger, J. and Kennedy, R. R., Temperature distribution and performance in balloon-sheet soil steaming, *Hortic. Res.*, 4, 25, 1964.

Greco, N., D'Addabbo, T., Stea, V., and Brondonisio, A., The synergism of soil solarization with fumigant nematicides and straw for the control of *Heterodera carotae* and *Ditylenchus dipsaci*, *Nematol. Mediterr.*, 20, 25, 1992.

Green, C. D. and Plumb, S., The effect of ultra-violet radiation on the invasion, survival and fertility of larvae of *Heterodera rostochiensis*, *Nematologica*, 13, 186, 1967.

Greenberger, A., Yogev, A., and Katan, J., Induced suppressiveness in solarized soils, *Phytopathology*, 77, 1663, 1987.

Grinstein, A. and Ausher, R., Soil solarization, in *Soil Solarization*, Katan, J. and DeVay, J. E., Eds., CRC Press, Boca Raton, FL, 1991, chap. 13.

Grinstein, A. and Hetzroni, A., The technology of soil solarization, in *Soil Solarization*, Katan, J. and DeVay, J. E., Eds., CRC Press, Boca Raton, FL, 1991, chap. 11.

Grinstein, A., Kritaman, G., Hetzroni, A., Gamliel, A., Mor, M., and Katan, J., The border effect of soil solarization, *Crop Prot.*, 14, 315, 1995.

Gruenzweig, J. M., Rabinowitch, H. D., and Katan, J., Physiological and developmental aspects of increased plant growth in solarized soil, *Ann. Appl. Biol.*, 122, 579, 1993.

Grooshevoy, S. E., Disinfection of seed-bed soil in cold frames by solar energy, *A. I. Mikoyan pan Soviet Sci. Res. Inst. Tob. and Indian Tob. Ind. (VITIM), Krasnodar*, Publ. 137, 31, 1939; (Russian, English summary in *Rev. Appl. Mycol.*, 18, 635, 1939).

Grooshevoy, S. E. and Levykh, P. M., The possibility of obtaining seed-bed soil free from infection in compost heaps, *A. I. Mikoyan pan Soviet Sci. Res. Inst. Tob. and Indian Tob. Ind. (VITIM), Krasnodar*, Publ. 141, 3, 1940; (Russian, English summary in *Rev. Appl. Mycol.*, 20, 87, 1941).

Hagan, H. R., Hawaiian pineapple field soil temperatures in relation to the nematode *Heterodera radicicola* (Greef) Muller, *Soil Sci.*, 36, 83, 1933.

Halloin, J. M., Effects of Soil solarization on sugarbeet production in Michigan, *J. Sugar Beet Res.*, 30, 151, 1993.

Hardison, J. R., Fire and flame for plant disease control, *Annu. Rev. Phytopathol.*, 14, 355, 1976.

Hartz, T. K. and Bogle, C. R., Response of tomato and watermelon to row solarization, *Appl. Agric. Res.*, 4, 15, 1989.

Hartz, T. K., Bogle, C. R., and Villalon, B., Response of pepper and muskmelon to row solarization, *HortScience*, 20, 699, 1985.

Hartz, T. K., Bogle, C. R., Bender, D. A., and Avila, F. A., Control of pink root disease in onion using solarization and fumigation, *J. Am. Soc. Hortic. Sci.*, 114, 587, 1989.

Hartz, T. K., DeVay, J. E., and Elmore, C. L., Solarization is an effective soil disinfestation technique for strawberry production, *HortScience*, 28, 104, 1993.

Heald, C. M., Classical nematode management practices, in *Vistas on Nematology*, Veech, J. A. and Dickson, D. W., Eds., Society of Nematologists, Hyattsville, MD, 1987, chap. 15.

Heald, C. M. and Robinson, A. F., Effects of soil solarization on *Rotylenchulus reniformis* in the lower Rio Grande Valley of Texas, *J. Nematol.*, 19, 93, 1987.

Heald, C. M. and Wayland, J. R., Ultra-high frequency electromagnetic energy as a means for nematode control, *Nematropica*, 5, 1, 1975.

Heald, C. M., Menges, R. M., and Wayland, J. R., Efficacy of ultra-high frequency (UHF) electromagnetic energy and soil fumigation on the control of the reniform nematode and common purslane among southern peas, *Plant Dis. Rep.*, 58, 985, 1974.

Heijina, B. J., Improved soil steaming, in *Jaarb. Tuinbouwtech., Inst. Tuinbouwtech.*, 1966, 81 (in Dutch).

Herbert, J. A. and Marx, D., Short-term control of Panama disease of bananas in South Africa, *Phytophylactica*, 22, 339, 1990.

Hesling, J. J., Nematode pests of mushrooms, in *Economic Nematology*, Webster, J. M., Ed., Academic Press, New York, 1972, chap. 18.

Hetzroni, A., Grinstein, A., Alper, Y., and Frankel, H., A continuous plastic film covering and welding machine for soil solarization, *Acta Hortic.*, 152, 259, 1983.

Heydayetullah, S. and Saha, J. C., Bacterial wilt disease of tomato, *Sci. Cult. (Calcutta)*, 7, 226, 1941.

Hoitink, H. A. J. and Fahy, P. C., Basis for the control of soilborne plant pathogens with composts, *Annu. Rev. Phytopathol.*, 24, 93, 1986.

Hoitink, H. A. J., Herr, L. J., and Schmitthenner, A. F., Survival of some plant pathogens during composting of hardwood tree bark, *Phytopathology*, 66, 1369, 1976.

Hoitink, H. A. J., Nelson, E. B., and Gordon, D. T., Composted bark controls soil pathogens of plants, *Ohio Rep.*, 67, 7, 1982.

Hollings, M. and Stone, O. M., The long-term survival of some plant viruses preserved by lyophilization, *Ann. Appl. Biol.*, 65, 411, 1970.

Hollis, J. P. and Rodriguez-Kabana, R., Rapid kill of nematodes in flooded soil, *Phytopathology*, 56, 1015, 1966.

Holtz, B. A. and Weinhold, A. R., *Thielaviopsis basicola* in San Joaquin Valley soils and the relationship between inoculum density and disease severity in cotton seedlings, *Plant Dis.*, 78, 986, 1994.

Horiuchi, S., Solarization for greenhouse crops in Japan, in *Soil Solarization*, DeVay, J. E., Stapleton, J. J., and Elmore, C. L., Eds., FAO Plant Production Paper 109, FAO, Rome, 1991, chap. 2.

Horsfall, J. G., Pasteurizing soil electrically to control damping-off, N. Y. *Agric. Exp. Stn. Bull.*, 651, 1, 1935.

Hoshino, H. M. and Godfrey, G. H., The thermal death point of *Heterodera radicicola* in relation to time, *Phytopathology*, 23, 260, 1933.

Ioannou, N. R., Schneider, R. W., and Grogan, R. G., Effect of flooding on the soil gas composition and the production of microsclerotia of *Verticillium dahliae* in the field, *Phytopathology*, 67, 651, 1977.

Jacobsohn, R., Greenberger, A., Katan, J., Levi, M., and Alon, H., Control of Egyptian broomrape *(Orobanche aegyptiaca)* and other weeds by means of solar heating of the soil by polyethylene mulching, *Weed Sci.*, 28, 312, 1980.

Jaffrin, A. and Makhlouf, S., Mechanism of light transmission through wet polymer films, *Acta Hortic.*, 281, 11, 1990.

Javed, M., Ahmad, R., Inam-ul-Haq, M., and Mukhtar, T., Effect of soil solarization on the population of root- knot nematode, *Meloidogyne incognita* and growth of tomato plants, *Pakistan J. Phytopathol.*, 6, 115, 1994.

Jimenez-Diaz, R. M., Bejarano, J., Blanco, M. A., Gomez, J., Gonzalez, R., and Melero, J. M., Control of Verticillium wilt and Fusarium wilt diseases by soil solarization in southern Spain, in *Soil Solarization*, DeVay, J. E., Stapleton, J. J., and Elmore, C. L., Eds., FAO Plant Production Paper 109, FAO, Rome, 1991, chap. 4.

Johnson, J., The influence of heated soils on seed germination and plant growth, *Soil Sci.*, 7, 1, 1919.

Johnson, J., Soil-steaming for disease control, *Soil Sci.*, 61, 83, 1946.

Johnson, S. R. and Berger, R. D., Nematode and soil fungi control in celery seedbeds on muck soil, *Plant Dis. Rep.*, 56, 661, 1972.

Johnston, W. J., Golob, C. T., and Sutton, J. W., Effect of temperature and postharvest field burning of Kentucky bluegrass on germination of sclerotia of *Claviceps purpurea*, *Plant Dis.*, 80, 766, 1996.

Juarez-Palacios, C., Felix-Gastelum, R., Wakeman, R. J., Paplomatas, E. J., and DeVay, J. E., Thermal sensitivity of three species of *Phytophthora* and the effect of soil solarization on their survival, *Plant Dis.*, 75, 1160, 1991.

Kaewruang W., Sivasithamparam, K., and Hardy, G. E., Effect of solarization of soil within plastic bags on root rot of gerbera (*Gerbera jamesonii* L.), *Plant Soil*, 120, 303, 1989.

Kamra., A. and Gaur, H. S., Control of plant parasitic nematodes by solarization of soil in polyethylene bags, *Ann. Plant Prot. Sci.*, 3, 90, 1995.

Kassaby, F. Y., Solar-heating soil for control of damping-off diseases, *Soil Biol. Biochem.*, 17, 429, 1985.

Katan, J., Solar pasteurization of soils for disease control: status and prospects, *Plant Dis.*, 64, 450, 1980.

Katan, J., Solar heating (solarization) of soil for control of soilborne pests, *Annu. Rev. Phytopathol.*, 19, 211, 1981.

Katan, J., Soil solarization, in *Innovative Approaches to Plant Dis., Management*, Chet, I., Ed., John Wiley & Sons, New York, 1987, 77.

Katan, J. and DeVay, J. E., Soil solarization: historical perspectives, principles and uses, in *Soil Solarization*, Katan, J. and DeVay, J. E., Eds., CRC Press, Boca Raton, FL, 1991, chap. 2.

Katan, J., Greenberger, A., Grinstein, A., and Alon, H., Solar heating by polyethylene mulching for the control of diseases caused by soilborne pathogens, *Phytopathology*, 76, 683, 1976.

Katan, J., Rotem, I., Finkel, Y., and Daniel, J., Solar heating of the soil for the control of pink root and other soilborne diseases in onions, *Phytoparasitica*, 8, 39, 1980.

Katan, J., Fishler, G., and Grinstein, A., Short- and long-term effects of soil solarization and crop sequence on Fusarium wilt and yield of cotton in Israel, *Phytopathology*, 73, 1215, 1983.

Katan, J., Grinstein, A., Greenberger, A., Yarden, O., and DeVay, J. E., The first decade of soil solarization (solar heating): a chronological bibliography, *Phytoparasitica*, 15, 229, 1987.

Katznelson, H. and Richardson, L. T., The microflora of the rhizosphere of tomato plants in relation to soil sterilization. III, *Can J. Res.*, 21, 249, 1943.

Keinath, A. P., Reductions in inoculum densities of *Rhizoctonia solani* and control of belly rot on pickling cucumber with solarization, *Plant Dis.*, 79, 1213, 1995.

Keinath, A. P., Soil amendment with cabbage residue and crop rotation to reduce gummy stem blight and increase growth and yield of watermelon, *Plant Dis.*, 80, 564, 1996.

Khandar, R. R. and Bhomik, T. P., Effect of different types of plastic tarp on the viability of sclerotia of Sclerotium rolfsii Sacc., a soil borne plant pathogen, Proc. XI Int. Congr. Uses of Plastics in Agric., E.127, 1990.

Kodama, T. and Fukui, T., Solar heating in closed plastic house for control of soilborne diseases. V. Applications for control of *Fusarium* wilt of strawberry, *Ann. Phytopathol. Soc. Jpn.*, 48, 570, 1982 (in Japanese, English summary).

Kreutzer, W. A., The reinfestation of treated soil, in *Ecology of Soilborne Pathogens*, Baker, K. F. and Snyder, W. C., Eds., University of California Press, Berkeley, 1965, 495.

Kulkarni, R. N., Kalra, A., and Ravindra, N. S., Integration of soil solarization with host resistance in the control of dieback and collar rot diseases of periwinkle, *Trop. Agric.*, 69, 217, 1992.

Lamberti, F. and Basile, M., Improvement in plastic technology for solar heating, in *Soil Solarization*, DeVay, J. E., Stapleton, J. J., and Elmore, C. L., Eds., FAO Plant Production Paper 109, FAO, Rome, 1991, chap. 8.

LaMondia, J. A. and Brodie, B. B., Control of *Globodera rostochiensis* by solar heat, *Plant Dis.*, 68, 474, 1984.

Lazarovits, G., Hawke, M. A., Tomlin, A. D., Olthof, Th. H. A., and Squire, S., Soil solarization to control *Verticillium dahliae* and *Pratylenchus penetrans* on potatoes in central Ontario, *Can. J. Plant Pathol.*, 13, 116, 1991.

Lemanova, N. B., Sultanova, O. D., and Zemshmann, A. Y., Sterilization of soil by radiation for raising super- elite grapevine planting material, *Sadov., Vinograd. Vinod. Moldav.*, 8, 38, 1983 (in Russian, with English summary).

Lewis, J. A. and Papvizas, G. C., Effect of sulfur-containing volatile compounds and vapors from cabbage decomposition on *Aphnomyces euteiches*, *Phytopathology*, 61, 208, 1975.

Lifshitz, R., Tabachnik, M., Katan, J., and Chet, I., The effect of sublethal heating on sclerotia of *Sclerotium rolfsii*, *Can. J. Microbiol.*, 29, 1607, 1983.

Lodha, S., Soil solarization, summer irrigation and amendments for control of *Fusarium oxysporum* f. sp. *cumini* and *Macrophomina phaseolina* in arid soils, *Crop Prot.*, 14, 215, 1995.

Lopez-Real, J. and Foster, M., Plant pathogen survival during composting of agricultural organic wastes, in *Composting of Agricultural and Other Wastes*, Gasser, J. K. R., Ed., Elsevier, London, 1985, 291.

Maas, P. W. T., Two important cases of nematode infestation in Surinam, in *Nematodes of Tropical Crops*, Peachey, J. E., Ed., Tech. Commun. No. 40, Commonwealth Agricultural Bureaux, St. Albans, U.K., 1969, 149.

Maas, P. W. T., Physical methods and quarantine, in *Principles and Practice of Nematode Control in Crops*, Brown, R. H. and Kerry, B. R., Eds., Academic Press, Orlando, FL, 1987, chap. 8.

Mahrer, Y. and Katan, J., Spatial soil temperature regime under transparent polethylene mulch: numerical and experimental studies, *Soil Sci.*, 131, 82, 1981.

Martyn, R. D. and Hartz, T. K., Use of soil solarization to control Fusarium wilt of watermelon, *Plant Dis.*, 70, 762, 1986.

Mayers, C. P. and Habeshaw, J. A., Depression of phagocytosis: a non-thermal effect of microwave radiation as a potential hazard to health, *Int. J. Radiat. Biol.*, 24, 449, 1973.

McGovern, R. J. and Begeman, J. P., Reduction of Phytophthora blight of Madagascar periwinkle in the landscape by soil solarization, *Proc. Fla. State Hortic. Soc.*, 108, 58, 1995.

McGovern, R. J. and Harper, F., Suppression of *Rhizoctonia solani* and *Fusarium* spp. in *Brassica oleracea* L. var. *acephala* by bed solarization and metam sodium in Bermuda, *Phytopathology* (Supplement), 86, 115, 1996.

McGovern, R. J., Vavrina, C. S., Obreza, T. A., and Capece, J. C., Reduction of Fusarium crown and root rot of tomato by combining soil solarization and metam sodium. Proc.1995 Int. Res. Conf. on Methyl Bromide Alternatives and Emissions Reductions, Methyl Bromide Alternatives Outreach, Fresno, CA, 34.1, 1995.

McSorley, R., Short-term effects of fire on the nematode community in a pine forest, *Pedobiologia*, 37, 39, 1993.

McSorley, R. and Duncan, L. W., Economic thresholds and nematode management, *Adv. Plant Pathol.*, 11, 147, 1995.

McSorley, R. and Parrado, J. L., The spiral nematode, *Helicotylenchus multicinctus*, on bananas in Florida and its control, *Proc. Fla. State Hortic. Soc.*, 96, 201, 1983.

McSorley, R. and Parrado, J. L., Application of soil solarization to Rockdale soils in a subtropical environment, *Nematropica*, 16, 125, 1986.

Meti, S. S. and Hosman, M. M., Effect of type and thickness of polyethylene on soil solarization, *Indian J. Agron.*, 39, 703, 1994.

Moore, W. D., Flooding as a means of destroying the sclerotia of *Sclerotinia sclerotiorum*, *Phytopathology*, 39, 920, 1949.

Moosman, A. and Koch, W., Soil disinfection by microwaves with special reference to weed control, Weed Control in Vegetable Production, Proc. Meet. EC Experts' Group, Stuttgart 28-31 Oct. 1986, 187, 1988.

Morgan, D. P., Liebman, J. A., Epstein, L., and Jimenez, M. J., Solarizing soil planted with cherry tomatoes vs. solarizing fallow ground for the control of Verticillium wilt, *Plant Dis.*, 75, 148, 1991.

Moura, M. L. R. and Palminha, J., A non-chemical method for the control of *Pyrenochaeta lycopersici* of tomato in the north of Portugal, *Acta Hortic.*, 366, 317, 1994.

Moyls, A. L., Heat transfer model for a heated cylinder of replant soil, *Appl. Eng. Agric.*, 10, 53, 1993.

Moyls, A. L. and Hocking, R. P., *In situ* soil steaming for the control of apple replant disease, *Appl. Eng. Agric.*, 10, 59, 1993.

Mueller, P. J. and Van Aartrijk, J., Flooding reduces the soil population of the stem nematode *Ditylenchus dipsaci* (Kuhn) in sandy soils, *Acta Hortic.*, 255, 261, 1989

Myers, R. F. and Dropkin, V. H., Impracticability of control of plant-parasitic nematodes with ionizing radiations, *Plant Dis. Rep.*, 43, 311, 1959.

Nederpel, L. J., Steriliation of glasshouse soil, Annu. Rep. Glasshouse Crops Res., Exp. Stn. Naaldwijk, 1972, 99.

Nederpel, L., Soil sterilization and pasteurization, in *Soil Disinfestation*, Mulder, D., Ed., Elsevier, Amsterdam, 1979, chap. 4.

Nelson, P. E. and Wilhelm, S., Thermal death range of *Verticillium albo-atrum*, *Phytopathology*, 48, 613, 1958.

Newcombe, M., Some effects of water and anaerobic conditions on *Fusarium oxysporum* in soils, *Trans. Br. Mycol. Soc.*, 43, 51, 1960.

Newhall, A. G., Control of root knot nematode in greenhouse soil sterilization. *Ohio Agric. Exp. Stn. Res. Bull.*, 451, 1, 1930.

Newhall, A. G., Experiments with new electric devices for pasteurizing soils, *Cornell Univ. Agric. Exp. Stn. Res. Bull.*, 731, 1, 1940.

Newhall, A. G., Disinfestation of soil by heat, flooding and fumigation, *Bot. Rev.*, 21, 189, 1955.

Newhall, A. G. and Nixon, M. W., Disinfesting soils by electric pasteurization, *Cornell Univ. Agric. Exp. Stn. Res. Bull.*, 636, 1, 1935.

Newhall, A. G. and Schroeder, W. T., New flash flame soil pasteurizer, *Cornell Univ. Agric. Exp. Stn. Res. Bull.*, 875, 1, 1951.

Nikitin, Y. P. and Nikitin, P. Y., Thyristor electrode soil sterilizer, *Mekhan. Electr. Sel. Khoz.*, 12, 16, 1990 (in Russian).

Noling, J. W., The use of hot water, broadspectrum fumigants, and soil solarization for nematode control, *J. Nematol.*, 26, 562, 1994.

Noling, J. W., Use of hot water for nematode control: a research perspective, Proc. 1995 Int. Res. Conf. on Methyl Bromide Alternatives and Emissions Reductions, Methyl Bromide Alternatives Outreach, Fresno, CA, 53.1, 1995a.

Noling, J. W., Physical methods of nematode control, *J. Nematol.*, 27, 512, 1995b.

Noling, J. W. and Becker, J. O., The challenge of research and extension to define and implement alternatives to methyl bromide, *J. Nematol.*, Suppl. 26, 573, 1994.

Norton, D. C., *Ecology of Plant-Parasitic Nematodes*, John Wiley & Sons, New York, 1978, chap. 2.

Noto, G., Soil solarization in greenhouse: effects on tomato crop, *Acta Hortic.*, 357, 237, 1994.

O'Bannon, J. H. and Good, J. M., Applications of microwave energy to control nematodes in soil, *J. Nematol.*, 3, 93, 1971.

Olsen, C. M. and Baker, K. F., Selective heat treatment of soil, and its effect on the inhibition of *Rhizoctonia solani* by *Bacillus subtilis*, *Phytopathology*, 58, 79, 1968.

Ou, L. T., Rothwell, D. F., and Mesa, M. V., Soil sterilization by 2450 MHz microwave radiation, *Soil Crop Sci. Soc. Fla. Proc.*, 44, 77, 1985.

Overman, A. J., Off-season land management, soil solarization and fumigation for tomato, *Soil Crop Sci. Soc. Fla. Proc.*, 44, 35, 1985.

Overman, A. J. and Jones, J. P., Soil solarization, reaction, and fumigation effects on double-cropped tomato under full-bed mulch, *Proc. Fla. State Hortic. Soc.*, 99, 315, 1986.

Overman, A. J., Howard, C. M., and Albregts, E. E., Soil solarization for strawberries, *Proc. Fla. State Hortic. Soc.*, 100, 236, 1987.

Palti, J., *Cultural Practices and Infectious Crop Diseases*, Vol 2, Springer-Verlag, Berlin, 1981.

Parker, F. E. and Vincent, J. M., Sterilization of peat by gamma radiation, *Plant Soil*, 61, 285, 1981.

Petrie, G. A., Long-term survival and sporulation of *Leptosphaeria maculans* (blackleg) on naturally infected rapeseed/canola stubble in Saskatchewan, *Can. Plant Dis. Surv.*, 75, 23, 1995.

Pflug, I. J. and Schmidt, C. F., Thermal destruction of microorganisms, in *Disinfection, Sterilization, and Preservation*, Lawrence, C. A. and Block, S. S., Eds., Lea & Febiger, Philadelphia, 1968, chap. 6.

Phillips, A. J. L., The effects of soil solarization on sclerotial populations of *Sclerotinia sclerotiorum*, *Plant Pathol.*, 39, 38, 1990.

Polizzi, G., Agosteo, G. E., and Cartia, G., Soil solarization for the control of *Phytophthora capsici* on pepper, *Acta Hortic.*, 366, 331, 1994.

Porter, I. J. and Merriman, P. R., Effects of solarization of soil on nematode and fungal pathogens at two sites in Victoria, *Soil Biol. Biochem.*, 15, 39, 1983.

Porter, I. J. and Merrimen, P. R., Evaluation of soil solarization for control of root diseases of row crops in Victoria, *Plant Pathol.*, 34, 108, 1985.

Porter, I. J., Merriman, P. R., and Keane, P. J., Integrated control of pink root *(Pyrenochaeta terrestris)* of onions by dazomet and soil solarization, *Aust. J. Agric. Res.*, 40, 861, 1989.

Porter, I. J., Merriman, P. R., and Keane, P. J., Soil solarization combined with low rates of fumigants controls clubroot of cauliflowers, caused by *Plasmodiophora brassicae* Woron, *Aust. J. Exp. Agric.*, 31, 843, 1991.

Pullman, G. S. and DeVay, J. E., Effect of flooding and paddy rice culture on the survival of *Verticillium dahliae* and incidence of Verticillium wilt in cotton, *Phytopathology*, 72, 1285, 1981.

Pullman, G. S., Devay, J. E., and Garber, R. H., Soil solarization and thermal death. A logarithmic relationship between time and temperature for four soilborne plant pathogens, *Phytopathology*, 71, 959, 1981a.

Pullman, G. S., DeVay, J. E., Garber, R. H., and Weinhold, A. R., Soil solarization: effects on Verticillium wilt of cotton and soilborne populations of *Verticillium dahliae, Pythium* spp., *Rhizoctonia solani*, and *Thielaviopsis basicola, Phytopathology*, 71, 954, 1981b.

Qiu, J., Wesyerdahul, B. B., Buchner, R. P., and Anderson, C. A., Refinement of hot water treatment for the management of *Aphelenchoides fragariae* in strawberry, *J. Nematol.*, Suppl. 25, 795, 1993.

Raj, H. and Kapoor, I. J., Soil solarization for the control of tomato wilt pathogen (*Fusarium oxysporum* Schl.), *Z. Pflanzenkr. Pflanzensch.*, 100, 652, 1993.

Ramirez-Villapudua, J. and Munnecke, D. E., Control of cabbage yellows (*Fusarium oxysporum* f. sp. *conglutinans*) by solar heating of field soils amended with dry cabbage residues, *Plant Dis.*, 71, 217, 1987.

Rattink, H., Disinfection of potting soil by means of gamma-radiation, *Meded. Fac. Landbouwwet, Rikjsuniv. Gent*, 47, 869, 1982.

Raviv, M., Reuveni, R., Medina, S., Shamir, Y., Durdevani, O., Shor, Y., and Schayer, R., Changes in tuff during prolonged cultivation which affects rose productivity, *Acta Hortic.*, 294, 99, 1991.

Raymundo, S. A. and Alcazar, J., Increasing efficiency of soil solarization in controlling root-knot nematodes by using two layers of plastic mulch, *J. Nematol.*, 18, 628, 1986.

Rhoades, H. L., Effect of fallowing and flooding on root knot in peat soil, *Plant Dis. Rep.*, 48, 303, 1964.

Rhoades, H. L., Effect of temperature on survival of *Meloidogyne incognita* in flooded and fallow muck soil, *Nematropica*, 12, 33, 1982.

Ristaino, J. B., Perry, K. B., and Lumsden, R. D., Effect of solarization and *Gliocladium virens* on *Sclerotium rolfsii*, soil microbiota, and the incidence of southern blight of tomato, *Phytopathology*, 81, 1117, 1991.

Rod, J., The use of soil solarization to control clubroot (*Plasmodiophora brassicae*), *Ochr. Rostl.*, 30, 183, 1994.

Rodriquez-Kabana, R., Organic and inorganic amendments to soil as nematode suppressants, *J. Nematol.*, 18,129, 1986.

Romano, F. and D'Errico, F. P., A rotary furnace for flame sterilization of soils, *Ann. Fac. Sci. Agric. Univ. Studi Napoli, Portici*, IV, 13, 2, 1979.

Rudd, W. N., Killing grubs in the soil, *Am. Flor.*, 9, 171, 1893.

Runia, W. T., A recent development in steam sterilization, *Acta Hortic.*, 152, 195, 1984.

Runia, W. T., Disinfestation of substrates used in protected culture, *Soilless Cult.*, 2, 35, 1986.

Runia, W. T., Elimination of root-infecting pathogens in recirculation water from closed cultivation systems by ultra-violet radiation, *Acta Hortic.*, 361, 361, 1994.

Saleh, H., Abu-Gharbieh, W. I., and Al-Banna, L., Augmentation of soil solarization effects by application of solar-heated water, *Nematol. Mediterr.*, 17, 127, 1989.

Salgado, S. M., Marban, M. N., and Zamudio, V., Comparacion de los efectos de agregados organicos, nematicidas y solarizacion en la incidencia de *Meloidogyne incognita* asociado al cultivo de frijol en Tecamachalco, Pue., Mem. XV Congr. Nac.Fitopatolo., Sociedad Mexicana de Fitopatología, Xalapa, Veracruz, Mexico, 1988, 10 (in Spanish).

Satour, M. M., El-Sherif, E. M., El-Ghareeb, L., El-Hada, S. A., and El-Wakil, H. R., Achievements of soil solarization in Egypt, in *Soil Solarization*, DeVay, J. E., Stapleton, J. J., and Elmore, C. L., Eds., FAO Plant Production Paper 109, FAO, Rome, 1991, chap. 7.

Scannavini, M., Antoniacci, L., Cobelli, L., and Brunelli, A., Esperienze di solarizzazione del terreno in Emilia- Romagna per il contenimento del marciume del colleto della lattuga, *Inform. Fitopatol.*, 43 (10), 30, 1993. (In Italian, with English summary).

Sewell, G. W. F., Wilson, J. F., and Martin, D. G., Machine picking in relation to progressive Verticillium wilt of the hop. II. The effect of composting on the infectivity of machine-picked hop waste, Annu. Rep. East Malling Res. Stn. 1961, 1962, 102.

Sharma, S. B. and Nene, Y. L., Effects of soil solarization on nematodes parasitic to chickpea and pigeonpea, *J. Nematol.*, Suppl. 22, 658, 1990.

Shaw, L. N. and Mitchell, D. J., A continuous soil pasteurizer for organic nursery potting mixes, *Trans. ASAE*, 33, 1978.

Sivakumar, M. and Marimuthu, T., Preliminary studies on the effect of solarization on phytonematodes of betelvine, *Indian J. Nematol.*, 17, 58, 1987.

Sivan, A. and Chet., I., Integrated control of fusarium crown and root rot of tomato with *Trichoderma harzianum* in combination with methyl bromide or soil solarization, *Crop Prot.*, 12, 380, 1993

Smith, W. H., *Recycling Composted Organic Materials in Florida*, Bull. BP-2, University of Florida, Institute of Food and Agricultural Sciences, Gainesville, FL, 1994.

Snyder, G. H., *Flooding as a Management Practice in the Everglades Agricultural Area*, University of Florida, Agric. Exp. Stn. Bull. No. 870, Institute of Food and Agricultural Sciences, Gainesville, FL, 1987.

Sonneveld, C. and Voogt, S. J., Studies on the manganese uptake of lettuce on steam-sterilized glass-house soils, *Plant Soil*, 42, 49, 1975.

Southey, J. F., Nematode pests of ornamental and bulb crops, in *Plant Parasitic Nematodes in Temperate Agriculture*, Evans, K., Trudgill, D. L., and Webster, J. M., Eds., CAB Int., Wallingford, U.K., 1993, chap. 12.

Sprau, F., Das verhalten von zysten des kartoffel-nematoden in klaranlagen, *Mitt. biol. BundAnst. Ld.-u. Forstw. Heft.*, 121, 39, 1967 (in German, cited in Bollen, 1985b).

Stapleton, J. J., Thermal inactivation of crop pests and pathogens and other soil changes caused by solarization, in *Soil Solarization*, DeVay, J. E., Stapleton, J. J., and Elmore, C. L., Eds., FAO Plant Production Paper 109, FAO, Rome, 1991, chap. 3.

Stapleton, J. J. and DeVay, J. E., Effect of soil solarization on populations of selected soilborne micro-organisms and growth of deciduous fruit tree seedlings, *Phytopathology*, 72, 323, 1982.

Stapleton, J. J. and DeVay, J. E., Response of phytoparasitic and free-living nematodes to soil solarization and 1,3-dichloropropene in California, *Phytopathology*, 73, 1429, 1983.

Stapleton, J. J. and DeVay, J. E., Thermal components of soil solarization as related to changes in soil and root microflora and increased growth response, *Phytopathology*, 74, 255, 1984.

Stapleton, J. J. and DeVay, J. E., Soil solarization: a non-chemical approach for management of plant pathogens and pests, *Crop Prot.*, 5, 190, 1986.

Stapleton, J. J. and Garza-Lopez, J. G., Mulching of soils with transparent (solarization) and black polyethylene films to increase growth of annual and perrenial crops in southwestern Mexico, *Trop. Agric.*, 65, 29, 1988.

Stapleton, J. J. and Heald, C. M., Management of phytopathogenic nematodes by soil solarization, in *Soil Solarization*, Katan, J. and DeVay, J. E., Eds., CRC Press, Boca Raton, FL, 1991, chap. 4.

Stapleton, J. J., Lear, B., and DeVay, J. E., Effect of combining soil solarization with certain nematicides on target and nontarget organisms and plant growth, *J. Nematol.*, Suppl. 1, 107, 1987.

Stapleton, J. J., Paplomatas, E. J., Wakeman, R. J., and DeVay, J. E., Establishment of apricot and almond trees using soil mulching with transparent (solarization) and black polyethylene film: effects on Verticilium wilt and tree health, *Plant Pathol.*, 42, 333, 1993.

Stephan, Z. A., Al-Maamoury, I. K., and Michbass, A. H., The efficacy of nematicides, solar heating and the fungus *Paecilomyces lilacinus* in controlling root-knot nematode *Meloidogyne javanica* in Iraq, in *Soil Solarization*, DeVay, J. E., Stapleton, J. J., and Elmore, C. L., Eds., FAO Plant Production Paper 109, FAO, Rome, 1991, chap. 9.

Stover, R. H., Flood-fallowing for erradication of *Fusarium oxysporum* f. *cubense*. II. Some factors involved in fungus survival, *Soil Sci.*, 77, 401, 1954.

Stover, R. H., Flood fallowing for erradication of *Fusarium oxysporum* f. *cubense*. III. Effect of oxygen on fungus survival, *Soil Sci.*, 80, 401, 1955.

Stover, R. H., Thornton, N. C., and Dunlap, V. C., Changes in the soil flora of banana lands flood-fallowed for the eradication of *Fusarium oxysporum* f. *cubense*, *Phytopathology*, 42, 476, 1952.

Stroetmann, I., Kampfer, P., and Dott, W., Untersuchngen zur effizienz von sterilisationsverfahren an unterschiedlichen boden, *Zbl. Hyg.*, 195, 11, 1994 (in German, with English summary).

Sundarum, T. K., Physiology and growth of thermophilic bacteria, in *Thermophiles: General, Molecular, and Applied Microbiology*, Brock, T. D., Ed., John Wiley & Sons, New York, 1986, 75.

Sztejnberg, A., Freeman, S., Chet, I., and Katan, J., Control of *Rosellinia necatrix* in soil and in apple orchard by solarization and *Trichoderma harzianum*, *Plant Dis.*, 71, 365, 1987.

Tavernetti, J. R., Characteristics of the resistance type soil sterilizer, *Agric. Eng.*, 16, 271, 1935.

Tavernetti, J. R., A continuous soil pasteurizer, *Agric. Eng.*, 23, 255, 1942.

Thames, W. H., Jr. and Stoner, W. N., A preliminary trial of lowland culture rice in rotation with vegetable crops as a means of reducing root-knot nematode infestations in the Everglades, *Plant Dis. Rep.*, 37, 187, 1953.

Thicoipe, J.-P. and Despujols, J., La solarisation limite la maladie des Taches orangees, *Infos*, 82, 31, 1992 (in French).

Thri Murty, V. S., Devadath, S., and Seshagiri Rao, C., Ecology of *Xanthomonas campestris* pv. *oryzae*, the incitant of bacterial blight of rice, *Indian. J. Agric. Sci.*, 52, 524, 1982

Tjamos, E. C., Control of *Pyrenochaeta lycopersici* by combined soil solarization and low dose of methyl bromide in Greece, *Acta Hortic.*, 152, 253, 1984.

Tjamos, E. C., Soil solarization in Greece, in *Soil Solarization*, Katan, J. and DeVay, J. E., Eds., CRC Press, Boca Raton, FL, 1991, chap. 14.

Tjamos, E. C. and Palomatas, E. J., Long-term effect of soil solarization in controlling Verticillium wilt of globe artichokes in Greece, *Plant Pathol.*, 37, 507, 1988.

Tjamos, E., Paplomatas, E., and Fountas, D., Control of Verticillium-wilt of artichokes by solar heating with a reduced dose of methyl bromide, in *Integrated and Biological Control in Protected Crops*, Proc. Meet. EC Experts' Group, Heraklion, 24-16, April, 1985, Cavalloro, R., Ed., A. A. Balkema, Rotterdam, 1987, 239.

Tjamos, E. C., Biris, D. A., and Paplomatas, E. J., Recovery of olive trees with Verticillium wilt after individual application of soil solarization in established olive groves, *Plant Dis.*, 75, 557, 1991.

Townsend, J. L., Gamma irradiation of *Hederodera schactii*, *Nematologica*, 13, 586, 1967.

Trivedi, P. C. and Barker, K. R., Nematode management by cultural practices, *Nematropica*, 16, 213, 1986.

Tu, C. C., Cheny, Y. H., and Hwang, S. C., Effects of solar energy and green manures on the control of southern blight of tomato, *Plant Prot. Bull. Taipei*, 33, 95, 1992.

Tu, J. C., Effect of soil flooding on the survival of *Fusarium oxysporum* f. sp. *pisi* and *Fusarium solani* f. sp. *pisi*, *Meded. Fac. Landbouwwet. Rijksuniv. Gent.*, 56, 431, 1991.

Tzortzakakis, E. and Gowen, S. R., Control of root-knot nematodes by integrating *Pasteuria penetrans* with oxamyl and solarization in protected crops in Crete, *J. Nematol.*, 26, 122, 1994.

Ulyanov, Y. A. and Zmeev, A. Y., Stationary electrode soil sterilizer of continuous action, *Sb. Nauchn. Rab. Sara. Sel'skokh. Inst.*, 119, 3, 1978 (in Russian).

Van Der Kroft, W. G. and Groen, J. V., Een nieuwe toepassing van electricteit ter bevordering van den plantgroei. *Ons Tijdschrift*, 13, 65, 1933.

Vannacci, G., Triolo, E., and Materazzi, A., Survival of *Sclerotinia minor* Jagger sclerotia in solarized soil, *Plant Soil*, 109, 49, 1988.

Van Schreven, D. A., Onderzoekingen met betrekking tot enkele plagen en ziekten van Vorstenlandsche Tabak, *Tijdschr. over Plantenziekten*, 54, 149, 1948 (in Dutch, English summary).

Van Wambeke, E., Wijsmans, J., Vanachter, A., and Van Assche, C., Alternatives for chemical crop growing substrate disinfestation, *Meded. Fac. Lanbouwwet. Rijksuniv. Gent*, 46, 961, 1981.

Van Wambeke, E., Wijsmans, J., and d'Hertefelt, P., Possibilities in microwave application for growing substrate disinfestation, *Acta Hortic.*, 152, 209, 1984.

Vela, G. R. and Wu, J. F., Mechanism of lethal action of 2450-MHz radiation on microorganisms, *Appl. Environ. Microbiol.*, 37, 550, 1979.

Verma, J. P., Nayak, M. L., and Singh, R. P., Survival of *Xanthomonas malvacearum* under north India conditions, *Indian Phytopathol.*, 30, 361, 1977.

Vidales, F. J. A., Munro, O. D., and Alcantar, J. L., Control de pathgenos del suelo mediante el uso de energia solar en al cultivo de melon (*Cucumis melo* L.), en el valle de Apatzingan, Mich., Mem. XIV Congr. Nac.Fitopatolo., Sociedad Mexicana de Fitopatologfa, Morelia, Mich., Mexico, 1987, 69 (in Spanish).

Walker, G. E. and Wachtel, M. F., The influence of soil solarization and non-fumigant nematicides on infection of *Meloidogyne javanica* by *Pasteuria penetrans*, *Nematologica*, 34, 477, 1988.

Wang, P. H. and Lin, Y. S., Ecology of pea wilt and root rot pathogens in drained paddy field, *Plant Prot. Bull. Taiwan*, 27, 317, 1985.

Wilcox, W. F. and Mircetich, S. M., Lack of host specificity among isolates of *Phytophthora megasperma*, *Phytopathology*, 77, 1132, 1987.

Wiley, B. B. and Westerberg, S. C., Survival of human pathogens in composted sewage, *Appl. Microbiol.*, 18, 994, 1969.

Wolf, D. C., Dao, T. H., Scott, H. D., and Lavy, T. L., Influence of sterilization methods on selected soil microbiological, physical, and chemical properties, *J. Environ. Qual.*, 18, 39, 1989.

Yuen, G. J. and Raabe, R. D., Effects of small-scale aerobic composting on survival of some plant pathogens, *Plant Dis.*, 68, 134, 1984.

Zakaria, M. A., Lockwood, J. L., and Filonow, A. B., Reduction in *Fusarium* population density in soil by volatile degradation products of oilseed meal amendments, *Phytopathology*, 70, 495, 1980.

Zentmeyer, G. A., *Phytophthora cinnamomi* and the Diseases It Causes, Monogr. 10, American Phytopathological Society, St. Paul, MN, 1980.

Part VI
Biotechnology

CHAPTER 13

Tissue Culture

Joyce M. Van Eck and Franzine D. Smith

1-8493-2627-3/97/$0.00+$.50

I. INTRODUCTION

In crop improvement programs for disease resistance, a number of strategies are used to acquire the desired trait. One source of plants with improved disease resistance can be material that is recovered from plant tissue culture. Plant tissue culture refers to the *in vitro* cultivation of all plant parts including single cells, tissues, and organs under aseptic conditions (Torres, 1989). These plant parts are cultured on media containing nutrients similar to those required by whole plants. In addition to these nutrients, the media may be supplemented with plant growth regulators, solidifying agents, amino acids, or other components. There are a number of articles which review the composition of various culture media (Gamborg et al., 1976; Dodds and Roberts, 1985). In addition to the appropriate culture medium, the plant material must be maintained in the appropriate environment. Attention should be given to temperature, light (intensity, quality, duration), and type of culture vessel.

The term plant tissue culture encompasses several different types of culture. Some specific types of culture include embryo culture, meristem culture, tissue or callus culture, and protoplast culture. A number of these culture methods have been used to improve disease resistance in many crops. It is not our intent to present an exhaustive review on this topic, but to give examples of how plant tissue culture has been used to generate pathogen-free plant material, create disease resistance for species where resistance does not exist, and transfer disease resistance genes between species where sexual incompatibilities impose crossing barriers.

II. MERISTEM CULTURE

A. Definition

Meristem culture involves the removal of the meristem with one to two leaf primordia and subsequent *in vitro* culturing of this explant. Many pathogens do not survive well in the actively dividing tissue of the apical dome, so this method is used to rid plant material of pathogens, especially viruses. The size of the explant affects its survival in culture and the presence or absence of the pathogen. The explant needs to be large enough to survive, but small enough to be pathogen free. The explant for meristem culture should be less than 0.1 mm long while that for shoot tip cultures range from 0.1 to 2 mm. Heat therapy or incorporation of an antiviral compound into tissue culture medium in combination with meristem culture has been used to increase rates of virus elimination.

B. Elimination of Fungi and Bacteria

Meristem tip culture in combination with culture indexing can be used to establish stock material of ornamental plants that are free of fungi and bacteria (Horst and Klopmeyer, 1993; Klopmeyer, 1995). Meristem or shoot tips are removed from greenhouse-grown plants and established in tissue culture. Systemic pathogens, if present, will grow out into the tissue culture medium. Those explants that remain free of fungi or bacteria are grown up, transferred to a certification greenhouse, and indexed an additional two times for systemic pathogens. Plant lines that remain free of the pathogens are used for propagation. Meristem culture has been applied to ornamental plants such as geraniums, New Guinea impatiens, and double-flowered impatiens and seed potatoes (Grimm et al., 1991; Horst and Klopmeyer, 1993; Klopmeyer, 1995). In seed potatoes, the bacterium *Erwinia carotovora* var. *atroseptica* (Eca) was eliminated with 50% success using meristem tip culture (Grimm et al., 1991). Treatment of the potatoes at high temperatures prior to meristem removal did not improve the eradication of Eca.

C. Elimination of Viruses

1. Meristem Culture

The meristem tip culture technique has been used to rid many plant species of viruses because it exploits the erratic distribution of viruses within plants. The cells in the meristem are actively dividing and the vascular tissue is not well developed, which makes transport of many viruses to this area very difficult. *Alstroemeria* mosaic virus was eliminated from several cultivars by meristem tip culture (Hakkaart and Versluijs, 1988). The success of this method for virus elimination was cultivar dependent with an elimination success rate from 0 to 100% in the cultivars tested. Leaf curl virus (gemini virus)-free sweet potato "Tainung 63" were produced in 75 to 80% of the meristem tips (0.3 mm) cultured (Green et al., 1992). Also, several varieties of dahlia were freed of Dahlia Mosaic Virus (DaMV) and DaMV/CMV complex in 80 to 90% of the plants derived from meristem cultures (Cai et al., 1988).

Often this method is sufficient to obtain pathogen-free plant material, but the addition of thermotherapy may improve recovery of virus-free plants. However, when the virus is heat stable as in the case of raspberry vein chlorosis virus (RVCV), the addition of thermotherapy did not increase the rate of virus-free plants recovered (66%) (Sobczykiewicz, 1989).

In some instances meristem culture is not effective in eliminating viruses either because the virus is able to multiply in the meristem or because the meristematic explants do not survive. For example, only larger meristems (0.51 to 1.0 mm) excised from a monopodial orchid hybrid survived, but these were all infected with cymbidium mosaic virus and odontoglossum ringspot virus (Lim et al., 1993).

2. Thermotherapy and Meristem Culture

Thermotherapy involves growing plants at the highest temperature tolerated (35 to 37°C) by the plant. In combination with meristem tip culture, it is an effective method for eliminating viruses that are heat sensitive. Heat treatment in the form of hot air instead of hot water is used most often when herbaceous material is being treated. Plants are treated either in pots or in tissue culture. During this period part of the plant may become virus free and then the shoot apex is removed and regenerated into a new plant. Heat treatment does not ensure virus elimination but plants can be indexed for viruses after treatment to determine its effectiveness.

Thermotherapy prior to meristem culture is a standard practice in some disease certification programs to produce virus-free plant material. *Pelargonium* flower break virus is heat sensitive and is eliminated from certified geranium stock plants by this method. Geraniums that are free of bacterial and fungal pathogens are grown at elevated temperature (16 h, 37°C; 8 h, 32°C) for 3 to 4 weeks. Meristems are then cultured and the resultant plants are indexed for the virus (Thorn-Horst et al., 1977).

Potatoes are plagued by a host of viruses that can affect yield. Meristem culture alone or in combination with thermotherapy has been used to address many of these viruses (Mellor and Stace-Smith, 1987). Potato viruses A and Y can be eliminated without prior heat treatment by culturing 0.3 mm meristem tips. PVX and PVS require heat treatments of six to eight weeks prior to culturing. Potato virus S (PVS) was eradicated from several potato cultivars in 90% of the meristem cultures when *in vitro* shoot tips were grown at 34°C for 1 to 2 months prior to taking meristem tips (Brown et al., 1988). This was an improvement over 11.4% and 41% eradication when potted plants were heat-treated prior to meristem excision (Stace-Smith and Mellor, 1968; Brown et al., 1988).

Peach cultivars (cvs. Hermosa and Summerset) free of prunus necrotic ringspot virus (PNRSV) were isolated after thermotherapy of *in vitro*-grown shoots (Stein et al., 1991). Alternating high (16 h, 38°C) and low (8 h, 28°C) temperatures were more effective in reducing the virus titer than a constant elevated temperature. After 22 days of thermotherapy, shoot apices (10 mm) were removed and the resultant plants were 90% (Hermosa) and 40% (Summerset) virus free.

In sweet potato, thermotherapy did not increase the percent of leaf curl virus-free plants derived from meristem culture (0.3 mm, 70 to 80%), however, it did improve virus elimination in plants derived from stem tips (1 cm) (Green et al., 1992). Virus elimination in stem tips increased from 0% elimination without thermotherapy to 50%.

3. Antiviral Compounds and Meristem Culture

Incorporation of an antiviral compound into tissue culture medium can eliminate viruses from plant tissue. Treatment of sweet potato stem sections (1 cm) with ribavirin incorporated in the culture medium increased the number of virus-free plants originating from stem tips as the concentration of ribavirin increased (Green et al., 1992). Another raspberry virus, raspberry bushy dwarf virus (RBDV) is heat stable and appears to be able to colonize the meristem because the combination of thermotherapy and meristem culture was not successful in eliminating it from red raspberry (Kudell and Buchenauer, 1989). However, growth of axillary bud cultures on medium containing antiviral compounds (ribavirin or dodecyl-*N*-methyl ephedrinium bromide) freed the plant material of RBDV.

D. Other Uses

Meristem culturing is also used as a method of micropropagation. A micropropagation method developed for ginger originates from meristems (Bhagyalakshmi and Singh, 1988). This source of pathogen-free plant material can replace seed pieces which are often systemically infected by *Pseudomonas solanacearum*, *Fusarium oxysporum* f. *zingiberi*, and *Pythium* sp.

E. Advantages and Disadvantages

Although meristem culture is not used to produce disease resistance, it does provide a source of pathogen-free planting material. This is an essential element in the control of plant diseases where chemical control of a pathogen during a growing season is either not available or too costly. Often, pathogen-free plants combined with good cultural practices and a tight sanitation program prevents reintroduction of pathogens and assures delivery of disease-free plants. One disadvantage to meristem culture is that because the explant is so small the culture often takes a relatively long time to grow. This problem can be eliminated in the case of some virus diseases by the addition of thermotherapy or antiviral compounds to meristem culturing. The addition of either thermotherapy or antiviral compounds may allow culturing of larger virus-free stem tips.

III. EMBRYO RESCUE

A. Definition

Embryo rescue is used to ensure the survival of immature embryos when wide crosses are made between distantly related plants. The incompatibility between distantly related plants often results in a breakdown of the endosperm, so embryos can be rescued by growing them on artificial medium which provides the neccessary nutrients. The immature embryos are grown in tissue culture until the plant can be transplanted to soil. Wide crosses are often attempted to transfer disease-resistant traits from wild species to cultivated varieties.

B. Disease Resistance

Embryo rescue is being used to introduce fungal, bacterial, and virus resistance traits into breeding material. Potato leaf roll virus (PLRV) is a serious pathogen, reducing potato yields in developing areas of the world. *Solanum etuberosum* is a wild nontuber-bearing potato which is resistant to PLRV (Chavez et al., 1988). Using embryo rescue, this trait was successfully introduced into a tuber-bearing *Solanum* (Chavez et al., 1988). Resistance in the progeny was expressed as a resistance to titer buildup of PLRV and not a resistance to infection.

A number of studies have been published on interspecific crosses made in barley with the intent of transferring insect or pathogen resistance. These studies are included in a recent review on embryo rescue (Sharma et al., 1996). *Hordeum bulbosum,* a wild relative of cultivated barley, is resistant to powdery mildew *(Erysiphe graminis* DC. f sp *hordei),* yellow rust (*Puccinia striiforms* WEST.), brown rust (*Puccinia hordei* OTTH.), scab (*Fusarium* spp.), yellow mosaic virus, net blotch *(Pyrenophre teres),* and scald *(Rhyncosporium secalis)* (Xu and Snape, 1989). The tetraploid hybrids resulting from crosses between *H. bulbosum* and the cultivated *H. vulgare* were resistant to powdery mildew and isolates of yellow rust and brown rust (Xu and Snape, 1989). This study showed the potential for transfering disease resistance from wild *H. bulbosum* to cultivated barley. In another study, crosses made with cultivated *H. vulgare* resulted in 11 triploid F_1 plants using embryo rescue (Xu and Kasha, 1992). After backcrossing, half of the resultant progeny were resistant to powdery mildew, indicating a single dominant gene for this trait.

Oryza minuta, a wild rice species, is resistant to the fungal disease rice blast *(Pyricularia grisea)* and bacterial blight *(Xanthomonas oryzae* pv. *oryzae),* two devastating diseases of rice worldwide. Crosses made with the cultivated rice *Oryza sativa* L. and *O. minuta* resulted in progeny that were resistant to bacterial blight or rice blast (Amante-Bordeos et al., 1992). Embryo rescue was necessary to recover hybrids and progeny from the backcross to *O. sativa.*

Other crops where interspecific hybridization between cultivated species and wild relatives has been successful due to embryo rescue include tomato *(Lycopersicon esculentum* and *L. peruviianum)* (Segeren et al., 1993), and lettuce (*Lactuca sativa* and *Lactuca* sp.) (Maisonneuve et al., 1995). These wild relatives are resistant to many diseases of these crops and could be used as a source of resistance in breeding programs. *L. peruviianum* is a source for tomato spotted wilt virus resistance and the wild *Lactuca* sp. used is resistant to *Bremia lactucae,* the causal agent of downy mildew.

Embryo rescue has been used to recover plants after intergeneric crosses. Hybrids from a cross between *Eruca sativa* and *Brassica campestris* were saved using embryo rescue (Agnihotri et al., 1990). *Eruca sativa,* a member of subtribe *Brassicineae,* is a source for resistance to white rust caused by the fungus *Albugo candida.*

C. Other Uses

In addition to disease resistance, other horticultural traits can be transferred using embryo rescue. Also, inherently weak embryos, a problem in some early ripening fruit varieties, can be saved with embryo rescue (Ramming, 1990). An early peach cultivar "Goldcrest" and an early nectarine cultivar "Mayfire" are the result of embryo rescue (Ramming and Tanner, 1987a,b). Hybridization between seedless grape varieties has also been possible using embryo rescue (Ramming and Emershad, 1984; Ramming, 1990).

D. Advantages and Disadvantages

The embryo rescue technique salvages plants that would not survive. This has major consequences for some plant breeders by shortening the breeding cycles. For example, some seedless grape varieties can be hybridized directly because hybrids can be rescued (Ramming, 1990). Without embryo rescue, an intermediate seeded female needs to be used followed by backcrossing.

Pathogen and pest resistance, in addition to other horticultural traits, can be transferred from wild species into breeding material. When desirable traits are not found in breeding material but are found in other species, embryo rescue may allow recovery of plants between species where crossing barriers exist due to sexual incompatibility. Also, hybrids in some intergeneric crosses can be saved with embryo rescue.

As with meristem culture, embryo rescue requires technical expertise, is time consuming, and tedious. However, it is one approach to consider when disease resistance is not available in breeding material.

IV. SOMACLONAL VARIATION

A. Background

During the tissue culture process, genetic variability can occur. This variability is called somaclonal variation, based on the term somaclones which is used in reference to regenerated plants. The variability may arise from preexisting mutations in the cells of the explant, however, a large part of the variation is induced during the tissue culture process (Lorz and Scowcroft, 1983; Orton, 1984). The variation can be at the genetic or molecular level resulting from changes in chromosomes (number, deletions, transposition, etc.) or methylation of DNA sequences in the nucleus, mitochondria, or chloroplasts (Phillips et al., 1990; Brown, 1991; van den Bulk, 1991). The extent of somaclonal variation can be determined by the explant source (Ostry et al., 1994), time in culture (Scowcroft and Larkin, 1985), and culture media components (Karp and Bright, 1985; Ostry et al., 1994). Variation can also be attributed to what is called epigenetic variation, which is a nonheritable change (Meins, 1983). Various sources of somaclonal variation and the possible uses in crop improvement have been described (Larkin and Scowcroft, 1981).

Since 1981, there have been numerous reports on somaclonal variation, some of which have been useful in crop improvement and some which have not. Traits such as salt and drought tolerance (Nabors and Dykes, 1985; Winicov, 1991), herbicide resistance (Taylor et al., 1989), frost tolerance (Dörffling et al., 1993), and disease resistance, which will be addressed in this section, have been observed in somaclonal variants.

B. Crops and Pathogens

One application of somaclonal variants has been as a source of disease resistance. The earliest report of somaclonal variation resulting in disease-resistant material was in sugar cane. Somaclones from several varieties were screened for resistance to Figi disease (viral) and Downy mildew *(Sclerospora sacchari)* (Heinz, 1973; Krishnamurthi, 1974; Krishnamurthi and Tlaskal, 1974). A number of variants resistant to both diseases were put into yield trials, and did not show a reduction in yield. Recently, somaclonal variants from sugar cane have been obtained that are resistant to Sugarcane Mosaic Virus (Oropeza and García, 1996). In field trials, the two resistant somaclones performed similarly to the mother plant.

From the first report on disease-resistant sugar cane until now, there are many examples of improved disease resistance in various crops via somaclonal variation. Source material for recovery of somaclonal variants has ranged from susceptible to nonsusceptible lines (Hammerschlag, 1990; Lineberger et al., 1990; Nagarathna et al., 1993). Plants recovered have ranged from very resistant to a particular pathogen, to more susceptible than the mother plant (Shoemaker et al., 1985; Donovan et al., 1994). Table 1 contains some examples of disease-resistant plants produced from somaclonal variation from 1991 until the present. For most crops developed between 1974 and 1994, see reviews by van den Bulk, 1991; Ahmed and Sagi, 1993; and Hammerschlag et al., 1995.

C. Screening Techniques

Screening for increased disease resistance via somaclonal variation has been evaluated *in vitro* at the cellular level or *in vivo* at the whole-plant level. There are advantages and disadvantages to each method, some of which are listed in Table 2. For *in vitro* selection, selection can be done on individual cells, callus, or even plants (van den Bulk, 1991; Ahmed and Sagi, 1993). Selection for increased resistance to a pathogen can be done with the toxic metabolite produced by the

Table 1 Disease-Resistant Plants from Somaclonal Variation

Crop	Pathogen	Ref.
Tomato	*Pseudomomonas solanacearum* E. F. Smith	Bobisud et al., 1996
	Alternaria solani	Rodeva et al., 1994
	Fusarium oxysporum f. sp. *radicus-lycopersici*	Rodeva et al., 1994
Potato	*Verticillium dahliae*	Sebastiani et al., 1994
Carrot	*Alternaria dauci*	Dugdale et al., 1993
Pearl millet	*Sclerospora graminicola* (Sacc. Schroet.	Nagarathna et al., 1993
Sugarcane	Sugarcane Mosaic Virus	Oropeza and García, 1996
Wheat	*Puccinia recondita* Roberge ex Desmaz. f. sp. *tritici* (Prt)	Oberthur et al., 1993
Apple	*Erwinia amylovora*	Donovan et al., 1994
Poplar	*Septoria musiva*	Ostry et al., 1994
Strawberry	*Fusarium oxysporum* f. sp. *fragariae*	Toyoda et al., 1991
Asparagus	*Fusarium oxysporum* f. sp. *asparagi*	Dan and Stephens, 1995
	Fusarium proliferatum	
Rice	*Xanthomonas oryzae*	Lihua et al., 1991
Celery	*Septoria apiicola* Speg.	Donovan et al., 1994

Table 2 Advantages and Disadvantages of Selection at the Cellular or Whole Plant Level

Selection at the Cellular Level		Selection at the Whole-Plant Level	
Advantages	**Disadvantages**	**Advantages**	**Disadvantages**
Able to screen a large population of uniform cells in limited space and time.	Resistance observed at the cellular level may not be expressed at the whole-plant level.	Resistance can be readily observed at the whole plant level.	Time and space limitations may not allow screening of large populations.
Can impose screening earlier in the system and limit the amount of material to be maintained.	Not able to observe undesirable traits that may have arisen due to somaclonal variation.	Can observe undesirable traits.	The need to carry along a large number of cultures that may not be resistant.
Easier control on the environmental conditions.	A selective agent is not always available for a disease	A selective agent is not necessary.	Not as easy to control environmental factors.

pathogen, with culture filtrates, or with the pathogen itself. Toxic metabolites produced by the pathogen and culture filtrates have been applied to individual cells or at the callus stage (Shahin and Spivey, 1986; Binarová et al., 1990; Sebastiani et al., 1994). One problem with using toxic metabolites is that toxins may not play a role in pathogenesis. Even the pathogen itself has been used for selection, however, there are numerous problems with this method (Daub, 1984). The pathogen may overgrow the cells, or on the other hand it may be difficult to grow the pathogen in culture.

Selection can also be done at the whole-plant level where no selection is applied *in vitro* and resistant plants have been recovered (Rodeva et al., 1994; Bobisud et al., 1996). There are studies of comparisons of selection *in vitro* vs. *in vivo*. For example, in an apple report regarding *Erwinia amylovora* resistance acquired by somaclonal variation, selection was done at both the *in vitro* level and *in vivo* on whole plants (Donovan et al., 1994). For *in vitro* selection, leaf disks were vacuum infiltrated with *E. amylovora* cells prior to regeneration. The assumption was that plants that regenerated would be resistant to *E. amylovora*. *In vitro* shoots were screened by inoculating the edges of cut leaves (Norelli et al., 1988). For whole plant selection, greenhouse-grown shoots were inoculated by inserting a hypodermic needle through the stem and filling the wound with a bacterial suspension (Norelli et al., 1984). Susceptible shoots developed typical wilting and blight of the shoot apex by 14 days. Higher frequencies of disease-resistant shoots were observed for greenhouse compared with *in vitro* inoculations. The leaf pretreatments had no significant effect on the frequency of somaclones recovered with increased resistance to the pathogen. Plants were recovered that had a greater resistance to fire blight than the parental material.

In contrast, in oil seed rape and tomato, there was no difference in the number of *Alternaria brassicicola*-resistant oil seed rape or *Fusarium*-resistant tomato plants recovered from *in vitro* and *in vivo* selection (MacDonald and Ingram, 1985, 1986; Shahin and Spivey, 1986). However, there are examples where selection *in vitro* does result in a higher frequency of more resistant somaclonal variants (Toyoda et al., 1989; Binarová et al., 1990).

D. Advantages and Disadvantages

Somaclonal variation can be a source of disease resistance in crops where genes for resistance are not available or are difficult to introgress in a breeding program. However, along with this improved resistance, other undesirable characteristics may be introduced. For example, in a program to develop bacterial wilt-resistant tomato somaclones, a number of undesirable traits were observed as compared to the source material (Bobisud et al., 1996). The source material significantly outyielded the resistant somaclones in total fruit weight/plant and total number of fruit/plant. Also, the fruit from the source plant were significantly greater in diameter. However, the major drawback was a reduction in yield in the somaclones. Although undesirable changes can occur, if these changes are not too extensive it may be possible to eliminate these traits by introducing the somaclones into a breeding program.

Another aspect that needs to be evaluated is the stability of the disease resistance. Because variation can result from nonheritable, epigenetic changes, an acquired resistance must be evaluated for stability. This may not be as great an issue in vegetatively propagated crops, but is an issue in seed-propagated crops. Stability of disease resistance acquired via somaclonal variation has been observed in downy mildew-resistant pearl millet (Nagarathna et al., 1993), *Fusarium* wilt-resistant strawberry (Toyoda et al., 1991), and leaf rust-resistant wheat (Oberthur et al., 1993).

So, although there are advantages and disadvantages to using somaclonal variation to acquire disease resistance, it can be a useful source of resistance. Researchers just need to be aware of the possibilities and structure their programs accordingly.

V. PROTOPLAST FUSION

A. Description of Method

During the 1960s, a method was developed for the isolation and culture of protoplasts, which are plant cells lacking a cell wall (Cocking, 1960). Protoplasts are isolated by enzymatic degradation of the cell wall with cellulases, pectinases, and hemicellulases. The integrity of the protoplasts is maintained by inclusion of an osmoticum in the isolation medium. If cultured in the appropriate medium, these protoplasts form new cell walls, then cell colonies, and possibly go on to regenerate plantlets (Takebe et al., 1971; Dodds and Roberts, 1985).

Aside from culturing the protoplasts for plant regeneration, the protoplasts can be fused, during which there is a mixing of cytoplasm. As a result of this mixing, the nuclei may remain separate or they may fuse, resulting in what are referred to as somatic hybrids. In some instances one nucleus completely disappears, but the cytoplasm remain mixed. This fusion product is referred to as a cybrid.

B. Fusion Methods

1. Polyethylene Glycol

The fusion of protoplasts can be facilitated in mainly two ways. One method involves the use of polyethylene glycol (PEG) (Puite et al., 1987). The PEG fusion method was introduced by Kao and Michayluk (1974). Since that first report, this fusion method has been used and modified for various crops (Thomas and Rose, 1988; Hansen and Earle, 1995).

2. *Electrofusion*

In a second fusion method, the fusion is induced by applying an electrical field which is sometimes called electrofusion. This method is not used as often as the PEG method, primarily due to the need for more elaborate equipment than is needed for the PEG method. However, electrofusion has been used successfully (Power et al., 1988; Hagimori et al., 1992).

C. Selection Strategies

There are various strategies to ensure selection of fusion products. Some of these include the lack of regeneration of one of the partners, treatment with metabolic inhibitors, X-rays, and gamma irradiation (Wright, 1978; Dudits et al., 1980; Hagimori et al., 1992; Gerdemann-Knorck et al., 1994). Treatment with X-rays or gamma irradiation of one of the donor parents at the appropriate level can result in transfer of only a few chromosomes, single chromosomes, or parts of chromosomes. This is referred to as asymmetric hybridization. By transferring only a part of the nuclear material, problems such as transfer of undesirable characteristics and fertility might be avoided.

Another selection strategy involves fusion of protoplasts from one partner that is antibiotic resistant as a result of transformation or by direct DNA uptake with protoplasts from a nonresistant partner (Bates et al., 1987; Gerdemann-Knorck et al., 1994). A dual selection system has been reported where both antibiotic and herbicide resistance were introduced into the protoplast donor parents (Masson et al., 1989). For reports on protoplast fusion and the various selection methods, see Gleba and Sytnik, 1984; Harms, 1985; and Rose et al., 1990.

D. Verification of Hybridity

Once plants regenerate from a fusion, there needs to be verification of the hybridity of these somatic hybrids. This would be especially true before introduction into a disease-screening or breeding program to allow for a more efficient use of the material. Depending on the nature of the fusion parents, verification might be done initially just by visualization (Narasimhulu et al., 1994). However, for most somatic hybrids, additional methods need to be utilized for confirmation such as isozyme markers (Waara et al., 1991; Hagimori et al., 1992; Gerdemann-Knorck et al., 1994) and testing at the chromosomal level or DNA level (Narasimhulu et al., 1994). At the DNA level there are a number of options including DNA content (Hansen and Earle, 1995), randomly amplified polymorphic DNA (RAPD) markers (Baird et al., 1992; Maisonneuve et al., 1995), restriction fragment length polymorphisms (RFLPs) (Hagimori et al., 1992), species-specific probes (Pehu et al., 1990), and ribosomal DNA probes (Pental et al., 1988; Rosen et al., 1988).

E. Use of Protoplast Fusions in Disease Resistance Programs

Sources for disease resistance in a breeding program may come from closely related species or from more distantly related species. The more distantly related a species, the more problems are encountered in crossing. Protoplast fusion is one of the methods that can be used to circumvent problems in introgressing disease resistance genes. By this method, factors that contribute to crossing barriers between species are avoided and viable hybrids have been recovered, even between distantly related species (Harms, 1985). Protoplast fusion can be used as long as plants can be regenerated from the crop of interest.

In lettuce, it is difficult to introgress disease resistance genes from wild species of lettuce due to sexual incompatibility between *Lactuca sativa* and wild *Lactuca* sp. (Maisonneuve et al., 1995). In a report by Maisonneuve et al. (1995) they used protoplast fusion as part of a program to overcome the crossing barriers. They recovered somatic hybrids between *L. sativa* and two wild species *(L. perennis* and *L. tatarica*). The hybrid nature was confirmed by RAPD analysis. Preliminary disease screening indicated the transfer of downy mildew resistance from *L. tatarica*. These plants were introduced into a backcross program.

Table 3 Disease-Resistant Plants from Protoplast Fusion

Species Used for Fusion	Disease	Ref.
Lactuca sativa wild *Lactuca* sp.	Downy mildew	Maisonneuve et al., 1995
Brassica oleracea *Raphanus sativus*	Club root	Hagimori et al., 1992
Brassica napus *Brassica nigra*	Blackleg Club root	Gerdemann-Knorck et al., 1994
Brassica oleracea *Brassica napus*	Black rot	Hansen and Earle, 1995
Solanum brevidens *Solanum tuberosum*	Bacterial soft rot	Austin et al., 1988

Wild relatives are not the only sources of disease resistance genes. For example, in *Brassica oleracea*, black rot, caused by *Xanthomonas campestris* pv. *campestris* (Pammel) Dowson is the most important disease of crucifers (Williams, 1980). Control measures include seed treatment, crop rotation, and general crop hygiene. However, Guo et al. (1994) identified very high resistance to *X. campestris* pv. *campestris* in certain *B. napus* lines, and in some lines there is very high resistance conferred by a dominant gene. In a few cases, sexual hybridization has been successful between *B. napus* and *B. oleracea*, however, it is difficult to recover progeny (Ayotte et al., 1987; Quazi, 1988). Therefore, Hansen and Earle (1995) used protoplast fusion to transfer black rot resistance from *B. napus* to rapid cycling *B. oleracea*. The hybrid nature of the somatic hybrids was confirmed by morphological studies, analysis of DNA content, and RAPD analysis. They recovered somatic hybrids that were highly resistant to *X. campestris* pv. *campestris* were fertile and set seed, and were successfully backcrossed to *B. oleracea* through embryo rescue.

Table 3 contains a list of some of the crops where protoplast fusion was used to transfer disease resistance. Any crop where crossing barriers exist is a candidate for protoplast fusion, provided that plants can be regenerated from the protoplasts.

F. Advantages and Disadvantages

The main advantage of protoplast fusion is that it provides a possible method of hybridization between two parents where crossing barriers may exist. These hybridizations may be interspecific, intraspecific, or intergeneric. However, somatic hybrids that result need to be evaluated for morphological characteristics and fertility. There may be undesirable morphological characteristics carried over from one of the parents or undesirable changes may result from somaclonal variation. As for fertility, there are reports of lower fertility in somatic hybrids, especially in intergeneric fusions (Rose et al., 1990). However, these problems may be resolved by taking an asymmetric hybridization approach where a limited amount of genetic material is transferred. Even though undesirable traits may appear, there is encouraging evidence that following backcrossing and restoration of the normal recipient phenotype, a trait like disease resistance can be maintained (Itoh et al., 1991; Larkin et al., 1993).

The main limitation for using protoplast fusion is the ability to regenerate plants from at least one of the partners. Although time and effort may need to be invested, protoplast fusion may be the only alternative for the transfer of disease resistance from one species to another.

VI. SUMMARY

All of the tissue culture methods discussed can be used to alleviate the use of chemicals to control diseases. Tissue culture may help to reduce the amount of pesticides needed to control diseases by generating disease-free plants through meristem culture, creating new sources of disease resistance as a result of somaclonal variation, or transfering disease resistance from other species by embryo rescue or protoplast fusion. For each of these methods there are pros and cons, but tissue culture should not be overlooked when an improvement in disease resistance is required, especially if there is no other viable means of disease control.

Tissue culture alone may be an asset in a crop improvement program, but even more important is tissue culture in combination with genetic engineering. If genes for disease resistance have been cloned, then tissue culture in combination with genetic engineering can be used to introduce the genes of interest. Genes of interest may be introduced by *Agrobacterium tumefaciens*, particle bombardment, tissue electroporation, microinjection, and direct gene transfer into protoplasts (Klee et al., 1987; Sanford et al., 1987; Paszkowski et al., 1989; D'Halluin et al., 1992). All of these gene transfer approaches are used in combination with some tissue culture method.

There are already examples of programs where disease resistance genes have been introduced into crops and improved resistance resulted. Genes for bacterial resistance have been introduced into flower crops and genes for virus resistance have been introduced into many crops (see reviews by Grumet, 1995 and Kuehnle et al., 1995). Genetic engineering is discussed in more detail in Chapter 14.

REFERENCES

Agnihotri, A., Gupta, V., Lakshmikumaran, M. S., Shivanna, K. R., Prakash, S., and Jagannathan, V., Production of *Eruca-Brassica* hybrids by embryo rescue, *Plant Breeding*, 104, 281, 1990.

Ahmed, K. Z. and Sagi, F., Use of somaclonal variation and *in vitro* selection for induction of plant disease-resistance: prospects and limitations, *Acta Phytopathol. Entomol. Hung.*, 28, 143, 1993.

Amante-Bordeos, A., Sitch, L. A., Nelson, R., Dalmacio, R. D., Oliva, N. P., Aswidinnoor, H., and Leung, H., Transfer of bacterial blight and blast resistance from the tetraploid wild rice *Oryza minuta* to cultivated rice, *Oryza sativa*, *Theor. Appl. Genet.*, 84, 345, 1992.

Austin, S., Lojkowska, E., Ehlenfeldt, M. K., Kelman, A., and Helgeson, J. P., Fertile interspecific somatic hybrids of *Solanum*: a novel source of resistance to *Erwinia* soft rot, *Phytopathology*, 78, 1216, 1988.

Ayotte, R., Harney, P. M., and Machado, V. S., The transfer of triazine resistance from *Brassica napus* L. to *B. oleracea* L. I. Production of F1 hybrids through embryo rescue, *Euphytica*, 36, 615, 1987.

Baird, E., Cooper-Bland, S., Waugh, R., DeMaine, M., and Powell, W., Molecular characterisation of inter- and intra-specific somatic hybrids of potato using randomly amplified polymorphic DNA (RAPD) markers, *Mol. Gen. Genet.*, 233, 469, 1992.

Bates, G. W., Hasenkampf, C. A., Contolini, C. L., and Piastuch, W. C., Asymmetric hybridisation in *Nicotiana* by fusion of irradiated protoplasts, *Theor. Appl. Genet.*, 74, 718, 1987.

Bhagyalakshmi and Singh, N. S., Meristem culture and micropropagation of a variety of ginger (*Zingiber officinale* Rosc.) with a high yield of oleoresin., *J. Hortic. Sci.*, 63(2), 321, 1988.

Binarová, P., Nedelnik, J., Fellner, M., and Nedbálková, B., Selection for resistance to filtrates of *Fusarium* spp. in embryogenic cell suspension culture of *Medicago sativa* L., *Plant Cell Tissue Org. Cult.*, 22, 191, 1990.

Bobisud, C. A., Martin, S. P., and Sekioka, T. T., Field testing bacterial wilt-resistant tomato somaclones, *J. Am. Soc. Hortic. Sci.*, 121(3), 384, 1996.

Brown, C. R., Kwiatkowski, S., Martin, M. W., and Thomas, P. E., Eradication of PVS from potato clones through excision of meristems from *in vitro*, heat-treated shoot tips, *Am. Potato J.*, 65, 633, 1988.

Brown, P. T. H., The spectrum of molecular changes associated with somaclonal variation, *Newslett, IAPTC*, 66, 14, 1991.

Cai, W. W., Tronchet, M., Larroque, N., Dorion, N., and Albouy, J., Production of virus-free dahlia by meristem-culture and virus detection through cDNA probes and ELISA, *Acta Hortic.*, 234, 421, 1988.

Chavez, R., Brown, C. R., and Iwanaga, M., Application of interspecific sesquiploidy to introgression of PLRV resistance from non-tuber-bearing *Solanum etuberosum* to cultivated potato germplasm, *Theor. Appl. Genet.*, 76, 497, 1988.

Chavez, R., Brown, C. R., and Iwanaga, M., Transfer of resistance to PLRV titer buildup from *Solanum etuberosum* to a tuber-bearing *Solanum gene* pool, *Theor. Appl. Genet.*, 76, 129, 1988.

Cocking, E. C., A method for the isolation of plant protoplasts and vacuoles, *Nature*, 187, 962, 1960.

D'Halluin, K., Bonne, E. H., and Bossut, M., Transgenic maize plants by tissue electroporation, *Plant Cell*, 4, 1495, 1992.

Dan, Y. and Stephens, C. T., The development of asparagus somaclones with high levels of resistance to *Fusarium oxysporum* f. sp. *asparagi* and *F. proliferatum*, *Plant Dis.*, 79, 923, 1995.

Daub, M. E., Tissue culture and the selection of resistance to pathogens, *Annu. Rev. Phytopathol.*, 24, 159, 1984.

Dodds, J. H. and Roberts, L. W., *Experiments in Plant Tissue Culture*, Cambridge University Press, Cambridge, 1985.

Donovan, A., Collin, H. A., Isaac, S., and Mortimer, A. M., Analysis of potential sources of variation in tissue culture derived celery plants, *Ann. Appl. Biol.*, 124, 383, 1994.

Donovan, A. M., Morgan, R., Valobra-Piagnani, C., Ridout, M. S., James, D. J., and Garrett, C. M. E., Assessment of somaclonal variation in apple. I. Resistance to the fire blight pathogen, *Erwinia amylovora*, *J. Hortic. Sci.*, 69(1), 105, 1994.

Dörffling, K., Dörffling, H., and Lessilich, G., *In vitro*-selection and regeneration of hydroxyproline resistant lines of winter wheat with increased proline content and increased frost tolerance, *J. Plant Physiol.*, 142, 222, 1993.

Dudits, D., Fejer, O., Hadlaczky, G., Koncz, C., Lazar, G. B., and Horvath, G., Intergeneric gene transfer mediated by plant protoplast fusion, *Mol. Gen. Genet.*, 179, 283, 1980.

Dugdale, L. J., Collin, H. A., Isaac, S., and Gill, J. J. B., Leaf blight resistance in carrot somaclones, *Acta Hortic.*, 336, 399, 1993.

Gamborg, O. L., Murashige, T., Thorpe, T. A., and Vasil, I. K., Plant tissue culture media, *In Vitro*, 12, 473, 1976.

Gerdemann-Knörck, M., Sacristan, M. D., Braatz, C., and Schieder, O., Utilization of asymmetric somatic hybridization for the transfer of disease resistance from *Brassica nigra* to *Brassica napus*, *Plant Breeding*, 113, 106, 1994.

Gleba, Y. Y. and Sytnik, K. M., *Protoplast Fusion, Genetic Engineering in Higher Plants*, Springer-Verlag, Berlin, 1984.

Green, S. K., Luo, C. Y., and Wu, S. F., Elimination of leafcurl virus of sweet potato by meristem tip culture, heat and ribavirin, *Plant Prot. Bull.*, 34, 1, 1992.

Grimm, F., Baumann, B., and Munzert, M., The influence of explant origin and heat treatment on plant regeneration and the elimination of *Erwinia carotovora* var. *atroseptica* (van Hall) Dye in potato meristem cultures, *Ber. Landwirtssch. Jahrbuch*, 68(2), 243, 1991.

Grumet, R., Genetic engineering for crop virus resistance, *HortScience*, 30, 449, 1995.

Guo, H., Dickson, M. H., and Hunter, J. E., *Brassica napus* sources of resistance to black rot in crucifers and inheritance of resistance, *HortScience*, 26, 1545, 1994.

Hagimori, M., Nagaoka, M., Kato, N., and Yoshikawa, H., Production and characterization of somatic hybrids between the Japanese radish and cauliflower, *Theor. Appl. Genet.*, 84, 819, 1992.

Hakkaart, F. A. and Versluijs, J. M. A., Virus elimination by meristem-tip culture from a range of *Alstroemeria* cultivars, *Neth. J. Plant Pathol.*, 94, 49, 1988.

Hammerschlag, F., Ritchie, D., Werner, D., Hashmil, G., Krusberg, L., Meyer, R., and Huettel, R., *In vitro* selection of disease resistance in fruit trees, *Acta Hortic.*, 392, 19, 1995.

Hammerschlag, F. A., Resistance responses of plants regenerated from peach callus cultures to *Xanthomonas campestris* pv. *pruni*, *J. Am. Soc. Hortic. Sci.*, 115, 1034, 1990.

Hansen, L. N. and Earle, E. D., Transfer of resistance to *Xanthomonas campestris* pv *campestris* into *Brassica oleracea* L. by protoplast fusion, *Theor. Appl. Genet.*, 91, 1293, 1995.

Harms, C. T., Hybridization by somatic cell fusion, in *Plant Protoplasts*, Fowke, L. C. and Constabel, F., Eds., CRC Press, Boca Raton, FL, 1985, 169.

Heinz, D. J., Sugar-cane improvement through induced mutations using vegetative propagules and cell culture techniques, in Proc. IAEA Symp. Induced Mutations in Vegetatively Propagated Plants, Vienna, 1973, 53.

Horst, R. K. and Klopmeyer, M. J., Controlling Viral Diseases, in *Geraniums IV The Grower's Manual*, White, J. W., Eds., Ball Publishing, Geneva, IL, 1993, 286.

Itoh, K., Iwabuchi, M., and Shimamoto, K., In situ hybridization with species-specific DNA probes gives evidence for asymmetric nature of *Brassica* hybrids obtained by X-ray fusion, *Theor. Appl. Genet.*, 81, 356, 1991.

Kao, K. N. and Michayluk, M. R., A method for high-frequency intergeneric fusion of plant protoplasts, *Planta*, 115, 355, 1974.
Karp, A. and Bright, S. W. J., On the causes and origins of somaclonal variation, *Oxford Surv. Plant Mol. Cell. Biol.*, 2, 199, 1985.
Klee, H., Horsch, R., and Rogers, S., Agrobacterium-mediated plant transformation and its further applications to plant biology, *Annu. Rev. Plant Physiol.*, 38, 467, 1987.
Klopmeyer, M. J., Certified Stock, in *New Guinea Impatiens — A Ball Guide*, Banner, W. and Klopmeyer, M. J., Ed., Ball Publishing, Geneva, IL, 1995, 55.
Krishnamurthi, M., Notes on disease resistance of tissue culture sub-clones and fusion of sugar cane protoplasts, *Sugarcane Breeders Newslett.*, 35, 24, 1974.
Krishnamurthi, M. and Tlaskal, J., Fiji disease resistant *Saccharum officinarum* var. Pindar sub-clones from tissue cultures, *Proc. XV Congr. Int. Soc. Sugar Cane Technol.*, 1, 130, 1974.
Kudell, A. R. and Buchenauer, H., Elimination of Raspberry Bushy Dwarf Virus from axillary bud cultures of red raspberry cv. Lloyd George by antiviral compounds, *J. Phytopathol.*, 124, 332, 1989.
Kuehnle, A. R., Chen, F., and Sugii, N., Novel approaches for genetic resistance to bacterial pathogens in flower crops, *HortScience*, 30, 456, 1995.
Larkin, P. J., Li, Y., Spindler, L. H., Tanner, G. J., and Banks, P. M., Disease resistance, cell culture and somatic recombination, *Acta Hortic.*, 336, 341, 1993.
Larkin, P. J. and Scowcroft, W. R., Somaclonal variation — a novel source of variability from cell cultures for plant improvement, *Theor. Appl. Genet.*, 60, 197, 1981.
Lihua, S., Jianming, S., Xuefeng, L., Zouwei, Z., Bangen, T., and Daguang, X., Rice somaclones resistant to *Xanthomonas oryzae*, in *Biotechnology in Agriculture and Forestry*, Bajaj, Y. P. S., Ed., Springer-Verlag, Berlin, 1991,
Lim, S. T., Wong, S. M., and Goh, C. J., Elimination of cymbidium mosaic virus and odontoglossum ringspot virus from orchids by meristem culture and thin section culture with chemotherapy, *Ann. Appl. Biol.*, 122, 289, 1993.
Lineberger, D. R., Sticklen, M. B., Pijut, P. M., Kroggel, M. A., Fink, C. V. M., and Domir, S. C., Use of protoplast, cell, and shoot tip culture in an elm germplasm improvement program, *Acta Hortic.*, 280, 247, 1990.
Lorz, H. and Scowcroft, W. R., Variability among plants and their progeny regenerated from protoplasts of SU/su hetrozygotes of *Nicotiana tabacum*, *Theor. Appl. Genet.*, 66, 67, 1983.
MacDonald, M. V. and Ingram, D. S., *In vitro* selection for resistance to *Alternaria brassicicola* in *Brassica napus* ssp. *oleifera* (winter oilseed rape) using partially purified culture filtrates, *Cruc. Newslett.*, 10, 97, 1985.
MacDonald, M. V. and Ingram, D. S., Towards the selection *in vitro* for resistance to *Alternaria brassicicola* (Schw.) wilts, in *Brassica napus* ssp. *oleifera* (Metzg.) Sinsk., winter oilseed rape, *New Phytol.*, 104, 621, 1986.
Maisonneuve, B., Chupeau, M. C., Bellec, Y., and Chupeau, Y., Sexual and somatic hybridization in the genus *Lactuca*, *Euphytica*, 85, 281, 1995.
Masson, J., Lancelin, C., Bellini, C., Lecerf, M., Guerche, P., and Pelletier, G., Selection of somatic hybrids between diploid clones of potato (*Solanum tuberosum* L.) transformed by direct gene transfer, *Theor. Appl. Genet.*, 78, 153, 1989.
Meins, F., Heritable variation in plant cell culture, *Annu. Rev. Plant Physiol.*, 34, 327, 1983.
Mellor, F. C. and Stace-Smith, R., Virus-free potatoes through meristem culture, in *Biotechnology in Agriculture and Forestry*, Vol. 3, Potato, Bajaj, Y. P. S., Ed., Springer-Verlag, Berlin, 1987, 30.
Nabors, M. W. and Dykes, T. A., Tissue culture of cereal cultivars with increased salt, drought, and acid tolerance, in *Biotechnology in International Agricultural Research*, International Rice Research Institute, Manila, 1985, 121.
Nagarathna, K. C., Shetty, S. A., Harinarayana, G., and Shetty, H. S., Selection for downy mildew resistance from the regenerants of pearl millet, *Plant Sci.*, 90, 53, 1993.
Narasimhulu, S. B., Kirti, P. B., Bhatt, S. R., Prakash, S., and Chopra, V. L., Intergeneric protoplast fusion between *Brassica carinata* and *Camelina sativa*, *Plant Cell Rep.*, 13, 657, 1994.
Norelli, J. L., Aldwinckle, H. S., and Beer, S. V., Differential host x pathogen interactions among cultivars of apple and strains of *Erwinia amylovora*, *Phytopathology*, 74, 136, 1984.
Norelli, J. L., Aldwinckle, H. S., and Beer, S. V., Virulence of *Erwinia amylovora* strains to *Malus* sp. Novole plants grown *in vitro* and in the glasshouse, *Phytopathology*, 78, 1292, 1988.

Oberthur, L. E., Harrison, S. A., Croughan, T. P., and Long, D. L., Inheritance of improved leaf rust resistance in somaclones of wheat, *Crop Sci.*, 33, 444, 1993.

Oropeza, M. and García, E. D., Somaclonal variants resistant to Sugarcane Mosaic Virus and their agronomic characterization, *In Vitro*, 32, 26, 1996.

Orton, T. J., Genetic variation in somatic tissues: method or madness?, *Adv. Plant Pathol.*, 2, 153, 1984.

Ostry, M., Hackett, W., Michler, C., Serres, R., and McCown, B., Influence of regeneration method and tissue source on the frequency of somatic variation in *Populus* to infection by *Septoria musiva*, *Plant Sci.*, 97, 209, 1994.

Paszkowski, J., Saul, M. W., and Potrykus, I., Plant gene vectors and genetic transformaiton: DNA-mediated direct gene transfer to plants, in *Cell Culture and Somatic Cell Genetics of Plants*, Schell, J. and Vasil, I. K., Eds., Academic Press, New York, 1989, 51.

Pehu, E., Thomas, M., Poutala, T., Karp, A., and Jones, M. G. K., Species-specific sequences in the genus *Solanum*: identification, characterization and application to study somatic hybrids of *S. brevidens* and *S. tuberosum*, *Theor. Appl. Genet.*, 80, 693, 1990.

Pental, D., Mukhopadhyay, A., Grover, A., and Pradham, A. K., A selection method for the synthesis of triploid hybrids by fusion of microspore protoplasts (n) with somatic cell protoplasts (2n), *Theor. Appl. Genet.*, 46, 237, 1988.

Phillips, R. L., Kaeppler, S. M., and Peschke, V. M., Do we understand somaclonal variation?, in *Progress in Plant Cellular and Molecular Biology*, Nijkamp, H. J. J., van der Plas, L. H. W., and van Aartrijk, J., Eds., Kluwer Acacemic, Dordrecht, 1990, 131.

Power, J. B., Davey, M. R., Freeman, J. P., Mulligan, B. J., and Cocking, E. C., Fusion and transformation of plant protoplasts, in *Methods for Plant Molecuolar Biology*, Weissbach, A. and Weissbach, H., Eds., Academic Press, London, 1988,

Puite, K. J., Dons, J. J. M., Huizing, H. J., Kool, A. J., Koornneef, M., and Krens, F. A., Progress in plant protoplast research, in Proc. 7th Int. Protoplast Symp., Wageningen, The Netherlands, 1987.

Quazi, M. H., Interspecific hybrids between *Brassica napus* L. and *Brassica oleracea* L. developed by embryo culture, *Theor. Appl. Genet.*, 75, 309, 1988.

Ramming, D. W., The use of embryo culture in fruit breeding, *HortScience*, 25(4), 393, 1990.

Ramming, D. W. and Emershad, R. L., Embryo culture of early-ripening seeded grape genotypes, *HortScience*, 19(3), 594, 1984.

Ramming, D. W. and Tanner, O., "Gold-Crest' peach, *Fruit Var. J.*, 41, 52, 1987a.

Ramming, D. W. and Tanner, O., 'Mayfire' nectarine, *Fruit Var. J.*, 41, 80, 1987b.

Rodeva, V. N., Stamova, L. I., and Stancheva, J. M., Comparative study of tomato somaclones for resistance to *Fusarium oxysporum* f.sp. *radicis-lycopersici* and *Alternaria solani* (Ellis & Martin) Jones and Grout, *Acta Hortic.*, 376, 411, 1994.

Rose, R. J., Thomas, M. R., and Fitter, J. T., The transfer of cytoplasmic and nuclear genomes by somatic hybridisation, *Aust. J. Plant Physiol.*, 17, 303, 1990.

Rosen, B., Hallden, C., and Heneen, W. K., Diploid *Brassica napus* somatic hybrids: characterisation of nuclear and organellar DNA, *Theor. Appl. Genet.*, 76, 197, 1988.

Sanford, J. C., Klein, T. M., and Wolf, E. D., Delivery of substances into cells and tissues using a particlc bombardment process, *J. Particle Sci. Technol.*, 5, 27, 1987.

Scowcroft, W. R. and Larkin, P. J., Somaclonal variation, cell selection and genotype improvement, in *Comprehensive Biotechnology: the Principles, Applications, and Regulations of Biotechnology in Industry, Agriculture, and Medicine*, Murray, M., Ed., Pergmon Press, Oxford, 1985, 153.

Sebastiani, L., Lenzi, A., Pugliesi, C., and Fambrini, M., Somaclonal variation for resistance to *Verticillium dahliae* in potato (*Solanum tuberosum* L.) plants regenerated from callus, *Euphytica*, 80, 5, 1994.

Segeren, M. I., Sondahl, M. R., Siqueira, W. J., Medina Filho, H. P., Nagai, H., and Lourençao, A. L., Tomato breeding. I. Embryo rescue of interspecific hybrids between *Lycopersicon esculentum* MILL. and *L. peruvianum* (L.) MILL., *Rev. Brasil Genet.*, 16(2), 367, 1993.

Shahin, E. A. and Spivey, R., A single dominant gene for *Fusarium* wilt resistance in protoplast-derived tomato plants, *Theor. Appl. Genet.*, 73, 164, 1986.

Sharma, D. R., Kaur, R., and Kumar, K., Embryo rescue in plants — a review, *Euphytica*, 89, 325, 1996.

Shoemaker, N. P., Swartz, H. J., and Galletta, G. J., Cultivar dependent variation in pathogen resistance due to tissue culture propagation of strawberries, *HortScience*, 20, 253, 1985.

Sobczykiewicz, D., *In vitro* culture of apical meristems of raspberry plants as an elimination method of a heat-stable raspberry vein chlorosis virus, *Fruit Sci. Rep.*, 16(4), 233, 1989.
Stace-Smith, R. and Mellor, F. C., Eradication of potato viruses X and S by thermotherapy and axillary bud culture, *Am. Potato J.*, 58, 199, 1968.
Stein, A., Spiegel, S., Faingersh, G., and Levy, S., Responses of micropropagated peach cultivars to thermotherapy for the elimination of prunus necrotic ringspot virus, *Ann. Appl. Biol.*, 119, 265, 1991.
Takebe, I., Labib, G., and Melchers, G., Regeneration of whole plants from isolated mesophyll protoplasts of tobacco, *Naturwissenschaften*, 58, 318, 1971.
Taylor, S. G., Shilling, D. G., and Quesenberry, K. H., *In vitro* selection for 2,4-D tolerance in red clover, *Theor. Appl. Genet.*, 78, 265, 1989.
Thomas, M. R. and Rose, R. J., Enrichment for *Nicotiana* heterokaryons after protoplast fusion and subsequent growth in agarose microdrops, *Planta*, 175, 396, 1988.
Thorn-Horst, H., Horst, R. K., Smith, S. H., and Oglevee, W. A., A virus-indexing tissue culture system for geraniums, *Florists' Rev.*, 160, 28, 1977.
Torres, K. C., *Tissue Culture Techniques for Horticultural Crops*, Van Nostrand Reinhold, New York, 1989.
Toyoda, H., Horikoshi, K., Yamano, Y., and Ouchi, S., Selection for Fusarium wilt disease resistance from regenerants derived from leaf callus of strawberry, *Plant Cell Rep.*, 10, 167, 1991.
Toyoda, H., Shimizu, K., Chatani, K., Kita, N., Matsuda, Y., and Ouchi, S., Selection of bacterial wilt-resistant tomato through tissue culture, *Plant Cell Rep.*, 8, 317, 1989.
van den Bulk, R. W., Application of cell and tissue culture and *in vitro* selection for disease resistance breeding — a review, *Euphytica*, 56, 269, 1991.
Waara, S., Wallin, A., and Eriksson, T., Production and analysis of intraspecific somatic hybrids of potato (*Solanum tuberosum* L.), *Plant Sci.*, 75, 107, 1991.
Williams, P. H., Black rot: a continuing threat to world crucifers, *Plant Dis.*, 64, 736, 1980.
Winicov, I., Characterization of salt tolerant alfalfa (*Medicago sativa* L.) plants regenerated from salt tolerant cell lines, *Plant Cell Rep.*, 10, 5461, 1991.
Wright, W. E., The isolation of heterokaryons and hybrids by a selective system using irreversible biochemical inhibitors, *Exp. Cell. Res.*, 112, 398, 1978.
Xu, J. and Kasha, K. J., Transfer of a dominant gene for powdery mildew resistance and DNA from *Hordeum bulbosum* into cultivated barley, *Theor. Appl. Genet.*, 84, 771, 1992.
Xu, J. and Snape, J. W., The resistance of *Hordeum bulbosum* and its hybrids with *H. vulgare* to common fungal pathogens, *Euphytica*, 41, 273, 1989.

CHAPTER 14

Genetic Engineering

Marc Fuchs and Dennis Gonsalves

1-8493-2627-3/97/$0.00+$.50

I. INTRODUCTION

Phytopathogenic bacteria, fungi, nematodes, and viruses can cause severe damage to crops by substantially reducing yield and the quality of marketable products, i.e., fruits, seeds, tubers, and flowers. Crop losses due to pathogens are very costly to agriculture worldwide. Losses reaching billions of dollars are reported yearly because of phytopathogenic agents. In developing countries, crop losses incited by pathogens are even more dramatic because they can be disastrous for subsistence farmers who rely greatly on their produce for survival.

Conventional strategies to control phytopathogenic bacteria, fungi, nematodes, and viruses are based predominantly on: (1) cultural practices, (2) the use of chemical compounds against bacteria, fungi, and nematodes, or against insect vectors in the case of viruses, and (3) the use of resistant crop varieties. Unfortunately, these control measures are not always available or effective. It is interesting to note that the majority of agronomically important crop plants lack effective resistance genes to control major diseases (Shah et al., 1995). Thus, there is a need for new sources of resistance genes and new environmentally sound approaches to develop disease-resistant crops. Recently, genetic engineering has opened new avenues for the development of disease-resistant crop plants.

A. Definition of Genetic Engineering

Genetic engineering allows the genetic modification of a plant by adding new traits through recombinant DNA technology. Genetic engineering makes possible the incorporation of a single gene or a combination of genes into the plant genome in order to confer traits such as disease resistance, herbicide tolerance, tolerance to salt, drought and cold, increased productivity, and increased nutritional value. This technology combines molecular biology, microbiology, tissue culture, cell transformation, plant selection and breeding techniques. Genetic engineering allows the development of genetically altered plants with defined traits and valuable properties without altering other desirable characteristics of the recipients. For example, genetic engineering is of special interest for the stable insertion of disease resistance genes into the genome of susceptible commercial cultivars as a method to protect such plants from phytopathogenic pests.

B. Rationale for the Use of Genetic Engineering to Control Plant Diseases

Cultural practices, i.e., crop rotation and delayed planting, can reduce the incidence of diseases caused by fungi, bacteria, nematodes, and viruses. However, changes in cultural practice are often not desired for economic reasons. Although they may allow growers to limit the incidence of pathogens for a short time, cultural practices may not be efficient in the long run, and not applicable in the case of monocultures. For example, in vineyards infected by grapevine fanleaf mepo virus (GFLV) that carry the nematode vector *Xiphinema index*, no control measures are effective to eradicate the degeneration disease of grapevines (Raski et al., 1993). In this case, complete elimination of GFLV-infected vines and a crop rotation with cereals or potatoes, or a 5- to 10-year fallow before replanting healthy material is recommended. However, this practice is economically undesirable or even impossible in highly priced vineyards.

Chemicals are widely used to control bacteria, fungi, and nematodes. In spite of their effectiveness at reducing the incidence of those phytopathogenic agents, public awareness and concern have been raised in recent years on their excessive use in agriculture and their potentially hazardous effects on the environment and human health. Concerns have been expressed in regard to the development of new pathogen populations that are resistant to the pesticides, the presence of chemical residues in food and water supplies, and needless to mention, the protection of the pesticide applicators. In the case of nematodes, control measures that are currently used are inadequate and rely mostly on environmentally hazardous nematicides. In some states in the U.S. and some countries in the European Union, the use of several nematicides has recently been limited or even prohibited.

Ideally, the use of resistant crop varieties is regarded as the most efficient strategy to control diseases caused by phytopathogenic bacteria, fungi, nematodes, and viruses. To develop disease-resistant crop plants, breeders use resistance genes from domesticated or related free-living species that are incorporated into the variety by sexual crosses. Host resistance genes have been used to control plant diseases for a long time. Despite a number of successes in enhancing disease resistance, conventional breeding has some limitations and pitfalls. For example, transfer of undesirable traits that are linked to the resistance genes of interest may create problems for conventional breeding. Secondary plant products, which are known to be involved in the plant defense mechanism system and are often present in abundance in free-living plants, are often undesirable in crop varieties because of their interference with digestion and food processing (Harms, 1992).

Genetic engineering offers great potential to complement conventional approaches for the control of plant diseases, and overcome some of their limitations.

C. Background on Gene Transfer Techniques for Plants

The first system to deliver foreign genes into the plant genome was developed in 1983 by eliminating the crown gall disease-causing genes from the T-DNA region of the Ti plasmid of *Agrobacterium tumefaciens*, a bacterium that can transfer a part of its Ti plasmid into the plant chromosomes, and by maintaining the DNA transfer functions intact (Zambryski et al., 1983). This system enabled researchers to substitute foreign genes for the tumor-causing genes, transfer these into the plant genome, and develop fertile plants containing foreign genes (Bevan et al., 1983; Fraley et al., 1983; Herrera-Estrella et al., 1983). Thus, plants that were resistant to the antibiotic kanamycin were produced by introducing into the plant genome the selectable bacterial gene coding neomycin phosphotransferase. This important breakthrough, which promoted the development of applied plant biotechnology, has become the most widely used gene transfer method in plants. Subsequently, the particle gun was developed to transfer DNA into cells by bombarding DNA-coated particles (Klein et al., 1987). The biolistic method has overcome the limitations encountered with some monocotyledonous crops that are not natural hosts for *A. tumefaciens*. Besides the *A. tumefaciens*- and biolistic-mediated transformation techniques, a few other gene transfer methods have been developed for plants (Walden and Wingender, 1995).

D. Objectives and Scope of this Review

This review will focus on the use of genetic engineering to develop agronomically important crops that are resistant to fungi, bacteria, nematodes, and viruses. In the first part of the review, we describe several examples of engineered resistance against these phytopathogenic agents with a special emphasis on resistance evaluation carried out under field conditions. In the second part of the review, we summarize environmental risk assessment data currently available for transgenic plants engineered for virus resistance. In the conclusion, we discuss the critical balance between risks and benefits. Considering that strategies used to engineer resistance to viruses were developed earlier than strategies used to engineer resistance to fungi, bacteria, and nematodes, this review focuses mainly on genetic engineering as a means to control diseases incited by viruses.

II. DEVELOPMENT OF GENETICALLY MODIFIED PLANTS TO CONTROL FUNGI, BACTERIA, AND NEMATODES

A. Rationale and Theory

The development of genetically modified plants that are resistant to diseases caused by fungi, bacteria, and nematodes has lagged behind that of virus-resistant crop plants. However, several examples of transgenic plants with high levels of resistance to fungi, bacteria, and nematodes have been recently reported. Cornelissen and Melchers (1993) discussed the different strategies used to engineer resistance to fungal diseases. Reinmann-Philipp and Beachy (1993) and Atkinson (1995) assessed the possibilities to engineer resistance to nematodes, and Herrera-Estrella and Simpson (1995) reviewed the different approaches used to engineer resistance to bacteria and fungi. In summary, to develop crop varieties that are resistant to bacteria, fungi, and nematodes, genetically modified plants have been engineered to express proteins that are either toxic to the pathogen, degrade cell walls, or are involved in the plant defense response to pathogen infection. Pathogen-derived resistance (Sanford and Johnston, 1985) has also been used as an approach to develop transgenic plants resistant to bacteria and fungi. Shah et al. (1995) reviewed several examples of genetically modified crops engineered for disease and insect resistance.

B. Current Status of Development

In spite of the limited number of reports on engineered resistance to phytopathogenic bacteria, fungi, and nematodes, some transgenic plants with enhanced tolerance have been developed for a few agronomically significant crops. Thereafter, we will focus on reports for these few crops and their field trials, even though these latter have not yet been extensive.

1. Control of Fungi

Oilseed rape, carrot, potato, rice, cucumber, and tomato plants are the first agronomically important crop plants engineered thus far for fungal resistance, but most of the early reports are on the model crop, tobacco. Several strategies have been used to develop transgenic plants exhibiting enhanced tolerance to fungal pathogens. These strategies are based on the use of foreign gene constructs encoding either single and multiple pathogenesis-related (PR) proteins that are involved in the plant defense response to pathogen infection, virus-derived antifungal toxins, plant-derived proteins, pathogen-encoded enzymes, bacterial ribonuclease, or ribosome-inactivating proteins.

Genes encoding PR proteins have been widely used in agronomically important crop plants. The first report on genetically modified plants engineered for resistance to a fungal pathogen was by Broglie et al. (1991), who evaluated oilseed rape plants expressing an endochitinase gene from bean for resistance to the soil-borne fungus *Rhizoctonia solani*. One transgenic line exhibited increased tolerance to the root-rot disease incited by *R. solani* upon growth in infested soil. Transgenic plants were more vigorous and showed a decrease from 75 to 45% in seed mortality compared to control plants.

Rice plants expressing a chitinase gene from rice were developed and shown to be resistant to *Rhizoctonia solani,* the causal agent of sheath blight disease (Lin et al., 1995). There was no delay in the development of lesions upon infection, however, both the number and size of the lesions were reduced in transgenic plants compared to control plants. Also, the lesions were confined to the lower leaves in some transgenic lines. The level of resistance correlated with the level of chitinase expression in the different lines tested. A similar approach was chosen by Xu et al. (1996) for the development of rice plants expressing the RC24 gene coding for a basic chitinase from rice.

Cucumber plants expressing genes encoding an acidic chitinase from petunia and basic chitinases from tobacco or bean were evaluated for resistance to *Altenaria cucumerina, Botrytis cinerea, Collectrichum lagenarium,* and *Rhizoctonia solani* in the growth chamber or in an *in vitro* detached leaf assay (Punja and Zhang, 1993). No difference in disease development was observed between transgenic and control plants. Carrot plants expressing an acidic chitinase from tobacco showed a significant lower rate in lesion development and in the final extent of the disease 7 days post-inoculation with *B. cinerea, R. solani,* and *Sclerotium rolfsii* in an *in vitro* assay using detached petioles. Interestingly, these transgenic carrot plants were as susceptible as control plants when inoculated with *Altenaria radicini* and *Thielaviopsis basicola.* In addition, carrot plants expressing the acidic chitinase from petunia did not exhibit any tolerance to the five fungal pathogens tested. So far, the use of chitinase genes has produced variable degree of resistance to fungal pathogens, and the resistance appears to be influenced greatly by the type and source of chitinase protein used, the characteristics of the fungal pathogen, and the target plant species.

Potato plants expressing a gene construct encoding osmotin, a basic PR-5 protein from tobacco, showed delay in the onset of symptom development upon inoculation with spore suspensions of *Phytophtora infestans,* the causal agent of late blight (Liu et al., 1994). Similarly, potato plants expressing a gene coding an osmotin-like protein from potato reduced the percentage of leaves showing sporulation (Zhu et al., 1996). In this case, tolerance to the fungus was greater during the early phase of infection.

Van den Elzen et al. (1993) reported on the development of tomato plants expressing four gene constructs coding for class I and class II chitinases and β-1,3-glucanases. These four genes were isolated from tobacco. Transgenic plants were resistant to the root fungal pathogen *Fusarium oxysporum* sp. *lycopersici.* Disease severity was significantly reduced in transgenic plants which were able to survive infection whereas control plants died. A correlation between class I antifungal protein expression and disease severity was observed. Interestingly, concurrent expression of these two genes resulted in substantially enhanced disease resistance while either gene alone did not provide any protection indicating a synergistic effect between the class I chitinase and β-1,3-glucanase used.

Besides genes encoding PR proteins, other gene constructs have been used in agronomically important crop plants genetically modified for resistance to fungal pathogens. Uchimiya et al. (1993) showed that rice plants expressing the bialaphos-resistant (bar) gene that were treated with this herbicide were not infected by *Rhizoctonia solani.* Similarly, rice plants expressing the bar gene showed a significant lower number of lesions upon inoculation with mycelia of *Magnaporthe grisea* when treated with bialaphos (Tada et al., 1996). The herbicide bialaphos has antibiotic activity against *R. solani* and *M. grisea,* which are the causal agents of sheath blight and blast, respectively.

Wu et al. (1995) reported on the development of potato plants containing a gene encoding glucose oxidase from *Aspergillus niger.* These plants expressed high levels of hydrogen peroxide by oxidization of glucose. Plants expressing elevated levels of hydrogen peroxide showed enhanced resistance to *Phytophtora infestans.* Transgenic plants showed a significant reduction in the percentage of diseased leaf areas (49 to 56% reduction in the size of necrotic lesions) and the number of dead petioles (18 to 100% reduction) upon infection by sporangia of *P. infestans.* Production of hydrogen peroxide was an absolute requirement for disease resistance. Significantly, resistance to *Verticillium dahliae* and *Erwinia carotovora* (see Section II.B.2) was also reported for these transgenic potato plants. *V. dahliae* incites early death or the wilt disease of potato. These data suggest that broad spectrum fungal resistance can be engineered in plants by manipulating active oxygen species. Given the fact that a large number of microbial pathogens are sensitive to hydrogen peroxide, and that resistance to the bacteria *Erwinia carotovora* was also exhibited by the same transgenic plants, broad spectrum disease-resistance against other plant pathogens, including bacteria, should be possible by engineering plants expressing high levels of hydrogen peroxide.

Strittmater et al. (1996) developed potato plants expressing a tandem gene construct coding barnase, a bacterial ribonuclease, and barstar, a specific inhibitor of barnase. These transgenic plants reduced the sporulation of *Phytophtora infestans* and substantially delayed and reduced the sporangiophore formation after leaf inoculation with fungal spores. In addition, the necrotic area on infected leaves was limited to the inoculation sites. This interesting approach was aimed at restricting the fungal pathogen at the inoculation site by a rapid and massive death of host cells.

Besides these reports on agronomically important crops, numerous studies have been carried out on the model crop, tobacco. Tobacco plants expressing genes encoding PR proteins (chitinase, β-1,3-glucanase or PR proteins of undefined function) (Alexander et al., 1993; Broglie et al., 1991; Glascock et al., 1992; Neuhaus et al., 1991; Vierheilig et al., 1993; Zhu et al., 1994) and antifungal substances (ribosome-inactivating proteins, stilbene-type phytoalexin, cystein-rich proteins, viral killer toxins) (Hain et al., 1993; Jach et al., 1995; Kinal et al., 1995; Logemann et al., 1992; Park et al., 1996; Terras et al., 1995) were evaluated for resistance against several fungal pathogens including *Rhizoctonia solani, Perenospora tabacina, Phytophtora parasitica* var. *nicotianae, Cercospora nicotianae, Botrytis cinerea, Alternaria longipes*, and *Ustilago maydis*. Variable degrees of resistance were reported ranging from reduced root colonization to reduced infected leaf area, reduced lesion number and size, delay in the onset of disease, and susceptibility. In general, the protection varied widely among the transgenic lines tested and appeared to be positively correlated with the amount of protein expressed. Tobacco plants were also engineered with multiple gene constructs in order to pyramid the effect of several antifungal proteins and evaluate synergism (Jach et al., 1995; Zhu et al., 1994). Interestingly, the combination of two genes (chitinase and β-1,3 glucanase, and chitinase and ribosome-inactivating protein) provided substantially greater protection than either gene alone (Jach et al., 1995; Zhu et al., 1994).

In general, the level of protection in the transgenic plants engineered so far for resistance against fungal pathogens was highly variable. Differences in degree of protection may be influenced by the specific activity of the enzymes or proteins, their localization and concentration within the plant cells, the characteristics of the fungal pathogen, the biochemical structure of the fungal cell walls, and the type of interaction between the plant and the pathogen. However, studies on synergism of antifungal proteins appear quite promising.

2. *Control of Bacteria*

Limited reports have been published on agronomically important crop plants genetically modified for resistance to bacterial pathogens. Among these are genetically modified potato and tomato plants and genetically modified apple rootstock. The majority of reports so far have used the model crop, tobacco. To achieve resistance to bacterial pathogens, several strategies have been employed. These strategies are based on the use of bactericidal genes of diverse origin, pathogen-derived genes encoding enzymes, genes encoding lytic peptides from the giant silk moth, genes encoding lysozyme from the bacteriophage T4, as well as plant genes encoding enzymes.

For agronomically important crops, potato plants expressing a gene encoding glucose oxidase from *Aspergillus niger* were produced by Wu et al. (1995). Transgenic plants were highly resistant to the soft rot disease incited by *Erwinia carotovora* ssp. *carotovora*. Tubers of transgenic plants inhibited the growth of *E. carotovora* resulting in limited tissue maceration, whereas tubers from control plants showed extensive rot. The production of hydrogen peroxide was an absolute requirement for disease resistance.

Potato plants expressing a gene construct coding for lysozyme, a bacteriolytic enzyme from the bacteriophage T4, displayed reduced susceptibility to *Erwinia carotovora* spp. *atroseptica* after tuber inoculation in the greenhouse (Düring et al., 1993). This bacterial pathogen causes soft rot and blackleg. Resistance was expressed as a reduction in tissue maceration of tuber discs from 100% for control plants to 65% for transgenic line T424 when inoculated with a high concentration of bacteria. Resistance was also expressed as capability to sprout from inoculated tuber tissue. A higher percentage (30 to 100%) of transgenic tubers produced shoots compared to control tubers (0 to 20%). A positive correlation between lysozyme accumulation and resistance was reported. The higher the expression of lysozyme, the better the resistance.

Norelli et al. (1994) developed an apple rootstock expressing a gene construct encoding attacin E, a lytic protein from the giant silk moth *Hyalophora cecropia*. This transgenic rootstock showed significant increased resistance to *Erwinia amylovora*, the causal agent of fire blight, under greenhouse conditions. Disease incidence, measured as percent of infected shoot length, was reduced by 50% in the transgenic plants compared to the susceptible Malling 26 rootstock parent 25 days post-inoculation. Transgenic plants showed a similar reaction to *E. amylovora* as Liberty, a fire blight-resistant scion cultivar produced by conventional breeding. Interestingly, a similar strategy based on the use of another lytic peptide from the giant silk moth was not successful (Allefs et al., 1995).

These authors showed that potato plants expressing a gene construct encoding a cecropin B precursor polypeptide were not resistant *to Erwinia carotovora* ssp. *atroseptica* and *Erwinia chrysanthemi*. Interestingly, no cecropin B could be detected by western blot in these transgenic plants, although transcription of the gene occurred at high rates. These data suggest that the peptide was likely degraded proteolytically in potato tissue, thus resulting in a concentration too low for a significant inhibition of bacterial cells.

Several other gene constructs have been used in the model crop, tobacco. Tobacco plants expressing gene constructs encoding a bacterial proton pump (Mittler et al., 1995), the A1 subunit of cholera toxin (CTX) (Beffa et al., 1995), α-thionin from barley (Carmona et al., 1993), and the lytic peptide cecropin from the giant silk moth (Hightower et al., 1994) were developed and evaluated for resistance against *Pseudomonas syringae* pv. *tabaci* and *P. syringae* pv. *phaseolicola*. Transgenic plants showed variable resistance ranging from substantial reduction of bacterial growth to reduced disease severity, reduced infected leaf surface, complete symptom inhibition, and susceptibility. In general, levels of resistance could be correlated to levels of protein expression in the different transgenic lines analyzed. Also, the pathogen-derived resistance strategy has been applied to protect plants from bacterial infection by expressing detoxifying enzymes (Anzai et al., 1989; De la Fuente et al., 1992). It will be interesting to see whether these approaches can provide useful levels of resistance in agronomically important crop plants.

The first report on map-based cloning and isolation of a plant resistance gene and its use to engineer resistance to a bacterial pathogen by transforming and complementing a susceptible crop variety was done by Martin et al. (1993). These authors transformed susceptible tomato plants with a 2.4-kb cDNA construct from tomato which had the *Pto* gene that confers resistance to *Pseudomonas syringae* pv. *tomato*. This cDNA clone contained a gene family encoding a putative serine-threonine protein kinase. Transgenic plants did not exhibit symptoms of bacterial speck disease upon infection by *P. syringae* pv. *tomato* and inhibited bacteria growth. Given the fact that several host resistance genes have been recently mapped or cloned (Michelmore, 1995), this approach is likely be used more often in the future.

3. Control of Nematodes

Genetic engineered resistance against nematodes has been the latest addition to the control of plant pests. Although several substances have been described as having potential nematicide activity (Atkinson et al., 1995; Reinmann-Philipp and Beachy, 1993) and progress has been made recently in engineered protection, the potential of genetic engineering to control nematodes has yet to be demonstrated on agronomically important crops.

Tobacco plants expressing the anti-sense of TobRB7, a root-specific gene that is expressed during root-knot nematode infection and encodes a putative water channel protein, have been developed (Opperman and Conkling, 1995). These plants were engineered to interfere with the nematode feeding sites. Transgenic plants exhibited a substantial reduction in galling and egg production upon infection. Field trials have shown that the feeding sites were limited, and that galling could be reduced by 70% in transgenic plants relative to control plants. Galls produced on transgenic plants were small in size and scarce compared to the large and clustered galls observed on control plants. Also, the protection was shown to be durable over time (Opperman and Conkling, 1995). Another approach used was based on the engineering of tobacco plants expressing the bacterial gene barnase that encodes a ribonuclease. Such an enzyme produced by transgenic plants was expected to disrupt the feeding behavior of the nematodes. One transgenic line showed promising levels of resistance in that it drastically reduced the number of eggs and no galls were observed (Opperman and Conkling, 1996).

Proteinase inhibitors have also been investigated as potential compounds that can interfere with the nematode feeding. Potato plants expressing the cowpea trypsin inhibitor have been shown to reduce the fecundity of *Meloidogyne* spp. (Atkinson et al., 1995). Experiments conducted on transformed tomato roots expressing a gene construct encoding a variant form of oryzacystatin, a cysteine proteinase inhibitor from rice, showed that the growth of *Globodera pallida* was substantially inhibited as well as the ability of females to produce eggs (Urwin et al., 1995). These results are promising, but further experiments need to be conducted using whole plants in the field to assess the agronomic value of this approach.

Tobacco plants expressing antibodies against the stylet secretions of the root-knot nematode *Meloidogyne incognita* were produced as an approach to engineer resistance against this nematode. Such plants were engineered to interfere with the feeding of the nematodes on roots. However, transgenic plants did not appear to have any deleterious effect on the reproduction nor the parasitism of *M. incognita* upon plant infection (Baum et al., 1996).

Although very limited, these examples of engineered resistance to nematodes demonstrate the feasibility and the potential of this approach. Further work is needed given the numerous environmental concerns with the use of nematicides and their limited efficacy.

C. Examples of Field Tests

The first report on extensive field trials of plants genetically engineered for resistance to fungal pathogens was on oilseed rape plants expressing an endochitinase gene from tomato (Grison et al., 1996). Chitinase catalyses the hydrolysis of chitin, a polymeric component of fungal cell walls. In field trials, transgenic plants expressing chitinase exhibited an increased tolerance to three fungal pathogens: *Cylindrosporium concentricum*, *Phoma lingam* and *Sclerotinia sclerotiorum*. Transgenic plants showed a reduction in the percentage of diseased leaves and foliar diseased area compared to control plants upon inoculation with *C. concentricum*. For *P. lingam*, one transgenic line showed a significant reduction of basal stem symptoms as reflected by a lower disease index. For *S. sclerotiorum*, transgenic plants were also efficient at reducing stem necrosis. Overall, different degrees of protection against the three fungal pathogens were observed with a symptom reduction ranging from 23 to 79%. This range in protection could be due to differences in the biochemical composition and cell wall structure among the three fungal species tested, the localization of the recombinant chitinase, and the type of interaction between the plant and the pathogen.

The second field test report of transgenic plants engineered for fungal resistance was conducted on carrot plants expressing gene constructs encoding combinations of class I and II chitinases, class I and II β-1,3-glucanases and/or osmotin, and a PR-5 protein (Tigelaar et al., 1996). These genetically modified plants were evaluated in the field for resistance to several fungal pathogens. Transgenic plants showed substantial resistance to *Altenaria dauci*, *Altenaria radicina*, *Cercospora carotae*, and *Erysiphe heraclei*.

It will be interesting to test whether agronomically significant transgenic crop plants that exhibit increased tolerance to fungal pathogens under laboratory conditions are capable of limiting in the field economically important diseases including sheath blight of rice and late blight of potato. Field trials will be needed to determine the potential of genetically modified plants in managing major diseases. Also, it will be interesting to see whether the delay in infection that was described for several transgenic lines in the laboratory provides practical control of the diseases in the field.

There is no published report yet on field experiments for plants genetically modified for enhanced tolerance to bacterial pathogens. For the few agronomically important crops that have been engineered for resistance to bacteria and tested under laboratory conditions, it will be interesting to investigate whether the antibacterial activity will be effective against natural infections in the field.

In the case of engineered resistance to nematodes, field trials have been conducted with transgenic tobacco plants (see Section II.B.3). It will be interesting to see whether the resistance obtained on tobacco can be translated to agronomically important crops.

In conclusion, despite significant and major breakthroughs in engineered protection against fungi, bacteria, and resistance in agronomically important crops, there are only limited examples of field trials. Resistance strategies shown to have some potential in laboratory and greenhouse tests in tobacco need to be evaluated on major crops. So far, genetically modified crop plants with resistance to fungi, bacteria, or nematodes have not yet reached commercial significance.

III. DEVELOPMENT OF GENETICALLY MODIFIED PLANTS TO CONTROL VIRUSES

A. Rationale and Theory

Plant viruses cause severe damage to numerous crops (Hull, 1994; Milne, 1987). For example, more than $1 billion losses are reported yearly for rice in Southeast Asia due to virus diseases (Herdt, 1991). Standard virus control measures include quarantine, use of virus-free certified plant material, and crop rotation. These approaches are important because they help to prevent the release of viruses and limit the incidence of diseases in areas where the virus(es) of concern are not yet endemic. In contrast, in areas where viruses are prevalent, management strategies such as suppression and cross protection are aimed at limiting the incidence of viruses or at protecting the crop against significant economic damage. Ideally, the use of virus-resistant crop varieties is the most valuable and effective approach to control virus diseases. The development of virus-resistant plants by conventional approaches has been successfully used over the years for many crops. However, most crop species have been recalcitrant to this strategy due to the lack of resistance genes, the complexity of these genes especially when multigenic traits are involved, and genetic barriers that make the introduction of resistance traits into commercial varieties difficult. Therefore, the development of effective methods for gene transfer to plants has resulted in major breakthroughs for the development of virus-resistant crop plants.

The majority of genetically modified crop plants engineered for virus resistance are developed based on the pathogen-derived resistance strategy (Sanford and Johnston, 1985). With this strategy, the resistant gene is derived from the viral genome itself and not from the genome of a plant or another organism. The pathogen-derived resistance approach to engineer virus resistance differs from the strategies used to engineer resistance to fungal, bacterial, and nematode diseases since in the latter case the genes that provide resistance are derived mainly from organisms other than the pathogen itself, although pathogen-derived resistance has also been used in a few cases. To develop virus-resistant plants, viral coding and noncoding sequences have been extensively used as full-length, truncated, or untranslatable constructs, and in a sense and antisense orientation. Viral gene constructs that have been used successfully to engineer virus resistance include the coat protein (CP) (Powell-Abel et al., 1986), the replicase (Golemboski et al., 1990), the protease (Maiti et al., 1993; Vardi et al., 1993), the cell-to-cell movement protein (Malyshenko et al., 1993), satellite RNAs (Harrison et al., 1987; Gerlach et al., 1987), and defective-interfering RNAs (Kollar et al., 1993; Stanley et al., 1990). Noncoding sequences from the 5′ (Nelson et al., 1993) or 3′ (Zaccomer et al., 1993) untranslated regions of viral genomes have also been used. Interestingly, virus resistance was also engineered recently using other strategies based on the use of genes coding antiviral proteins including ribosome-inactivating proteins from pokeweed (Lodge et al., 1993), from *Dianthus cayophyllus* (Hong et al., 1996), from root tubers of *Trichosanthes kirilowii* (Lam et al., 1996), rat 2′-5′ oligoadenylate synthetase gene (Truve et al., 1993), human 2-5A synthetase gene and a 2-5A dependent RNase L (Mitra et al., 1996; Ogawa et al., 1996), antibodies (Tavladoraki et al., 1993; Voss et al., 1995), and defense-related compounds (Herbers et al., 1996). Also, ribozymes have been successfully used in tobacco plants (Nakamura et al., 1995, 1996).

B. Current Status of Development

Since the first report of engineered resistance to a plant virus by Powell-Abel et al. (1986), who showed that tobacco plants expressing the CP gene of TMV were protected from infections by this virus, pathogen-derived resistance has been applied to numerous viruses and crop plants (Baulcombe, 1994; Beachy, 1993a,b; Beachy et al., 1990; Fitchen and Beachy, 1993; Grumet,

1990, 1994; Hackland et al., 1994; Hanley-Bowdoin and Hemenway, 1992; Hull and Davies, 1992; Lomonossoff, 1995; Scholthof et al., 1993; Surtevant and Beachy, 1993; Wilson, 1993). By far, a larger number of transgenic crops engineered for virus resistance have been evaluated in the field than transgenic plants engineered to control bacteria, fungi, and nematodes combined. Most examples of virus resistance are based on coat protein-mediated protection (Beachy et al., 1990). Table 1 lists transgenic crop plants of agronomic significance that have been extensively evaluated for virus resistance under field or greenhouse conditions.

The first virus-resistant genetically modified crop plant, Freedom II, was deregulated in the U.S. in 1994 (Medley, 1994) and released commercially in the spring of 1995 by the Asgrow Seed Company. Freedom II, a summer squash which is resistant to zucchini yellow mosaic potyvirus (ZYMV) and watermelon mosaic potyvirus 2 (WMV 2), was experimentally tested under the code name ZW-20 (Tricoli et al., 1995; Fuchs and Gonsalves, 1995b). Transgenic squash line CZW-3 is the second virus-resistant transgenic product deregulated in the U.S. (Acord, 1996). This squash line is resistant to ZYMV, WMV 2, and cucumber mosaic cucumo virus (CMV) (Tricoli et al., 1995). Papaya is the second virus-resistant transgenic crop that recently received exemption status from APHIS/USDA (Strating, 1996). Transgenic papaya lines 55-1 and 63-1 are resistant to papaya ringspot potyvirus (PRSV). Undoubtedly, more virus-resistant crops will be deregulated in the near future.

C. Examples of Field Tests

In this section, we summarize the reports on virus-resistant transgenic crops that have been extensively evaluated in the field. The first field studies were conducted in the late 1980s on potato plants engineered for resistance to potato potexvirus X (PVX) and/or potato potyvirus Y (PVY), tomato plants engineered for resistance to tobacco (TMV) and tomato mosaic (ToMV) tobamoviruses, and cucumber plants engineered for resistance to cucumber mosaic cucumovirus (CMV). More recent field tests conducted in the 1990s were done on squash plants engineered for resistance to ZYMV, WMV 2, and/or CMV, cantaloupe plants engineered for resistance to the same three viruses, and papaya plants engineered for resistance to PRSV. All of these examples illustrate the impact that virus-resistant transgenic plants can have on disease management.

The first report of a field trial was by Nelson et al. (1988) who evaluated tomato plants expressing the CP gene of TMV for resistance to TMV. Transgenic plants were highly resistant to mechanical inoculation by TMV, with only 5% of the plants being symptomatic compared to 99% of the control plants. Inoculated transgenic and uninoculated control plants showed identical fruit yield, indicating that genetic engineering did not alter the horticultural performance of the transgenic tomato plants. All the transgenic plants were resistant to TMV strain U1 and to the more severe strain PV230, however, they were not resistant to infection by ToMV. Subsequently, tomato plants expressing the CP gene of ToMV were developed to engineer resistance to ToMV (Sanders et al., 1992). Transgenic line 4174 displayed high levels of resistance to ToMV and also to TMV strain U1. Tomato plants with resistance to TMV and ToMV should be useful in greenhouses and plastic houses where these two viruses can be mechanically transmitted from plant to plant at high ratio.

Fuchs et al. (1996) evaluated CMV resistance of tomato plants expressing the CP gene of CMV strain WL. CMV-WL belongs to CMV subgroup II. Considering that transgenic plants displayed high levels of resistance to CMV strains from the two CMV subgroups I and II (Xue et al., 1994; Provvidenti and Gonsalves, 1995), plants were challenged in the field with the subgroup I CMV strain Fny. None of the mechanically inoculated transgenic plants became infected, whereas infection occurred in 80% of the control plants. Transgenic plants had a 17-fold higher fruit yield than inoculated control plants. Such high levels of resistance to CMV had not been reported previously in tomato. Unfortunately, the low rate of natural CMV spread by aphid vectors from tomato to tomato plant observed in New York did not allow a critical evaluation of resistance against natural aphid inoculations. Providing that resistance holds against aphid inoculations, these transgenic tomato plants will be especially valuable because CMV causes severe losses in tomato fields worldwide and host genes for resistance have not been identified.

Table 1 Agronomically Significant Virus-Resistant Genetically Engineered Crops

Crop	Resistance to	Viral Transgene Construct	Resistance Evaluation	Status of Development	Ref.
			Vegetables		
Tomato	TMV	CP	Field	Experimental	Nelson et al., 1988
			Greenhouse	Experimental	Motoyoshi and Ugaki, 1993
	ToMV	CP	Field	Experimental	Sanders et al., 1992
	CMV	CP	Field	Experimental	Fuchs et al., 1996
			Greenhouse	Experimental	Xue et al., 1994
					Provvidenti and Gonsalves, 1995
	TSWV	NP	Greenhouse	Experimental	Kim et al., 1994
					Ultzen et al., 1995
	CMV, TSWV	CP, NP	Greenhouse	Experimental	Gonsalves et al., 1993
	TYLCV	CP (VI)	Greenhouse Insect-proof cages	Experimental	Kunik et al., 1994
Potato	PVX	CP	Field	Experimental	Jonkedijk et al., 1992
	PVY	CP	Field	Experimental	Malnoë et al., 1994
	PVX, PVY	CP, CP	Field	Experimental	Kaniewski et al., 1990
			Greenhouse	Experimental	Lawson et al., 1990
	PVS, PVM	CP	Greenhouse	Experimental	MacKenzie et al., 1991
	PLRV	CP	Greenhouse	Experimental	Barker et al., 1992, 1994
					del Vas et al., 1993
					Kawchuk et al., 1990, 1991
					van der Wilk et al., 1991
					Brown et al., 1995
					Presting et al., 1995
		MP	Greenhouse	Experimental	Tacke et al., 1996
Squash	CMV	CP	Field	Experimental	Tricoli et al., 1995
	ZYMV	CP	Field	Experimental	Tricoli et al., 1995
	WMV 2	CP	Field	Experimental	Tricoli et al., 1995
	CMV, WMV 2	CP, CP	Field	Experimental	Tricoli et al., 1995
	ZYMV, WMV 2	CP, CP	Field	**Commercial**	Tricoli et al., 1995
					Fuchs and Gonsalves, 1995b
					Arce-Ochoa et al., 1995
					Clough and Hamm, 1995
	CMV, ZYMV, WMV 2	CP, CP, CP	Field	**Deregulation**	Tricoli et al., 1995
					Arce-Ochoa et al., 1995
					Clough and Hamm, 1995
					Fuchs and Gonsalves, 1993
	SqMV	CP	Field, Greenhouse	Experimental	Pang et al., 1995
Melon	ZYMV	CP	Greenhouse	Experimental	Fang and Grumet, 1993
	CMV	CP	Greenhouse	Experimental	Gonsalves et al., 1994
					Yoshioka et al., 1993
	CMV, ZYMV, WMV 2	CP, CP, CP	Field	Experimental	Clough and Hamm, 1995
					Fuchs et al., 1997
Cucumber	CMV	CP	Field	Experimental	Gonsalves et al., 1992
Lettuce	LMV	CP	Greenhouse	Experimental	Zerbini et al., 1995
	TSWV	NP	Greenhouse	Experimental	Pang et al., 1996
			Fruits		
Papaya	PRSV	CP	Field	**Deregulation**	Lius et al., 1997
			Greenhouse		Fitch et al., 1992
					Tennant et al., 1994
Plum	PPV	CP	Greenhouse and quarantine enclosures	Experimental	Scorza and Ravelonandro, 1996

Table 1 (continued) Agronomically Significant Virus-Resistant Genetically Engineered Crops

Crop	Resistance to	Viral Transgene Construct	Resistance Evaluation	Status of Development	Ref.
			Cereals		
Corn	MDMV MCMV	CP	Growth chamber Experimental	Experimental	Murray et al., 1993
Rice	RSV	CP	Greenhouse	Experimental	Hayakawa et al., 1992
			Ornamentals		
Petunia	CMV	Satellite RNA	Greenhouse	Experimental	Kim et al., 1995
			Forage		
Alfalfa	AlMV	CP	Greenhouse	Experimental	Hill et al., 1991

Jongedijk et al. (1992) developed potato plants expressing the CP gene of PVX and extensively evaluated 12 different transgenic lines to infection by PVX. Two lines (MGE-32 and MGE-44) were highly resistant with no infected tubers, whereas the other lines showed moderate resistance or susceptibility.

Malnoë et al. (1994) analyzed the resistance of potato plants expressing the CP gene of PVY to natural spread by aphid vectors. Transgenic line Bt6 was immune to the homologous PVY strain N and significantly reduced the incidence (23%) of heterologous PVY strain O. In addition, analysis of transmission of PVY strain O in tubers of transgenic plants that were infected by aphids showed that the presence of this latter strain was not detected in the majority (59 to 72%) of progeny.

Lawson et al. (1990) and Kaniewski et al. (1990) were the first to demonstrate that resistance to two viruses could be achieved in a single plant line by genetic engineering. These authors evaluated potato plants expressing the CP genes of PVX and PVY. Although several transgenic lines were tested in the field, line 303 was the only one that was highly resistant to mixed infections by the two viruses (Kaniewski et al., 1990). This study demonstrated the feasibility and the usefulness of multiple CP gene constructs to control mixed virus infections.

Cucumber plants expressing the CP gene of CMV strain C were evaluated for resistance against mechanical and aphid inoculations by CMV (Gonsalves et al., 1992). Transgenic plants significantly reduced the incidence of CMV with a lower percentage of symptomatic plants than control plants. Transgenic line T48 showed comparable degree of resistance to Marketmore, a popular conventional resistant cultivar.

Squash plants containing single or multiple CP gene constructs of CMV, ZYMV, and/or WMV 2 were evaluated for resistance to single or mixed infections (Tricoli et al., 1995). Transgenic lines showed variable resistance ranging from no symptom development, to delay in the onset of symptoms, reduced symptom severity, and susceptibility. For example, line C-82 was highly resistant to CMV, line W-164 was highly resistant to WMV 2, and line Z-33 was highly resistant to ZYMV with no symptomatic plants for any of these three lines. For transgenic lines expressing more than one CP gene, line CW-14 was highly resistant to CMV and WMV 2, and line ZW-20 was highly resistant to ZYMV and WMV 2. Interestingly, some virus-resistant plants of line ZW-20 had lost the neomycin phosphotransferase II (NPT II) marker gene. Fuchs and Gonsalves (1995b) highlighted the value of transgenic line ZW-20 in controlling mixed infections by ZYMV and WMV 2. In addition to lines with resistance to two viruses, transgenic line CZW-3 expressing three CP genes was highly resistant to mixed infections by CMV, ZYMV, and WMV 2. This was the first report of engineered resistance against three different viruses using CP-mediated protection. Development of transgenic squash lines ZW-20 and CZW-3 is significant because of the difficulties in obtaining squash varieties with multiple resistance genes and similar levels of resistance to mixed infections by conventional breeding. The resistance of transgenic lines ZW-20 and CZW-3 was confirmed by field testing in different geographic locations (Arce-Ochoa et al., 1995; Clough and Hamm, 1995; Fuchs and Gonsalves, 1993). Transgenic line ZW-20 was used to develop Freedom II, the first virus-resistant crop to be deregulated for commercial release in the U.S. in 1994. Recently, transgenic line CZW-3 received exemption status from APHIS/USDA (Acord, 1996).

Cantaloupe plants expressing the CP genes of CMV, ZYMV, and WMV 2 were evaluated for resistance against single and mixed infections (Fuchs et al., 1997). Across all trials, transgenic plants exhibited resistance against single and mixed infections of CMV, ZYMV, and WMV 2. Homozygous plants showed localized symptoms only late in the growing season, and hemizygous plants displayed a significant delay in the onset of disease. In contrast, control plants developed severe systemic symptoms early in the growing season. Both transgenic homozygous and hemizygous plants substantially reduced the incidence of mixed infections, grew vigorously, and yielded significantly more marketable fruits than control plants.

In 1992, papaya plants expressing the CP gene of PRSV strain HA 55-1 (Fitch et al., 1992) were evaluated in the field for resistance to PRSV on the Oahu island in Hawaii. Transgenic papaya line 55-1 was completely resistant to PRSV even after 18 months (Lius et al., 1996). A subsequent field trial was established in the Puna area on the Hawaii island in 1995, and transgenic plants 55-1 have shown excellent resistance so far. It thus appears that transgenic line 55-1 would be suitable to control PRSV in Hawaii, especially since the genetically engineered cultivar is the popular "Sunset" cultivar which is a sib of "Sunrise", the most widely grown papaya cultivar worldwide. Transgenic line 55-1, which was named "SunUp", and transgenic line 63-1 were recently deregulated by APHIS/USDA (Strating, 1996). These two lines are expected to clear regulatory status from EPA and FDA in 1997. Deregulation and subsequent commercialization of PRSV-resistant transgenic papaya plants will provide Hawaii with an urgently needed control strategy for a disease that is currently devastating the papaya industry. This may be the first case where a biotechnology product will have a significant and timely impact in controlling a virus disease that threatens the economic survival of an industry. Similar efforts to develop transgenic papaya that are resistant to strains of PRSV in other parts of the world are underway.

Pathogen-derived resistance continuous to be used to engineer virus resistance in numerous other crops. Thus, one can expect that more reports of field trials will appear in the near future and that a number of these crops will be deregulated and eventually commercialized.

D. Epidemiological Considerations

Extensive field evaluations of virus-resistant genetically engineered crop plants showed that they have the ability to exclude viruses from a crop. Also, transgenic plants restrict the amount of virus-infected tissues and lower virus titers, thus reducing the spread of viruses by limiting the availability of virus inoculum. In other words, virus-resistant transgenic plants are unlikely to serve as secondary virus source for aphid spread. This important epidemiological consideration has been experimentally demonstrated recently in our laboratory using transgenic squash line ZW-20. We wanted to assess whether transgenic ZW-20 plants that develop localized chlorotic dots or blotches upon infection by ZYMV and WMV 2 (Fuchs and Gonsalves, 1995b) could serve as secondary virus source for aphid transmission. Interestingly, ZYMV and WMV 2 can be recovered from those localized chlorotic dots and blotches but not from the surrounding leaf tissues. We evaluated the spatial relationships among ZW-20 plants that were field tested in replicated commercial-type fields by using geostatistical analysis (Klas et al., 1994). Modeling of our field data showed that semivariograms constructed for transgenic squash plants were overall flat and with no apparent fit to any theoretical semivariogram models. These results indicated the absence of dependence between transgenic plants, suggesting that ZW-20 plants did not serve as a virus source (Klas et al., 1994). Subsequent experiments carried out in screen enclosures showed that transgenic ZW-20 plants lowered the probability of transmission of ZYMV and WMV 2 by reducing the efficiency of acquisition by vectors.

IV. ENVIRONMENTAL SAFETY ASSESSMENT OF GENETICALLY MODIFIED PLANTS RESISTANT TO VIRUSES

A. Environmental Concerns

Numerous articles and reviews have been published on the potential environmental risks regarding the commercial release of plants engineered for disease resistance in general (Ahl-Goy and Duesing, 1996; Dale, 1992, 1994; Darmency, 1994; Gates, 1995; Gliddon, 1994; Keeler, 1995; Linder and Schmitt, 1994; Mellon and Rissler, 1995; Rissler and Mellon, 1993; Rogers and Parkes, 1995; Thacker, 1993; Tiedje et al., 1989), and for virus-resistant transgenic crop plants (de Zoeten, 1991; Falk and Bruening, 1994; Hull, 1989, 1990; Tepfer, 1993; Wilson, 1993) in particular. There has been a significant number of papers published on the potential risks that virus-resistant genetically modified plants may have on the environment. However, very limited scientific information is available, if any, based on risk assessment experiments conducted in the field. Nonetheless, field experiments are the only realistic way to assess environmental risks. In addition, results from greenhouse experiments do not always directly translate into similar field results. Obtaining scientific data from field experiments are critical if one is concerned with tracking long-term environmental effects of the commercial release of virus-resistant transgenic plants. In addition, this type of information should be published to help the scientific community inform regulatory agencies and the general public using a systematic scientific approach and not solely speculations and predictions.

Two major concerns have been raised with the release of virus-resistant transgenic crop plants into the environment. First, the development or creation of modified viruses, and second, the escape of transgenes. The first concern is related to the presence and expression of viral sequences in plant chromosomes. Constitutive expression of viral sequences occurs in genetically engineered plants, but does not occur naturally in current crops. Therefore, when virus-resistant transgenic plants are released on a large scale, these viral sequences constitute a reservoir of virus-derived components, i.e., transcripts derived from virus genes and proteins encoded by virus genes. These viral products which provide virus resistance may also change the properties of existing viruses and trigger the development of new viruses through complementation, transencapsidation, and recombination.

1. Recombination

RNA recombination between a transgene mRNA and a challenging viral RNA is a concern in transgenic crop plants because it can lead to the development of novel viruses. Expression of a viral transgene, in many cases the CP gene, confers virus resistance to a transgenic plant, but the mRNAs of the transgene may be available for recombination with the RNA of a challenging virus. The process which involves the exchange of template RNA species during virus replication can result in a chimeric product that combines two originally distinct RNA species. The likelihood of recombination between a transgene transcript and an infecting virus should not be high if the plant resists infection. However, virus isolates and viruses that are able to overcome the resistance can increase the probability of recombination. Consequently, when virus-resistant transgenic plants are commercially released, the opportunity for recombination exists because transgenic plants are still susceptible to infection by viruses other than the targeted one(s) and viral transgenes are available for recombination with challenging viruses.

2. Transencapsidation

Transencapsidation in transgenic crop plants is a concern because the majority of agronomically important transgenic crops engineered so far for virus resistance contain viral CP genes (see Table 1), and some lines express high levels of CP. Through transencapsidation, the genome of a challenging virus can be partially or completely encapsidated by CP subunits produced by genetically modified plants. Thus, transgenic plants expressing viral CP genes can drive the development of newly coated virus particles with possible new properties. Transencapsidation can have significant consequences given the different functions of the CP during the virus replication cycle. It is well documented that the CP is not only involved in protection of the virus genome from degradation by host nucleases, but is also implicated in RNA binding, systemic movement, establishing host range, symptomatology, and seed transmissibility. Also, the CP determines the antigenic properties and the vector transmission specificity. This is particularly important for insect-borne viruses for which the CP plays a critical role in virus transmission (Atreya et al., 1990; Chen and Francki, 1990; Harrison and Murant, 1984; Murant et al., 1988; Pirone, 1991). Through transencapsidation, newly coated virus particles can have altered vector specificity and expanded host range. This phenomenon is of particular concern if a transgenic plant becomes infected by a virus isolate to which the genetically modified plants do not display resistance. However, acquisition of a different CP is only effective in the transgenic plant or in one transmission from that plant. This is a single-generation event and should not be sufficient to develop an epidemic. Thus, transencapsidation is unlikely to be an environmental risk.

3. Complementation

Through complementation, a viral gene product, i.e., the CP or the movement protein produced by a transgenic plant could supply a functional protein to a challenging virus that is deficient in the synthesis of this protein or that produces a dysfunctional protein. For example, a virus that carries a deficient movement protein and is thus restricted to the inoculation sites, could translocate to uninoculated leaves by complementation with a functional movement protein that is expressed by transgenic plants. Through complementation, existing viruses may acquire new properties. However, complementation is unlikely to cause any environmental hazards.

4. Synergism

Interactions between co-infecting unrelated viruses can sometimes be synergistic leading to an increase in virus titer and/or an increase in disease severity. Thus, transgenic plants expressing virus-derived sequences may increase virus titers and symptoms severity when challenged by unrelated viruses. Synergism, however, does not modify existing viruses nor create novel viruses with new characteristics such as expanded host range, and altered vector transmissibility. Therefore, synergism is of very limited, if any, environmental concern.

5. Gene Flow

The second environmental concern is related to the fact that transgenic plants express foreign genes that could escape into wild crop relatives. If a virus-resistant transgenic crop is cultivated in physical proximity to a virus-susceptible free-living crop relative, these two species being capable of interbreeding, hybrids can develop by pollen flow. Such hybrids can acquire the viral transgenes and subsequently become virus resistant. A weedy crop relative may show enhanced weediness due to the acquisition of the virus resistance trait. Usually, pollen reaches neighbor plants that are within a short distance from a donor plant. However, under favorable conditions, pollen can also move long distances by insects and be transported by the wind. For several agronomically important crops, bees are major pollinators and pollen dispersal agents. Thus, with pollen flow, there is a possibility of gene flow to wild species with subsequent reduction of genetic diversity. At this point of time, it seems that flow of transgenes from transgenic plants to compatible species is inevitable (Kareiva et al., 1994). Therefore, the evaluation of risks would be based on whether the resulting plants that acquire the transgenes will pose environmental risks.

6. Considerations

One has to remember that complementation, synergistic interactions, transencapsidation, and recombination are not new phenomena that emerged with the use of transgenic plants. They have already been demonstrated to occur when nontransformed plants are subject to mixed virus infections. Recombination was described for numerous viruses that belong to several plant virus families including caulimoviruses (Howell et al., 1981; Lebeurier et al., 1982), bromoviruses (Allison et al., 1990; Bujarski and Kaesberg, 1986), carmoviruses (Cascone et al., 1990), tobamoviruses (Beck and Dawson, 1990), luteoviruses (Gibbs and Cooper, 1995; Rathjen et al., 1994), tobraviruses (Robinson et al., 1987), tombusviruses (White and Morris, 1994), nepoviruses (Le Gall et al., 1995), potyviruses (Cervera et al., 1993), cucumoviruses (Fernandez-Cuartero et al., 1994), hordeiviruses (Edwards et al., 1992), and alfalfa mosaic alfamovirus (Huisman et al., 1989; van der Kuyl et al., 1991). The involvement of RNA recombination in virus diversity and evolution has been discussed (Lai, 1992; Simon and Bujarski, 1994). Also, transencapsidation is known to occur with potyviruses (Bourdin and Lecoq, 1991), luteoviruses (Creamer and Falk, 1990; Hu et al., 1988; Wen and Lister, 1991), and it has been suspected with cucumoviruses (Perry and Francki, 1992). Similarly, complementation has been reported in co-infected plants for a replication-defective TMV mutant (Ogawa et al., 1991), and for a cell-to-cell movement-defective cowpea mosaic comovirus mutant (Talianski et al., 1993). In general, complementation for movement is accomplished by several related or unrelated viruses (Malyshenko et al., 1989). Synergistic interactions have been documented for potato leafroll luteovirus and PVY (Barker, 1987), soybean mosaic potyvirus and bean pod mottle como virus (Calvert and Ghabrial, 1983), maize chlorotic mottle machlomo virus and maize dwarf mosaic poty virus (Goldberg and Brakke, 1987), CMV and ZYMV (Poolpol and Inouye, 1986), PVX and PVY (Damirdagh and Ross, 1967; Rochow and Ross, 1955), PVX and other potyviruses (Bowman Vance et al., 1995), and potato aucuba mosaic potexvirus and PVY (Xu and AbouHaidar, 1996) among others.

Similarly, gene flow between a crop and weed populations has probably occurred since the beginning of organized farming. Pollen flow has been documented for several crop plants and weed relatives. For example, gene flow has been described for some crops including squash (Kirkpatrick and Wilson, 1988; Wilson, 1990), sugar beet (Boudry et al., 1993; Raybould et al., 1996), carrot (Wijnheijmer et al., 1989), potato (Eijlander and Stiekema, 1994; Quiros et al., 1992), rice (Langevin et al., 1990), corn (Doebley, 1990), and sunflower (Arias and Rieseberg, 1994). Kirkpatrick and Wilson (1988) demonstrated pollen movement and gene flow between domesticated *Cucurbita pepo* L. ssp. *ovifera* (L.) squash and the wild relative Texas gourd *C. pepo* L. spp. *ovifera* var. *texana* by analyzing progeny of experimental populations. Interestingly, the frequency of crop-weed hybrids was low (5%) relative to the high levels of pollen flow and the high activity of large populations of bees.

Therefore, for environmental risk assessment, the key question is to assess whether transgenic plants will increase the likelihood of complementation, synergism, transencapsidation, and recombination. Another critical issue is to assess whether the introgression of virus resistance into related plant species will enhance the development of undesirable traits such as increased weediness.

B. Risks of Development of New Viruses

Several laboratories have initiated work to assess the development of modified viruses or the creation of novel viruses through complementation, transencapsidation and recombination. Most experiments have been conducted in the laboratory and essentially with tobacco as a model crop.

1. Examples of Recombination

Recombination has been demonstrated to occur for transgene transcripts and viral RNAs using transgenic tobacco plants and several viruses that belong to different families. Lommel and Xiong (1991) showed that the RNA of a red clover necrotic mosaic dianthovirus (RCNMV) isolate defective in cell-to-cell movement recombined with a transgene encoding the movement protein of RCNMV that was expressed in transgenic *Nicotiana benthamiana*. Subsequently, Xiong and colleagues challenge-inoculated these transgenic plants with RNA 1 of RCNMV and showed that recombination between the transgene transcript and the inoculated RNA occurred. Such recombination events restored the sequences necessary for viral replication of the transcript, developed a functional viral RNA, and reestablished the bipartite genome (summary in Allison, 1996).

Cauliflower mosaic caulimovirus (CaMV) was shown to acquire gene VI of CaMV that was expressed in transgenic *Brassica napus* (Gal et al., 1992). Gene VI of CaMV determines systemic infection in solanaceous species and also acts as a translational transactivator. To assess the effect of selection pressure on the development of recombinants, *Nicotiana bigelovii* plants expressing gene VI from CaMV were challenged under moderate and high selection pressure. High selection pressure was achieved by using a CaMV strain that does not naturally infect *N. bigelovii*. Under these conditions, systemic infection of transgenic plants was observed and viable recombinants were recovered (Schoelz and Wintermantel, 1993). Low selection pressure was achieved by using a CaMV strain that infects *N. bigelovii*. In that case, recombinants were also isolated but at a very low frequency (Wintermantel and Schoelz, 1996). In competition assays, recombinant viruses showed enhanced fitness relative to wild-type viruses because they acquired the ability to systemically infect *N. bigelovii* and other solanaceous plants (Schoelz and Wintermantel, 1993; Wintermantel and Schoelz, 1996).

Greene and Allison (1994) showed that a cowpea chlorotic mottle bromovirus (CCMV) isolate that contained a defective CP gene was able to acquire a functional gene by recombination with the CP gene from CCMV that was expressed in transgenic *Nicotiana benthamiana*. Interestingly, deletions in the 3′ untranslated region that is adjacent to the CCMV transgene construct abolished the occurrence of recombination events (Gene and Allison, 1996). In subsequent studies, Allison and collegues mixed infected *N. benthamiana* plants with two related bromoviruses, brome mosaic virus (BMV) and CCMV, and showed that recombination was not detected within the CP genes of the viruses recovered from plants mixed-infected with BMV and CCMV. When plants were infected with recombinant viruses in which the CP genes were exchanged, recombination was detected readily in the recovered viral RNAs (summary in Allison, 1996). The rate of recombination was similar to that seen previously in transgenic plants (Greene and Allison, 1994). These data are very important to risk assessment since they provide direct evidence that transgenic plants did not increase the likelihood of recombination.

In general, all the previous examples illustrate that RNA recombination can occur with transgenic plants expressing viral sequences leading to the development of novel virus entities with new properties. Interestingly, recombination was not detected in the case of tobacco plants containing the 3′ noncoding sequence of cherry leafroll nepovirus (Cooper et al., 1994). Since most of the recombination evidence obtained so far comes from laboratory experiments, there is still need to assess the risk of recombination in relation to the release of transgenic plants into the environment. First of all, recombination is well known to occur for several families of plant viruses (see Section IV.A.1), therefore it is not surprising that recombination has been also detected with transgenic plants that express sequences derived from viruses known to recombine propensely. Using computer-assisted predictions of RNA recombination, Revers et al. (1996) speculated that recombination is likely to be common for potyviruses, the largest group of plant viruses.

2. *Examples of Transencapsidation*

Transencapsidation was recently demonstrated to occur in transgenic tobacco plants expressing the CP of TMV (Osbourn et al., 1990; Holt and Beachy, 1991), the CP of plum pox potyvirus (Lecoq et al., 1993), and the CP of AlMV (Candelier-Harvey and Hull, 1993), and for potato plants expressing the CP of PVY (Farinelli et al., 1992). These studies showed that the CP expressed in transgenic plants was able to encapsidate the RNA genome of the challenging virus. Transencapsidated viruses were mutant TMV isolates (Osbourn et al., 1990), synthetic mutant TMV RNAs (Holt and Beachy, 1991), ZYMV (Lecoq et al., 1993), CMV (Candelier-Harvey and Hull, 1993), and another strain of PVY that was identifiable by monoclonal antibodies (Farinelli et al., 1992). More importantly, transencapsidation was shown to mediate the systemic spread of TMV mutants deficient in this function (Osbourn et al., 1990; Holt and Beachy, 1991), and the spread of an aphid nontransmissible strain of ZYMV (Lecoq et al., 1993).

In our laboratory, we assessed possible changes in virus-vector specificity through transencapsidation by carrying out experiments in the field. Transgenic plants expressing the CP gene of the aphid transmissible CMV strain WL were challenge-inoculated with the aphid nontransmissible CMV strain C (Fuchs et al., 1994; Fuchs and Gonsalves, 1995a). These two CMV strains belong to two distinct CMV subgroups I and II that can be differentiated by their serological properties and nucleotide homologies (~80% between the CP genes). CMV-C belongs to subgroup I and CMV-WL belongs to subgroup II. By using several different lines of transgenic squash, melon, and tomato, we analyzed whether the level of CP expression and the degree of resistance to infections by CMV-C would affect the spread of CMV-C. Our field experiments were carried out essentially with two melon lines that expressed high levels of CP and were not resistant to CMV infections, although they displayed a delay in the onset of disease. Spread of CMV-C was monitored on healthy nontransformed plants under field conditions with indigenous aphid populations as natural vectors, and under greenhouse and screen enclosures with the green peach aphid *Myzus persicae*. Virus spread was analyzed by visual monitoring of symptom development, ELISA, a reverse transcription-polymerase chain reaction (RT-PCR) assay combined with restriction fragment length polymorphism (RFLP), and a RT-PCR assay combined with sequencing. The RT-PCR RFLP assay enabled us to classify the CMV isolates in their CMV subgroup, and the second RT-PCR assay enabled us to sequence a region of the CP gene that includes the nucleotide triplet in position 502–504. This triplet codes for the CP amino acid residue 168 which is a phenylalanine (Phe) for CMV-C. Residue Phe 168 is unique to CMV-C because all the other CMV strains carry a tyrosine or a cysteine at this position in the CP. Our results showed that none of the 444 inoculated transgenic plants containing the CP gene of CMV-WL mediated the spread of CMV-C to a total of 1119 potential recipient nontransformed plants.

3. Examples of Complementation

Complementation was demonstrated with tobacco plants expressing the TMV cell-to-cell movement protein. These plants were able to complement the movement of TMV mutants defective in cell-to-cell spread (Deom et al., 1987; Lapidot et al., 1993). Similar results were obtained with transgenic potato plants expressing the movement protein of TMV (Valkonen et al., 1995). Van Dun et al. (1988) showed that *Nicotiana tabacum* plants expressing the P1 replicase gene of AlMV were able to complement AlMV lacking that gene. Similarly, Hanley-Bowdoin et al. (1990) demonstrated that a tomato golden mosaic geminivirus (TGMV) Al1 replicase gene expressed in *Nicotiana benthamiana* plants could functionally complement a TGMV variant carrying a lethal mutation in the Al1 open reading frame. Schoelz et al. (1991) showed that transgenic *Nicotiana bigelovii* expressing gene VI of CaMV were able to complement the systemic movement of a CaMV strain defective in long-distance movement. Complementation was also described in transgenic *Nicotiana tabacum* plants expressing the movement protein of CMV (Kaplan et al., 1995). Transgenic plants complemented the long-distance movement of a cucumovirus pseudorecombinant (carrying RNAs 1 and 2 of tomato aspermy virus and RNA 3 of CMV strain M) that was defective for this function. Also, transgenic plants complemented the cell-to-cell movement but not the long-distance spread of BMV and peanut stunt cucumovirus, but not of unrelated viruses.

Recently, Li and Carrington (1995) showed that active NIb RNA polymerase from tobacco etch potyvirus (TEV) expressed in transgenic *N. tabacum* cells complemented NIb-defective TEV mutants.

4. Examples of Synergism

Recently, it was shown that transgenic plants can display synergism-like characteristics. Tobacco plants expressing the 5′-terminal third of the tobacco vein mottling potyvirus or TEV genome showed symptoms resembling those of PVX/potyvirus synergism upon infection by PVX (Bowman Vance et al., 1995).

C. Risks of Transgene Spread and Introgression into Wild Crop Relatives

1. Transgene Flow and Introgression into Wild Crop Relatives

No report has been published yet on gene flow between a transgenic crop that is resistant to a bacteria, fungus, nematode, or virus and a free-living crop relative. To our knowledge, the first study on transgene escape into a free-living relative was recently published by Mikkelsen et al. (1996) who studied the introgression of a transgene conferring tolerance to the herbicide glufosinate from transgenic oilseed rape, *Brassica napus*, into the wild relative *Brassica campestris*. These two *Brassica* species are known to spontaneously hybridize in nature. Interspecific herbicide tolerant *B. napus x B. campestris* hybrid plants were grown in the field with *B. campestris*. Backcrossing occurred leading to the development of fertile plants. The progeny of the first backcross was screened for herbicide tolerance. Out of 4000 seeds from 32 interspecific hybrids, 865 herbicide-tolerant plants were randomly selected and further grown to flowering. Subsequently, out of 44 plants that showed a *B. campestris*-type morphology, 7 plants had highly fertile pollen (>90%) and a number of chromosomes similar to *B. campestris*. A second backcross was made with four of these plants showing *B. campestris* characteristics. Overall, seeds of this second backcross had a pronounced dormancy and low viability (1 to 12%). Of the 416 progenies obtained, 175 (42%) were herbicide tolerant. This study clearly showed that fertile herbicide-tolerant weed-like plants can be produced under field conditions as early as the first-backcross generation. In addition, two generations of hybridization and backcrossing were enough to recover herbicide-tolerant weeds upon herbicide selection.

In our laboratory, we have carried out experiments to assess gene flow using two squash species, the virus-resistant genetically modified squash (*Cucurbita pepo* ssp. *ovifera* var. *ovifera*) line CZW-3 (Tricoli et al., 1995) and the wild relative *Cucurbita pepo* ssp. *ovifera* var. *texana*, known as Texas gourd and referred to hereafter as *C. texana*. *C. texana* is a free-living squash species that is found in 12 states in the south-central U.S. and upper Mississippi drainage area. It can be a localized weed problem of soybean and cotton fields in the flood plains of the Red River Valley in Arkansas (Weidemann and Templeton, 1988), in some cotton fields in Mississippi (Andres, 1995), and in other areas in Arkansas, Louisiana, Illinois, and Kentucky (Wilson et al., 1994). *C. texana* can easily hybridize with cultivated squash. The ecology of *C. texana* and the development of hybrids between *C. texana* and cultivated squash is well documented (Kirkpatrick and Wilson, 1988; Wilson, 1990; Wilson et al., 1994).

Since gene flow has been demonstrated to occur naturally between cultivated nontransformed squash and *C. texana* squash and vice versa (Kirkpatrick and Wilson, 1988; Wilson, 1990), it is conceivable that CP genes from virus-resistant transgenic squash plants will escape into *C. texana* plants that grow naturally in the vicinity of commercial fields of transgenic squash plants. The chances for CP gene flow between genetically modified virus-resistant squash and *C. texana* is much higher now because transgenic squash line Freedom II, which is resistant to ZYMV and WMV 2, has been recently cleared for commercial use, and other transgenic squash lines such as CZW-3 are in the process of being commercialized.

2. Development of Hybrids between Transgenic and Free-Living Squash Plants

We developed hybrid plants between transgenic and *C. texana* squash by hand pollination in the greenhouse. For these crop/wild hybrid plants, transgenic squash were used as the staminate parent and wild squash as the pistillate parent to mimic pollen flow from transgenic to wild squash plants. The transgenic parent used was squash line CZW-3 which contains the CP genes of CMV, ZYMV, and WMV 2, and the NPT II marker gene (Tricoli et al., 1995). Given that virus resistance correlated well with the expression of the NPT II gene in transgenic line CZW-3 (Tricoli et al., 1995; Fuchs and Gonsalves, unpublished results), detecting the protein expressed by the NPT II gene by enzyme-linked immunosorbent assay (ELISA) was used as a simple, fast, and reliable way to indirectly monitor the presence of the CP genes.

3. Will CP Genes Confer Virus Resistance in Free-Living Squash Plants?

We evaluated the fitness of hybrids between transgenic squash CZW-3 and wild *C. texana* squash under field conditions of low and high disease incidence. High disease incidence was achieved by mechanically inoculating susceptible plants and transplanting them to the field sites along with the test plants. These inoculated plants served as the primary virus source for natural transmissions by aphids. Low disease pressure was achieved by transplanting only uninoculated plants to the field sites. In addition, plants that became infected by aphids early in the growing season were quickly rogued to minimize the incidence of viruses. Resistance was evaluated by visual monitoring of symptom development, ELISA, and fruit analysis.

Our data showed that the transgenic hybrids were symptomless or had local chlorotic lesions while all of the nontransgenic hybrid and *C. texana* plants became infected 33 days after transplanting and showed severe symptoms that consisted of mosaic and distortion, and plants were stunted (Figure 1). The local chlorotic lesions observed on transgenic hybrids were similar to those described for the virus-resistant transgenic parent CZW-3 (Fuchs and Gonsalves, 1993) and for transgenic squash line ZW-20 (Fuchs and Gonsalves, 1995b). Similar symptoms developed on plants in plots with low and high disease incidence, except that the rate of virus infection was lower in the former plots. A few transgenic hybrid and CZW-3 plants developed severe symptoms consisting of mosaic, but only late in the growing season (Figure 1). These symptoms could be attributed to PRSV, to which these two genotypes are not resistant. Resistance of the hybrid plants was confirmed by ELISA. Transgenic hybrid plants, as well as transgenic CZW-3 plants, did not react positive in ELISA for CMV, ZYMV, and WMV-2, whereas all nontransgenic hybrids and *C. texana* plants gave positive reactions (Table 2). A few transgenic hybrid and CZW-3 plants reacted positive in ELISA for PRSV.

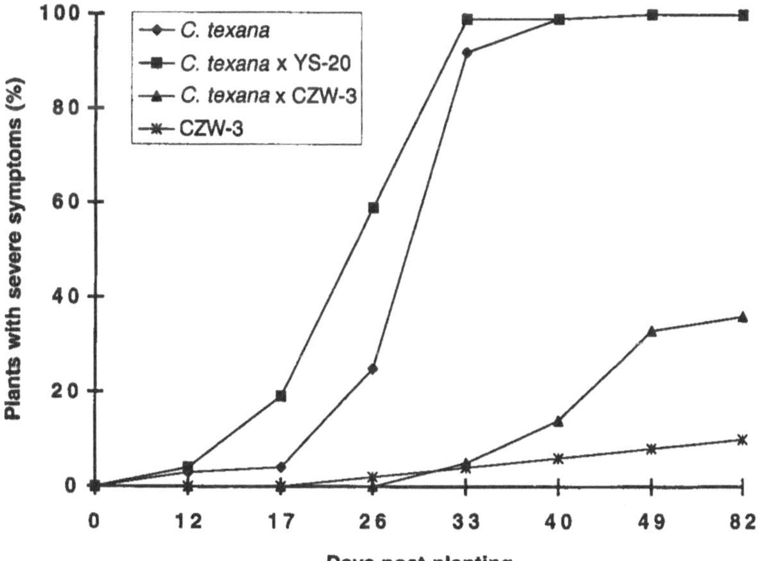

Figure 1 Disease progress curve on aphid-inoculated squash plants under field conditions of high disease incidence. The transgenic genotypes tested were transgenic *Cucurbita pepo* cv. Pavo CZW-3 expressing the CP genes of ZYMV, WMV 2, and CMV (CZW-3), and hybrids of *Cucurbita texana* x transgenic CZW-3 (C. texana x CZW-3). The nontransformed genotypes were hybrids of *C. texana* x nontransformed *C. pepo* Pavo cv. YS-20 (C. texana x YS-20), and *C. texana* (C. texana). The cumulative number of plants with severe symptoms was determined throughout the growing season for each genotype.

Table 2 **Virus Detection by ELISA on the Transgenic Squash *Cucurbita pepo* cv. Pavo CZW-3, Hybrids between *Cucurbita texana* × Transgenic CZW-3, Hybrids between *C. texana* × Nontransformed cv. YS-20, and the Wild Squash *C. texana* Evaluated in a Field Trial Conducted in 1994 Under High Disease Incidence**

		Percent ELISA Positive Plants[a]								
Genotype	**n[b]**	**+**	**C**	**Z**	**W**	**CZ**	**CW**	**ZW**	**CZW**	**P**
Nontransformed										
C. texana	138	100	3	7	9	3	2	63	5	8
C. texana × YS-20	53	100	4	6	8	0	9	58	6	9
Transgenic										
C. texana × CZW-3	42	36	0	0	0	0	0	0	0	36
CZW-3	48	19	0	0	0	0	0	0	0	19

[a] ELISA was performed on leaf samples collected 63 days post-planting (September 21). ELISA (+), CMV (C), ZYMV (Z), WMV 2 (W), and PRSV (P) positive reactions were discriminated.

[b] Number of plants tested for each genotype. Some *C. pepo* cv. Pavo plants were mechanically inoculated with ZYMV or WMV 2 on July 15, prior to their transplanting on July 20. These plants served as a virus source for secondary transmissions by aphids.

Table 3 **Fruit and Seed Analysis of Transgenic Squash *Cucurbita pepo* cv. Pavo CZW-3, Hybrids between *C. texana* and Transgenic CZW-3, Hybrids between *C. texana* and Nontransformed *C. pepo* cv. YS-20, and *C. texana* Plants Evaluated in a Field Trial Conducted in 1994 Under High Disease Incidence**

		Fruits		
Genotype	**n[a]**	**No.**	**Symptomatic (%)[b]**	**No. Viable Seeds/Plant**
Nontransformed				
C. texana	138	173	98	0
C. texana × YS-20	53	81	100	78
Transgenic				
C. texana × CZW-3	42	106	54	168
CZW-3	48	53	15	243

[a] Number of plants tested for each genotype. Squash seedlings were transplanted to the field site on July 20. Data were collected for fruits on October 10 and for seeds on October 26.

[b] Symptoms on fruits (distortions, discolorations, green dots, knobs) were visually monitored.

4. *Will CP Genes Affect the Fitness of Free-Living Squash Plants?*

As described above, transgenic hybrids between transgenic squash CZW-3 and wild *C. texana* squash showed virus resistance. Fitness was further evaluated by analyzing the fruits produced by each genotype for seed viability (Table 3). In the plots with high disease incidence, transgenic hybrid and CZW-3 plants produced many and symptomless fruits whereas the majority of the other plants had very few fruits, but almost all were symptomatic. Fruits from transgenic hybrids contained significantly more viable seeds than fruits from *C. texana*. In contrast, all genotypes, including *C. texana*, that were tested in the plots with low disease incidence, produced many healthy-appearing fruits. *C. texana* plants produced substantially more fruits than the hybrids, but smaller, and fruits of *C. texana* contained similar numbers of seeds relative to those of the fruits produced by the hybrids (Data not shown). The few fruits of the two transgenic genotypes were symptomatic due to infection by PRSV.

In a separate set of experiments designated to evaluate the introgression of the CP genes through pollen flow, virus-resistant hybrid plants were grown in the field with *C. texana* plants under conditions of low and high disease incidence. At the end of the growing season, we analyzed the progeny of *C. texana* for the presence of the CP genes by screening germinating seeds by ELISA. We found that a high number of *C. texana* seeds contained the transgenes, indicating that gene introgression occurred under field conditions of low disease pressure, but not under conditions of high disease pressure. Under high disease pressure, viruses had severely affected the development of the *C. texana* plants, impairing their flowering and fruiting capabilities.

5. Will Hybrids that Acquire CP Genes Become More Weedy and Invasive?

It is possible that virus resistance could enhance weediness and invasiveness in wild populations if the disease is widespread and endemic. Interestingly, Bartsch et al. (1996) conducted a survey of wild beet populations in the estuary of the river Po in northeastern Italy. These authors found numerous wild beet plants at six different locations, but none was infected by beet necrotic yellow vein furovirus (BNYVV). These results suggested that even if BNYVV-resistant wild beet plants were to develop, they would unlikely become more invasive. Our field trials did not address the question of weediness because *C. texana* does not occur naturally in New York. However, preliminary data suggest that *C. texana* plants are not readily infected by viruses in their natural ecosystems (Quemada, 1995). More extensive surveys of virus incidence in natural populations of *C. texana* must be done to assess the potential of viruses in controlling the evolution of these free-living plants.

V. BENEFITS OF GENETICALLY MODIFIED PLANTS RESISTANT TO VIRUSES

A. Benefits

Virus-resistant transgenic crop plants offer numerous benefits to agriculture. These benefits are economical, epidemiological, and environmental. First, such plants are particularly valuable in cases where virus-resistant cultivars are not available or where resistance genes have not been identified. In that case, engineered protection may be the only approach to develop virus-resistant plants. Second, plants that are genetically modified for virus resistance increase yield and improve crop quality. This can be critical, especially in developing countries where virus diseases can reduce the already limited food supply. Third, virus-resistant transgenic plants should reduce the application of insecticides to control vectors. This would minimize the chances of contaminating the groundwater supply. Fourth, virus-resistant transgenic plants also reduce sources of viruses for secondary spread by displaying resistance, which may reduce epidemics in neighboring crops that are not resistant. Fifth, genetic engineering has increased the potential for combining multiple virus resistance traits in a plant, something that was often difficult to achieve by conventional breeding. Lastly, the transgenes can be monitored easily with molecular tools which makes it easier to carry out gene flow survey experiments.

B. Commercialization

Freedom II is the first commercially available virus-resistant and disease-resistant transgenic crop released in the U.S. Freedom II is a summer squash line that expresses the CP genes of ZYMV and WMV 2, and is resistant to these two potyviruses. This line was experimentally evaluated under the code name ZW-20 (Tricoli et al., 1995; Fuchs and Gonsalves, 1995b). It received exemption status by USDA-APHIS in 1994 (Medley, 1994) and was released by the Asgrow Seed Company in the spring of 1995. Others squash cultivars have been subsequently developed from the ZW-20 genotype (Tricoli et al., 1995).

Transgenic tobacco, tomato, potato, and oilseed rape plants, all genetically modified for virus resistance, are currently cultivated on thousands of hectares at different locations throughout the People's Republic of China (Plafker, 1994). However, no data have been reported yet on the field performance of these plants.

C. Deregulation

Recently, transgenic summer squash line CZW-3 (Tricoli et al., 1995) and transgenic papaya lines 55-1 and 63-1 (Fitch et al., 1993; Lius et al., 1996; Tennant et al., 1994) have entered the process of deregulation. These crop plants are likely to receive exemption status for commercial release sometime in the near future. Transgenic squash line CZW-3 expresses the CP genes of CMV, ZYMV, and WMV 2 and is highly resistant to these three viruses. This line has been recently approved by APHIS-USDA for deregulation (Acord, 1996). Transgenic papaya lines 55-1 and 63-1 express the CP gene of PRSV strain HA 55-1, and are highly resistant to PRSV isolates from Hawaii. These two lines were recently exempted by APHIS-USDA (Strating, 1996). More virus-resistant agronomically significant crop plants should be deregulated in the near future.

VI. DISCUSSION: DO RISKS OUTWEIGH BENEFITS?

The development of genetically engineered crops for disease control have raised concerns on their potential risk to the environment and human health. These include (1) the toxicity of the transgene product to nontarget species, (2) the potential invasiveness of the transgenic crop variety itself, (3) the development of resistance to toxins expressed by transgenic plants especially in the case of bacteria and fungi, (4) the development of new pathogens, in particular when pathogen-derived resistance is applied, and (5) the development of weeds through the transfer of resistance to noncrop species.

As mentioned previously, RNA recombination has been raised as a concern for the release of virus-resistant genetically modified crops. The critical question is to assess whether these transgenic plants provide an enhanced opportunity for RNA recombination. It is known that plant viruses exchange genetic information during mixed infections when they replicate simultaneously in cells of nontransgenic plants. Recombination between transgenic mRNAs and infecting viruses has not yet been shown to occur in the field with agronomically significant transgenic crops, although it has been demonstrated in the laboratory with transgenic tobacco and several viruses for which recombination is well documented. So far, it does not appear from the data currently available that the widespread use of transgenic plants expressing viral sequences will increase the likelihood of recombination. There are several reasons for this. First, it appears that viral sequences involved in recombination, i.e., common untranslated 5′ and 3′ sequences of some multipartite viruses, do not provide practical levels of virus resistance. Second, plants are commonly exposed to infections by multiple viruses and viral RNA sequences. Therefore, opportunities for recombination are already widespread in infected weeds and crop plants. Third, virus-infected plants produce much higher concentrations of viral RNA than transgenic plants. Fourth, genetic variations among virus strains already occur widely in nature. In addition, RNA recombination appears to occur more frequently for some virus genera (luteo, caulimo, tobra, and potyviruses) than others (como, nepo, cucumoviruses), and computer-assisted sequence comparisons suggest that recombination may be common for viruses in the same genus but unlikely between viruses of different genera. Fifth, if recombination occurs between transgenic mRNAs and infecting viruses, it is likely to be rare and probably no more frequent than natural recombinations or mutations. Also, the likelihood of the recombinants to survive appears low.

Transencapsidation is another risk issue related to the release of virus-resistant transgenic crops expressing CP genes. Nevertheless, transencapsidation is known to occur when viruses co-infect nontransgenic plants. Therefore, the critical question is to assess whether transgenic plants will increase its occurrence. Although demonstrated to occur in the laboratory mainly with the model crop tobacco (Osbourn et al., 1990; Holt and Beachy, 1991; Candelier-Harvey and Hull, 1993; Lecoq et al., 1993), transencapsidation has not been described yet in the field with agronomically significant crops. In the case where a change in virus-vector specificity has been documented (Lecoq et al., 1993), the experimental conditions do not represent a realistic system because the two viruses involved in transencapsidation have a distinct host range. ZYMV and

PPV are unlikely to interact in nature since PPV infects stone fruit trees and ZYMV infects cucurbits, therefore these two viruses are unlikely to co-infect the same plants. Thus far, it seems that transgenic plants are unlikely to enhance the occurrence of transencapsidation. There are several reasons for this. First, the level of CP expression is often very low in transgenic plants. This is the case for Freedom II, the first virus-resistant transgenic crop commercially available (Fuchs and Gonsalves, 1995b). Second, mixed virus infections are common already in nature in crop plants and can be expected at numerous field sites where a diversity of vectors (aphids, beetles, thrips, whiteflies, leafhoppers, nematodes) occur readily or where a single vector carries many different viruses. Therefore, the chances of transencapsidation are already high in co-infected plants. Third, infected plants contain much higher amounts of viral CP than transgenic plants. Therefore, chances of transencapsidation should be lower in transgenic plants than in mixed-infected plants.

Another environmental concern raised by large-scale cultivation of transgenic crops is the movement of transgenes into related weedy species. Engineered traits, like any of the other conventional traits, can be transferred from crops to sexually compatible weed species (Mikkelsen et al., 1996). If the transgene confers a selective advantage, i.e., disease resistance, it may enhance the fitness of the recipient weed and develop a more weedy free-living species that could become invasive and eventually more difficult to control.

Some studies address risk assessment issues of genetically modified crops for particular environments. For example, Kapteijns (1993) investigated potato, beet, oilseed rape, and corn in the Netherlands and showed that only transgenic beet and oilseed rape should be highly examined because these two crops can hybridize with wild species in this country, whereas potato and corn are isolated from any related wild flora. Similarly, Raybould and Gray (1993) analyzed the probability of gene flow from crops to wild species in the U.K. These authors concluded that hybridization should be an issue for a limited number of crop species including cabbage, carrot, poplar, pine, plums, apple, ryegrass, white clover, and lucerne.

Ahl-Goy and Duesing (1996) assessed the potential of gene transfer to wild relatives in Europe and concluded that the majority (91%) of field trials conducted so far with transgenic crop species had limited, if any, impact on the environment and the remaining 9% of field trials had low potential environmental impact. However, others have reported that small-scale and confined field trials, like most of those performed so far in Europe and the U.S., contribute little to environmental risk assessment (Regal, 1994; Mellon and Rissler, 1995). Thus, commercial-scale tests have been recommended (Stone, 1994). Tiedje et al. (1989) formulated early on recommendations to develop environmentally sound products of genetic engineering. Dale (1992) analyzed the likelihood of transgene transfer from domesticated crops to wild species. As more transgenic crops are being approved for commercial release, gene flow has become an important issue (Kling, 1996). Keeler (1995) examined the characteristics of major cultivated crop plants and weed species and predicted that genetically engineered plants are unlikely to create significantly weedy plants for most crops, although Williamson (1994) mentioned that it is not yet possible to predict which plant species can become invasive in the U.K. Linder and Schmitt (1994) suggested methods to evaluate the relative fitness of crop-wild hybrids and to determine the likelihood of backcrossing. Rogers and Parkes (1995) extensively evaluated the factors that influence the dispersal and persistence of transgenes in the environment. Schmitt and Linder (1994) analyzed the potential ecological effects of transgene escape. The impact of hybrids between crop and wild species was examined by Darmency (1994). Andow (1994) suggested the use of mathematical models to identify plants that may have adverse environmental potential.

Risks have to be balanced against benefits. Scientifically based risk assessment data are important because they generate important information that can be used in making decisions and setting policies on the deployment of genetically engineered crops. A number of risk assessment studies have been performed in laboratories. However, experiments carried out in confined conditions have only limited value for assessing environmental risks. Field experiments that are under way should yield very valuable information. Interestingly, field data so far do not seem to indicate that risks outweigh benefits. In contrast, they tend to show that genetic engineering appears to be an environmentally sound approach for disease control. Looking at the impact on agriculture, virus-resistant genetically modified crops offer potential benefits like no other strategy used so far to control viral diseases.

VII. CONCLUSIONS

Remarkable progress has been made in developing disease-resistant crops through genetic engineering. Pathogen-derived resistance has proven to be a very effective approach for the development of virus-resistant transgenic plants. Field trials with virus-resistant vegetable crops such as tomato and cucurbits, and fruit crops such as papaya, conclusively show the effectiveness of genetically engineered plants at producing marketable fruits under heavy disease pressure. Numerous publications have also been reported on genetically engineered plants for resistance to bacteria, fungi, or nematodes. Although evaluation of these plants is primarily at the laboratory or greenhouse stage, we can expect reports on results of field trials in the near future.

The benefits of genetically engineered plants for disease resistance and thus for agriculture are obvious, but legitimate concerns have been expressed regarding the potential environmental risks of these plants. Assessments are being made on the risks of virus-resistant plants that can occur through the phenomena of recombination, transencapsidation, or complementation, all which could lead to enhanced virus spread, and the creation of more deleterious or even novel viruses. It has been shown in the laboratory that these phenomena can occur when the invading virus interacts with transgenes or their products. Furthermore, these phenomena have been shown to occur in nature during co-infection of nontransgenic plants by viruses. Thus, the question is their frequency of occurrence and whether their occurrence involving transgenes will prove harmful to the environment. Introgression of transgenes conferring virus resistance to free-living species is another risk that may enhance weediness, for example, of the latter.

Risk assessment studies with disease-resistant plants are relatively recent and largely limited to laboratory or confined studies. This is understandable given the fact that virus-resistant plants have been extensively field tested only in the last 7 years. APHIS/USDA, however, has determined that two virus-resistant transgenic squash lines and two virus-resistant papaya lines should have no significant impact on the environment, and thus they were exempted from regulations. These crops express virus CP genes that could be involved in recombination, transencapsidation, and complementation with invading viruses. With squash, the introgression of CP transgenes into the free-living squash does occur, but it is not deemed to be risky to the environment. Will this be a general trend for similarly created transgenic plants of other crops? Although data so far suggest that virus-resistant transgenic plants will not pose undue environmental risks, each case will be examined individually and their risks weighed against the benefits. Risk assessment experiments under field conditions are underway and it is hoped that such studies will help in striking the right balance.

In summary, genetic engineering using recombinant DNA techniques has made the development of disease-resistant transgenic plants a practical reality. We can expect many more advances in developing effective disease-resistant plants. The proper deployment of these crops will provide tremendous benefits to agriculture and humankind.

ACKNOWLEDGMENT

We thank Dr. L. M. Yepes for critically reading the manuscript.

REFERENCES

Acord, B. R., Availability of determination of nonregulated status for squash line genetically engineered for virus resistance, *Fed. Regist.*, 61, 33485, 1996.

Ahl-Goy, P. and Duesing, J. H., Assessing the environmental impact of gene transfer to wild relatives, *Bio/Technology*, 14, 39, 1996.

Alexander, D., Goodman, R. M., Gut-Rella, M., Glascock, C., Weymann, K., Friedrich, L., Maddox, D., Ahl-Goy, P., Luntz, T., Ward, E., and Ryals, J., Increased tolerance to two oomycete pathogens in transgenic tobacco expressing pathogenesis-related protein 1a, *Proc. Natl. Acad. Sci. U.S.A.*, 90, 7327, 1993.

Allefs, S. J. H. M., Florack, D. E. A., Hoogendoorn, C., and Stiekema, W. J., *Erwinia* soft rot resistance of potato cultivars transformed with a gene construct coding for antimicrobial peptide cecropin B is not altered, *Am. Potato J.*, 72, 437, 1995.

Allison, R. F., Update on virus recombination in transgenic crops, in Nat. Biol. Imp. Ass. Pro./Inf. Syst. Bio., August issue, 1996, presented at 8th Symp. on Environmental Releases of Biotechnology Products: Risk Assessment Methods and Research Progress, Ottawa, June 23 to 26, 1996.

Allison, R., Thompson, C., and Ahlquist, P., Regeneration of functional RNA virus genome by recombination between deletion mutants for cowpea chlorotic mottle virus 3a and coat protein genes for systemic infection, *Proc. Natl. Acad. Sci. U.S.A.*, 87, 1820, 1990.

Andow, D. A., Community response to transgenic plant release: using mathematical theory to predict effects of transgenic plants, *Mol. Ecol.*, 3, 65, 1994.

Andres, C. T., "Texas gourd" becomes a weed problem in Mississippi cotton fields, *Cucurbit Network News*, 2, 4, 1995.

Anzai, H., Yoneyama, K., and Yamaguchi, I., Transgenic tobacco resistant to a bacterial disease by the detoxification of a pathogenic toxin, *Mol. Gen. Genet.*, 219, 492, 1989.

Arce-Ochoa, J. P., Dainello, F., Pike, L. M., and Drews, D., Field performance comparison of two transgenic summer squash hybrids to their parental hybrid line, *HortScience*, 30, 492, 1995.

Arias, D. M. and Rieseberg, L. H., Gene flow between cultivated and wild sunflowers, *Theor. Appl. Genet.*, 89, 655, 1994.

Atkinson, H. J., Urwin, P. E., Hansen, E., and McPherson, M. J., Designs for engineered resistance to root-parasitic nematodes, *Trends Biotechnol.*, 13, 369, 1995.

Atreya, C., D., Raccah, B., and Pirone, T. P., A point mutation in the coat protein abolishes aphid transmissibility of a potyvirus, *Virology*, 178, 161, 1990.

Barker, H., Invasion of non-phloem tissue in *Nicotiana clevelandii* by potato leafroll luteovirus is enhanced in plants also infected with potato Y potyvirus, *J. Gen. Virol.*, 68, 1223, 1987.

Barker, H., Reavy, B., Kumar, A., Webster, K. D., and Mayo, M. A., Restricted virus multiplication in potatoes transformed with the coat protein gene of potato leafroll luteovirus: similarities with a type of host gene-mediated resistance, *Ann. Appl. Biol.*, 120, 55, 1992.

Barker, H., Webster, K. D., Jolly, C. A., Reavy, B., Kumar, A., and Mayo, M. A., Enhancement of resistance to potato leafroll virus multiplication in potato by combining the effects of host genes and transgenes, *Mol. Plant-Microbe Interact.*, 7, 528, 1994.

Bartsch, D., Schmidt, M., Pohl-Orf, M., Haag, C., and Schuphan, I., Competitiveness of transgenic sugar beet resistant to beet necrotic yellow vein virus and potential impact on wild beet populations, *Mol. Ecol.*, 5, 199, 1996.

Baulcombe, D., Novel strategies for engineering virus resistance in plants, *Curr. Opinion Biotechnol.*, 5, 117, 1994.

Baum, T. J., Hiatt, A., Parrott, W. A., Pratt L. H., and Hussey, R. S., Expression in tobacco of a functional monoclonal antibody specific to stylet secretions of the root-knot nematode, *Mol. Plant-Microbe Interact.*, 9, 382, 1996.

Beachy, R. N., Ed., Transgenic resistance to plant viruses, *Semin. Virol.*, 4, 327, 1993a.

Beachy, R. N., Virus resistance through expression of coat protein genes, in *Biotechnology in Plant Dis., Control*, Chet, I., Ed., Wiley-Liss, New York, 1993b, 89.

Beachy R. N., Loesch-Fries, S., and Tumer, N. E., Coat protein-mediated resistance against virus infection, *Annu. Rev. Phytopathol.*, 28, 451, 1990.

Beck, D. L. and Dawson, W. O., Deletion of repeated sequences from tobacco mosaic virus mutants with two coat protein genes, *Virology*, 177, 462, 1990.

Beffa, R., Szell, M., Meuwly, P., Pay, A., Vögeli-Lange, R., Métraux, J-P., Neuhaus, G., Meins, F., and Nagy, F., Cholera toxin elevates pathogen resistance and induces pathogenesis-related gene expression in tobacco, *EMBO J.*, 14, 5753, 1995.

Bevan, M. W., Flavell, R. B., and Chilton, M. D., A chimaeric antibiotic resistance gene as selectable marker for plant cell transformation, *Nature*, 304, 184, 1983.

Boudry, P., Morchen, M., Saumitou-Laprade, P., Vernet, P., and van Dijk, H., The origin and evolution of weed beets: consequences for the breeding and release of herbicide-resistant transgenic sugar beets, *Theor. Appl. Genet.*, 87, 471, 1993.
Bourdin, D. and Lecoq, H., Evidence that heteroencapsidation between two potyviruses is involved in aphid transmission of a non-aphid transmissible isolate from mixed infections, *Phytopathology*, 81, 1459, 1991.
Bowman Vance, V., Berger, P. H., Carrington, J. C., Hunt, A. G., and Shi, X. M., 5' Proximal potyviral sequences mediate potato virus X/potyviral synergistic disease in transgenic tobacco, *Virology*, 206, 583, 1995.
Broglie, K., Chet, I., Holliday, M., Cressman, R., Biddle, P., Knowlton, S., Mauvais, C. J., and Broglie, R., Transgenic plants with enhanced resistance to the fungal pathogen *Rhizoctonia solani*, *Science*, 254, 1194, 1991.
Brown, C. R., Smith, O. P., Damsteegt, V. D., Yang, C. P., Fox, L., and Thomas, P. E., Suppression of PLRV titer in transgenic Russet Burbank and Ranger Russet, *Am. Potato J.*, 72, 589, 1995.
Bujarski, J., J. and Kaesberg, P., Genetic recombination between RNA components of a multipartite plant virus, *Nature*, 321, 528, 1986.
Calvert, L. A. and Ghabrial, S. A., Enhancement by soybean mosaic virus of bean pod mottle virus titer in doubly infected soybean, *Phytopathology*, 73, 992, 1983.
Candelier-Harvey, P. and Hull, R., Cucumber mosaic virus genome is encapsidated in alfalfa mosaic virus coat protein expressed in transgenic tobacco plants, *Transgenic Res.*, 2, 277, 1993.
Carmona, M. J., Molina, A., Fernández, J. A., López-Fando, J. J., and Garcia-Olmedo, F., Expression of the α-thionin gene from barley in tobacco confers enhanced resistance to bacterial pathogens, *Plant J.*, 3, 457, 1993.
Cascone, P. J., Carpenter, C. D., Li, X. H., and Simon, A. E., RNA recombination between satellite RNAs of turnip crinkle virus, *EMBO J.*, 9, 1709, 1990.
Cervera, M. T., Riechmann, J. L., Martin, M. T., and Garcia, J. A., 3'-Terminal sequence of the plum pox virus PS and o6 isolates: evidence for RNA recombination within the potyvirus group, *J. Gen. Virol.*, 74, 1709, 1993.
Chen, B. and Francki, R. I. B., Cucumovirus transmission of the aphid *Myzus persicae* is determined solely by the viral coat protein, *J. Gen. Virol.*, 71, 939, 1990.
Clough, G. H. and Hamm, P. B., Coat protein transgenic resistance to watermelon mosaic and zucchini yellows mosaic virus in squash and cantaloupe, *Plant Dis.*, 79, 1107, 1995.
Cooper, J. I., Edwards, N. L., Rosenwasser, O., and Scott, N. W., Transgenic resistance genes from nepoviruses: efficacy and other properties, *N.Z. J. Crop Hortic. Sci.*, 22, 129, 1994.
Cornelissen, B. J. C. and Melchers, L. S., Strategies for control of fungal diseases with transgenic plants, *Plant Physiol.*, 101, 709, 1993.
Creamer, R. and Falk, B., W., Direct detection of transencapsidated barley yellow dwarf luteoviruses in doubly infected plants, *J. Gen. Virol.*, 71, 211, 1990.
Dale, P., J., Spread of engineered genes to wild relatives, *Plant Physiol.*, 100, 13, 1992.
Dale, P. J., The impact of hybrids between genetically modified crop plants and their related species: general considerations, *Mol. Ecol.*, 3, 31, 1994.
Damirdagh, I. S. and Ross, A. F., A marked synergistic interaction of potato viruses X and Y in inoculated leaves of tobacco, *Virology*, 31, 296, 1967.
Darmency, H., The impact of hybrids between genetically modified crop plants and their related species: introgression and weediness, *Mol. Ecol.*, 3, 37, 1994.
De la Fuente-Martínez, J. M., Mosqueda-Cano, G., Alvarez-Morales, A., and Herrera-Estrella, L., Expression of a bacterial phaseotoxin resistant ornithyl transcarbamylase in transgenic tobacco confers resistance *to Pseudomonas syringae* pv. *phaseolicola*, *Bio/Technology*, 10, 905, 1992.
del Vas, M., Ceriani, M. F., Collavita, M., Butzonich, I., and Bopp, H. E., Analysis of transgenic potato plants expressing potato leafroll virus coat protein gene, *Plant Physiol.*, 102, 174, 1993.
Deom, C. M., Oliver, M. J., and Beachy, R. N., The 30-kilodalton gene product of tobacco mosaic virus potentiates virus movement, *Science*, 237, 389, 1987.
de Zoeten, G. A., Risk assessment: do we let history repeat itself?, *Phytopathology*, 81, 585, 1991.
Doebley, J., Molecular evidence for gene flow among *Zea* species, *BioScience*, 40, 443, 1990.
Düring, K., Porsch, P., Fladung, M., and Lörz, H., Transgenic potato plants resistant to the phytopathogenic bacterium *Erwinia carotovora*, *Plant J.*, 34, 587, 1993.

Edwards, M. C., Petty, I. T. D., and Jackson, A. O., RNA recombination in the genome of barley stripe mosaic virus, *Virology*, 189, 389, 1992.

Eijlander, R. and Stiekema, W. J., Biological containment of potato *(Solanum tuberosum)*: outcrossing to the related wild species black nightshade *(Solanum nigrum)* and bittersweet *(Solanum dulcamara), Sex. Plant Reprod.*, 7, 29, 1994.

Falk, B. W. and Bruening, G., Will transgenic crops generate new viruses and new diseases?, *Science*, 263, 1395, 1994.

Fang, G. and Grumet, R., Genetic engineering of potyvirus resistance using constructs derived from the zucchini yellow mosaic virus coat protein gene, *Mol. Plant-Microbe Interact.*, 6, 358, 1993.

Farinelli, L., Malnöe, P., and Collet, G. F., Heterologous encapsidation of potato virus Y strain O (PVY^O) with the transgenic coat protein of PVY strain N (PVY^N) in *Solanum tuberosum* cv. Bintje, *Bio/Technology*, 10, 1020, 1992.

Fernandez-Cuartero, B., Burgyan, J., Aranda, M. A., Salanki, K., Moriones, E., and Garcia-Arenal, F., Increase in the relative fitness of a plant virus RNA associated with its recombinant nature, *Virology*, 203, 373, 1994.

Fitch, M. M. M., Manshardt, R. M., Gonsalves, D., Slightom, J. L., and Sanford, J. C., Virus resistant papaya plants derived from tissues bombarded with the coat protein gene of papaya ringspot virus, *Bio/Technology*, 10, 1466, 1992.

Fitchen, J. H. and Beachy, R. N., Genetically engineered protection against viruses in transgenic plants, *Annu. Rev. Microbiol.*, 47, 739, 1993.

Fraley, R. T., Rogers, S. C., Horsch, R. B., Sanders, P. R., Flick, J. S., Adams, S. P., Bittner, M. L., Brand, L. A., Fink, C. L., Fry, J. S., Galluppi, G. R., Goldberg, S. B., Hoffmann, N. L., and Woo, S. C., Expression of bacterial genes in plant cells, *Proc. Natl. Acad. Sci. U.S.A.*, 80, 4803, 1983.

Fuchs, M. and Gonsalves, D., Unpublished results, 1993.

Fuchs, M. and Gonsalves, D., Risk assessment of virus spread and gene introgression using transgenic vegetable crops expressing viral coat protein genes, *Phytopathology*, 85, 1138, 1995a.

Fuchs, M. and Gonsalves, D., Resistance of transgenic hybrid squash ZW-20 expressing the coat protein genes of zucchini yellow mosaic virus and watermelon mosaic virus 2 to mixed infections by both potyviruses, *Bio/Technology*, 13, 1466, 1995b.

Fuchs, M., Klas, F. E., and Gonsalves, D., Resistance evaluation of transgenic plants expressing virus coat protein genes and risk assessment of virus spread, *Phytopathology*, 84, 1371, 1994.

Fuchs, M., McFerson, J., Tricoli, D. M., McMaster, J. R., Deng, R. Z., Boeshore, M. L., Reynolds, J. F., Russell, P. F., Quemada, H. D., and Gonsalves, D., Cantaloupe line CZW-30 containing coat protein genes of cucumber mosaic virus, zucchini yellow mosaic virus, and watermelon mosaic virus-2 is resistant to these aphid-borne viruses in the field, *Mol. Breeding*, 1997, in press.

Fuchs, M., Provvidenti, R., Slightom, J. L., and Gonsalves D., Evaluation of transgenic tomato plants expressing the coat protein gene of cucumber mosaic virus strain WL under field condition, *Plant Dis.*, 80, 270, 1996.

Gal, S., Pisan, B., Hohn, T., Grimsley, N., and Hohn, B., Agroinfection of transgenic plants leads to viable cauliflower mosaic virus by intermolecular recombination, *Virology*, 187, 525, 1992.

Gates, P., The environmental impact of genetically engineered crops, *Biotechnol. Genet. Eng. Rev.*, 13, 181, 1995.

Gerlach, W. L., llewellyn, D., and Haseloff, J., Construction of a disease resistant gene using the satellite RNA of tobacco ringspot virus, *Nature*, 328, 802, 1987.

Gibbs, M. J. and Cooper, J. I., A recombinational event in the history of luteoviruses probably by base-pairing between the genomes of two distinct viruses, *Virology*, 206, 1129, 1995.

Glascock, C. B., Kamoun, S., Tyler, B., Ryals, J., and Alexander, D., Transgenic tobacco resistant to *Phytophtora parasitica*: analysis of expression and activity of a novel pathogenesis-related (PR) protein, SAR 8.2, *J. Cell. Biochem.*, 16F, 215, 1992.

Gliddon, C., The impact of hybrids between genetically modified crop plants and their related species: biological models and theoretical perspectives, *Mol. Ecol.*, 3, 41, 1994.

Goldberg, K. and Brakke, M. K., Concentration of maize chlorotic mottle virus increased in mixed infections with maize dwarf mosaic virus, strain B, *Phytopathology*, 77, 162, 1987.

Golemboski, D. B., Lomonossoff, G. P., and Zaitlin, M., Plants transformed with tobacco mosaic virus nonstructural gene sequence are resistant to the virus, *Proc. Natl. Acad. Sci. U.S.A.*, 87, 6311, 1990.

Gonsalves, C., Xue, B., Pang, S. Z., Provvidenti, R., Slightom, J. L., and Gonsalves, D., Breeding multiple virus-resistant transgenic tomatoes, in Proc. 6th Int. Congr. Pathology, Montreal, July 28 to August 6, 1993, 190.

Gonsalves, C., Xue, B., Yepes, M., Fuchs, M., Ling, K., Namba, S., Chee, P., Slightom, J. L., and Gonsalves, D., Transferring cucumber mosaic virus-white leaf strain coat protein gene into *Cucumis melo* L. and evaluating transgenic plants for protection against infections, *J. Am. Soc. Hortic. Sci.*, 119, 345, 1994.

Gonsalves, D., Chee, P., Provvidenti, R., Seem, R., and Slightom, J. L., Comparison of coat protein-mediated and genetically-derived resistance in cucumbers to infection by cucumber mosaic virus under filed conditions with natural challenge inoculations by vectors, *Bio/Technology*, 10, 1562, 1992.

Greene, A. E. and Allison, R. F., Recombination between viral RNA and transgenic plant transcript, *Science*, 263, 1423, 1994.

Greene, A. E. and Allison, R. F., Deletions in the 3′ untranslated region of cowpea chlorotic mottle virus transgene reduce recovery of recombinant viruses in transgenic plants, *Virology*, 225, 231, 1996.

Grison, R., Grezes-Besset, B., Schneider, M., Lucante, N., Olsen, L., Leguay, J.-J., and Toppan, A., Field tolerance to fungal pathogens in *Brassica napus* constitutively expressing a chimeric chitinase gene, *Nat. Biotechnol.*, 14, 643, 1996.

Grumet, R., Genetically engineered plant virus resistance, *HortScience*, 25, 508, 1990.

Grumet, R., Development of virus resistant plants via genetic engineering, *Plant Breeding Rev.*, 12, 47, 1994.

Hackland, A. F., Rybicki, E. P., and Thomson, J. A., Coat protein-mediated resistance in transgenic plants, *Arch. Virol.*, 139, 1, 1994.

Hain, R., Reif, H.-J., Krause, E., Langerbartels, R., Kindl, H., Vorman, B., Wiese, W., Schmelzer, E., Schreier, P. H., Stöcker, R. H., and Stenzel, K., Disease resistance results from foreign phytoalexin expression in a novel plant, *Nature*, 361, 153, 1993.

Hanley-Bowdoin, L., Elmer, J. S., and Rogers, S. G., Expression of a functional replication protein from tomato golden mosaic virus in transgenic tobacco plants, *Proc. Natl. Acad. Sci. U.S.A.*, 87, 1446, 1990.

Hanley-Bowdoin, L. and Hemenway, C., Expression of plant viral genes in transgenic plants, in *Genetic Engineering with Plant Viruses*, CRC Press, Boca Raton, FL, 1992, 251.

Harms, C. T., Engineering genetic disease resistance into crops: biotechnological approaches to crop protection, *Crop Prot.*, 11, 291, 1992.

Harrison, B. D., Mayo, M. A., and Baulcombe, D. C., Virus resistance in transgenic plants that express cucumber mosaic virus satellite RNA, *Nature*, 328, 799, 1987.

Harrison, B. D. and Murant, A. F., Involvement of virus-coded proteins in transmission of plant viruses by vectors, in *Vectors in Virus Biology*, Mayo, M. A. and Harrap, K. A., Eds., Academic Press, London, 1984, 1.

Hayakawa, T., Zhu, Y., Itoh, K., Kimura, Y., Izawa, T., Shimamoto, K., and Toriyama, S., Genetically engineered rice resistant to rice stripe virus, an insect-transmitted virus, *Proc. Natl. Acad. Sci. U.S.A.*, 89, 9865, 1992.

Herbers, K., Meuwly, P., Frommer, W. B., Métraux, J. P., and Sonnewald, U., Systemic acquired resistance mediated by the ectopic expression of invertase: possible hexose sensing in the secretory pathway, *Plant Cell*, 8, 793, 1996.

Herdt, R. W., Research priorities for rice biotechnology, in *Rice Biotechnology*, Khush, G. S. and Toenissen, G. H., Eds., CAB International, Wallingford, U.K., 1991, 19.

Herrera-Estrella, L., Depicker, A., van Montagu, M., and Schell, J., Expression of chimaeric genes transferred into plant cells using a Ti-plasmid-derived vector, *Nature*, 303, 209, 1983.

Herrera-Estrella, L. and Simpson, L., Genetically engineered resistance to bacterial and fungal pathogens, *W. J. Microbiol. Biotechnol.*, 11, 383, 1995.

Hightower, R., Baden, C., Penzes, E., and Dunsmuir, P., The expression of cecropin peptide in transgenic tobacco does not confer resistance to *Pseudomonas syringae* pv *tabaci*, *Plant Cell Rep.*, 13, 295, 1994.

Hill, K. K., Jarvis-Egan, N., Halk, E. L., Krahn, K. J., Liao, L. W., Mathewson, R. S., Merlo, D. J., Nelson, S. E., Rashka, K. E., and Loesch-Fries, L., The development of virus-resistant alfalfa, *Medicago sativa* L., *Bio/Technology*, 9, 373, 1991.

Hong, Y., Saunders, K., Hartley, M. R., and Stanley, J., Resistance to geminivirus infection by virus-induced expression of dianthin in transgenic plants, *Virology*, 220, 119, 1996.

Holt, C. A. and Beachy, R. N., *In vivo* complementation of infectious transcripts from mutant tobacco mosaic virus cDNAs in transgenic plants, *Virology*, 181, 109, 1991.

Howell, S. H., Walker, L. L., and Walden, R. M., Rescue of *in vitro* generated mutants of cauliflower mosaic virus genome in infected plants, *Nature*, 293, 483, 1981.

Hu, J. S., Rochow, W. F., Palukaitis, P., and Dietert, R. R., Phenotypic mixing: mechanism of dependent transmission for two related isolates of barley yellow mosaic virus, *Phytopathology*, 78, 1326, 1988.

Huisman, M. J., Cornelissen, B. J. C., Groenendijk, C. F. M., Bol, J. F., and Van Vlotten-Dotting, L., Alfalfa mosaic virus temperature-sensitive mutants. V. The nucleotide sequence of TBTS 7 RNA3 shows limited nucleotide changes and evidence for heterologous recombination, *Virology*, 171, 409, 1989.

Hull, R., Non-conventional resistance to viruses in plants. Concepts and risks, in *Proc. Symp. 19th Stadler Genetics*, Gustafson, J. P., Ed., Columbia, 1989, 289.

Hull, R., The use and misuse of viruses in cloning and expression in plants, in *Recognition and Response in Plant-Virus Interactions*, Fraser, R. S. S., Ed., Springer Verlag, Berlin, 1990, 443.

Hull, R., Resistance to plant viruses: obtaining genes by non-conventional approaches, *Euphytica*, 75, 195, 1994.

Hull, R. and Davies, J. W., Approaches to nonconventional control of plant virus diseases, *Crit. Rev. Plant Sci.*, 11, 17, 1992.

Jach, G., Görnhardt, B., Mundy, J., Logemann, J., Pinsdorf, E., Leah, R., Schell, J., and Maas, C., Enhanced quantitative resistance against fungal disease by combinatorial expression of different barley antifungal proteins in transgenic tobacco, *Plant J.*, 8, 97, 1995.

Jongedijk, E., de Schutter, A. A. J. M., Stolte, T., van den Elzen, P. J. M., and Cornelissen, B. J. C., Increased resistance to potato virus X and preservation of cultivar properties in transgenic potato under field conditions, *Bio/Technology*, 10, 422, 1992.

Kaniewski, W., Lawson, C., Sammons, B., Haley, L., Hart, J., Delannay, X., and Tumer, N. E., Field resistance of transgenic Russet Burbank potato to effects of infection by potato virus X and potato virus Y, *Bio/Technology*, 8, 750, 1990.

Kaplan, I. B., Shintaku, M. H., Li, Q., Zhang, L., Marsh, L., and Palukaitis, P., Complementation of virus movement in transgenic tobacco expressing the cucumber mosaic virus 3a gene, *Virology*, 209, 188, 1995.

Kapteijns, A. J. A. M., Risk assessment of genetically modified crops. Potential of four arable crops to hybridize with the wild flora, *Euphytica*, 66, 145, 1993.

Kareiva, P., Morris, W., and Jacobi, C. M., Studying and managing the risk of cross-fertilization between transgenic crops and wild relatives, *Mol. Ecol.*, 3, 15, 1994.

Kawchuk, L. M., Martin, R. R., and McPherson, J., Resistance in transgenic potato expressing the potato leafroll virus coat protein gene, *Mol. Plant-Microbe Interact.*, 3, 301, 1990.

Kawchuk, L. M., Martin, R. R., and McPherson, J., Sense and antisense RNA-mediated resistance to potato leafroll virus in Russet Burbank potato plants, *Mol. Plant-Microbe Interact.*, 4, 247, 1991.

Keeler, K. H., Can genetically engineered crops become weeds, *Bio/Technology*, 7, 1134, 1995.

Kim, J. W., Sun, S. S. M., and German, T. L., Disease resistance in tobacco and tomato plants transformed with the tomato spotted wilt virus nucleocapsid gene, *Plant Dis.*, 78, 615, 1994.

Kim, S. J., Paek, K. H., and Kim, B. D., Delay of disease development in transgenic petunia plants expressing cucumber mosaic virus I17N-satellite RNA, *J. Am. Soc. Hortic. Sci.*, 120, 353, 1995.

Kinal, H., Park, C.-M., Berry, J. O., Koltin, Y., and Bruenn J. A., Processing and secretion of a virally encoded antifungal toxin in transgenic tobacco plants: evidence for a Kex2p pathway in plants, *Plant Cell*, 7, 677, 1995.

Kirkpatrick, K. J. and Wilson, H., Interspecific gene flow in *Cucurbita*: *C. texana* vs. *C. pepo*, *Am. J. Bot.*, 75, 519, 1988.

Klas, F. E., Fuchs, M., and Gonsalves, D., Spatial analysis as a tool to evaluate virus resistance in a transgenic crop, *Phytopathology*, 84, 1372, 1994.

Klein, T. M., Wolf, E. D., and Sanford, J. C., High velocity microprojectiles for delivering nucleic acids into living cells, *Nature*, 327, 70, 1987.

Kling, J., Could transgenic supercrops one day breed superweeds?, Science, 274, 180, 1996.

Kollar, A., Dalmay, T., and Burgyan, J., Defective interfering RNA mediated resistance against cymbidium ringspot tombusvirus in transgenic plants, *Virology*, 193, 313, 1993.

Kunik, T., Salomon, R., Zamir, D., Navot, N., Zeidan, M., Michelson, I., Gafni, Y., and Czosnek, H., Transgenic tomato plants expressing the tomato yellow leaf curl virus capsid protein are resistant to the virus, *Bio/Technology*, 12, 500, 1994.

Lai, M. M. C., RNA recombination in animal and plant viruses, *Microbiol. Rev.*, 56, 61, 1992.

Lam, Y. H., Wong, Y. S., Wang, B., Wong, R. N. S., Yeung, H. W., and Shaw, P. C., Use of trichosanthin to reduce infection by turnip mosaic virus, *Plant Sci.*, 114, 111, 1996.

Langevin, S. A., Clay, K., and Grace, J. B., The incidence and effects of hybridization between cultivated rice and its related weed red rice, *Evolution*, 44, 1000, 1990.

Lapidot, M., Gafny, R., Ding, B., Wolf, S., Lucas, W. J., and Beachy, R. N., A dysfunctional movement protein of tobacco mosaic virus that partially modifies the plasmodesmata and limits virus spread in transgenic plants, *Plant J.*, 4, 959, 1993.

Lawson, C., Kaniewski, W., Haley, L., Rozmann, R., Newall, C., Sanders, P., and Tumer, N. E., Engineered resistance to mixed virus infection in a potato cultivar: resistance to potato virus X and potato virus Y in transgenic Russet Burbank, *Bio/Technology*, 8, 127, 1990.

Lebeurier, G., Hirth, L., Hohn, B., and Hohn, T., *In vivo* recombination of cauliflower mosaic virus DNA, *Proc. Natl. Acad. Sci. U.S.A.*, 79, 1932, 1982.

Lecoq, H., Ravelonandro, M., Wipf-Scheibel, C., Monsion, M., Raccah, B., and Dunez, J., Aphid transmission of a non-aphid transmissible strain of zucchini yellow mosaic virus from transgenic plants expressing the capsid protein of plum pox potyvirus, *Mol. Plant-Microbe Interact.*, 6, 403, 1993.

Le Gall, O., Candresse, O., and Dunez, J., Transfer of the 3′ non-translated region of grapevine chrome mosaic virus RNA-1 by recombination to tomato black ring virus RNA-2 in pseudorecombinant isolates, *J. Gen. Virol.*, 76, 1285, 1995.

Li, X. H. and Carrington, J. C., Complementation of tobacco etch potyvirus mutants by active RNA polymerase expressed in transgenic cells, *Proc. Natl. Acad. Sci. U.S.A.*, 92, 457, 1995.

Lin, W., Anuratha, C. S., Datta, K., Potrykus, I., Muthukrishnan, S., and Datta, S. K., Genetic engineering of rice for resistance to sheath blight, *Bio/Technology*, 13, 686, 1995.

Linder, C. R. and Schmitt, J., Assessing the risks of transgene escape through time and crop-wild hybrid persistence, *Mol. Ecol.*, 3, 23, 1994.

Liu, D., Kashchandra, G., Hasegawa, P. M., and Bressan, R. A., Osmotin overexpression in potato delays development of disease symptoms, *Proc. Natl. Acad. Sci. U.S.A.*, 91, 1888, 1994.

Lius, S., Manshardt, R. M., Fitch, M. M. M., Slightom, J. L., Sanford, J. C., and Gonsalves, D., Pathogen-derived resistance provides papaya with effective protection against papaya ringspot virus, *Mol. Breeding*, 1997, in press.

Lodge, J. K., Kaniewski, W. K., and Tumer, N. E., Broad-spectrum virus resistance in transgenic plants expressing pokeweed antiviral protein, *Proc. Natl. Acad. Sci. U.S.A.*, 90, 7089, 1993.

Logemann, J., Jach, G., Tommerup, H., Mundy, J., and Schell, J., Expression of a barley ribosome-inactivating protein leads to increased fungal protection in transgenic tobacco plants, *Bio/Technology*, 10, 305, 1992.

Lommel, S. A. and Xiong, Z., Reconstitution of a functional red clover necrotic mosaic virus by recombinational rescue of the cell-to-cell movement gene expressed in a transgenic plant, *J. Cell. Biochem.*, 15A, 151, 1991.

Lomonossoff, G. P., Pathogen-derived resistance to plant viruses, *Annu. Rev. Phytopathol.*, 33, 323, 1995.

MacKenzie, D. J., Tremaine, J. H., and McPherson, J., Genetically engineered resistance to potato virus S in potato cultivar Russet Burbank, *Mol. Plant-Microbe Interact.*, 4, 95, 1991.

Maiti, I. B., Murphy, J. F., Shaw, J. G., and Hunt, A. G., Plants that express a potyvirus proteinase gene are resistant to virus infection, *Proc. Natl. Acad. Sci. U.S.A.*, 90, 6110, 1993.

Malnoë, P., Farinelli, L., Collet, G. F., and Reust, W., Small-scale field tests with transgenic potato, cv. Bintje, to test resistance to primary and secondary infections with potato virus Y, *Plant Mol. Biol.*, 25, 963, 1994.

Malyshenko, S. I., Kondakova, O. A., Nazarova, J. V., Kaplan, I. B., Taliansky, M. E., and Atabekov, J. G., Reduction of tobacco mosaic virus accumulation in transgenic plants producing non functional viral transport proteins, *J. Gen. Virol.*, 74, 1149, 1993.

Malyshenko, S. I., Kondakova, O. A., Taliansky, M. E., and Atabekov, J. G., Plant virus transport function: complementation by helper viruses is nonspecific, *J. Gen. Virol.*, 70, 2751, 1989.

Martin, G. B., Brommenschenkel, S. H., Chunwongse, J., Frary, A., Ganal, M. W., Spivey, R., Wu, T., Earle, E. D., and Tanksley, S. D., Map-based cloning of a protein kinase gene conferring disease resistance in tomato, *Science*, 262, 1432, 1993.

Medley, T. L., Availability of determination of nonregulated status for virus resistant squash, *Fed. Regist.*, 59, 64187, 1994.

Mellon, M. and Rissler, J., Transgenic crops: USDA data on small-scale tests contribute little to commercial risk assessment, *Bio/Technology*, 13, 96, 1995.

Michelmore, R., Molecular approaches to manipulation of disease resistance genes, *Annu. Rev. Phytopathol.*, 15, 393, 1995.

Mikkelsen, R. T., Andersen, B., and Jorgensen, R. B., The risk of crop transgene spread, *Nature*, 380, 31, 1996.

Milne, R. G., The economic impact of filamentous viruses, in *The Plant Viruses*, Milne, R. G., Ed., Plenum Press, New York, 1987, 331.

Mitra, A., Higgins, D. W., Langenberg, W. G., Nie, H., Sengupta, D. N., and Silverman, R. H., A mammalian 2-5 A system functions as an antiviral pathway in transgenic plants, *Proc. Natl. Acad. Sci. U.S.A.*, 93, 6780, 1996.

Mittler, R., Shulaev, V., and Lam E., Coordinated activation of programd cell death and defense mechanisms in transgenic tobacco plants expressing a bacterial proton pump, *Plant Cell*, 7, 29, 1995.

Motoyoshi, F. and Ugaki, M., Production of transgenic tomato plants with specific TMV resistance, *Jpn. Agric. Res. Q.*, 27, 122, 1993.

Murant, A. F., Raccah, B., and Pirone, T. P., Transmission by vectors, in *The Plant Viruses*, Milne, R. G., Ed., Plenum Press, New York, 1988, 237.

Murray, L. E., Elliott, L. G., Capitant, S. A., West, J. A., Hanson, K. K., Scarafia, L., Johnston, S., DeLuca-Flaherty, C., Nichols, S., Cunanan, D., Dietrich, P. S., Mettler, I. J., Dewald, S., Warnick, D. A., Rhodes, C., Sinibaldi, R. M., and Brunke, K. J., Transgenic corn plants expressing MDMV strain B coat protein are resistant to mixed infections of maize dwarf mosaic virus and maize chlorotic mottle virus, *Bio/Technology*, 11, 1559, 1993.

Nakamura, S., Honkura, R., Ugaki, M., Ohshima, M., and Ohashi, Y., Resistance to viral infection in transgenic plants expressing ribozymes designed against cucumber mosaic virus RNA3, *Ann. Phytopathol. Soc. Jpn.*, 61, 53, 1995.

Nakamura, S., Iwai, T., Honkura, R., Ugaki, M., and Ohasho, Y., Degradation of transgene RNA in virus-resistant transgenic plants expressing catalytic antisense RNA against virus genome, *Ann. Phtytopathol. Soc. Jpn.*, 62, 156, 1996.

Nelson, R. S., McCormick, S. M., Delannay, X., Dube, P., Layton, J., Anderson, E. J., Kaniewska, M., Proksch, R. K., Horsch, R. B., Rogers, S. G., Fraley, R. T., and Beachy, R. N., Virus tolerance, plant growth, and field performance of transgenic tomato plants expressing coat protein from tobacco mosaic virus, *Bio/Technology*, 6, 403, 1988.

Nelson, R. S., Roth, D. A., and Johnson, J. D., Tobacco mosaic virus infection of transgenic *Nicotiana tabacum* plants is inhibited by antisense constructs directed at the 5′ region of viral RNA, *Gene*, 127, 227, 1993.

Neuhaus, J.-M., Ahl-Goy, P., Hinz, U., Flores, S., and Meins, F., High-level expression of a tobacco chitinase gene in *Nicotiana sylvestris*. Susceptibility of transgenic plants to *Cercospora nicotianae* infection, *Plant Mol. Biol.*, 16, 141, 1991.

Norelli, J. L., Aldwinckle, H. S., Destéfano-Beltrán, L., and Jaynes, J. M., Transgenic 'Malling 26' apple expressing the attacin E gene has increased resistance to *Erwinia amylovora*, *Euphytica*, 77, 123, 1994.

Ogawa, T., Hori, T., and Ishida, I., Virus-induced cell death in plants expressing the mammalian 2′,5′ oligoadenylate system, *Nature Biotechnology*, 14, 1566, 1996.

Ogawa, T., Watanabe, Y., Meshi, T., and Okada, Y., Transcomplementation of virus-encoded replicase components of tobacco mosaic virus, *Virology*, 185, 580, 1991.

Opperman, C. H. and Conkling, M. A., Field evaluations of transgenic root-knot nematode resistant tobacco, *J. Nematol.*, 27, 513, 1995.

Opperman, C. H. and Conkling, M. A., Genetic engineering resistance to root-knot nematodes, *Can. J. Plant Pathol.*, 18, 97, 1996.

Osbourn, J. K., Sarkar, S., and Wilson, T. M. A., Complementation of coat protein-defective TMV mutants in transgenic tobacco plants expressing TMV coat protein, *Virology*, 179, 921, 1990.
Pang, S. Z., Jan, F. J., Carney, K., Stout, J., Tricoli, D. M., Quemada, H. D., and Gonsalves, D., Post-transcriptional transgene silencing and consequent tospovirus resistance in transgenic lettuce are affected by transgene dosage and plant development, *Plant J.*, 9, 899, 1996.
Pang, S. Z., Jan, F. J., Tricoli, D. M., Hu, J. S., Fuchs, M., Quemada, H. D., and Gonsalves, D., Unpublished results, 1995.
Park, C-M., Berry, J. O., and Bruenn, J. A., High-level secretion of a virally encoded anti-fungal toxin in transgenic tobacco plants, *Plant Mol. Biol.*, 20, 359, 1996.
Perry, K. L. and Francki, R. I. B., Insect-mediated transmission of mixed and reassorted cucumovirus genomic RNAs, *J. Gen. Virol.*, 73, 2105, 1992.
Pirone, T. P., Viral genes and gene product that determine insect transmissibility, *Semin. Virol.*, 2, 81, 1991.
Plafker, T., First biotech safety rules don't deter Chinese efforts, *Science*, 266, 966, 1994.
Poolpol, P. and Inouye, T., Enhancement of cucumber mosaic virus multiplication by zucchini yellow mosaic virus in doubly infected cucumber plants, *Ann. Phytopathol. Soc. Jpn.*, 52, 22, 1986.
Powell-Abel, P., Nelson, R. S., De, B., Hoffmann, N., Rogers, S. G., Fraley, R. T., and Beachy, R. N., Delay of disease development in transgenic plants that express the tobacco mosaic virus coat protein gene, *Science*, 232, 738, 1986.
Presting, G. G., Smith, O. P., and Brown, C. R., Resistance to potato leafroll virus in potato plants transformed with the coat protein gene or with vector control constructs, *Phytopathology*, 85, 436, 1995.
Provvidenti, R. and Gonsalves, D., Inheritance of resistance to cucumber mosaic virus in a transgenic tomato line with the coat protein gene of the white leaf strain, *J. Hered.*, 86, 85, 1995.
Punja, Z. K. and Zhang, Y.-Y., Plant chitinases and their role in resistance to fungal diseases, *J. Nematol.*, 25, 526, 1993.
Quemada, H. D., Personal communication, 1995.
Quiros, C. F., Ortega, R., Van Raansdonk, L., Herrera-Montoya, M., Cisneros, P., Schmidt, E., and Brush, S. B., Increase of potato genetic resources in their center of diversity: the role of natural outcrossing and selection by the Andean farmer, *Genet. Resour. Crop Evol.*, 39, 107, 1992.
Raski, D. J., Goheen, A. C., Lider, L. A., and Meredith, C. P., Strategies against grapevine fanleaf virus and its nematode vector, *Plant Dis.*, 67, 335, 1993.
Rathjen, J. P., Karageorgos, L. E., Habili, N., Waterhouse, P. M., and Symons, R. H., Soybean dwarf luteovirus contains the third variant genome type in the luteovirus group, *Virology*, 198, 671, 1994.
Raybould, A. F., Goudet, J., Mogg, R. J., Gliddon, C. J., and Gray, A. J., Genetic structure of a linear population of *Beta vulgaris* ssp. maritima (sea beet) revealed by isozyme and RFLP analysis, *Heredity*, 76, 111, 1996.
Raybould, A. F. and Gray, A. J., Genetically modified crops and hybridization with wild relatives: a U.K. perspective, *J. Appl. Ecol.*, 30, 199, 1993.
Regal, P. J., Scientific principles for ecologically based risk assessment of transgenic organisms, *Mol. Ecol.*, 3, 5, 1994.
Reinmann-Philipp, U. and Beachy, R. N., Plant resistance to virus diseases through genetic engineering: can a similar approach control plant-parasitic nematodes?, *J. Nematol.*, 25, 541, 1993.
Revers, F., Le Gall., O., Candresse, T., Le Romancer, M., and Dunez, J., Frequent occurrence of recombinant potyvirus isolates, *J. Gen. Virol.*, 77, 1953, 1996.
Rissler, J. and Mellon M., Perils admidst the promise. Ecological risks of transgenic crops in a global market, *Union of Concern Scientists*, 92, 1993.
Robinson, D. J., Hamilton, W. D. O., Harrison, B. D., and Baulcombe, D. C., Two anomalous tobravirus isolates: evidence for RNA recombination in nature, *J. Gen. Virol.*, 70, 935, 1987.
Rochow, W. F. and Ross, A. F., Virus multiplication in plants doubly infected by potato viruses X and Y, *Virology*, 1, 10, 1955.
Rogers, H. J. and Parkes, H. C., Transgenic plants and the environment, *J. Exp. Bot.*, 46, 467, 1995.
Sanders, P. R., Sammons, B., Kaniewski, W., Haley, L., Layton, J., Lavallee, B. J., Delannay, X., and Tumer, N. E., Field resistance of transgenic tomatoes expressing the tobacco mosaic virus or tomato mosaic virus coat protein genes, *Phytopathology*, 82, 683, 1992.
Sanford, J. C. and Johnston, S. A., The concept of parasite-derived resistance — Deriving resistance genes from the parasite's own genome, *J. Theor. Biol.*, 113, 395, 1985.

Schmitt, J. and Linder, C. R., Will escaped transgenes lead to ecological release?, *Mol. Ecol.*, 3, 71, 1994.

Schoelz, J. E., Goldberg, K. B., and Kiernan, J., Expression of cauliflower mosaic virus (CaMV) gene VI in transgenic *Nicotiana bigelovii* complements a strain of CaMV defective in long distance movement in nontransformed *N. bigelovii*, *Mol. Plant-Microbe Interact.*, 4, 350, 1991.

Schoelz, J. E. and Wintermantel, W. M., Expansion of viral host range through complementation and recombination in transgenic plants, *Plant Cell*, 5, 1669, 1993.

Scholthof, K. B., Scholthof, H. B., and Jackson, A. O., Control of plant virus diseases by pathogen-derived resistance in transgenic plants, *Plant Physiol.*, 102, 7, 1993.

Scorza R. and Ravelonandro, M., Personal communication, 1996.

Shah, D. M., Rommens, C. M. T., and Beachy, R. N., Resistance to diseases and insects in transgenic plants: progress and applications to agriculture, *Trends Biotechnol.*, 13, 362, 1995.

Simon, A. E. and Bujarski, J. J., RNA-RNA recombination and evolution in virus-infected plants, *Annu. Rev. Phytopathol.*, 32, 337, 1994.

Stanley, J., Frischmuth, T., and Ellwood, S., Defective viral DNA ameliorates symptoms of geminivirus infection in transgenic plants, *Proc. Natl. Acad. Sci. U.S.A.*, 87, 6291, 1990.

Stone, R., Large scale are next test for transgenic crop safety, *Science*, 266, 1472, 1994.

Strating, A., Availability of determination of nonregulated status for papaya lines genetically engineered for virus resistance, *Fed. Regist.*, 61, 48663, 1996.

Strittmater, G., Janssens, J., Opsomer, C., and Botterman, J., Inhibition of fungal disease development in plants by engineering controlled cell death, *Bio/Technology*, 13, 1085, 1996.

Sturtevant, A. P. and Beachy, R. N., Virus resistance in transgenic plants: coat protein -mediated resistance, in *Transgenic Plants: Fundamentals and Applications*, Hiatt, A., Ed., Marcel Dekker, New York, 1993, 93.

Tacke, E., Salamini, F., and Rohde, W., Genetic engineering of potato for broad-spectrum protection against virus infection, *Nature Biotechnology*, 14, 1597, 1996.

Tada, T., Kanzaki, H., Uchimiya, H., and Nakamura, I., Decreased symptoms of rice blast disease on leaves of *bar*-expressing transgenic rice plants following treatment with bialaphos, *Mol. Plant-Microbe Interact.*, 9, 758, 1996.

Talianski, M. E., de Jager, C. P., Wellink, J., van Lent, J. W. M., and Goldbach, R. W., Defective cell-to-cell movement of cowpea mosaic virus mutant N123 is efficiently complemented by sunn-hemp mosaic virus, *J. Gen. Virol.*, 74, 1895, 1993.

Tavladoraki, P., Benvenuto, E., Rinca, S., De Martinis, D., Cattaneo, A., and Galeffi, P., Transgenic plants expressing a functional single chain Fv antibody are specifically protected from virus attack, *Nature*, 366, 469, 1993.

Tennant, P. F., Gonsalves, C., Ling, K. S., Fitch, M., Manshardt, R., Slightom, J. L., and Gonsalves, D., Coat protein gene transgenic papaya and classically cross protected papaya show differential protection against papaya ringspot virus isolates, *Phytopathology*, 84, 1359, 1994.

Tepfer, M., Viral genes and transgenic plants. What are the potential environmental risks?, *Bio/Technology*, 11, 1125, 1993.

Terras, F. R., Eggermont, K., Kovaleva, V., Raikhel, N. V., Osbron, R. W., Kester, A., Rees, S. B., Torrekens, S., Van Leuven, F., Vanderleyden, J., Cammue, B. P. A., and Broekaret, W. F., Small cystein-rich antifungal proteins from radish: their role in host defense, *Plant Cell*, 7, 573, 1995.

Thacker, J. R. M., Transgenic crop plants and pest control, *Sci. Prog.*, 77, 207,1993.

Tiedje, J. M., Colwell, R. K., Grossman, Y. L., Hodson, R. E., Lenski, R. E., Mack, R. N., and Regal, P. J., The planned introduction of genetically engineered organisms: ecological considerations and recommendations, *Ecology*, 70, 298, 1989.

Tigelaar, H., Stuiver, M. H., Molendijk, L., Troost-van Deventer, E., Sela-Buurlage, M. B., Plooster, L., Sijbolts, F., Custers, J., Apotheker-de-Groot, M., and Melchers, L. S., Broad spectrum fungal resistance in transgenic carrot plants, presented at the 88th APS Meeting, Indianapolis, July 27 to 31, 1996.

Tricoli, D., Carney, K. J., Russell, P. F., McMaster, J. R., Groff, D. W., Hadden, K. C., Himmell, P. T., Hubbard, J. P., Boeshore, M. L., Reynolds, J. F., and Quemada, H. D., Field evaluation of transgenic squash containing single or multiple virus coat protein gene constructs for resistance to cucumber mosaic virus, watermelon mosaic virus 2, and zucchini yellow mosaic virus, *Bio/Technology*, 13, 1458, 1995.

Truve, E., Aaspollu, A., Honkanen, J., Puska, R., Mehto, M., Hassi, A., Teeri, T. H., Kelve, M., Seppänen, P., and Saarma, M., Transgenic potato plants expressing mammalian 2'-5' oligoadenylate synthetase are protected from potato virus X infection under field conditions, *Bio/Technology*, 11, 1048, 1993.

Uchimiya, H., Iwata, M., Nojiri, C., Samarajeewa, P. K., Takamatsu, S., Ooba, S., Amzai, H., Christensen, A. H., Quail, P. H., and Toki, S., Bailiffs treatment of transgenic rice expressing a *bar* gene prevents infection by the sheath blight pathogen *(Rhizoctonia solani)*, *Bio/Technology*, 11, 835, 1993.

Ultzen, T., Gielen, J., Venema, F., Westerbroek, A., De Haan, P., Tan, M., Schram, A., Van Grinsven, M., and Goldbach, R., Resistance to tomato spotted wilt virus in transgenic tomato hybrids, *Euphytica*, 85, 159, 1995.

Urwin, P. E., Atkinson, H. J., Waller, D. A., and McPherson, M. J., Engineered oryzacystatin-I expressed in transgenic hairy roots confers resistance to *Globodera pallida*, *Plant J.*, 8, 121, 1995.

Valkonen, Koivu, K., Slack, S. A., and Pehu, E., Modified resistance of *Solanum brevidens* to potato Y potyvirus and tobacco mosaic tobamovirus following genetic transformation and explant regeneration, *Plant Sci.*, 106, 71, 1995.

van den Elzen, P. J. M., Jongedijk, E., Melchers, L. S., and Cornelissen, B. J. C., Virus and fungal resistance: from laboratory to field, *Philos. Trans. R. Soc. London B*, 342, 271, 1993.

van der Kuyl, A. C., Neeleman, L., and Bol, J. F., Complementation and recombination between alfalfa mosaic virus RNA3 mutants in tobacco plants, *Virology*, 183, 731, 1991.

van der Wilk, F., Willink, D. P. L., Huisman, M. J., Huttinga, H., and Goldbach, R., Expression of the potato leafroll luteovirus coat protein gene in transgenic potato plants inhibits viral infection, *Plant Mol. Biol.*, 17, 431, 1991.

van Dun, C. M. P., van Vloten-Doting, L., and Bol, J. F., Expression of alfalfa mosaic virus cDNA1 and 2 in transgenic tobacco plants, *Virology*, 163, 572, 1988.

Vardi, E., Sela, I., Edelbaum, O., Livneh, O., Kuznetsova, L., and Stram, Y., Plants transformed with a cistron of potato virus Y protease (Nia) are resistant to virus infection, *Proc. Natl. Acad. Sci. U.S.A.*, 90, 7513, 1993.

Vierheilig, H., Alt, M., Neuhaus, J.-M., Boller, T., and Wiemken, A., Colonization of transgenic *Nicotiana sylvestris* plants, expressing different forms of *Nicotiana tabacum* chitinase, by the root pathogen *Rhizoctonia solani* and by the mycorrhizal symbiont *Glomus mosseae*, *Mol. Plant-Microbe Interact.*, 6, 261, 1993.

Voss, A., Niersbach, M., Hain, R., Hirsch, H. J., Liao, Y. C., Kreuzaler, F., and Fischer, R., Reduced virus infectivity in *N. tabacum* secreting a TMV-specific full-size antibody, *Mol. Breeding*, 1, 39, 1995.

Walden, R. and Wingender, R., Gene-transfer and plant-regeneration techniques, *Trends Biotechnol.*, 13, 324, 1995.

Weidemann, G. J. and Templeton, G. E., Efficacy and soil persistence of *Fusarium solani* f. sp. *cucurbitae* for control of Texas gourd *(Cucurbita texana)*, *Plant Dis.*, 72, 36, 1988.

Wen, F. and Lister, R., M., Heterologous encapsidation in mixed infections among four isolates of barley yellow dwarf virus, *J. Gen. Virol.*, 72, 2217, 1991.

White, K. A. and Morris, T. J., Recombination between defective tombusvirus RNAs generates functional hybrid genomes, *Proc. Natl. Acad. Sci. U.S.A.*, 91, 3642, 1994.

Wijnheijmer, E. H. M., Brandenburg, W. A., and Ter Borg, S. J., Interactions between wild and cultivated carrots *(Daucus carota* L.) in the Netherlands, *Euphytica*, 40, 147, 1989.

Williamson, M., Community response to transgenic plant release: predictions from British experience of invasive plants and feral crop plants, *Mol. Ecol.*, 3, 75, 1994.

Wilson, H. D., Gene flow in squash species, *BioScience*, 40, 449, 1990.

Wilson, H. D., Lira, R., and Rodriguez, I., Crop-weed gene flow: *Cucurbita argyrosperma* Huber and *C. fraterna* L. H. Bailey (Cucurbitaceae), *Econ. Bot.*, 48, 293, 1994.

Wilson, T. M. A., Strategies to protect crop plants against viruses: pathogen-derived resistance blossoms, *Proc. Natl. Acad. Sci. U.S.A.*, 90, 3134, 1993.

Wintermantel, W. M. and Schoelz, J. E., Isolation of recombinant viruses between cauliflower mosaic virus and a viral gene in transgenic plants under conditions of moderate selection pressure, *Virology*, 223, 156, 1996.

Wu, G., Schortt, B. J., Lawrence, E. B., Levine, E. B., Fitzsimmons, K. C., and Shah, D. M., Disease resistance conferred by expression of a gene encoding H_2O_2-generating glucose oxidase in transgenic potato plants, *Plant Cell*, 7, 1357, 1995.

Xu, H. and AbouHaidar, M. G., Synergistic relationship between potato aucuba mosaic virus and potato virus Y strain O, *Can. J. Plant Pathol.*, 18, 102, 1996.

Xu, Y., Zhu, Q., Panbangred, W., Shirasu, K., and Lamb, C., Regulation, expression and function of a new basic chitinase gene in rice (*Oryza sativa* L.), *Plant Mol. Biol.*, 30, 387, 1996.

Xue, B., Gonsalves, C., Provvidenti, R., Slightom, J. L., Fuchs, M., and Gonsalves, D., Development of transgenic tomato expressing a high level of resistance to cucumber mosaic virus strains of subgroup I and II, *Plant Dis.*, 78, 1038, 1994.

Yoshioka, K., Hanada, K., Harada, T., Minobe, Y., and Oosawa, K., Virus resistance in transgenic melon plants that express the cucumber mosaic virus coat protein gene and in their progeny, *Jpn. J. Breeding*, 43, 629, 1993.

Zaccomer, B., Cellier, F., Boyer, J. C., Haenni, A. L., and Tepfer, M., Transgenic plants that express genes including the 3′ untranslated region of the turnip yellow mosaic virus (TYMV) genome are partially protected against TYMV infection, *Gene*, 136, 87, 1993.

Zambryski, P., Herrera-Estrella, L., De Block, M., van Montagu, M., and Schell, J., Ti plasmid vector for the introduction of DNA into plant cells without alteration of their normal regeneration capacity, *EMBO J.*, 2, 2143, 1983.

Zerbini, F. M., Michelmore, R. W., and Gilbertson, R. L., Resistance to LMV infection in lettuce due to the expression of different forms of the LMV coat protein gene, *Phytopathology*, 85, 1138, 1995.

Zhu, B., Chen, T. H. H., and Li, P. H., Analysis of late-blight disease resistance and freezing tolerance in transgenic potato plants expressing sense and antisense genes for an osmotin-like protein, *Planta*, 198, 70, 1996.

Zhu, G., Maher, E. A., Masoud, S., Dixon, R. A., and Lamb, C. J., Enhanced protection against fungal attack by constitutive co-expression of chitinase and glucanase genes in transgenic tobacco, *Bio/Technology*, 12, 807, 1994.

Index

C

D

E

G

H

I

J

K

L

M

N

O

Q

R

S

T

U

V

W

T

U

V

W

X

Y

Z